Semiconducting Polymers

ACS SYMPOSIUM SERIES **735**

Semiconducting Polymers

Applications, Properties, and Synthesis

Bing R. Hsieh, EDITOR
Xerox Corporation

Yen Wei, EDITOR
Drexel University

American Chemical Society, Washington, DC

Library of Congress Cataloging-in-Publication Data

Semiconducting polymers : applications, properties, and synthesis / Bing R. Hsieh, Yen Wei, editor.

p. cm.—(ACS symposium series : 735)

Includes bibliographical references and index.

ISBN 0–8412–3612–7

1. Polymers—Electric properties congresses. 2. Organic semiconductors Congresses.

I. Hsieh, Bing R. II. II. Wei, Yen. III. Series

QD382.S4S46 1999
620.1′92042972—dc21

99–28602
CIP

The paper used in this publication meets the minimum requirements of American National Standard for Information Sciences—Permanence of Paper for Printer Library Materials, ANSI Z39.48-94 1984.

Distributed by Oxford University Press

PRINTED IN THE UNITED STATES OF AMERICA

Foreword

THE ACS SYMPOSIUM SERIES was first published in 1974 to provide a mechanism for publishing symposia quickly in book form. The purpose of the series is to publish timely, comprehensive books developed from ACS sponsored symposia based on current scientific research. Occasionally, books are developed from symposia sponsored by other organizations when the topic is of keen interest to the chemistry audience.

Before agreeing to publish a book, the proposed table of contents is reviewed for appropriate and comprehensive coverage and for interest to the audience. Some papers may be excluded in order to better focus the book; others may be added to provide comprehensiveness. When appropriate, overview or introductory chapters are added. Drafts of chapters are peer-reviewed prior to final acceptance or rejection, and manuscripts are prepared in camera-ready format.

As a rule, only original research papers and original review papers are included in the volumes. Verbatim reproductions of previously published papers are not accepted.

ACS BOOKS DEPARTMENT

Contents

INDEXES

Dedication

This book is dedicated to
Professor Alan G. MacDiarmid
and to Our Families

Preface

Semiconducting polymers refer to polymeric materials with electrical conductivity in the range of 100 to 10^{-12} (Ohm cm)$^{-1}$. There are at least four major classes of semiconducting polymers: filled polymers, ionically conducting polymers, conjugated polymers, and charge transfer polymers. The first two classes have had a wide range of commercial applications; the last two have not been as successful. Conducting conjugated polymers and charge transfer polymers constitute two important subgroups in semiconducting organics, a field initiated by the discovery of electrical conductivity in molecular charge transfer complexes in 1954. A similar observation was made in iodine doped polymers in 1967. The discovery of metallic conductivity in doped polyacetylene in 1977 generated tremendous excitement in conducting conjugated polymers. After the discovery of superconducting molecular charge transfer complexes in 1980 and fullerenes in 1985, the field of semiconducting organics had finally evolved from phenomenological pursuit to application-driven research. Today, exploration of semiconducting polymers as active constituents in electronic devices has become the major focal point.

The international symposium on "Semiconducting Polymers" was held as part of the 215th ACS National Meeting in Dallas on March 29–30, 1998. "Electroplastics for Plastic Electronics" was used as the symposium logo to heighten the application-driven orientation of electroactive polymer research and to suggest the arrival of the plastic electronics era. The ultimate goal of the field is to develop a wide range of polymeric materials with electronic properties of molecular solids and processability and mechanical properties of conventional polymers. The symposium consisted of a poster session and four oral sessions: (1) filled polymers and ionically conducting polymers; (2) conducting polymers; (3) charge transfer polymers; and (4) polymer LEDs. The session on conducting polymers was dedicated to Alan G. MacDiramid for his pioneering contributions to conducting polymers. The symposium attracted an excellent attendance of 100–150 people, reflecting a strong interest in the field. Selected papers of the symposium were highlighted in an article entitled, "Plastics with an Electrical Bend," in the April 13, 1998 issue of *Chemical and Engineering News* (pages 41–46).

This book comprises a collection of papers presented in the symposium as well as several invited review papers. This volume is organized into three sections: Properties (Chapters 2–11); Device Applications (Chapters 11–19), and Synthesis and Fabrication (Chapters 20–27). This is in accordance the subtitle of the book, "Applications, Properties, and Synthesis," because they are three key, inseparable elements in today's materials research.

We thank Mary Galvin for help in organizing the meeting. We are very grateful to the contributing authors for accepting the burden of contributing to this volume. We thank Anne Wilson of the American Chemical Society Books Department for smooth cooperation and patient support. We also thank the following organizations for their financial support of the symposium: ACS Division of Polymer Chemistry, Inc., ACS Petroleum Research Fund, National Science Foundation (Materials Research Program), Hewlett-Packard Company, 3M, and Xerox Corporation.

BING R. HSIEH
The Wilson Center for Research and Technology
Xerox Corporation
800 Phillips Road, 114–22D
Webster, NY 14580

YEN WEI
Department of Chemistry
Drexel University
32nd and Chestnut Streets
Philadelphia, PA 19104

Chapter 1

Electroplastics for Plastic Electronics

Bing R. Hsieh [1] and Yen Wei [2]

[1] Xerox Corporation, The Wilson Center for Research and Technology, 800 Phillips Road, 114-39D, Webster, NY 14580 (bhsieh@crt.xerox.com)
[2] Department of Chemistry, Drexel University, 32nd and Chestnut Streets, Philadelphia, PA 19104 (weiyen@duvm.ocs.drexel.edu)

Four major classes of semiconducting electroplastics, namely filled polymers, ionically conducting polymers, charge transfer polymers, and conjugated conducting polymers are reviewed. Their applications in plastics electronics such as plastic displays, plastic transistors, plastic lasers, plastic photodiodes, and plastic fuel cells are highlighted. Although much progress has been made in the development of electroplastics, we are still in critical need of new materials with improved properties and functionality to truly fulfill our quest of all-plastic electronics. In particular, we need printable electroplastics with (1) a wide range of controlled conductivity, (2) with high electron mobility (n-type materials), and (3) blue light emission. The fundamental requirements for such new electroplastics are availability, processability, and reliability.

Electroplastics refer to electroactive polymers with charge transporting properties. The majority of electroplastics are semiconducting with electrical conductivity in the range of somewhere from 10^{-12} to 100 $(\text{Ohm cm})^{-1}$ [1]. There are at least four major classes of semiconducting polymers: filled polymers, ionically conducting polymers, charge transfer polymers and conjugated conducting polymers.

Filled polymers are polymers loaded with conductive fillers such as carbon black, graphite fiber, metal particles, or metal oxide particles [2-6]. Among the four classes of semiconducting polymers, filled polymers have the longest history and broadest application in electronic devices. Conducting filled polymers were invented in 1930 for the prevention of corona discharge and have been used in advanced printed circuitries. The dominant use of filled polymers as semiconducting materials can be attributed to their ease of processing and wide range of electrical properties. Being inhomogeneous is an inherent weakness of filled polymer systems, which have three phases, namely the polymer, the filler, and the interface. Such inhomogeneity tends to result in problems such as lack of reproducibility, heavy process dependency, steep percolation threshold in conductivity, or weak dielectric strength, to name a few. Controlling the quality of filler dispersion in a polymer matrix is the most critical and challenging technical issue in filled polymers.

Although ionic polymers or ionomers have been known for more than 30 years [7, 8], their ionic conductivity was not investigated until 1975 [9]. Since then, ionically conducting polymers or polymer electrolytes have had a wide range of commerical electronic applications [10-13], including rechargeable batteries, fuel cells [14], and polymer light-emitting devices [15]. The drive toward these important commercial applications has propelled our understanding of ionically conducting polymers significantly. Ionic polymers also have the advantages of being highly processible and abundantly available. However, ionic conductivity is highly sensitive to humidity. This is related to the ionic conduction mechanism which requires the dissociation of the opposite ionic charges followed by solvation. Solvation is generally affected by the polar polymer matrix and by the presence of polar solvents, especially water. As a result, most ionic conductors become highly insulating upon drying. In many practical ionic polymer systems, one sign of ionic species are attached to the polymer chains while the counter ions are mobile. This can prevent rapid depletion of ionic species in the systems. Materials with both ionic and electronic conductivity are known and may have novel electronic applications [16]. It should be noted that some ionically conducting polymers can also be viewed as filled polymers when the ionic additives are not completely soluble in a polymer matrix.

Conducting conjugated polymers and charge transfer polymers constitute two important subgroups in semiconducting organics [17-20], a field initiated by the discovery of electrical conductivity in molecular charge transfer complexes in the 1950s [21]. A similar observation was made in iodine doped poly(vinyl carbazole) in the mid-1960s [22]. The discovery of metallic conductivity in doped polyacetylene in 1977 generated tremendous excitement [23] and launched the field of synthetic metals. For the next 10 years, the field focused on searching for new materials with metallic conductivity, without seriously addressing critical application parameters such as availability, processability, and reliability. After the discovery of superconducting molecular charge transfer complexes in 1980 for tetramethylselenafulvalene [24] and in 1986 for fullerene [25], the field of semiconducting organics finally evolved from phenomenological pursuit to application driven research [26]. Today, exploration of semiconducting polymers as active constituents in electronic devices has become the major focal point. The ultimate challenge of the field is to develop intrinsically conducting polymers with electronic properties like molecular systems as well as processability and mechanical properties like conventional polymers.

Several practical approaches have emerged recently from both the conjugated polymer and the charge transfer polymer fronts. For example, in the conjugated polymer front, polyaniline and modified polyaniline are being used as conductive fillers to give conducting filled polymers [27, 28]. Water soluble stable conducting polythiophenes (Baytron) have been commercialized by Bayer Corporation [29]. This breakthrough has inspired much work in the area of water soluble conducting polymers [30]. Ionic conductivity may also be present in such water soluble conducting polymers. It has also been shown that condensation polymers having conjugated oligomeric segments of well-defined structures, such as an aniline trimer, showed desirable electroactivity as well as processability and mechanical properties [31].

Because of their commercial use in xerographic photoreceptors, charge transport polymers have become the most established semiconducting organic systems [32]. The first commercial organic photoreceptor, introduced by IBM in 1972, was a single layer device based on a charge transfer polymer, namely trinitrofluorenone doped poly(vinyl carbozole). Most aromatic organic materials, including conjugated polymers such as poly(p-phenylene vinylene) (PPV), have since been evaluated for photoreceptor applications [33]. Charge transport polymers, instead of charge transfer polymers, are being used in today's multi-layer photoreceptors. Most of the charge transport polymers such as poly(vinyl carbazole) (PVK), triarylamine doped polycarbonate and polysilanes are hole transporting or p-type materials [34]. Although several electron transport materials have been reported, their charge mobility of 10^{-6} cm^2 V^{-1} s^{-1} is about three order of magnitude lower than that for hole transport materials [35]. The lack of high mobility electron transport materials is a critical weakness associated with semiconducting organics. The world of semiconducting organics will likely remain p-type if high mobility electron transport materials are not realized.

The concept of adding an oxidant (such as $SbCl_5$ and tri-(p-bromophenyl)aminium hexachloroantimonate) to a charge transport polymer (such as PVK) to give a semiconducting charge transfer polymer was first demonstrated in the 1970s [36]. Different oxidant and charge transport polymer combinations have been developed to give various semiconducting polymers [37-39] and have been used as contact modification layers in organic light-emitting diodes [40]. For example, cation radical salts of N,N,N',N'-tetra-p-tolyl-4,4'-biphenyldiamine were introduced and used in combination with triarylmine doped polycarbonate charge transport polymers to give semiconducting polymers [39a]. The conductivity of the polymers could be tuned readily by adjusting the concentrations of the charge transport group and the oxidant. This semiconducting polymer is close to ideal because it has many advantages, including a wide range of highly stable, tunable, homogeneous conductivity [10^{-12}-10^{-5} (Ohm cm)$^{-1}$], excellent wear resistance, and high dielectric strength. The only problem is that higher conductivity is not yet attainable. Further work in this direction may result in more highly conducting polymers in order to expand the range of applications.

The discovery of electroluminescence in PPV in 1990 [41] ushered in the era of plastic electronics. Since then doping conjugated polymers into their conducting forms is no longer of prime interest. The use of neutral or undoped conjugated polymers as well as other electroplastics as active materials for semiconductor device applications, such as light-emitting devices [42-43], photo-diodes [44], transistors [45], integrated display [46], plastic fuel cells [47], plastic lasers [48], and integrated circuits [49], has become the dominant theme of the field. Highly stable plastic LEDs have been demonstrated and will likely be commercialized in the near future [50]. PPV and its derivatives have been the materials of choice for plastic LEDs because of their availability, processability, relatively high efficiency, and a wide range of color emission from orange to green [51]. There has been a great deal of interest in blue emitting plastic LEDs because they may enable full color display through a simple energy down conversion scheme. Although poly(9,9-

dialkylfluorenes) are promising blue emitting polymers [52], we are still in need of blue emitting polymers with improved functional performance.

Reducing the cost of electronic devices is the driving force behind plastic electronics. Ease of processing of plastic devices can offer the most significant reduction in cost. For example, devices with electroactive micropatterns have been fabricated by screen printing, ink jet printing, or lay-by-lay deposition process [53-55].

In summary, much progress has been made in the development of electroplastics for plastic electronics. We still need electroplastics with (1) a wide range of controlled conductivity, (2) high electron mobility (n-type materials), and (3) blue light emission. Preferably, the materials should be printable and have well defined structures. The fundamental requirements for electroplastics are availability, processability and reliability. Scientists and engineers are creating new device concepts and then looking for the most promising materials solutions. This system level approach toward materials research is a positive development and will result in a greater impact.

References:

1. C. C. Ku and R. Liepins, *Electrical Properties of Polymers: Chemical Principles*; Hanser Publishers: Munich, 1987.
2. R. H. Norman, *Conductive Rubbers and Plastics*: Elsevier, Amsterdam, 1970.
3. R. Landauer, "Electrical Conductivity in Inhomogenious Media" in *Electrical Transport and Optical Properties of Inhomogenious Media*; J. C. Garland and D. B. Taner, Eds., American Institute of Physics: New York, 1978.
4. E. K. Sichel Ed., *Carbon Black Polymer Composites*; Dekker: New York, 1982.
5. S. K. Bhattacharya, *Metal-Filled Polymers: Properties and Applications*, Dekker: New York, 1986.
6. F. Carmona, *Physica A*, **157**, 461 (1989).
7. A. Eisenberg, *Ion-Containing Polymers*; Academic Press: New York, 1977.
8. A. Eisenberg and J.-S. Kim, *Introduction to Ionomers*; Wiley: New York, 1998.
9. P. V. Wright, *Br. Polym. J.*, **7**, 319 (1975).
10. M. Watanable and N. Ogata, *Br. Polym. J.*, **20**, 181 (1988).
11. M. Armand, *Adv. Mater.*, **2**, 278 (1990).
12. J. R. MacCallum and C. A. Vincent Eds., *Polyelectrolytes*; Elsvier: New York, 1989.
13. D. E. Katsoulis, *Chem. Rev.* **98**, 359 (1998).
14. (a) M. Jacoby "Taking Charge of The 21st Century" in Chemical and Engineering News, August 3, 1998, page 37. (b) A. Salzhauer, "Plastic Batteries All Charging up and Waiting to Go" Technology Review, July-August, 1998, page 60. (c) M. Winter, J. O. Besenhard, M. E. Spahr and P. Novak, *Adv. Mater*, **10**, 725 (1997).
15. (a) Q. Pei, G. Yu, C. Zhang, Y. Yang, A. J. Heeger, *Science*, **269**, 1086 (1995). (b) Q. Pei, Y. Yang, G. Yu, C. Zhang, A. J. Heeger, *J. Am. Chem. Soc.*, **118**, 3922 (1996).
16. (a) M. G. Minett, J. R. Owen, *Solid State Ionics*, **28/30**, 1193 (1988); (b) T. Nakamura, T. Akutagawa, K. Honda, A. E. Underhill, A. T. Coomber, R. H. Friend, *Nature*, **394**, 159 (1998).
17. F. Gutmann and L. E. Lyons, *Organic Semiconductors*, Wiley: New York, 1964.
18. H. Meier, *Organic Semiconductors: Dark and Photoconductivity of Organic Solids*, Verlag Chemie: Weinheim, 1974.

19. J.-P. Farges Ed., *Organic Conductors: Fundamental and Applications*, Marcel Dekker: New York, 1994.
20. H. S. Nalwa Ed., *Handbook of Orgnaic Conductive Molecules and Polymers*, Wiley: Chichester, 1997.
21. H. Akamatu, H. Inokutchi, Y. Matsunaga, *Nature*, **173**, 168 (1954).
22. A. H. Hermann and A. Rembaum, *J. Polym. Sci. C*, 109 (1967).
23. H. Shirakawa, E. J. Louis, A. G. MacDiarmid, C. K. Chiang, A. J. Heeger, *J. Chem. Soc. Chem. Commun*, 578, (1977).
24. D. Jerome, M. Mazaud, M. Ribault, K. Bechgaard, *J. Phys. Lett.* **41**:L95 (1980).
25. (a) A. F. Hebard, M. J. Rosseinsky, R. C. Haddon, D. W. Murphy, S. H. Glarum, T. T. Palstra, A. P. Ramirez, A. R. Kortan, *Nature*, **350**, 600 (1991); (b) Z. Iqbal, R. H. Baughman, B. L. Ramakrishna, S. Khare, N. S. Murthy, H. J. Bornemann, and D. E. Morris, *Science*, **254**, 826 (1986).
26. See references 41-49.
27. Anand, J.; Palaniappan, S.; Sathyanarayana, D. N. "Conducting Polyaniline Blends and Composites" *Prog. Polym. Sci.*, **23**, 993 (1998).
28. B. Wessling, *Synth. Met.* **93**, 143 (1998).
29. (a) F. Jonas, G. Heywang, W. Schmidtberg, J. Heinze, M. Dietrich US 4,959,430 (1990) (b) F. Jonas, G. Heywang, W. Schmidtberg, J. Heinze, M. Dietrich, US Patent 4,987,042 (1991). (c) F. Jonas, A. Karbach, B. Muys, E. van Thillo, E. Wehrmann, K. Rolf, A. Elschner, R. Dujardin US Patent 5,766,515 (1998).
30. M. T. Nguyen, M. Leclerc and A.F. Diaz, *Trends in Polym. Sci.*, **3(6)**, 186 (1995).
31. Y. Wei, C. Yang and T. Ding, *Tetrahedron Lett.*, **37**, 731 (1996). (b) Y. Wei, C. Yang, G. Wei, T. Ding and G. Feng, *Synth. Met.*, **84**, 289 (1997). (c) Z. Y. Wang, C. Yang, J. P. Gao, J. Lin, Y. Wei and S. Li, *Macromolecules*, **31**, 2702 (1998).
32. P. M. Borsenberger and D. S. Weiss, *Organic Xerographic Photoreceptors* (Marcel Dekker, New York, 1998.
33. B. R. Hsieh in *Polymer Materials Encyclopedia* J. C. Salamone Ed., CRC Press, vol. 9, p6537 (1996), and references cited therein.
34. M. Abkowitz, H. Bassler, M. Stolka, *Philosophical Magazine B*, **63**, 201, (1991).
35. P. M. Borsenberge et al. *J. Appl. Phys.* **68**, 4100 (1990).
36. (a) Y. Yamamoto, S. Kanda, S. Kusabayashi, T. Nogaito, K. Ito, H. Mikawa, *Bull. Chem. Soc. Jpn.*, **38**, 2015 (1965). (b) R. H. Partridge, UK. Patent 74/44704 (1974), Chem. Abstr. 86, P11261 (1974) (c) H. Block, M. A. Cowd, Walker, S. M. *Polymer*, **18**, 781 (1977). (d) H. Block, S. M. Bowker, S. M. Walker, *Polymer*, **19**, 531 (1978).
37. (a) J. Mort, S. Grammatica, D. J. Sandman, and A. Troup, *J. Electronic Mater*, **9**, 411 (1980). (b) W. W. Limburg, D. M. Pai, J. M. Pearson, "Semiconductive Organic Compositions" U.S. Patent 4,338,222 (1982).
38. M. Fukushima, M. Aramata, S. Mori US Patent 5,549,851 (1996).
39. (a) B. R. Hsieh US Patent 5,853,906 (1998). (b) J. Mort, B. R. Hsieh, M. A. Machonkin, J. Mammino, US Patent 5,834,080 (1998). (c) B. R. Hsieh, J. Mort, M. A. Machonkin, J. R. Ewing U.S. Patent 5,587,224 (1996).
40. (a) R. H. Partridge, Polymer, 24, 733 (1983). (b) R. H. Partridge, Polymer, 24, 739 (1983). (c) R. H. Partridge, Polymer, 24, 748 (1983). (d) R. H. Partridge, Polymer, 24, 755 (1983). (e) A. Yamamori, C. Adachi, T. Koyama, Y. Taniguchi, Appl. Phys. Lett., 72, 2147 (1998).
41. J. H. Burroughes, D. D. C. Bradley, A. R. Brown, R. N. Marks, K. Mackay, R. H. Friend, P. L. Burn, A. B. Holmes, *Nature* **347**, 539 (1990).

42. (a) A. Kraft, A. C. Grimsdale, A. B. Homles, *Angew. Chem. Int. Ed.* **37**, 402 (1998). (b) B. R. Hsieh Ed. *Organic Light-Emitting Materials and Devices*, Macrom. Symp. Vol 125, 1998. (c) S. Miyata and H. S. Nalwa, Eds. *Organic Electroluminescence Materials and Devices*, Gordon and Breach, Amsterdam, 1997.
43. J. A. Rogers, Z. Bao, L. Dhar, *Appl. Phys. Lett.*, **73**, 294 (1998).
44. M. Granstrom, K. Petritsch, A. C. Arias, A. Lux, M. R. Andersson, R. H. Friend, *Nature*, **395**, 257 (1998) and references cited therein.
45. (a) A. R. Brown, A. Pomp, C. M. Hart, D. M. de Leeuw, *Science*, **270**, 972 (1995). (b) J. A. Rogers, Z. Bao, V. R. Raju, *Appl. Phys. Lett.*, **72**, 2716 (1998). (c) C. D. Dimitrakopoulos, S. Purushothaman, J. Kymissis, A. Callegari, and J. M. Shaw, *Scienc*e, **283**, 822 (1999).
46. (a) A. Dodabalapur, Z. Bao, A. Makhija, J. G. Laquindanum, V. R. Raju, Y. Feng, H. E. Katz, J. Rogers, *Appl. Phys. Lett.*, **73**, 142 (1998). (b) H. Sirringhaus, N. Tessler, R. H. Friend, *Science*, **280**, 1741 (1998).
47. P. Searson, J. G. Killian, H. Sarker, J. Giaccai, Y. Gofer, T. O. Poehler, U.S. Patent 5,733,683 (1998) and references cited therein.
48. J. A. Rogers, M. Meier, A. Dodabalapur, *Appl. Phys. Lett.*, **73**, 1766 (1998).
49. C. J. Drury, C. M. J. Mutsaers, C. M. Matters, D. M. de Leeuw, *Appl. Phys. Lett.*, **73**, 108 (1998).
50. D. B. Roitman, H. Antoniadis, J. Sheats, F. Pourmirzaie, *Optoelectronics World*, July issue, 163 (1998).
51. B.R. Hsieh, Y. Yu, E.W. Forsythe, G.M. Schaaf and W.A. Feld, , *J. Am. Chem. Soc.* **120**, 231 (1998).
52. (a) E. P. Woo, W. R. Shang, M. Inbasekaran, G. R. Roof, U.S. Patent 5,708,130 (1998). (b) M. Kreyenschmidt, G. Klaerner, T. Fuhrer, J. Ashenhurst, S. Karg, W. D. Chen, V. Y. Lee, J. C. Scott, R. D. Miller, *Macromolecules*, **31**, 1099 (1998). (c) G. Klarner, M. H. Davey, W.-D. Chen, J. C. Scott, R. D. Miller, *Adv. Mater.* **10**, 993 (1998).
53. (a) S.-C. Chang, J. Bharathan, Y. Yang, R. Helgeson, F. Wudl, M. B. Ramey, J. R. Reynolds, *Appl. Phys. Lett.*, **73**, 2561 (1998). (b) J. Bharathan and Y. Yang, *Appl. Phys. Lett.*, **72**, 2660, (1998).
54. Z. Bao, Y. Feng, A. Dodabalapur, V. R. Raju, A. J. Lovinger, *Chem. Mater.*, *9*, 1299 (1997).
55. (a) G. Decher, Science, 277, 1232 (1998). (b) O. Onitsuka, A. C. Fou, M. Ferreira, B. R. Hsieh, M. F. Rubner, *J. Appl. Phys.*, **80**, 4067 (1996).

PROPERTIES

Chapter 2

Imaging and Electrical Resistivity Measurements of Disordered Carbon–Black–Polymer Composites

Michael B. Heaney

Huladyne Research, 160 Waverley Street, Palo Alto, CA 94301–1138 (mheaney@alum.mit.edu)

This chapter reviews some of the work done on disordered carbon-black–polymer composites by my collaborators and myself over the past several years. These composites have widespread commercial applications. A qualitative analysis of a transmission electron microscope image is presented. Quantitative analyses of scanning probe microscope images and dc electrical resistivity data are presented. The resistivity and linear expansion of a typical composite between 25 and 180°C are measured and analyzed. The scaling theory of percolation provides a good explanation of most of our data.

Disordered carbon-black–polymer composites have many common uses in modern technology. These uses include inks, automobile tires, reinforced plastics, wire and cable sheaths, antistatic shielding, resettable fuses, and self-regulating heaters. As an immediate example, the inked letters on this page consist of a disordered carbon-black–polymer composite bound to the surface of the paper. Despite these widespread applications, many of the important physical properties of these composites are not well understood. There is a long history of experimental and theoretical work on disordered carbon-black–polymer composites [*1*]. One of the most exciting recent advances in the field has been the application of the scaling theory of percolation [*2*] to these composites. This is based on the many similarities between the percolative transition in a disordered conductor–insulator composite and the thermodynamic phase transition common in many materials.

This chapter is not intended as a comprehensive review of the subject. Many other groups and individuals have made significant contributions to this field.

Composite Fabrication

There are many techniques for making carbon-black–polymer composites. Perhaps the simplest technique is to add carbon-black powder and polymer powder together in a bowl and stir briskly by hand. The resulting mixture will have many of the unusual physical properties that make this composite useful. However, in this powder form the composite is awkward to handle and measure. More convenient composites can be made by mixing carbon-black powder into molten wax and allowing the wax to harden, or by mixing carbon-black powder into an epoxy and curing the epoxy. More durable and useful composites are made by mixing the carbon-black powder with a molten high molecular weight polymer such as polyethylene, then cooling the composite to room temperature. For small amounts of composite this is most easily done in a "batch mixer," which is essentially a heated bowl with rotating mixing blades [*3*]. After mixing, the composite can be remelted and molded into the desired shape. For larger amounts of composite, it is more convenient to use a "continuous mixer," where the carbon-black powder and polymer powder are fed continuously into one end of a heated tube containing a rotating screw [*3*]. The rotating screw mixes the carbon-black and molten polymer, while simultaneously pushing the mixture towards the further end of the tube. The mixed composite is extruded out the further end of the tube, and can be molded into the desired shape during or after this extrusion. Most of the composites discussed in this chapter were made using a batch mixer and a hand-operated pneumatic compression mold.

Composite Imaging.

It is often useful to image the microscopic structure of the carbon-black–polymer composite. Microscopic images can reveal if the carbon-black and polymer are mixed uniformly, if the carbon-black has flocculated or formed nonrandom structures, if there are voids in the composite, or if there is anything else unusual. The most common imaging techniques are optical microscopy, scanning electron microscopy, and transmission electron microscopy. Viswanathan and I have recently used a new imaging technique, based on scanning probe microscopy, to study the microscopic structure of these composites [*4*].

Transmission Electron Microscopy. Figure 1 shows a transmission electron microscope (TEM) image of a carbon-black–polymer composite sample. The dark shapes in Figure 1 are carbon-black. The diffuse gray areas are polymer. The bulk composite is electrically conductive with a dc resistivity of about 10^4 Ω cm. The imaged sample is a microtomed slice of the bulk sample, with mean thickness less than 1 μm.

Carbon-blacks have structure on several different length scales [*5*]. Carbon atoms tend to bond in graphitic platelets a few nm in diameter and a few Å in thickness. These graphitic platelets, observable in higher magnification TEM images, are bonded at their edges to other graphitic platelets to form spheroidal structures of mean diameter 80 nm. These spheroidal structures are termed "particles," and have an onion-like structure. Many individual particles are visible in Figure 1. Some of these particles are fused together with other particles, forming larger structures termed "aggregates" with mean diameters of 200 nm. Many aggregates are also visible in Figure 1. The shape of these aggregates can vary greatly. Some aggregates appear spheroidal, others appear rod-like, while yet others appear to have a random fractal-like structure.

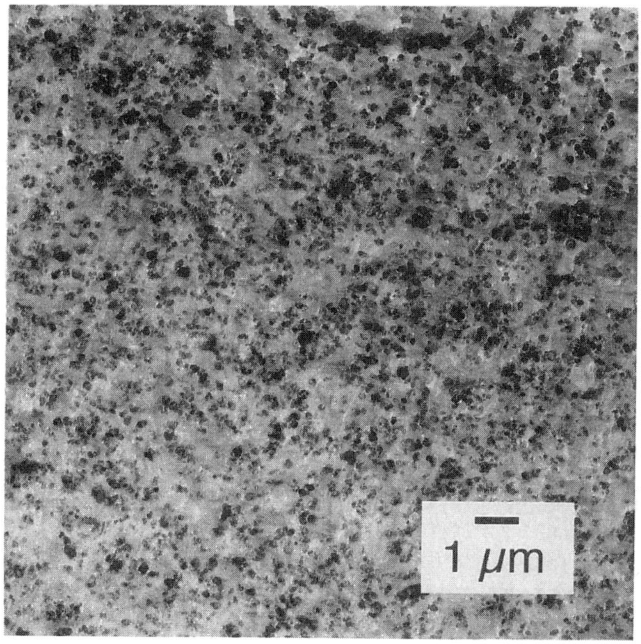

Figure 1. Transmission electron microscope image of a carbon-black–polymer composite. The bulk resistivity of this composite is approximately 10^4 Ω cm. (Reproduced with permission from ref. 17. Copyright 1997 Elsevier Science.)

The adjacent fused carbon-black particles in an aggregate usually share some graphitic planes. These shared graphitic planes make the aggregate a robust structure, which is usually not extensively broken down by mechanical or thermal processing. These aggregates tend to loosely adhere to other aggregates, forming even larger structures termed "agglomerates" with diameters up to 1 μm. Some potential agglomerates are visible in Figure 1. However, it is not possible to confirm which carbon particles are in direct contact from a two-dimensional image. Agglomerates can usually be broken down by mechanical and thermal processing.

The carbon-black aggregates are assumed to fall into three possible categories, in accordance with the Skal-Shklovskii-de Gennes (SSDG) model [*6*]. These three categories are: (a) the carbon-black aggregate is electrically connected to the conductive backbone and can carry dc current, (b) the carbon-black aggregate is electrically connected to the conductive backbone but is part of a dead end pathway and cannot carry dc current, or (c) the carbon-black aggregate is not electrically connected to the conductive backbone, but exists as isolated aggregates or agglomerates. These three classes of carbon-black are represented by the different patterned circles in Figure 2. The continuous conductive pathways control the important dc electrical properties of the composite. The TEM image of Figure 1 shows essentially all of the carbon-black present in the sample slice. From this image, it is not possible to conclusively identify the continuous pathways. A key reason is the difficulty of distinguishing if two apparently touching carbon-black aggregates are actually in good electrical contact. A second reason is that the continuous pathways are three-dimensional in nature, and probably meander in and out of the sample plane. It may be possible to overcome these problems by imaging successive slices of the same sample, then digitally reconstructing a three-dimensional image from this set of two-dimensional images. However, it could still be difficult to conclusively distinguish the continuous pathways. One problem is that during the microtoming process some carbon-black aggregates are pulled out of the sample, while others are pressed into the sample. This would make reconstruction of the original three-dimensional structure challenging.

Scanning Probe Microscopy. The scanning probe microscope (SPM) is a commercially available instrument (Nanoscope III, from Digital Instruments) that offers a relatively new means to distinguish continuous conductive pathways in disordered carbon-black–polymer composites. Figure 2 is a schematic illustration of how the SPM can be used to image carbon-black–polymer composites.

The carbon-black–polymer composite sample is attached to a metal substrate with conductive silver paint. The metal substrate is held at an electrical potential of 3 V relative to the conductive tip. The SPM tip makes two interleaved scans of the composite surface. During the first scan the SPM is operated as a tapping mode atomic force microscope (TMAFM). In this mode the probe tip makes physical contact with the composite and measures the topography of the surface. Figure 3 shows the TMAFM image of an area on the surface of the carbon-black–polymer composite sample. The darker areas represent hills, while the lighter areas represent valleys on the surface. There are some relatively straight diagonal lines visible in the image. These are probably imprints on the composite surface of polishing scratches

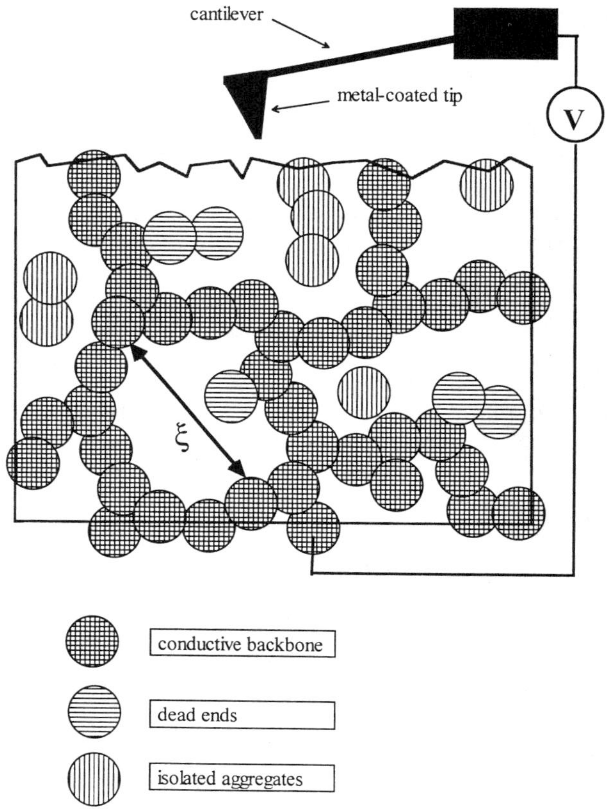

Figure 2. The Skal-Shklovskii-DeGennes model for the carbon-black–polymer composite, and the scanning probe microscopy technique used to image the conductive network at the surface.

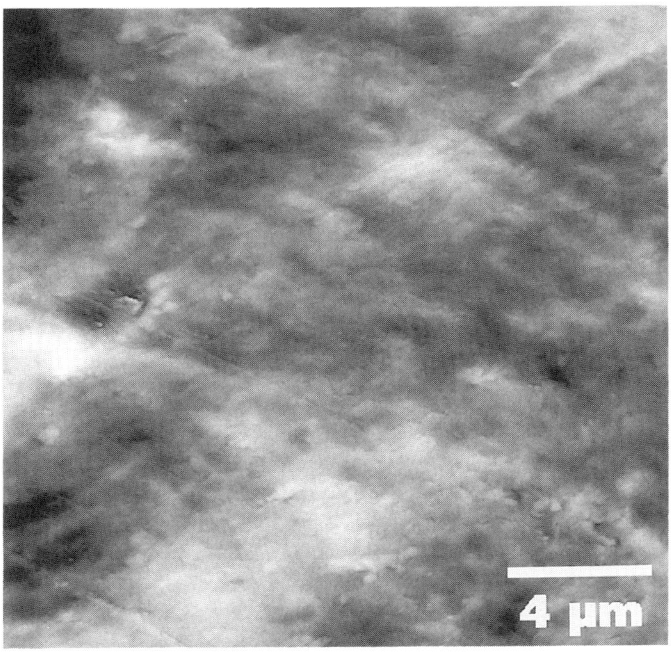

Figure 3. The topography of a particular area of the carbon-black–polymer composite surface, as imaged in the "tapping mode atomic force microscope" mode of the scanning probe microscope. (Reproduced with permission from ref. 4. Copyright 1995 American Physical Society.)

on the mold used to form the composite sheet. There are some features visible which may be carbon-black aggregates, but it is not possible to say so conclusively.

During the second scan the SPM is operated as an electric force microscope (EFM). In this mode the probe tip never makes physical contact with the composite surface. Instead, the probe tip scans at a constant distance above the surface, moving in and out as needed to follow the surface topography. This is done by using the topographic data stored from the earlier TMAFM scan. A dc voltage is applied between the conductive probe tip and the bottom surface of the sample during the second scan. The deflection of the conductive tip is recorded as the tip scans across the surface. It is important to note that electric force microscopy (EFM) is very different than scanning tunneling microscopy (STM). In EFM mode the tip is relatively far from the sample surface, and no significant electric current (tunneling or otherwise) flows between the tip and the sample. In EFM mode the electrostatic deflection is measured. In STM mode the tip is relatively close to the sample surface, and a significant tunneling current flows between the tip and the sample. In STM mode the tunneling current is measured.

Figure 4 shows the electric force microscope (EFM) image of the carbon-black–polymer composite. Note that Figs. 3 and 4 are images of exactly the same area of the composite surface, taken concurrently, with the same probe tip. The dark areas in Figure 4 correspond to conducting regions, while the light areas correspond to insulating regions. There is a striking contrast between the TMAFM image of Figure 3 and the EFM image of Figure 4. Fairly distinct dark shapes are visible in the EFM image. These dark shapes have a size and structure consistent with that of the carbon-black aggregates visible in the TEM image of Figure 1. Note also that some of the diagonal lines are visible in both Figures 3 and 4. Evidently some of the topography can influence the EFM image. There is also some possible correlation between the dark shapes (elevated hills) observed in Figure 3 and the dark shapes (conductive areas) observed in Figure 4. This may be a similar effect, or a real correlation caused by carbon-black aggregates protruding from the surface. The correlation is clearly not present for most of the dark shapes, suggesting the EFM mode is giving an independent image of a different property of the composite surface.

Image Analysis. The image of the conductive network shown in Figure 4 can be quantitatively analyzed. This is done by a two-point correlation analysis of the binarized image. Figure 5 shows a binarized image of Figure 4. Conductive areas are represented by black pixels, and nonconductive areas are represented by white pixels.

Twenty five black pixels are chosen in Figure 5, such that the pixels are approximately equally spaced from each other. The chosen black pixels are each circled in Figure 5 for identification. A series of boxes of varying size L are then constructed, where L is the length of one side of the box, centered around each chosen pixel. Note that the boxes must be centered around a conductive (black) pixel, since we intend to measure the correlations of the conductive network with itself. A few such boxes are shown around the centermost chosen pixel in Figure 5. For each box size L, we measure the area contained within the box that is conductive (black). This area is defined as the "mass" $M(L)$ for the box of size L. The ratio of the conductive area to the total area within the box is then defined as the "density" $\rho_d(L)$ of the conductive area:

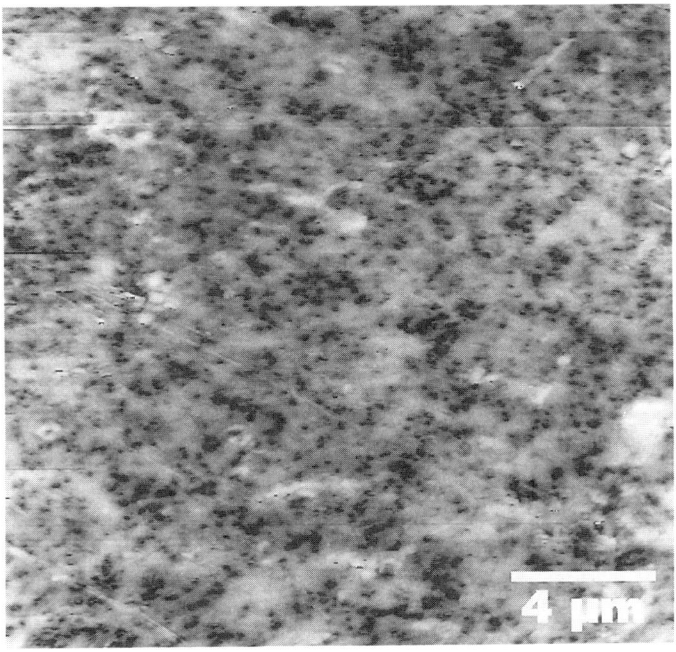

Figure 4. The electric field potential of the same area of the carbon-black–polymer composite surface shown in Figure 3, as imaged in the "electric force microscope" mode of the scanning probe microscope. (Reproduced with permission from ref. 4. Copyright 1995 American Physical Society.)

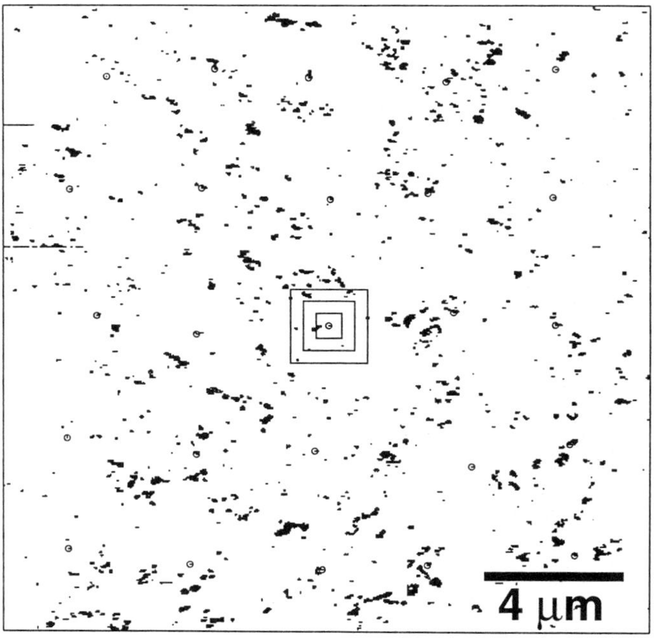

Figure 5. The binarized image of Figure 4. The conductive regions are represented by black pixels, while the insulating regions are represented by white pixels. The 25 chosen black pixels are each circled. Some example boxes are shown around the centermost chosen black pixel. (Reproduced with permission from ref. 4. Copyright 1995 American Physical Society.)

$$\rho_d(L) \equiv \frac{M(L)}{L^2}. \tag{1}$$

The maximum box size L is chosen such that the overlaps between the largest boxes around adjacent center pixels are negligible. Allowing significant overlap would convolve the structure analysis at small length scales with the structure analysis at large length scales, causing spurious results (Kapitulnik, A., Stanford University, personal communication, 1995).

Figure 6 shows a plot of $\rho_d(L)$ vs. L for the binarized image shown in Figure 5. The data points are the mean values of $\rho_d(L)$, averaged over the 25 different boxes of side L for each value of L. The three representative error bars are the one sigma standard deviations calculated from the 25 different boxes of side L. The solid lines are fits to the data, and will be discussed shortly.

The point-to-point scatter in the mean values of $\rho_d(L)$ is significantly smaller than the size of the error bars. This suggests that the statistical variations are non-Gaussian, and contain information about the long-range correlations of the conductive network (Bergman, D.J., Tel Aviv University, personal communication, 1996). There is currently no quantitative theoretical understanding of this phenomenon.

The mean density of conductive area in Figure 6 shows three distinct regimes. For length scales less than about 0.6 μm the density of conductive area decreases most steeply as a function of L. For length scales between about 0.6 μm and 3 μm the density of conductive area decreases less steeply as a function of L. For length scales above about 3 μm the density of conductive area is approximately constant as a function of L. This behavior is very similar to behavior observed earlier in quantitative analyses of two-dimensional percolative systems [*7*], and is consistent with the scaling theory of percolation. The scaling theory of percolation predicts that the conductive network in a percolative system near the percolation threshold will be a fractal object with the conductive "mass" obeying the equation [*2*]

$$M(L) \propto L^{D'}, \tag{2}$$

where D' is the fractal dimensionality of the conductive network. This is a generalization of conventional Euclidean scaling behavior to non-integral dimensions. The sloped solid lines in Figure 6 are weighted nonlinear least-squares fits of the data to equation 2. The slope of the fitted line below 0.6 μm corresponds to a fractal dimension $D' = 0.9 \pm 0.4$. The slope of the fitted line between 0.8 μm and 2 μm corresponds to a fractal dimension $D' = 1.6 \pm 0.6$. Above 3 μm the density is approximately constant, corresponding to a fractal dimensionality of $D' = 2.0$.

A generalization of Euclidean scaling relations to fractal geometries implies that a fractal object of dimension D embedded in a d-dimensional space and cut by a surface of dimensionality d_s will show a fractal dimensionality of $D' = D - (d - d_s)$. This allows us to estimate the fractal dimension of the conductive network as $D = 1.9 \pm 0.4$ below 0.3 μm, $D = 2.6 \pm 0.6$ between 0.8 μm and 2 μm, and $D \approx 3$ above 2 μm.

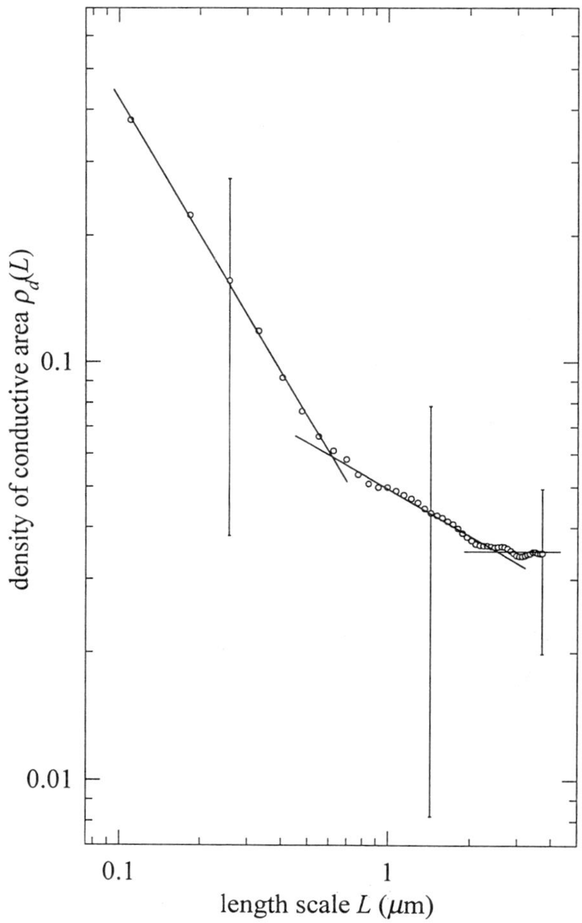

Figure 6. A log-log plot of the average density of the conductive regions $\rho_d(L)$ versus the length scale L for the data shown in Figure 5. The minimum conductive unit is a square of side L = 0.037 μm. (Reproduced with permission from ref. 4. Copyright 1995 American Physical Society.)

Below 0.3 μm we are probing the structure of the carbon-black aggregates themselves. Carbon-black aggregates are known to have fractal-like structures [*5*]. Separate electron microscopy image analysis of similar carbon-blacks yields a fractal dimensionality in good agreement with our measured value [*8*]. However, the resolution of Figure 5 at small length scales is not very high, so this value should be regarded as only a rough estimate. At intermediate length scales we are probing the structure of the conductive network itself: the structure of the individual conductive areas has been "scaled away." The scaling theory of percolation [*2*] predicts a fractal dimensionality of D_{th} = 2.53, consistent with our measured value D = 2.6 ± 0.6. Note that the estimated experimental uncertainty is relatively large. This is because we implicitly assumed Gaussian errors in the weighted nonlinear least-squares fit, which is probably too conservative. The real uncertainties in the fitted values of the fractal dimensions are probably much smaller. A better theoretical understanding of the non-Gaussian nature of the longer-range correlations would be needed to make a more definitive analysis.

The binarization threshold used to obtain Figure 5 is somewhat arbitrary, and the chosen threshold is on the "light" side. This threshold is chosen because it makes the intermediate range in Figure 6 more pronounced. Various other binarization thresholds were tried as a test of the data analysis technique. Changing the binarization threshold such that there is up to three times as much total conductive area present in Figure 5 makes no difference in the fitted values for the fractal dimensionalities in the same three regimes. However, the "breaks" between the three regimes are less obvious for this "darker" threshold.

Figure 6 shows that the average density of the conductive network at large length scales is about $\rho_d(L) \approx 0.035$. This means that only about 3.5% of the surface is conductive. This particular composite sample has a carbon-black concentration (volume percent) of 19.78%. Therefore only about 18% of the carbon-black present in the composite is part of the conductive backbone. The other 82% of the carbon-black is presumably in dead ends and isolated aggregates, and can not carry dc current through the sample. This is consistent with the SSDG model of percolation [*6*]: near the percolation threshold, only a small fraction of the conductive units are actually part of the conductive backbone. Most of the conductive units in a percolative system near but above percolation are not part of the conductive backbone and can not carry any dc current.

Bulk Resistivity Measurements.

The electrical properties of disordered carbon-black–polymer composites are critical to many of their commercial applications. Many research groups have used scaling laws to explain the electrical resistivity of disordered conductor–insulator composites. Some results [*9*] appear to obey scaling laws with a critical exponent in good agreement with the theoretical three-dimensional percolation value [*2*] t_{th}= 2.0. Other results [*10*] show agreement with a scaling law, but with a critical exponent in disagreement with t_{th}= 2.0. Understanding these differences in behavior is one of the major unresolved questions in the field. To address this issue, I measured the resistivity of a series of carbon-black–polymer composites with many samples close to the percolation threshold [*11*]. This is the regime where scaling behavior should be most evident.

Sample Fabrication. The composites are fabricated from commercially available carbon-black and polymer. The carbon-black resistivity is of order 10^{-2} Ω cm and the polymer resistivity is of order 10^{18} Ω cm. The carbon-black consists of 200 nm mean diameter aggregates composed of smaller fused semispherical particles of 80 nm mean diameter [*5*]. The polymer is high density polyethylene with a melting point of 130°C. The preweighed carbon-black and powdered polymer are mechanically mixed and then melt compounded in a batch mixer at 150°C. Fourteen different mixtures were made, with carbon-black concentrations (volume fractions) ranging from about 0.18 to 0.40. Each compound is compression molded into three slabs of nominal thickness 0.025 cm and then cut into samples of nominal width 1.25 cm and nominal length 14 cm, giving a total of about 30 samples at each concentration. The concentration of carbon-black in the composite is determined by thermogravimetric analysis of four representative pieces removed from regions near the edges of the three slabs.

Measurement Technique. Rectangular stripes of silver paint are applied to the surface of each sample at regularly spaced intervals to form electrical contacts. Wires are attached to the samples by spring-loaded metal clips placed on the silver-paint electrodes. A comparison of two-point and four-point measurements showed that contact resistance between the sample and the silver-paint electrodes can be significant, especially for composites close to the percolation threshold. All measurements reported here are made with a four-point technique [*12*] in the linear (ohmic) range of the resistance versus voltage characteristic, and show no significant time dependence.

Resistivity vs. Concentration. Figure 7 shows the dependence of the carbon-black–polymer composite resistivity on the concentration (volume fraction) of carbon-black at room temperature. The data points are the mean resistivity of the approximately 30 samples measured at each carbon-black concentration. The error bars are not shown. At zero carbon-black concentration the measured resistivity is about 10^{18} Ω cm, which is consistent with the expected resistivity for this polymer. The resistivity of the composites gradually decreases as the carbon-black concentration increases, until a critical point (the "percolation threshold") at $p \approx 0.17$. The composite resistivity drops abruptly by about 12 decades, then decreases more gradually as the carbon-black concentration approaches $p \approx 0.40$. This general behavior is analogous to a phase transition [*2*]. The two phases of the carbon-black–polymer composite are an insulating phase and a conducting phase, the analog to temperature is carbon-black concentration, and the analog to the critical temperature of the phase transition is the percolation threshold. This is analogous to a second order phase transition because there is no "latent heat" associated with the transition. It was suggested (see [*13*] for a review) that the analogy could extend further, such that the key physical properties of the composite may obey universal scaling laws, similar to the scaling laws observed in thermodynamic phase transitions.

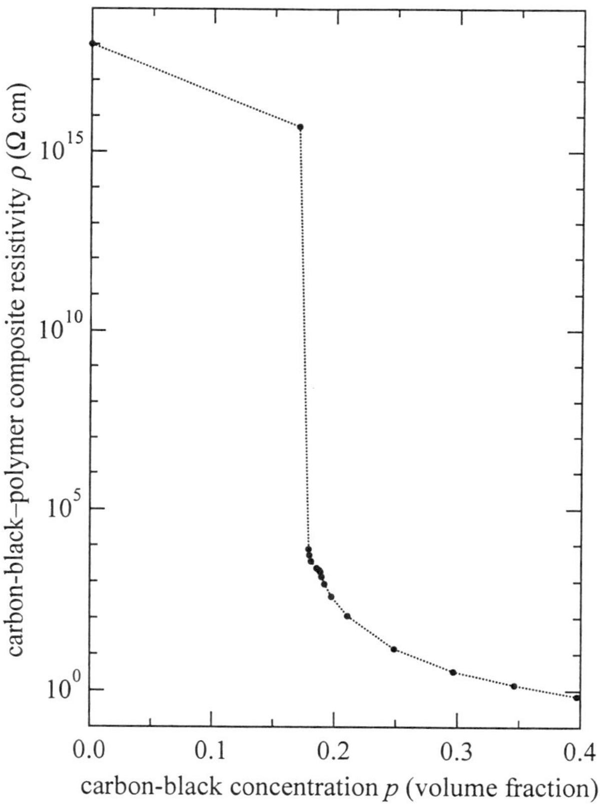

Figure 7. The bulk dc resistivity of a carbon-black–polymer composite versus the carbon-black concentration, measured at room temperature. The percolation threshold occurs at a carbon-black concentration of $p \approx 0.17$. The dashed line connects adjacent data points, and has no other physical significance.

The concept of "scaling" means that many of the key physical properties of a physical system near a percolation threshold or phase transition will fall on a straight line when plotted on reduced axes, with the slope of that line giving the value of the critical exponent associated with the particular physical property plotted. The concept of "universality" means that the resistivity (for example) of composites made from different types of conducting particles (with different microscopic structures) and different types of insulating matrices will all fall on the same straight line, when plotted on reduced axes. All of the details of the microscopic structure of the carbon-black, the mixing procedure, the thermal history of the composite, etc., are included in the two phenomenological parameters ρ_0 and p_c. Abeles, Pinch, and Gittelman [*9*] published the first attempt to fit a scaling law to percolation data in a disordered conductor–insulator composite.

Figure 8 shows a fit of the data of Figure 7 above the percolation threshold to the scaling law [*2*]

$$\rho = \rho_0 \left(\frac{p - p_c}{1 - p_c} \right)^{-t}, \tag{3}$$

where ρ is the carbon-black–polymer composite resistivity (Ω cm), ρ_0 is the resistivity scale factor (Ω cm), p is the carbon-black concentration (volume fraction), p_c is the percolation threshold (volume fraction), and t is the conductivity critical exponent. The weighted nonlinear least-squares fit gives a conductivity critical exponent $t = 2.9 \pm 0.1$, a percolation threshold $p_c = 0.170 \pm 0.001$, and a resistivity scale factor $\rho_0 = 15 \pm 2$ mΩ cm. The reduced chi-square for the fit is 1.3 ± 0.4, indicating a good fit.

The fitted value $t = 2.9 \pm 0.1$ for the conductivity critical exponent is in good agreement with the theoretical value [*2*] $t_{th} = 3$ for mean-field percolation and clearly inconsistent with the theoretical value [*2*] $t_{th} = 2$ for three-dimensional percolation.

Figure 8 shows the data and fit plotted on "reduced axes," which effectively normalize out the two phenomenological parameters, the resistivity scale factor and the percolation threshold. On these axes a scaling law such as equation 3 predicts a straight line, with a slope equal to the value of the critical exponent. Plotting the data and fit on reduced axes allows a better qualitative estimate of the quality of the fit, and also facilitates comparisons between different physical systems where the resistivity scale factor and percolation threshold can be very different.

Resistivity Fluctuations vs. Concentration. Figure 9 shows the relative fluctuations in resistivity versus the carbon-black concentration, plotted on reduced axes. The relative fluctuations in resistivity $\delta\rho/\rho$ are calculated from the one sigma standard deviations of the sample-to-sample variations in resistivity for the approximately 30 samples at each concentration, divided by the mean resistivity at each concentration. The one sigma error bars on the relative fluctuations in resistivity are estimated from the variations in the scatter in resistivity between the three slabs made at each concentration. The solid line is a fit to the scaling law

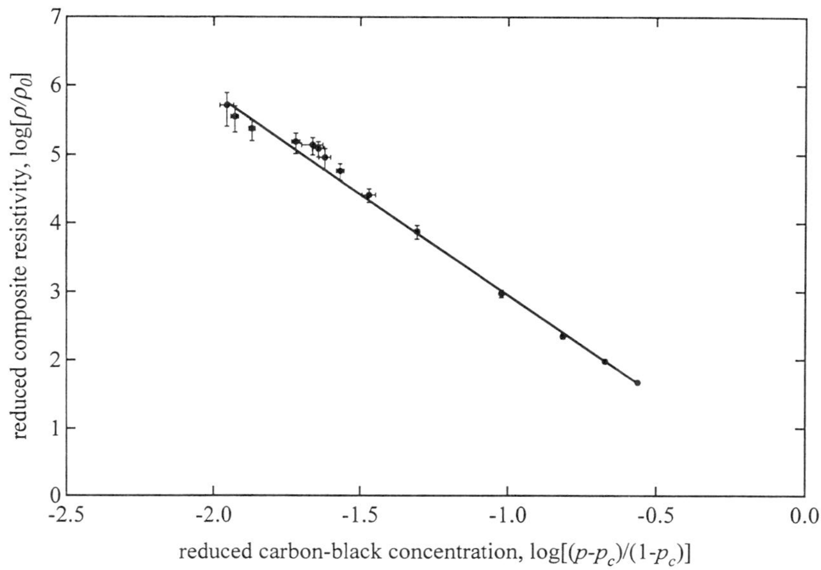

Figure 8. The composite resistivity versus carbon-black concentration above the percolation threshold, plotted on reduced axes. The solid line is a fit to the scaling law of Eq. 3. (Reproduced with permission from ref. 11. Copyright 1995 American Physical Society.)

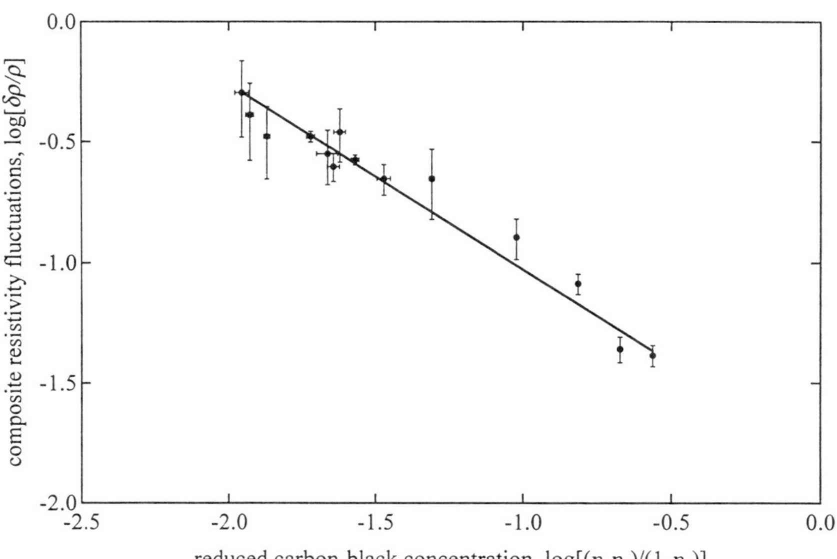

Figure 9. The sample-to-sample fluctuations in composite resistivity versus carbon-black concentration above the percolation threshold, plotted on reduced axes. The solid line is a fit to the scaling law of Eq. 4. (Reproduced with permission from ref. 11. Copyright 1995 American Physical Society.)

$$\frac{\delta\rho}{\rho} = a\left(\frac{p-p_c}{1-p_c}\right)^{-b}, \tag{4}$$

where a is a scale factor, p is the carbon-black concentration, p_c is the percolation threshold (volume fraction), and b is a critical exponent. The percolation threshold is taken as the value found from the resistivity versus concentration fit. This gives a more reliable value and reduces the number of free parameters. The weighted nonlinear least-squares fit gives a scale factor $a = 0.016 \pm 0.002$ and an exponent $b = 0.77 \pm 0.03$. The reduced chi-square for the fit is 1.0 ± 0.4, indicating a good fit.

The data appear to obey a scaling law, but the physical interpretation is not completely clear. Carmona, Prudhon, and Barreau [*10*] had suggested that the relative fluctuations in resistivity in similar composites are due to fluctuations in the correlation length. A random-walk argument in three dimensions predicts the relative fluctuations in resistivity for a statistical ensemble of cubic samples of side L containing a percolative network with a correlation length of ξ is given by [*14*]

$$\left(\frac{\delta\rho}{\rho}\right)^2 \propto \left(\frac{\xi}{L}\right)^3. \tag{5}$$

This can be generalized to non-cubic geometries by substituting the sample volume V in place of the cube volume L^3. Note that the volumes of the composite samples in this experiment are the same for the different carbon-black concentrations. This facilitates the data analysis, and makes any scaling behavior immediately evident in the raw data.

The scaling theory of percolation predicts that the correlation length ξ will obey a scaling law of form [*2*]

$$\xi = \xi_0\left(\frac{p-p_c}{1-p_c}\right)^{-\nu}, \tag{6}$$

where ξ_0 is the correlation-length scale factor, p is the carbon-black concentration (volume fraction), p_c is the percolation threshold (volume fraction), and ν is the correlation-length critical exponent. Combining Eqs. 5 and 6 gives

$$\frac{\delta\rho}{\rho} \propto \left(\frac{\xi_0^{\,3}}{V}\right)^{1/2}\left(\frac{p-p_c}{1-p_c}\right)^{-3\nu/2}. \tag{7}$$

The results of the fit to equation 4 then imply a correlation-length critical exponent $\nu = 0.51 \pm 0.02$. This is in good agreement with the theoretical mean-field value [*2*] of $\nu_{th} = 1/2$ and in clear disagreement with the theoretical three-dimensional percolation value [*2*] of $\nu_{th} = 0.88$. It is not well understood why this particular composite appears to obey mean-field percolation and not three-dimensional percolation. A number of other disordered conductor–insulator composites [*10*] also give critical exponents consistent with mean-field percolation and inconsistent with three-dimensional percolation.

A potential problem with the interpretation of the resistivity fluctuation data is that the fitted value for the correlation-length scaling factor gives a value $\xi_0 \propto 360 \pm 30$ μm. This is over two decades larger than the estimated size of the carbon-black aggregates present in Figure 1, and is larger than the sample thickness. Similar anomalously large values of the correlation-length scale factor were also observed by Carmona, Prudhon, and Barreau [*10*] in their experiments.

There are two reasons why the fitted correlation-length scale factor might be larger than expected in this interpretation. First, scaling laws such as equation 6 are only asymptotic relations, and are known to contain nonuniversal prefactors. Second, since equation 7 is only a proportionality, the correlation-length scale factor is just *proportional to* 360 μm. In the absence of quantitative knowledge of the nonuniversal prefactor in equation 6 and the exact constant that should multiply the right hand side of equation 7, it is not possible to determine conclusively if this model is correct or not. There are also several other possible interpretations of this data [*11*].

Resistivity vs. Temperature.

Several of the common commercial uses of disordered carbon-black–polymer composites rely on their resistivity versus temperature behavior above room temperature. In an attempt to better understand this behavior, I measured simultaneously the resistance and linear expansion of a composite sample as a function of temperature from 25 to 180°C [*15*]. The composite sample is made from the same carbon-black and polymer shown in Figure 1, but the concentration (volume fraction) of carbon-black is nominally $p = 0.35$. The composite sample has the shape of a disk with diameter 1.27 cm and thickness 500 μm. Electrical contact is made by laminating metal foil on the top and bottom flat faces of the composite disk. Separate experiments showed that these type of electrical contacts produced no significant contact resistance. Figure 10a shows the composite resistance increasing gradually at first as temperature is raised above room temperature. The resistance then increases dramatically between 120 and 133°C. Above 133°C the resistance increases at a rate between the two earlier regimes. The monotonic increase in resistance with temperature from 25 to 180°C is known as the positive temperature coefficient of resistance (PTC) effect. The precipitous increase in resistance between 120 and 133°C is known as the PTC anomaly. The PTC effect in disordered conductor–polymer composites was first observed by Gerald Pearson at Bell Labs in 1939 [*16*]. He discussed possible applications of the PTC effect to the regulation and control of electrical current.

The polymer in this composite is high density polyethylene, which is semicrystalline with a crystallinity of about 75%. The crystalline melting point is about 130°C. This corresponds well to the abrupt increase in resistance observed in Figure 10a. The width of the polymer melting transition also corresponds well with the width of the abrupt increase in resistivity. The polymer melting is a second order phase transition, and shows hysteresis. When this composite is cooled from 180 to 20°C the polymer does not recrystallize at the same temperature at which it melted (130°C); it recrystallizes at a lower temperature. The resistance vs. temperature behavior shows the same hysteresis. This suggests the PTC anomaly in these composites is driven by the polymer melting transition.

Figure 10b shows the linear expansion of the composite disk, measured in the direction perpendicular to the plane of the disk. Note that both axes are now linear. The composite thickness increases almost linearly at first, then increases sharply between 120 and 133°C, then increases linearly above 133°C at a rate between that of the two earlier regimes. The abrupt increase in composite thickness occurs at the melting point of the polymer, with a width similar to the width of the polymer melting transition.

Figure 10c shows the resistance versus thickness of the sample. Note that there is a one-to-one correspondence between the data points in Figs. 10a, 10b, and 10c. There is a slight kink in the data at a thickness of about 630 μm. This thickness corresponds to a temperature of 133°C, which is the upper end of the PTC anomaly in Figure 10a. For the several data points immediately below 133°C the resistance tended to drift upwards with time. For the several data points immediately above 133°C the resistance tended to drift downwards with time. This suggests the kink may be caused by minor temporal instabilities in the resistance of the composite at the completion of polymer melting.

Note that there is no longer an anomalous, abrupt increase in the resistance of the composite in Figure 10c. This suggests that the mechanism causing the PTC effect is the same below, during, and above the polymer melting point. The exact nature of the mechanism is still not fully understood.

The smooth and gradual nature of the increase of sample resistance with thickness suggests a new approach to understanding the PTC effect and the PTC anomaly in this and similar materials. Figure 10c suggests that there is no "critical point" for the electronic conduction mechanism at the polymer melting point. This rules out several proposed theoretical models for the PTC effect in this class of composites [*15*]. The resistance versus thickness behavior appears to be more fundamental for understanding the PTC effect than the resistance versus temperature behavior. Developing a theoretical understanding of the resistance versus thickness behavior should be the first and primary step in understanding the PTC effect in this class of composites. The entire resistance versus temperature behavior, including the PTC anomaly, can then be understood by measurements of the physical expansion of the composite.

Acknowledgments

Special thanks to Derek Leong for assistance with TEM imaging, Aharon Kapitulnik for much useful advice and help, Marty McKenzie for instruction in thermomechanical analysis, Vijay Reddy for assistance with thermogravimetric analysis, Craig Prater for advice on scanning probe microscopy, and David J. Bergman for theoretical guidance.

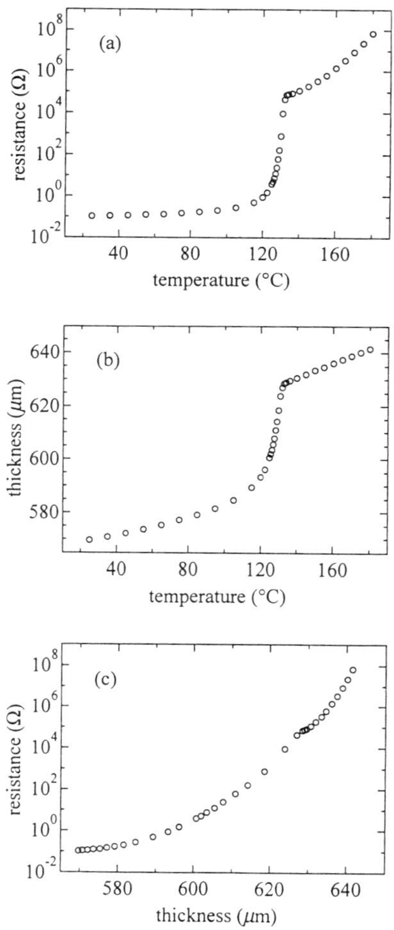

Figure 10. Simultaneous measurements of (a) the dc resistance versus temperature, (b) the sample thickness versus temperature, and (c) the dc resistance versus thickness of a disk-shaped sample of a carbon-black–polymer composite. (Reproduced with permission from ref. 15. Copyright 1996 American Institute of Physics.)

References

1. *Carbon Black-Polymer Composites: the Physics of Electrically Conducting Composites;* Sichel, E.K., Ed.; Plastics Engineering; Marcel Dekker, Inc.: New York, NY, 1982; Vol. 3, and references therein.
2. Sahimi, M.; *Applications of Percolation Theory*; Taylor & Francis: Bristol, PA, 1994. Stauffer, D., and Aharony, A.; *Introduction to Percolation Theory, Second Edition;* Taylor & Francis: Bristol, PA, 1991. Zallen, R.; *The Physics of Amorphous Solids*; John Wiley & Sons: New York, NY, 1983.
3. Rauwendaal, C.; *Polymer Mixing: A Self-Study Guide*; Hanser Gardner: Cincinnati, OH, 1998.
4. Viswanathan, R., and Heaney, M.B., *Phys. Rev. Lett.* **1995**, *75*, 4433; **1996**, *76*, 3661(E).
5. *Carbon Black, Second Edition;* Donnet, J.-B., Bansal, R.C., and Wang, M.-J., Eds.; Marcel Dekker, Inc.: New York, NY, **1993**.
6. Skal, A.S., and Shklovskii, B.I.; *Sov. Phys.-Semicond.* **1975**, *8*, 1029. De Gennes, P.G.; *J. Physique* **1976**, *37*, L1.
7. Kapitulnik, A., and Deutscher, G.; *Phys. Rev. Lett*. **1982**, *49*, 1444. Voss, R.F., Laibowitz, R.B., and Allessandrini, E.I.; *Phys. Rev. Lett.* **1982**, *49*, 1441.
8. Ehrburger-Dolle, F., and Tence, M.; *Carbon* **1990**, *28*, 448.
9. Van der Putten, D., Moonen, J.T., Brom, H.B., Brokken-Zijp, J.C.M., and Michels, M.A.J.; *Phys. Rev. Lett.* **1992**, *69*, 494. Carmona, F., and Mouney, C.; *J. Mat. Sci.* **1992**, *27*, 1322. Chakrabarty, R.K., Bardhan, K.K., and Basu, A.; *Phys. Rev.* **1991, *B*** *44*, 6773. Benguigui, L., Yacubowicz, J., and Narkis, M.; *J. Polym. Sci.* **1987**, *B 25*, 127. Song, Y., Noh, T.W., Lee, S.-I., and Gaines, J.R.; *Phys. Rev.* **1986**, *B 33*, 904. Lee, S.-I., Song, Y., Noh, T.W., Chen, X.-D., and Gaines, J.R.; *Phys. Rev.* **1986**, *B 34*, 6719. Chen, C.C., and Chou, Y.C.; *Phys. Rev. Lett.* **1985**, *54*, 2529. Balberg, I., and Bozowski, S.; *Solid State Commun.* **1982**, *44*, 551. Abeles, B., Pinch, H.L., and Gittleman, J.I.; *Phys. Rev. Lett.* **1975**, *35*, 247.
10. McLachlan, D.S., Rosenbaum, R., Albers, A., Eytan, G., Grammatika, N., Hurwits, G., Pickup, J., and Zaken, E.; *J. Phys. Condens. Matter* **1993**, *5*, 4829. Carmona, F., Valot, E., Servant, L., and Ricci, M.; *J. Phys. I France* **1992**, *2*, 503. Carmona, F., and Mouney, C.; *J. Mat. Sci.* **1992**, *27*, 1322. McLachlan, D.S., Blaskiewicz, M., and Newnham, R.E.; *J. Am. Ceram. Soc.* **1990**, *73*, 2187. Quivy, A., Deltour, R., Jansen, A.G.M., and Wyder, P.; *Phys. Rev.* **1989**, *B 39*, 1026. Carmona, F., and El Amarti, A.; *Phys. Rev.* **1987**, *B 35*, 3284. Balberg, I.; *Phys. Rev. Lett.* **1987**, *59*, 1305. Lee, S.-I., Song, Y., Noh, T.W., Chen, X.-D., and Gaines, J.R.; *Phys. Rev.* **1986**, *B 34*, 6719. Carmona, F., Prudhon, P., and Barreau, F.; *Solid State Commun.* **1984**, *51*, 255. Pike, G.E. in *Electrical Transport and Optical Properties of Inhomogeneous Media*; Garland, J.C., and Tanner, D.B., Eds.; AIP Conf. Proc. 40; AIP; New York, 1978, p. 366.
11. Heaney, M.B.; *Phys. Rev.* **1995**, *B 52*, 12477.
12. Heaney, M.B., in *The Measurement, Instrumentation and Sensors Handbook;* Webster, J.G., Ed.; CRC Press: Boca Raton, FL, 1998.
13. Stanley, H.E.; *Introduction to Phase Transitions and Critical Phenomena*; Oxford University Press: New York, 1971.

14. Straley, J.P., in *Electrical Transport and Optical Properties of Inhomogeneous Media*; Garland, J.C., and Tanner, D.B., Eds.; AIP Conf. Proc. 40; AIP; New York, 1978, p. 118.
15. Heaney, M.B.; *Appl. Phys. Lett.* **1996**, *69*, 2602.
16. Pearson, G.; US Patent No. 2,258,958 (issued 14 October 1941); UK Patent Specification No. 541,222 (issued 18 November 1941).
17. Heaney, M.B.; *Physica* 1997, *A 241*, 296.

Chapter 3

The Effect of Dopant Counter Ion on the Conductivity and Morphology of Polyaniline–Nylon Blends

Alan R. Hopkins [1,4], Paul G. Rasmussen [1], Rafil A. Basheer [2], B. K. Annis [3], and G. D. Wignall [3]

[1] Center for Macromolecular Science and Engineering and Department of Chemistry, The University of Michigan, Ann Arbor, MI 48109–1055
[2] Polymers Department, General Motors Research and Development Center, Warren, MI 48090–9055
[3] Chemical and Analytical Sciences Division, Oak Ridge National Laboratory, Oak Ridge, TN 37831–6197

We have prepared a series of nylon / polyaniline blends using the solvent hexafluoroisopropanol (HFIP), which is an excellent solvent for polyaniline emeraldine base (PANI-EB), polyaniline doped with various sulfonic acids (PANI-ES) and for high molecular weight nylon 6 and nylon 12. It was observed that conductivity and morphology of the blends varied with the compatibility of the sulfonic acid anion with the nylon. Methanesulfonic acid, butane sulfonic acid dodecylbenzene sulfonic acid and camphor sulfonic acid were used as PANI dopants and the PANI-ES / nylon blends were characterized by electrical conductivity (room and low temperature) and transmission electron microscopy. The results of these various measurements and the conclusions which can be drawn regarding morphology and conductivity of the blends, will be reported.

Polyanilines are of particular current interest primarily due to their relative ease of synthesis, low cost and stable conductivity in air. The insulating, polyaniline emeraldine base (PANI-EB) form becomes electrically conducting by preferential protonation or 'doping' imine nitrogen sites to yield an electrically conducting polyaniline emeraldine salt (PANI-ES). Conductivity values of polyanilines range from insulating to 300 S/cm and are dopant concentration sensitive (1). Like most other organic conducting polymers, the PANI-ES form suffers from limited solubility in common solvents. Cao etc. al. (2) recently found that this limited solubility can be overcome by functionalizing the dopant counter-anion with polar and non-polar analogs to promote solubility in common solvents. These functionalized anions act as both flexiblizers and surfactants to the stiff polyaniline backbone to result in a soluble and processable form of PANI-ES.

[4] Current address: 2511 S.W. 35th Place, 54, Gainesville, FL 32608.

Background

Blending of polymers offers a means of engineering into one material certain combinations of desired properties exhibited individually by the component polymers. As previously demonstrated by MacDiarmid et. al. (3) polyanilines show high electrical conductivity. Their utilization in many applications is limited, however, due to their inherent brittleness. Blending polyaniline with a commercial insulating thermoplastic has the advantage of combining the plastic properties of the host material and the electrical conductivity of polyaniline.

Insulating polymers possessing many desirable mechanical properties may be rendered conductive by mixing them with conductive particles such as carbon black, metal powders, flakes or fibers of metal coated particles. The level of electrical conductivity in these heterogeneous materials depends primarily on the concentration and geometries of the conductive filler particles. The observed critical filler content which marks the onset of a sharp increase in conductivity (e.g. the percolation threshold) in the majority of these composites falls between 15-25 % (vol/vol) (4). This percolation threshold has been shown both theoretically and experimentally to be well described by a system of small spheres packed in three dimensions.

Recently, Cao et. al. discovered that after doping and complexing with sulfonic acid having a functionalized counter-anion (e.g. camphor sulfonic acid), the resulting PANI-ES can be co-dissolved in m-cresol in various ratios with an insulating host polymer (e.g. poly(methylmethacrylate), PMMA which form robust, transparent conducting films when cast from solution. Since then, other approaches have been introduced such as blending of PANI-ES through electrochemical (5) and chemical (6) polymerization of aniline monomer on polymer substrates; coating of PANI-ES polymer surfaces (7), dispersing of polyaniline in the melt of a thermoplastic polymer (8), blending undoped PANI with poly(vinyl pyrrolidone) then co-dissolving both polymers in N-methyl pyrrolidone (NMP) and subsequently doping in an acidic media (9), dispersing of PANI-ES in a melt of thermoplastic polymer and blending the non-conducting form of PANI with polystyrene sulfonic acid (10). The electrical conductivities of the blends obtained from these approaches are relatively low and the amount of PANI-ES necessary to provide high conductivity in the blend approaches 16 % (vol/vol).

The processing of doped polyaniline and host polymer in a single solvent as developed by Cao et. al. (11) gave high electrical conductivities at relatively low loading levels of polyaniline salt. Specifically, polyaniline, optimally doped with camphor sulfonic acid and blended with PMMA (e.g. PANI-0.5-HCSA / PMMA) yielded electrical conductivities on the order of 1 S/cm with only ≈ 0.3 % (vol/vol) loading of the polyaniline in the PMMA (12). Polyaniline salt, when dispersed with PMMA in the melt, requires significantly higher loading fractions of salt (≈ 9-10 % (vol/vol)) before showing the onset of electrical conductivity (8).

To account for the large difference in electrical conductivity thresholds, Cao et. al. (11) investigated the morphology of the solution cast polyaniline blends. They reported (12) that the PANI-0.5-HCSA / PMMA blend, upon casting from m-cresol solution, forms a multiphase system. The low onset of electrical conductivity in the blend was apparently due to the fractal-like network formation of the phase-separated salt in the host PMMA during solvent (e.g. m-cresol) evaporation (13). The formation of this network, was confirmed by Reghu et. al. (14) using TEM. These blends exhibit a pseudo-percolation conduction behavior in that no sharp increase in conductivity (e.g. no conventional percolation threshold) is observed. The conductivity rises smoothly from the insulating state with increasing polyaniline salt concentrations.

The 'non-conventional' relationship between electrical conductivity and polyaniline salt content in the solution cast blend was interpreted by many authors

(15), (16), (17) and (18) as evidence of a phase-separated structure with conducting networks or pathways of pure polyaniline salt forming at low loading levels. It is difficult to pinpoint the conditions and molecular interactions that control the formation of the salt network. Interactions between solvent and polymer and between polymers; rate of crystallization of respective blend components and kinetic effects of solvent evaporation from blend solution are all thought to be operative in forming these conductive pathways.

Most polyaniline blends described in the literature were processed from m-cresol, a solvent which is both acidic and high boiling. Secondary effects, such as hydrolysis and a high residual solvent content after drying dilutes the effect of the salt counter-anion on the host polymer and gives a distorted picture of the blend system. By using hexafluoroisopropanol (HFIP) as co-solvent for both salt and host polymer, these secondary effects can be eliminated and a 'cleaner' blend is produced from which the role of the salt in the blend can be studied without interference from residual solvent.

In this work, we describe the effect of the dopant counter-anion functionality on the network morphology. We explore the blend morphologies of PANI-0.5-HMSA, PANI-0.5-HDBSA and PANI-0.5-HCSA salts in nylon 6 and nylon 12. The methane sulfonic acid (HMSA) and dodecylbenzene sulfonic acid (HDBSA) dopants were chosen for this study because they differ greatly in polarity. (Dodecylbenzene sulfonic acid has a longer hydrophobic tail compared to methane sulfonic acid). In addition, camphorsulfonic acid dopant (HCSA) has, in addition to the sulfonate groups, an extra functionality (carbonyl) to hydrogen bond with the host and neighboring salt polymer. The PANI-0.5-HCSA salt was selected since it was observed that it yields the highest electrical conductivity of all the surveyed polyaniline salts. We have selected two aliphatic polyamides as the insulating hosts where the repeat frequency of amide groups between the hydrocarbon backbone is the variable.

We have previously reported (19) the electronic and vibrational spectra, solution viscosity and electronic properties of PANI-EB and PANI-ES in HFIP solvent. We are able to correlate the change in hydrodynamic volume with a concomitant red shift in the UV-Vis / near IR spectra and thus verify the "coil" to "expanded coil conformation hypothesis" (20). We have also found that high quality films can be cast from HFIP that are relatively solvent free, free standing and exhibit high electrical conductivities which depends on which dopant is used. However, not matter what solvent is used, pure polyaniline films are quite brittle and lack the physical properties necessary for many applications.

Thus in this study, we report on the use of HFIP as a blending solvent for PANI-ES with commercial high molecular weight nylons. We observe that the onset of electrical conductivity occurs at low volume fractions of the conductive component, compared to metal filled composites (4). This is primarily due to the unique morphology of the PANI-ES that is formed within the insulating host polymer. PANI-ES forms a network structure at volume fractions of < 1 % (vol/vol) (21). The mechanism behind the formation of this immiscible polymer network bay be due, in part, to the polarity differences (induced by the functionality of the dopant counter-anion) in the doped polyaniline and the insulating host polymer. This phase separation of PANI-ES and an insulating matrix into a cellular-like network is central to the idea of obtaining high conductivities at low salt loading fractions.

Experimental

Synthesis: Polyaniline emeraldine base (PANI-EB) was prepared via the FeCl3 method as described elsewhere (19).

Safety Information: Although HFIP is not identified as a carcinogen or teratogen, its solvent power towards skin, mucous membrane of the eyes and nose, combined with its high volatility, require that it be handled with considerable caution. See standard safety data tabulations such as MSDS or similar for details.

Doping: Doped polyaniline solutions were prepared in HFIP by a solution doping method (22). Solutions turned from blue / brown to forest green, characteristic of doped polyaniline. Molar doping of undoped polyaniline is calculated for polyaniline emeraldine salt from the mole ratio y = (moles of dopant) / (moles of phenyl-NH), determined by elemental analysis. Optimally doped polyaniline has the value of y = 0.5. Dopants used were: camphor sulfonic acid (HSCA), (Aldrich); methane sulfonic acid (HMSA), Aldrich; and dodecyl benzene sulfonic acid (HDBSA), TCI America. Nylon 6 and 12 (Aldrich) were vacuum dried before solution blending. All PANI-ES solutions were filtered with a 0.50 μm filter.

Film preparation: All filtered, doped polyaniline salt and blend solutions were solution cast onto a Teflon coated glass substrate. They were covered with a glass dish to allow for slow evaporation (over a period of 24 hours) of the solvent at room temperature. HFIP, with a boiling point of 60 °C, allows gentle stripping at room temperature. All films were peeled off the Teflon tape substrate and dried in a dynamic vacuum oven at 75 °C for approximately 24 hours. From fluorine elemental analysis, percent residual HFIP solvent in films was less than 0.5% (wt/wt). This casting process from HFIP produced robust, free standing and solvent free polyaniline / nylon blend films.

Results and Discussion

Effect of Dopant Counter-anion and Polyamide Host on Blend Electrical Conductivity. Electrical conductivity has been shown to be sensitive to the morphology of a conducting polymer (20). Conductivity is used to address the nature of the polyaniline salt morphology in the insulating polyamide matrix. In an attempt to modify this morphology (and resulting conductivity) we have evaluated six optimally doped polyaniline salt / polyamide blends to better understand the role of the counter-anion on PANI-0.5-ES's morphology and resulting conductivity.

Figure 1 shows electrical conductivity as a function of volume fraction of the polyaniline in blends with nylon 6 or with nylon 12. It is seen that the rate of increase in conductivity varies depending on the composition of the blend. Both nylon 6 and nylon 12 blends show low onsets of electrical conductivity at less than 4 % (vol/vol) of conductive component which is several times lower than the percolation threshold for globular conducting particles in an insulating matrix (23). In addition, the conductivity for all polyaniline / polyamide blends increases at fast rate as the concentration of conducting polyaniline increases above the value marking the onset of conductivity. None of the PANI-ES / nylon blends (Figure 1) exhibit any clear conductivity saturation but rather a continuous increase in electrical conductivity. At 50-60 % salt content, the polyaniline blend conductivity value becomes more like the respective pure polyaniline salt, reflecting the nature of the salt component rather than the blend. The absence of any data above 60 % blend in Figure 1 is due to extremely brittle nature of the film samples which prohibited accurate electrical conductivity measurements.

Since low concentrations are technologically more important, we restricted work to these blend concentrations. At these lower levels, the electrical conductivity of the polyaniline blends is quite sensitive to salt levels. Figure 2 shows the effect on conductivity when the host nylon is changed from nylon 6 to nylon 12. As seen in Figure 2-a, ≈0.5 % concentration of PANI-0.5-HMSA salt in nylon 6 shows a conductivity of ≈ 10-4 S/cm, whereas a similar concentration value of the same salt in nylon 12 has a conductivity of ≈ 10-5 S/cm . The level of conductivity above the

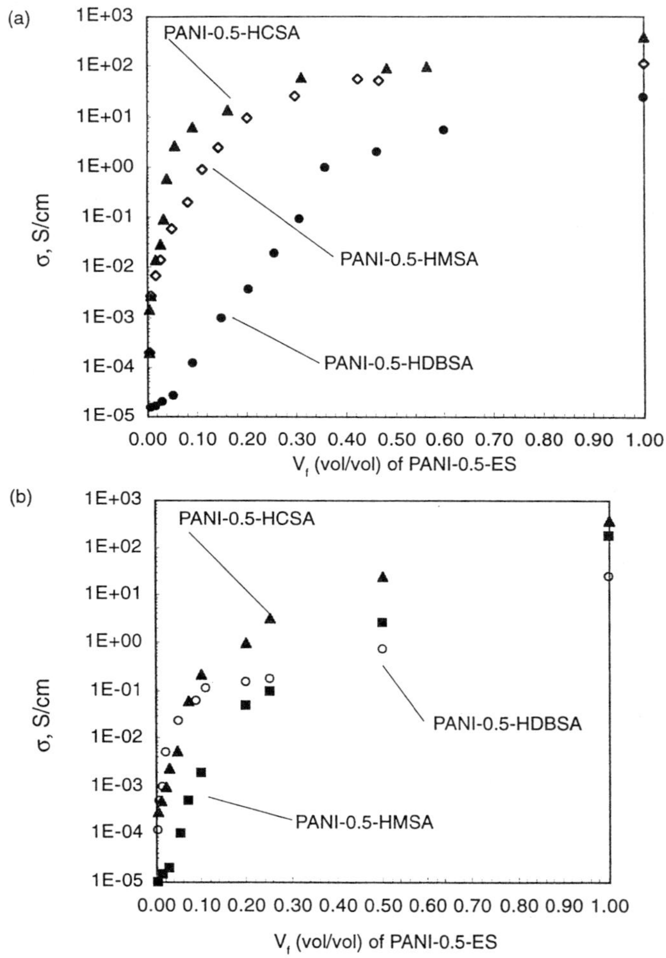

Figure 1. d.c. Electrical conductivity, σ, versus volume fraction (vol/vol) of PANI-0.5-HCSA; PANI-0.5-HMSA and PANI-0.5-HDBSA salts solution blended with (a) nylon 6 and (b) nylon 12.

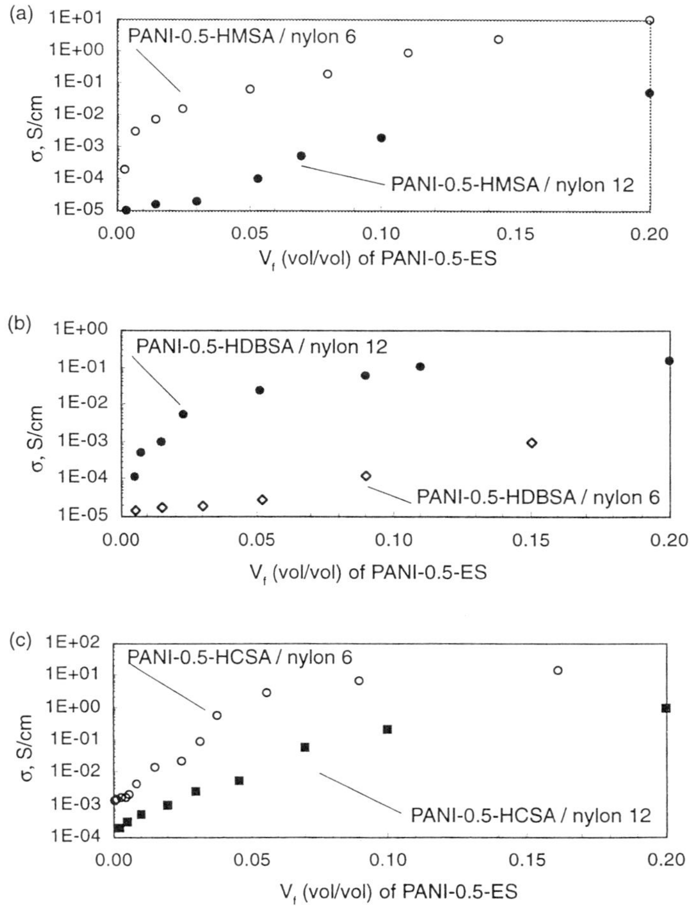

Figure 2. Summary of d.c. Electrical conductivity, σ, versus volume fraction (vol/vol) of (a) PANI-0.5-HMSA / nylon; (b) PANI-0.5-HDBSA / nylon and (c) PANI-0.5-HCSA / nylon blends.

onset value in PANI-0.5-HMSA / nylon 6 blends increase approximately three orders of magnitude to ≈0.5 S/cm at 8 % (vol/vol) loading, compared to an increase of only one order of magnitude to a value of 10-4 S/cm at the same salt loading level for the nylon 12 blend. At the same concentration, blending the less polar salt (e.g. PANI-0.5-HDBSA) in nylon 12 (Figure 2-b), which has a larger hydrocarbon segment between amide groups, shows a higher conductivity than the PANI-0.5-HMSA / nylon 12 blend.

The relationship between conductivity and volume fraction of PANI-0.5-HCSA in nylon 6 or nylon 12 (Figure 2-c) shows behavior similar to that of the other blends surveyed. However, the electrical conductivity in PANI-0.5-HCSA blends is seen to increase at a higher level compared to the PANI-0.5-HMSA and the PANI-0.5-HDBSA systems. This behavior suggests that the more highly conducting salt (i.e. PANI-0.5-HCSA) is more efficiently dispersed in the both nylons, resulting in very low onset thresholds and higher blend conductivities.

It is important to note the differences in the electrical conductivity behavior at low salt loading fractions. By blending a polar salt such as PANI-0.5-HMSA with nylon 6, with its shorter hydrocarbon segments than the nylon 12, percolation of the conducting salt occurs at a lower volume fraction. We suggest that the polarity matching, along with other secondary interactions between the salt and the nylon acts to better disperse the phase separated conducting pathways. The sensitivity of electrical conductivity to this degree of dispersion implies that salt morphology is determined by the degree of interactions between polyaniline salt and host nylon polymer.

Effect of Dopant Counter-Anion and Polyamide Host on Blend Morphology. The low onset value of electrical conductivity in all the blends suggests a branch-like structure of the polyaniline phases in the nylon. The biphasic nature of these materials was established by a series of experiments using transmission electron microscopy (TEM)

A continuous, multi-connected (dark) network is seen in Figure 3-b for the PANI-0.5-HMSA / nylon 6 blends at the 0.5 % loading level. As the concentration increases to 1.5 % (Figure 4-b) the density of these networks increases. Further increase in salt concentration to 5 % shows a convoluted mass of individual networks due to their increase density per square unit of the TEM micrograph. The formation of these conducting pathways at 1.5 % provides evidence that the low onset of electrical conductivity is due to the formation of these conductive pathways. The observed exponential increase in conductivity (Figure 3-a) is consistent with the growth of the conductive network phase seen in Figures 3-b and 3-c. The lower onset of electrical conductivity for the less polar salt, PANI-0.5-HDBSA in the relatively polar nylon 6 implies a less developed network (Figure 4-b) in terms of its morphology, compared to that of the PANI-0.5-HMSA / nylon 6 blend. As the PANI-0.5-HDBSA salt concentration increases to 1.5 % (Figure 4-c), the polyaniline shows signs of semi-connected aggregated domains embedded in the nylon 6 matrix. Further increases in salt concentration show gross phase separation (Figure 4-d),where the dark areas in the optical micrograph are represented by PANI-0.5-HDBSA. The absence of a well-developed and connected network is consistent with the observed slow increase in conductivity in these blends.

The highly conducting PANI-0.5-HCSA salt morphology (Figure 5-b) in nylon 6 shows initial growth of a morphology in which the connective nature of the network increases with increasing concentration. Further salt loading to 4 % (Figure 5-c) shows an extremely dense network. Here again, the formation of a fibrillar network is consistent with the low onset and exponential increase in electrical conductivity of the PANI-0.5-HCSA / nylon 6 blend system.

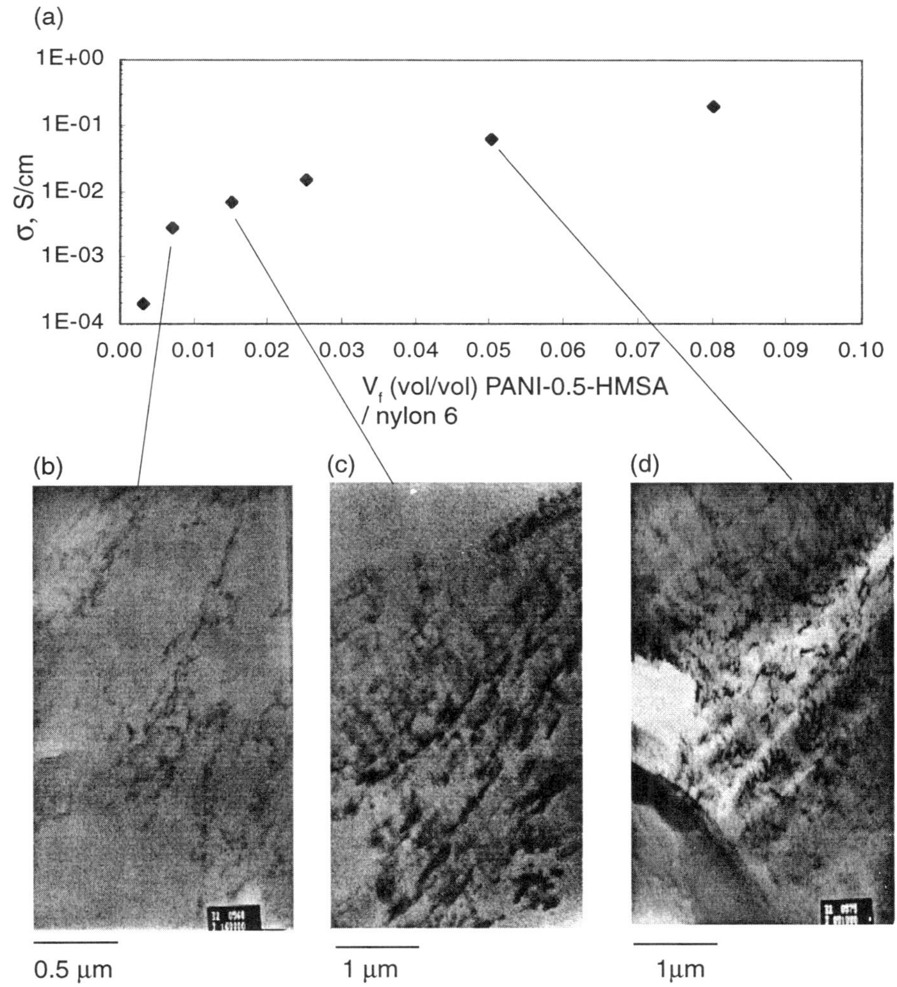

Figure 3. (a) Onset of electrical conductivity for x % (vol/vol) of PANI-0.5-HMSA / nylon 6 blend system. (b), (c) and (d) TEM micrographs of 0.5 %, 1.5 % and 5 % salt blends, respectively. Dark areas in micrographs represent stained polyaniline salt imbedded in polyamide.

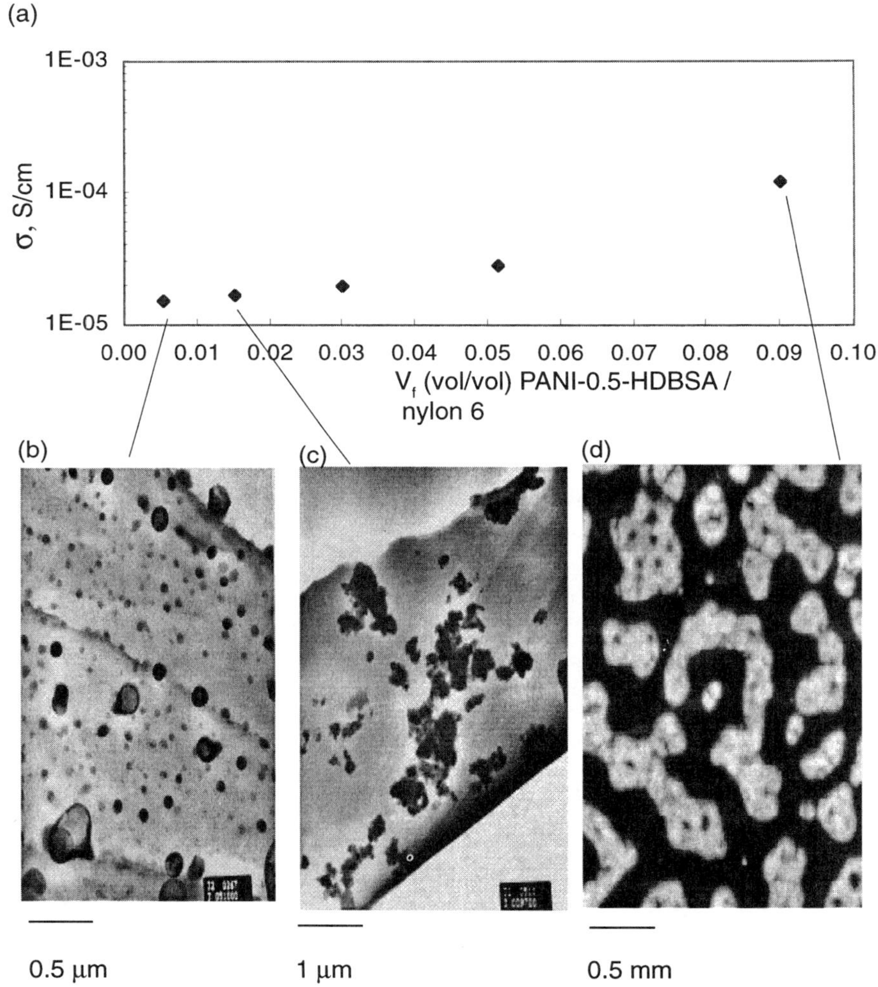

Figure 4. (a) Onset of electrical conductivity for x % (vol/vol) of PANI-0.5-HDBSA / nylon 6 blend system. (b) and (c) TEM micrographs of 0.5 % and 1.5 % salt blends, respectively. (d) Optical micrograph of 10 % PANI-0.5-HDBSA / nylon 6 blends.

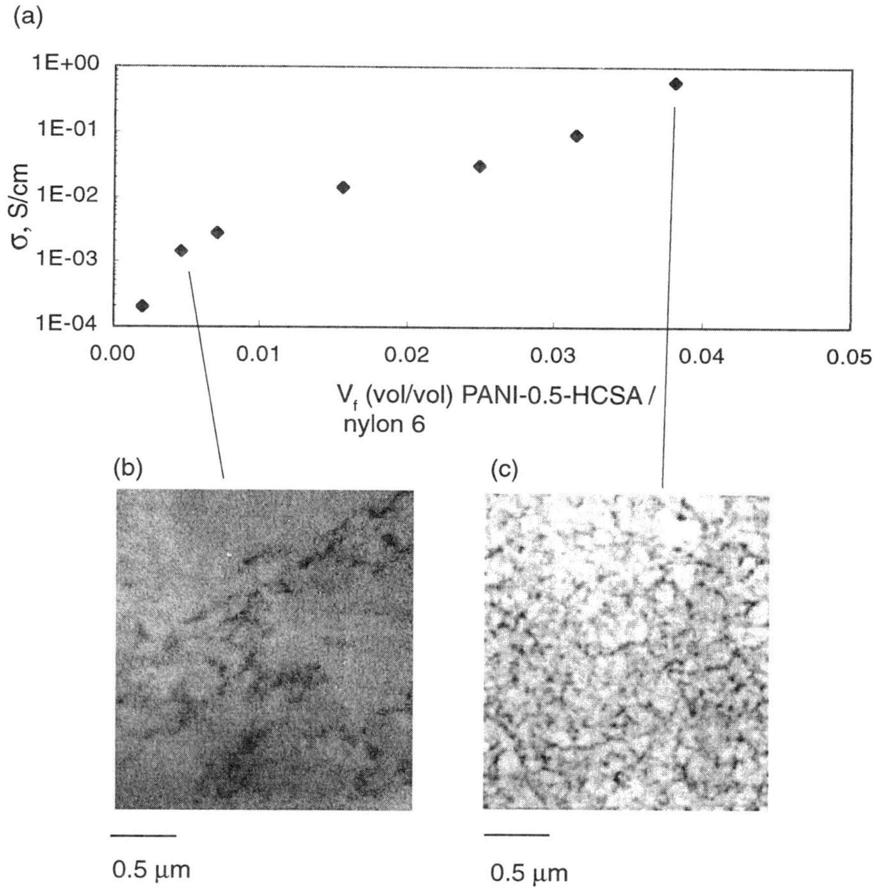

Figure 5. (a) Onset of electrical conductivity for x % (vol/vol) of PANI-0.5-HCSA / nylon 6 blend system. (b) and (c) TEM micrographs of 0.5 % and4 % salt blends, respectively.

When the three different salts are separately blended with a less polar nylon such as nylon 12, the morphology is seen to be fractal at lower salt concentrations for the case of PANI-0.5-HDBSA containing blends (Figure 7-b and c) than the PANI-0.5-HMSA and PANI-0.5-HCSA / nylon 12 blends (Figure 6-b, c, d and Figure 8-a, respectively). The lower onset of electrical conductivity of PANI-0.5-HDBSA / nylon 12 compared to the PANI-0.5-HMSA and PANI-0.5-HCSA containing blends reflects the presence of the fractal conducting pathways in the PANI-0.5-HDBSA blend.

It is interesting to note that the PANI-0.5-HCSA blends at ≈ 1 % show in nylon 12 a discontinuous network, but a more fibrillar-like network when blended with nylon 6. This more aggregated morphology in the nylon 12 is reflected in the more gradual growth versus a more continuous increase in electrical conductivity as a function of salt loading. (Compare 4.6-a to 4.7-a.)

In summary, all the polyaniline blends studied exhibit a conductivity which rises smoothly and rapidly from the insulating state with increasing polyaniline concentration. However, the onset of conduction seems to be dependent on the nature of the conductive pathways that are present at low loading fractions. Polyaniline salts with the more polar counter-anions (e.g. MSA-) blended with the more polar polyamide (e.g. nylon 6) show signs of a continuous, multi-connected network whereas spherical salt domains are characteristic of a more nonpolar PANI-0.5-HDBSA blended with nylon 6. The threshold for electrical conductivity is sensitive to the morphological structure of the polyaniline / nylon blends.

Conduction Behavior of Polyaniline Blends. The characteristic transport properties of the PANI-ES network in a matrix of polyamide may be followed by observing the temperature dependence of conductivity. The details of the possible mechanism of conduction have been previously described (24) . In summary, the transport dependence of conductivity of the different polyaniline salts displayed a behavior best described by a variable range hopping between localized domains in three dimensions at low temperature and a nearest neighbor hopping conduction at higher temperatures, as seen in Figure 9. In Figure 10, we give a representative example (PANI-0.5-HMSA / nylon) of the temperature dependence of conductivity for blends of polyaniline in nylon 6 and in nylon 12 insulating hosts. It is clear that unlike the polyaniline salts, the conductivity data follow a hopping transport in three dimension at all temperatures (σ vs $T^{-1/4}$). The slope of the temperature dependence increases with dilution, indicating increased disorder (slope = T_0 which is inversely proportional to localization length). The electrical conductivity is specially sensitive to the existence of insulating barriers in a heterogeneous system. This sensitivity is much more pronounced at dilution levels closer to the percolation threshold (≈ 0.5 %). However, blends containing low polyaniline salt content were more resistive, as the temperature is lowered, which made it difficult to cover a reasonable temperature range due to instrument limitations.

At a concentration closer to the percolation threshold, manifestation of the affects which are due to the network morphology and counter-anion on the conductivity may be revealed. From Figure 10, it is seen that at high concentration of PANI-ES (10 % and 25 %), the behavior of conductivity temperature dependence of PANI-0.5-HMSA in nylon 6 and in nylon 12 are similar. As the concentration of PANI-ES lowered, the data became more sensitive, demonstrating greater temperature dependence (larger slope) for nylon 6 blend than that for nylon 12 blends. This behavior is opposite to expectation since the room temperature conductivity of PANI-0.5-HMSA salt in nylon 6 is higher than that in nylon 12. As pointed out earlier, the higher conductivity of PANI-0.5-HMSA / nylon 6 blend is due to a more network type morphology compared to blends with nylon 12. Better network structure means few barriers to conduction and higher charge mobility. We explain

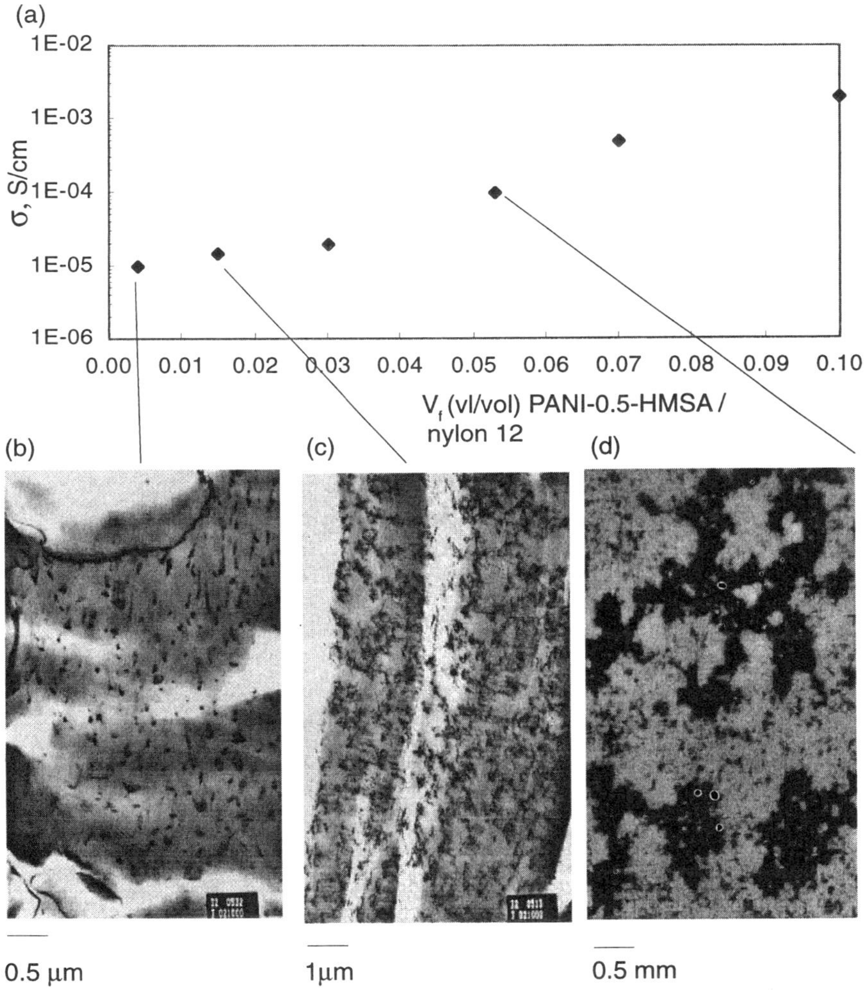

Figure 6. (a) Onset of electrical conductivity for x % (vol/vol) of PANI-0.5-HMSA / nylon 12 blend system. (b) and (c) TEM micrographs of 0.5 % and 1.5 % salt blends, respectively. (d) Optical micrograph of 5 % PANI-0.5-HMSA / nylon 12 blends.

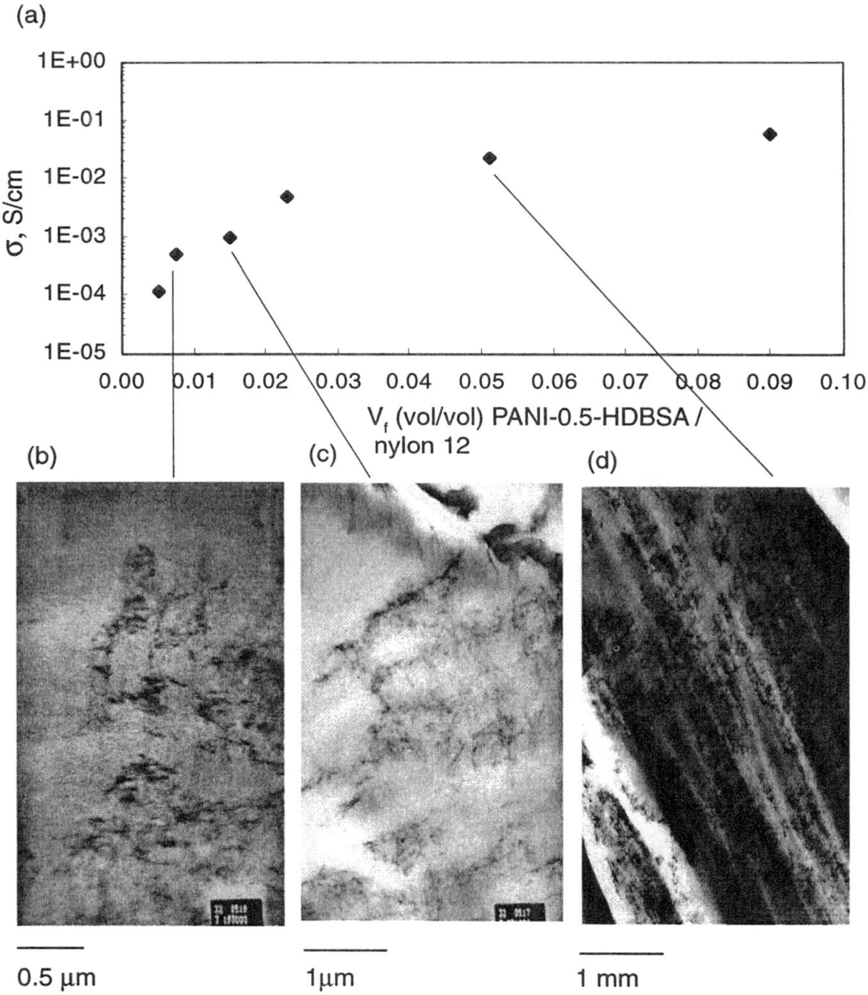

Figure 7. (a) Onset of electrical conductivity for x % (vol/vol) of PANI-0.5-HDBSA / nylon 12 blend system. (b), (c) and (d) TEM micrographs of 0.5 %,1.5 % and 5 % salt blends, respectively.

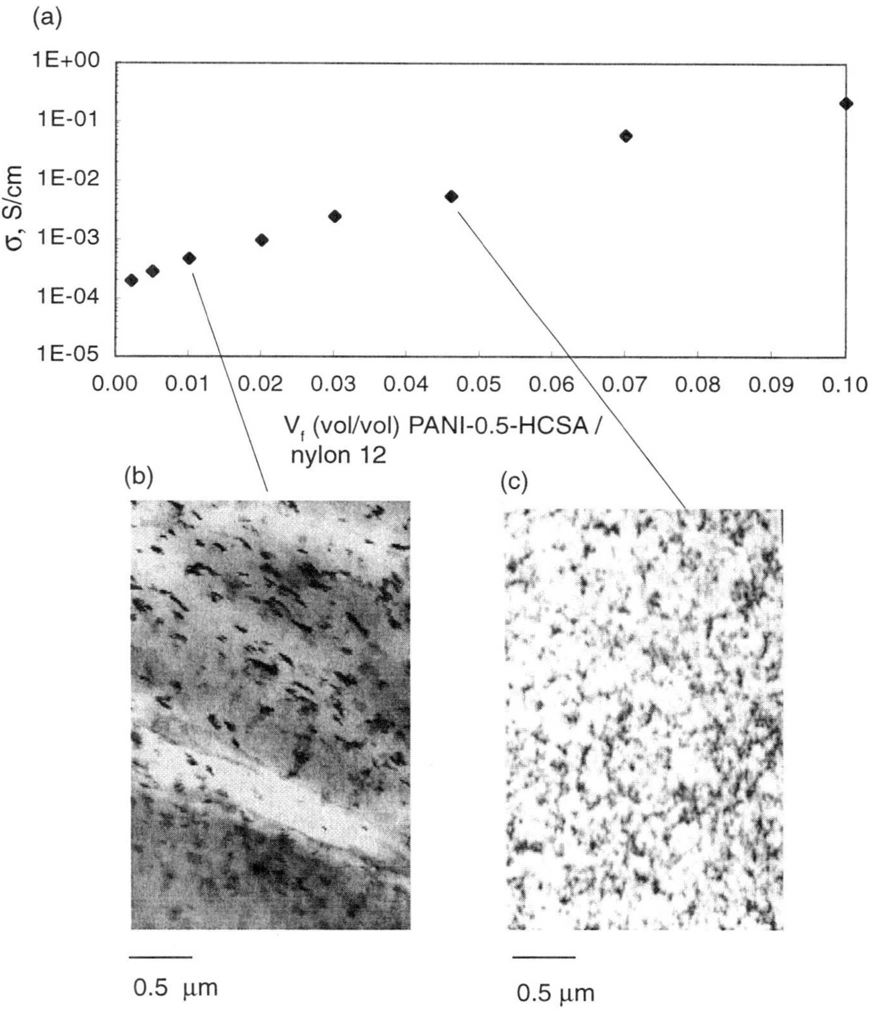

Figure 8. (a) Onset of electrical conductivity for x % (vol / vol) of PANI-0.5-HCSA / nylon 12 blend system. (b) and (c) TEM micrographs of 0.7 % and 5% salt blends.

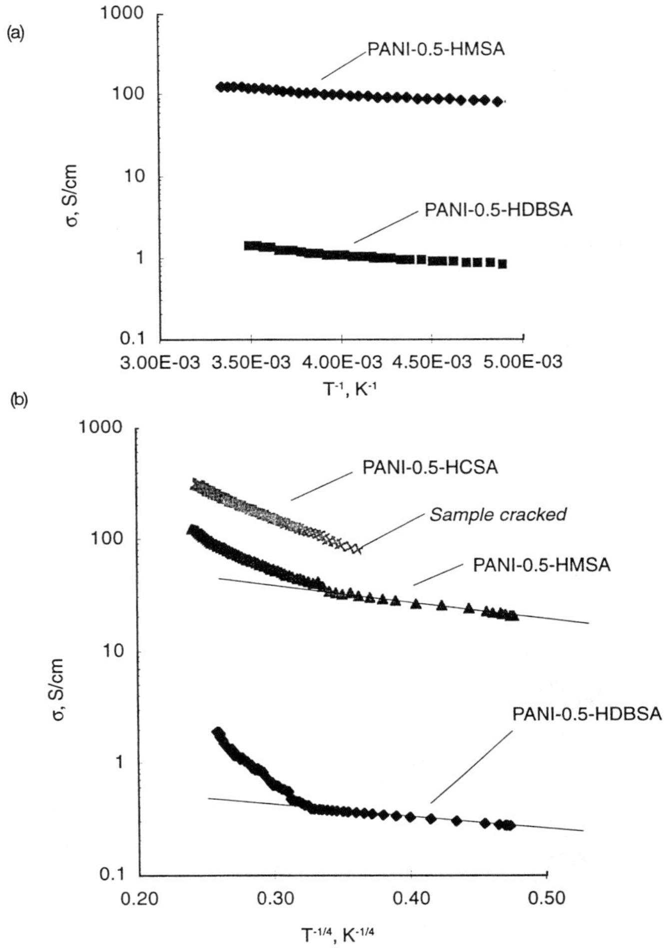

Figure 9. Temperature dependence of the d.c. conductivity of optimally doped polyaniline doped with HCSA , HMSA and HDBSA against (a) T^{-1} and (b) $T^{-1/4}$

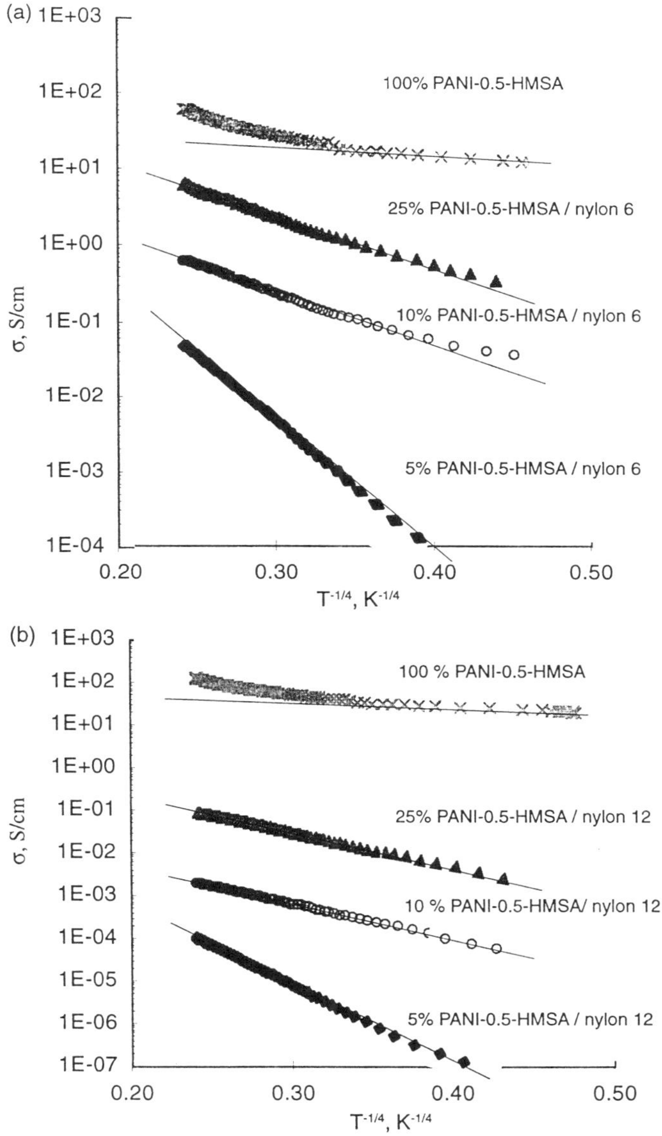

Figure 10. Temperature dependence of the d.c. conductivity behavior of PANI-0.5-HMSA / (a) nylon 6 and (b) nylon 12 blends against $T^{-1/4}$.

the unexpected greater temperature dependence of conductivity in the case of the blend containing nylon 6 in terms of lower carrier concentration arising from the a fraction of HMSA acid being present in the nylon 6 matrix and therefor not available for doping. In other words, the conductivity of PANI-0.5-HMSA salt in nylon 6 blend is not optimally doped. Holland etc. al. (25) recently studied the temperature dependence of conductivity of PANI-0.5-HCSA salt, with different doping levels and observed that salts that are not optimally doped have stronger temperature dependence than the optimally doped salts. They further concluded that the acid added to protonate the sample did not completely react with polyaniline and that an excess acid is required to obtain optimally doped salt. Although the carrier mobility in PANI-0.5-HMSA / nylon 6 blends is higher than with nylon 12 blends due to better fractal network in the former, the carrier concentration is lower leading to stronger temperature dependence at lower temperatures.

In blends containing polyaniline doped with HDBSA, no significant variation in the temperature dependence behavior between PANI-0.5-HDBSA / nylon 6 and PANI-0.5-HDBSA / nylon 12 is observed, although a weaker temperature dependence is expected for the latter blend. In this case, a larger fraction of the doping acid is expected to remain in the nylon 12 matrix rather than in nylon 6. This would result in lower carrier concentration for the PANI-0.5-HDBSA in the nylon 12 host and a stronger than expected temperature dependence of conductivity.

Conclusion

The threshold value for electrical conductivity is shown to be sensitive to the morphological structure of the salt network. Measurable electrical conductivity of these blends shows very low values at approximately 0.5 % by volume of the salt. At this threshold value for electrical conductivity, a fractal network is established through the nylon host. Lower onsets of electrical conduction in the PANI-0.5-HMSA / nylon 6, PANI-0.5-HDBSA / nylon 12 and PANI-0.5-HCSA / nylon 6 blends paralleled a finer, more branched network whereas the lower conductivities accompanied a coarser, more globular morphology in the PANI-0.5-HMSA/ nylon 12, PANI-0.5-HDBSA / nylon 6 blends. At concentrations less than 5 % (vol/vol) a conducting network forms throughout the nylon sample. As salt level increases above the threshold value, the network becomes dense and convoluted and the electrical conductivity becomes less sensitive to dopant anion functionality.

Transport in conducting polyaniline / nylon blends is observed to be independent on the composition, dopant anion, host matrix and structure of the conducting phase. The microscopic transport in all salt networks remains unchanged by the dilution of the salt in the nylon matrix. The conductivity follows a temperature dependence characteristic of generalized variable range hopping mechanism with χ = -1/4. The slope of the temperature dependence increases with dilution, indicating increased disorder in the polyaniline salt.

Acknowledgment

This research was in part supported by a grant to Paul G. Rasmussen from the donors of the Petroleum Research Fund, administered by the American Chemical Society and by the Division of the Materials Sciences, Office of Basic Energy Sciences, Department of Energy, under contract DE-AC05-96OR22464.

References

(1) Hopkins, A. R.; Rasmussen, P. G.; Basheer. R. A.; Annis, B.K.; Wignall, G. D. ACS Polymer Preprints, **1997**, 38, 1, 427
(2) (a) Cao, Y.; Smith, P.; Heeger, A.J. *Synth. Metals*, **1992**, 48, 91.
(3) Huang, W.S.; Humphrey, B.D.; MacDiarmid, A.G. *J. Chem. Soc. Faraday Trans*. **1986**, 82, 2385
(4) (a) Zallen, R.; The Physics of Amorphous Solids, Wiley, New York, **1983**. (b) Hotta, S.; Rughooputh, S.D.D.V.; Heeger, A.J.; *Synth. Met*. **1987**, 22, 79.
(5) (a) Dalas, E.; Kalitsis, J.; Sakkopoulos, S.; Vitoratos, E.; Koutsoukos, P.G. *Synth. Met*. **1991**, 41, 963. (b) Chen, S.A.; Fang, W.G., *Macromolecules* **1991**, 24, 1242.
(6) Hsu, C.H. *Synth. Met*. **1991**, 41, 671.
(7) (a) Gregory, R.V.; Kimrell W.C.; Kuhn, H.H. *Synth. Met*. **1989**, 28, C823. (b) Genies, E.M.; Petrescu, C.; Olmedo, L. *Synth. Met*. **1991**, 41, 665.
(8) (a) Wessling, B.; Volk, H.; *Synth. Met*., **1987**, 18, 671. (b) Paoli, M.A.; Duek, E.A.; Rodrigues, M.A.; *Synth. Met*. **1991**, 41, 973. (c) Wessling, B. *Kunststoffe*, **1986**, 76, 930. (d) Ezquerra, T.A.; Kremer, F.; Mohommadi, M.; Rueche, J.; Wegner, G.; Wessling, B.; *Synth. Met*. **1989**, 28, C83.
(9) Stockton, W. B. Ph. D. thesis, Massachusetts Institute of Technology, **1995**, 12-24.
(10) Ghosh, S.; Chandra, R. *J. Applied Polymer Science*, **1990**, 40, 1049.
(11) (a) Cao, Y.; Smith, P.; Heeger, A.J. *Synth. Met*.. **1992**, 91, 48. (b) Cao, Y.; Smith, P.; Heeger, A.J. *Appl. Phys. Lett*..
(12) Yan, C.Y.; Cao, Y.; Smith, P.; Heeger, A.J., *J. Polym. Prepr*. (*Am. Chem. Soc., Div. Polym. Chem*.) **1993**, 1, 790.
(13) (a) Cao, Y. Smith, P. Heeger, A.J. *Synth. Met*. **1992**, 91, 48. (b) Yang, C.Y., Cao, Y.; Smith, P.; Heeger, AJ. *ibid*. **1993**, 53, 293.
(14) Reghu, M.; Yoon, C.O.; Yang, C.Y.; Moses, D.; Heeger, A.J.; Cao, Y.; *Macromolecules*, **1993**, 26, 7245.
(15) Minto, C. D. G.; Vaughhan, A. S. *Synth. Met*. **1996**, 81, 81-86.
(16) Heeger, A. J. *Trends in Polymers* **1995**, 2, 39-47.
(17) Banerjee, P.; Mandal, B.M. *Macromolecules* **1995**, 28, 3940-3943.
(18) Cao, Y.; Smith, P.; Heeger, A. J. *Synth. Met*.. **1992**, 48, 91-97.
(19) Hopkins, A. R.; Rasmussen, P. G.; Basheer. R. A.; *Macromolecules*. **1996**, 29, 7838-7846.
(20) Xia, Y.; MacDiarmid, A.G.; Epstein, A. J. *Macromolecules* **1994**, 27, 7212-7214.
(21) Reghu, M.; Yoon, C.O.; Yang, C.Y.; Moses, D.; Smith, Paul; Heeger, A. J.; Cao, Y. *Phys. Rev. B*. **50**, 19, 13931-13941.
(22) Hopkins, A. R.; Rasmussen, P.G.; Basheer, R.A. *Macromolecules*, 29,(1996) 7838.
(23) Cao, Y.; Smith, P.; Heeger, A.J. *Synth. Met*. **1993**, 3514-3519.
(24) Hopkins, A. R. *Ph. D. Thesis*, The University of Michigan, December 1997
(25) Holland, E.R.; Pomfret, S.J.; Adams, P. N.; Monkman, A.P. *J. Phys. Condens. Matter* **1996**, 8, 2991-3002. (25) Hopkins, A. R.; Rasmussen, P. G.; Basheer. R. A.; Annis, B.K.; Wignall, G. D. ACS Polymer Preprints, **1997**, 38, 1, 427.

Chapter 4

Dependence of Electrical and Optical Properties on the Morphology and Microstructure of Polyaniline

S. S. Hardaker [1], K. Eaiprasertsak [1], J. Yon [1], R. V. Gregory [1], G. X. Tessema [2], R. Ou [3], C. Cha [3], and R. J. Samuels [3]

[1] Center for Advanced Engineering Fibers and Films and School of Textiles, Fiber, and Polymer Science, and [2] Department of Physics and Astronomy, Clemson University, Clemson, SC 29634
[3] School of Chemical Engineering, Georgia Institute of Technology, Atlanta, GA 30332

Since the discovery of electroactive polymers in the 1970's extensive work has been carried out to utilize these materials in a variety of applications normally not thought to be germane to organic polymers. Anticipated applications have ranged from "organic wires" made from fibers of intrinsically conductive organic polymers to semiconductor devices, and light emitting diodes.[1] Use of these polymers for battery electrodes and actuators is also finding applications in a variety of engineered high technology materials. Since the early work of Shrikawa, MacDiarmid and Heeger, countless new polymers for electronic, and recently photonic, applications have been synthesized.[2]

While many of the descriptions of charge transport in these polymers is based on our knowledge of the solid state physics and chemistry governing transport mechanisms in metals and inorganic semiconductors, few models describing this phenomena in polymers have been put forth. Most of our understanding of these materials is based upon the knowledge base already established for well-ordered and glassy metals, and on the crystalline state of inorganic semiconductors. Organic polymers are, however, quite different than most of their inorganic counter-parts. Organic polymers will have a variety of differing structures depending not only on the chemical repeat unit of the mers but also on the interactive forces between the polymer chains. These forces will determine the final polymer morphology. In addition the processing of these polymers will have a substantial effect on the polymer's final morphology and structure, crystallinity, density, elastic properties, solubility, strength, bend modulus, stability, and a variety of other properties. An example of a processing related effect on structure is the degradation of molecular weight due to shearing in a processing stage, exposure to high temperatures, or other conditions the polymer might be exposed to while being processed into an end item. Although the literature is rich with detailed accounts of these effects on the mechanical properties of organic polymers, such as the polyamides and polyesters used in fiber production, there are very few papers devoted to these effects in electroactive polymers. Even fewer papers discuss the effect of morphology development during processing on the electrical and optical properties of these polymers. This contribution will seek to demonstrate the effect of various formation methodologies on the resultant electrical and optical properties.

Polyaniline (PANI) is a polymer of particular interest due to its facile processing capability in its non-conductive emeraldine base form. This chapter will consider the effects of polymer orientation on the refractive index in all three dimensions of polyaniline (PANI) films formed by various casting techniques. These changes in the refractive index can be related to the orientation of the polymer molecular chains and therefore the molecular structure of the formed polymer. Additionally we will discuss the effect of PANI films prepared from solutions containing differing concentrations of a reducing agent, which reduces the emeraldine base (EB) to the lecoemeraldine base (LEB) form of the polymer resulting in significantly different morphologies. Determination of the thermopower of films formed from PANI in different oxidation states will be shown to be indicative of the charge transport mechanisms at different temperatures.

The effects of differences in polymer structure and morphology on final properties will vary from polymer to polymer. Morphological effects on final electrical, optical, and mechanical properties of the formed polymer will in general hold for all organic polymers whether or not they are electroactive. Although the observed changes reported here are specific to PANI the effect of these differences must be considered for all electroactive polymeric materials.

Experimental

There are many different experimental methods described in the literature regarding the chemical and electrochemical synthesis of polyaniline.[3] For the purposes of this writing we will only discuss the effect of morphological changes of PANI films formed from chemically synthesized polyaniline Synthesis conditions will affect the properties of the final polymer product due to changes in molecular weight and interaction of the growing polymer with either itself or neighboring polymer chains and possibly other molecular species present in the reaction mixture. Normally polyaniline is synthesized under acidic conditions leading to a doped electrically conductive powder. Subsequent dedoping of polymer results in the emeraldine base (EB) form that is soluble in several different solvents. Much work in the literature is reported on solvation of the EB form in N-methyl-2-pyrrolidinone (NMP). We have found and reported in previous work that di-methyl-propylene urea (DMPU) is in fact a better solvent for the PANI base forms than is (NMP) for processing purposes.[4] By better solvent we mean that the polymer in DMPU has a higher intrinsic viscosity and viscoelastic properties consistent with better solvation. The solvent from which PANI is processed will have a significant influence on the morphology of the solid state polymer. If for example a film is cast from a poor solvent, where the polymer/polymer interactions are strong resulting in a tight ball chain configuration, the mechanical and electrical properties will be considerably different than if the chains were in an expanded coil conformation by processing in a much better solvent. Methods of synthesis, dedoping, reduction of EB to LEB, and solvation for the study described in this chapter are outlined below.

Spin Coated and Drawn Films. Polyaniline was chemically synthesized by addition of an ammonium persulfate solution (180g $(NH_4)_2S_2O_8$ in 1.0M HCl) to a stirred aniline solution (73.6g in 1.0M HCl). The oxidizing solution was added over two hours under constant stirring and the reaction mixture was held at about 0^oC. After two additional hours, the reaction mixture was filtered and washed with 6.5 liters of water, followed by deprotonation in excess 3 wt% NH_4OH. The deprotonated base polymer was filtered and washed with 6.0 liters of water. The polymer was washed in one liter of methanol, filtered and dried. Molecular weights were estimated with gel permeation chromatography (DMSO, 0.2%LiBr, 75^oC, polyvinyl pyridine standard) as Mn=35000, Mw=83000, Mw/Mn=2.4. N,N'-dimethyl propylene urea (DMPU) was chosen as the processing solvent for the preparation of films via spin coating as this solvent inhibits the formation of gels, allowing stable solutions of PANI to be prepared for solution processing.[5] A 7.5 wt % solution of PANI was spin-coated onto glass substrates with a spin time of two minutes and a spin speed of 750 rpm. The films were dried at 130^oC for 90 seconds and allowed to cool to room temperature. The films were removed from the substrates by submersing in water. The spin coated samples were hand drawn at 100^oC to varying extension ratios to a maximum of 1.63 at about 5 cm/min.

Oxidation State Studies. For the oxidation state studies polyaniline was chemically synthesized by addition of an ammonium persulfate solution (in 1.0 M HCl) to a stirred aniline solution (in 1.0 M HCl) over a two and one half-hour period. The temperature of the reaction mixture was controlled at -30 °C. LiCl (6.0 M) was present to prohibit freezing. For this reaction, the aniline:oxidant molar ratio was 2:1. The total reaction time was 24 hours after which the resultant emeraldine salt was washed with 6.0 L of deionized water. The emeraldine salt was deprotonated by stirring in 3 wt% NH_4OH for 25 hours and washed with 6.0 L deionized water. Oligomer was removed by stirring the emeraldine base in methanol for 45 minutes followed by rinsing with 6.0 L methanol. Gel permeation chromatography (polystyrene standards, N-methyl-2-pyrrolidinone/0.05 M LiBr as eluent, 80 °C) was employed to determine the molecular weight of the emeraldine base (Mn=32,000 and Mw=60,000).

Solutions (9 wt%) of the emeraldine base were prepared in N,N'-dimethyl propylene urea (DMPU) containing varying amounts of phenyl hydrazine. The molar ratios of phenyl hydrazine to emeraldine base (based on a four ring repeat unit) were R = 0.0, 0.36, 0.71, and 1.11. These solutions were spin coated onto silicon wafers at a spinning speed of 500 rpm and a spinning time of 30 seconds on a Headway Research Inc., PM101DT-R790 spin coater. These films were then placed on a heating plate for 10 minutes at ~75 °C. The films were removed from the substrate by submersion in deionized water and annealed at 200 °C for 20 minutes under a vacuum. I_2 vapor was used to dope the films.

Differential scanning calorimetry (DSC) was carried out on a TA Instruments, Model 2920 at a heating rate of 20 °C per minute under nitrogen, with indium as a calibration standard. For the electrical measurements, samples were mounted across two copper blocks using carbon paint to make electrical contacts, with a silicon diode mounted near the sample to measure temperature. The resistivity (using a standard four probe technique) was measured as a dipstick containing the sample was slowly lowered into a liquid nitrogen dewar. After the temperature dependence of the resistivity was measured, the sample was slowly withdrawn from the dewar and the thermoelectric power was measured. The two copper mounting blocks were controlled at two different temperatures and the gradient measured by a Au-Fe(0.07%)-Cu thermocouple anchored near the ends of the sample. Both the DC conductivity and thermoelectric power were measured in the temperature range between 77 and 300 K.

Although PANI is easily synthesized, precaution must be taken to minimize exposure to aniline in the polymerization process. Under certain conditions aniline will react with itself to form the highly carcinogenic oligomer benzidine. This oligomer results from a head-to-head coupling of the aniline monomer. Benzidine once formed and subsequently incorporated into the polymer chain as a defect in the normal head to tail polymerization of aniline is not considered to be dangerous. One must still be concerned with the possibility that unreacted aniline may well be present in the formed PANI and an oligomerization reaction forming benzidine may occur in the polymer. Recent work in our laboratory clearly demonstrates that benzidine easily partitions into emeraldine base polyaniline.[6] A detailed discussion of the incorporation of benzidine in formed PANI and in the starting reaction mixtures may be found in reference 6.

Optical Properties of PANI Films

The three dimensional refractive indices of the freestanding PANI films were characterized with a modified prism-wave guide coupler (Metricon PC-2010) at 1550nm. The technique was previously developed and tested on a range of different freestanding and spin coated polymer films[7,8] In this work Nz is the refractive index along the optical symmetry axis, Ny is the refractive index perpendicular to the symmetry axis in the film plane, while Nx is the refractive index normal to the film plane. The average refractive index, Nav = (Nz+Ny+Nx)/3, is proportional to the density and hence is a measure of the crystallinity. The birefringence is defined as: Nzy = Nz - Ny, Nzx = Nz - Nx, and Nyx = Ny - Nx and is a measure of molecular orientation.

In order to obtain accurate values of the principal refractive indices in the plane of the sample, it is best to measure the refractive index as a function of the angle, ϕ, around the film. By introducing a large number, 16 in our case, of angular measurements into the optical indicatrix equation, a precise identification is obtained of the symmetry axis direction and of both principal refractive indices in the plane of the film. The optical indicatrix equation in the plane of the film has the linear form:

$$1/N\phi 2 = 1/Ny2 + (1/Nz2 - 1/Ny2)\cos 2\phi \quad (1)$$

Experimentally the above equation is slightly modified to:

$$1/N\phi 2 = 1/Ny2 + (1/Nz2 - 1/Ny2)\cos 2(\phi - \beta) \quad (2)$$

where ß, is the angle between the assumed reference direction and the optical symmetry axis direction. To find the exact direction of the optical symmetry axis, a linear least squares computer subroutine is used. [9] Figure 1 shows a linear indicatrix plot of the in-plane refractive indices of the PANI film cast from DMPU and stretched to a draw ratio of 1.63.

This technique eliminates alignment errors in the refractometer that can easily be missed by the more conventional technique of simply measuring only along and perpendicular to the assumed symmetry axis direction. A further advantage is this measurement technique also yields 16 Nx values to average, increasing the reliability of that important parameter. All refractive index measurements made in our laboratory utilized the angular indicatrix approach.

Effect of Film Orientation. The effect on the three dimensional refractive indices of fabricating film by spin coating is shown in Figure 2. In spin coating a polymer droplet is deposited onto a spinning surface and is spread out uniformly from the center. Figure 2 shows that the resulting undrawn film orientation is random in the film plane, Nz = Ny, but the principle refractive index in the thickness direction, Nx, is different. The spin coated film has a more planar-like structure. In contrast the DMPU cast undrawn film before drawing was isotropic.

With extension of the freestanding spin coated film changes in orientation occur in all three film planes. Nz increases with increasing draw ratio as the random molecules align increasingly in the draw direction. Ny decreases with increasing draw, as does Nx. The rate of decrease in Ny is greater than that of Nx so that the two values approach each other at the highest draw ratio. Thus the spin coated film goes from a planar to almost a uniaxial structure with increasing extension. The average refractive index remains constant indicating there is no phase change with extension. [10]

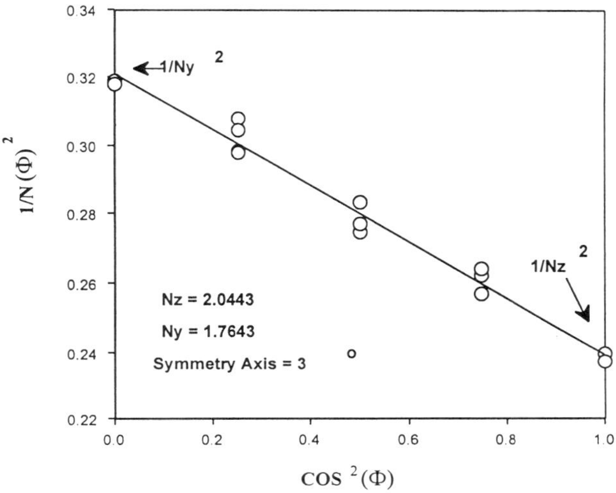

Figure 1. Linear refractive index indicatrix plot for spin coated and drawn film with a draw ratio of 1.63.

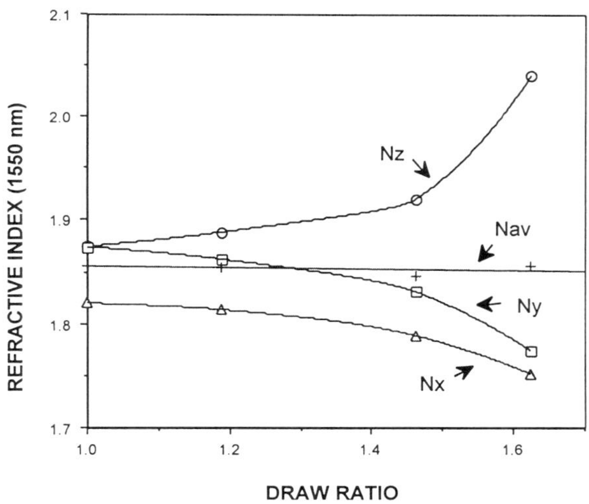

Figure 2. Refractive index of DMPU spin coated films as a function of fabrication draw ratio.

The new near infrared wave guide coupling technique can be used for the determination of the three dimensional refractive indices of films opaque in the visible region. This technique is appropriate for other electroactive polymer systems providing a window exists in the visible or near infrared spectrum where measurements can be made. Applying this technique to PANI films has clearly shown that spin coated films show some planar character in the thickness direction. Unidirectional stretching of freestanding spin-coated PANI film leads to increased orientation in the deformation direction, and conversion from a planar to a more uniaxial structure. Similar studies have been carried out on PANI films formed from NMP solutions.[11]

Although these studies elucidate the effect of morphological changes of the optical properties of PANI films which have been cast and subsequently oriented, the technique can be applied to other electroactive polymers as well to determine the effect of orientation and subsequent morphological changes on the formed polymer's optical properties.

Effect of Structural Morphology on the Electrical Properties of PANI Films

The chemical structure of polyaniline in various oxidation states is shown in Figure 3. The fully reduced form is referred to as leucoemeraldine base (LEB) while the fully oxidized form is called pernigraniline (PNB). Both LEB and PNB will shift towards the EB form upon exposure to an oxidizing environment.[2] Although it is well known that the oxidation state of polyaniline is an important characteristic of this polymer, there are few reports of its influence on the development of morphology in fibers and films. Previous work has shown that both films and fibers produced from solutions of leucoemeraldine base in N,N'-dimethyl propylene urea (DMPU) exhibit crystallinity and a melting transition.[12]

Electrical transport properties of chemically prepared polyaniline show a strong temperature dependence. In general, depending on the sample preparation and dopant species, the polymers may be metallic, insulating, or be in the critical region for a metal-insulator transition. [13,14] It is also known that disorder plays a predominant role in the transport process. The disorder can be homogeneous on a molecular scale, mesoscopic heterogeneity's with highly doped conducting clusters, or crystalline islands separated by insulating regions. The DC conductivity is thus influenced to a great deal by the extent of amorphous regions. However, because of the absence of a measuring current flow, the thermoelectric power is less dependent on these regions. Several models have been proposed for describing the temperature dependence and transport properties of charge carriers, including quasi-one dimensional variable range hopping (Q-1D VRH), charging energy limited tunneling for granular metals, and three dimensional variable range hopping with a Coulomb gap.[15,16,17]

In this section we will elucidate the interconnection between structure, both molecular (i.e. oxidation state) and morphological, and electrical transport properties. The techniques employed for this study are differential scanning calorimetry for

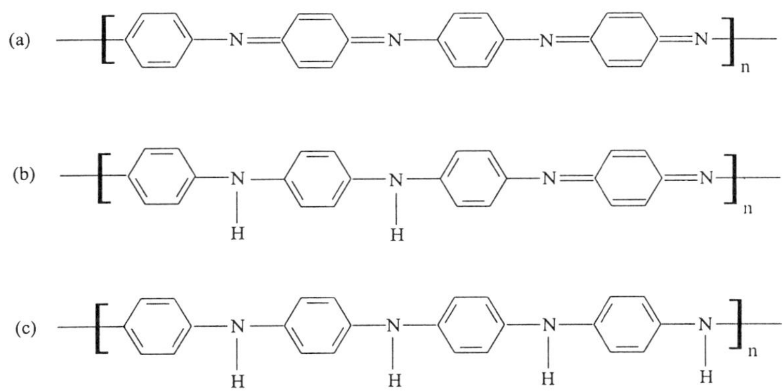

Figure 3. Three oxidation states of polyaniline base: a) fully oxidized pernigraniline base (PNB), b) partially oxidized emeraldine base (EB), and c) fully reduced leucoemeraldine base (LEB).

characterization of thermal transitions, and temperature dependence of conductivity and thermoelectric power.

Films were spin coated from solutions containing molar ratios R=1.11, 0.71, 0.36, and 0.00 of phenyl hydrazine, a reducing agent, to emeraldine base (based on a four ring repeat unit). As one might expect, the appearance of the films depended strongly upon the level of oxidation. The fully reduced films were a grey color with a poorly reflecting surface indicative of the leucoemeraldine base. With increasing oxidation, the films were more reflective and had a dark grey color, while the unreduced films were quite reflective and had the characteristic coppery color of emeraldine base films.

This optical behavior is attributed to the presence of a crystalline phase (probably less than ~15%) in the leucoemeraldine base form of the films which causes a rough surface to form. The surfaces of the films which were against the substrates are much more reflective. As reported earlier, an endotherm was observed from differential scanning calorimetry consistent with melting of the crystallites.[13]

For the present study, DSC was chosen to investigate the effect that oxidation state has on the morphological development of the prepared films. A discussion of thermal transitions in the emeraldine base form of polyaniline may be found in the literature.[18,19] Figure 4 presents the DSC results on four annealed films. For the emeraldine base film (R = 0.00), as the temperature increases a broad exotherm centered at ~300 °C was observed, attributed to oxidation or crosslinking and characteristic of emeraldine base. With increasing R (i.e. more reduction) one can see the development of an endothermic peak between ~300 and ~385 °C, depending on oxidation state. As a greater proportion of the repeat units are reduced (as R increases), more of the chain segments can crystallize, which is accompanied by an increase in the magnitude of the endotherm. Another effect is that the endotherm becomes sharper and shifts to higher temperatures with increasing R, believed to be due to increasing crystallinity in conjunction with more perfect ordering in the crystallites.

When treated with I_2 vapor, the reduced segments are oxidatively doped to the conducting state. Figure 5 presents the results of the electrical conductivity temperature dependence as a function film oxidation level prior to doping. The room temperature conductivity of the R = 0.00 film doped with 1.0 M HCl was measured to be 0.22 S/cm. With increasing R, the conductivity of the I_2 doped films increases, and can be interpreted in terms of the quasi-1D variable range hopping mechanism (Equation 3):[16,14]

$$\sigma_{dc} = \sigma_o \exp\left[-\left(\frac{T_o}{T}\right)^{\gamma}\right] \qquad (3)$$

here $\gamma = ½$, and $T_o = 485$ K for the R = 0.36 film and $T_o = 137$ K for the R = 0.71 film. If fit to Equation 1, the R = 1.11 (most reduced) film exhibited a T_o of 70 K. A decreasing T_o can be attributed to more intra- and interchain ordering resulting in

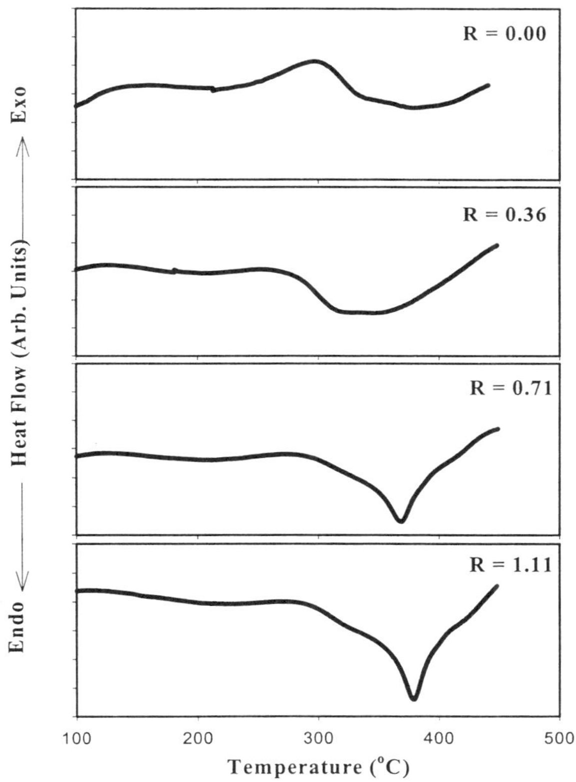

Figure 4. Differential scanning calorimetry of polyaniline spin coated films as a function of chemical reduction: R = 0.00: ~emeraldine base, R = 1.11: ~leucoemeraldine base.

greater charge delocalization, consistent with the morphological picture obtained from the DSC results. However, for the R = 1.11 film, a slightly better fit is obtained when plotted against 1/T, indicative of a semiconductor mechanism. This is in fact born out by thermoelectric power measurements on this film in which the thermoelectric power is proportional to 1/T. Figure 6 presents the thermoelectric power for the R = 0.36 film as a function of 1/T. A prominent feature of this figure is the transition at ~150 K, which is consistent with polaron scattering by phenyl ring vibrational modes as reported by Pratt et.al.[20] This temperature corresponds to a β-transition observed by dynamic mechanical thermal analysis and temperature dependence of the anisotropy in σ_{dc} indicating a lack of phenyl ring rotation below 150 K.[21] The change in slope at the 150 K temperature is consistent with a change from a semiconductor transport mechanism to a variable range hopping mechanism. Ongoing work suggests that there may be another transition in the thermoelectric power spectrum around 250 K. This transition has yet to be quantified.

This work demonstrates that there is a clear connection between morphological structure as evidenced by the DSC results and electrical transport properties. With increasing chemical reduction, there is less hindrance to crystallization and sharper, higher temperature endotherms are obtained from the DSC, characteristic of larger, more perfect crystalline structures. This behavior is confirmed by the temperature dependence of conductivity in which T_o decreases with increasing chemical reduction. Unlike the other two samples (R = 0.36, 0.71), the R = 1.11 film exhibits the 1/T dependence of thermoelectric power indicative of a semiconductor transport mechanism. Although the LEB form of PANI slowly reoxidizes back toward the EB form the above study combined with the optical studies suggest that the structure formed when processed in the LEB form is maintained after reoxidation and subsequent doping. This is consistent with the results obtained for fiber spinning from LEB where the formed fibers demonstrate superior mechanical properties when compared with fibers spun from EB systems.[5]

Concluding Remarks

This chapter reports on the effect of morphological structure on the determined electrical and optical properties of polyaniline when processed by different methodologies. Although the information reported is specific to PANI the concept of changes in structural properties of electroactive polymers due to processing or pre-processing steps is not unique to polyaniline but rather extends to all synthetic organic polymers. Such morphological changes affect not only the mechanical properties of the formed polymer but also the electrical transport mechanisms as well as their optical properties. When these polymers are considered for device applications one must consider the route by which the materials are prepared and the resulting structure as this will have demonstrative effects on the polymers, thermal and electrical transport, optical characteristics, failure mechanisms, and efficiency as hole or electron injectors.

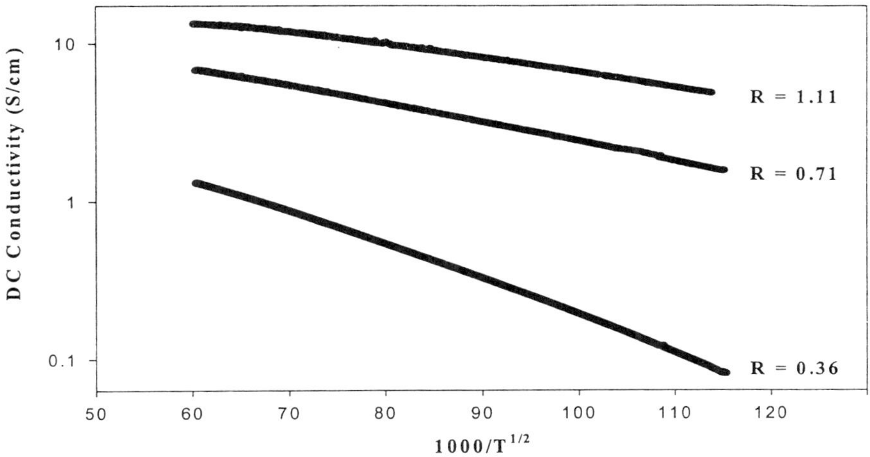

Figure 5. DC conductivity of iodine doped polyaniline films as a function of R, the molar ratio of reducing agent to emeraldine base (based on a four ring repeat unit) of the solution used to prepare films.

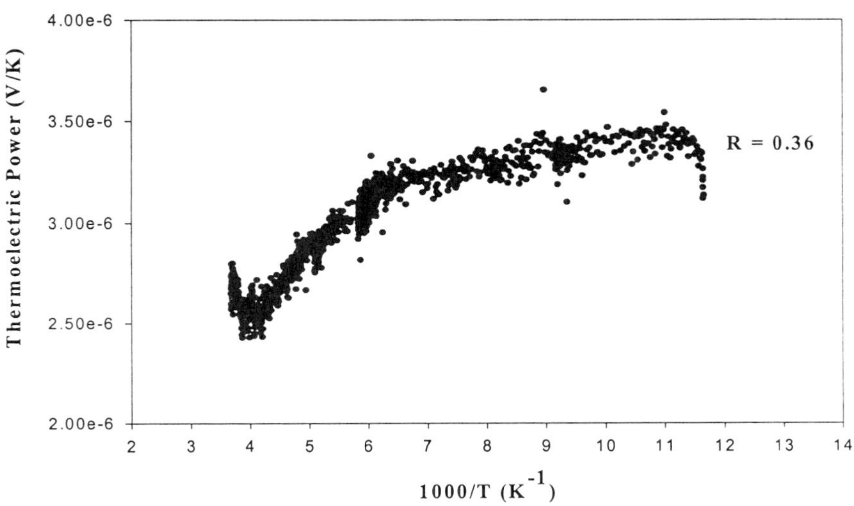

Figure 6. Thermoelectric power of a polyaniline spin coated film. R = 0.36 is the molar ratio of reducing agent to emeraldine base (based on a four ring repeat unit) of the solution used to prepare film.

References

[1] Proceedings of the *"International conference on the Science oand Technology of Synthetic Metals"* Snowbird Utah, USA July 1996

[2] A. G. MacDiarmid, **Conjugated Polymers and Related Materials**, edited by W. R. Salanek, I. Lundstrom, and B. Ranby, Oxford University Press, New York, **1993**, pp. 73 - 98.

[3] Genies, E. M., Boyl, A., Lapkowski, M., Tsintavis, C.; *Synthetic Metals*, **1990** *36*, 139

[4] R. Jain and R.V. Gregory, *Synthetic Metals*, **1995**, *74*, 263

[5] A.P. Chacko, S. S. Hardaker, B. Huang, and R. V. Gregory, *Mat. Res. Soc. Symp. Proc.*, **1995**, *413*, 503; S.S. Hardaker, A.P. Chacko, B. Huang, and R. V. Gregory, *SPE-ANTEC '96*, **1996**, *Vol. II,* 1358.

[6] Watters, M. J. *"Determination of Trace quantities of Benzidine in Emeraldine-Base Polyaniline"* M.S. Thesis, Clemson University, 1995

[7] Hardaker, S., C.Y. Cha, S. Moghazy and R. J. Samuels, in *"Advances in Polyimide Science and Technology,"* C. Feger, M. Khojasteh and M. Htoo, eds. Lancaster, Pa., Technomic Pub. Co., **1993**, 571.

[8] Hardaker, S., Moghazy, S., Cha, C. and R.J. Samuels *J. Polymer Sci., Part B: Polymer Physics*, **1993**, *31*, 1951.

[9] R.E. Pepper and R.J. Samuels, *Encyclopedia of Polymer Science and Engineering*, H.F.Mark, N.M.Bikales, C.G.Overberger, G.Menges, and J.I. Kroschwitz, Eds., **1988**, *14*, 261.

[10] C. Cha, S. S. Hardaker, R .V. Gregory and R .J. Samuels, *Synthetic Metals*, **1997**, *84*, 743.

[11] R. Ou, T. Liu, H. Wang, S. S. Hardaker, L. Ding, B. Mattes. R. V. Gregory, and R. J. Samuels; *SPE-ANTEC'98*, **1998**, *Vol. II*, 1351

[12] A. P. Chacko, S. S. Hardaker, R. V. Gregory, and T. W. Hanks, *Polymer,* **1998**, *39(14)*, 3289

[13] H. K. Chaudhari and D. S. Kelkar, *J. Appl. Poly. Sci.*, **1996**, *62*, 15

[14] J. Joo, V. N. Prigodin, Y. G. Min, A. G. MacDiarmid, and A. J. Epstein, *Phys. Rev. B*, **1994**, *50*, 12226

[15] E. P. Nakmedov, V. N. Prigodin, and A. N. Samukhin, *Sov. Phys. Solid. Stat.*, **1989**, *31*, 368

[16] B. Abeles, P. Sheng, M. D. Coutts, and Y. Arie, *Adv. Phys.*, **1975**, *24*, 407

[17] B. I. Shklovskii and A. L. Efros, *Electronic Properties of Doped Semiconductors*, Springer-Verlag, New York, 1984.

[18] Y. Wei, G. W. Jang, K. F. Hsueh, E. M. Scherr, A. G. MacDiarmid, and A. J. Epstein, *Polymer*, **1992**, 33, 314

[19] A. J. Milton and A. P. Monkman, *J. Phys. D: Appl. Phys.*, **1993**, *26*, 1468

[20] F. L. Pratt, S. J. Blundell, W. Hayes, K. Nugamine, K. Ishida, and A. P. Monkman, *Phys. Rev. Lett.*, **1997**, *79*, 2855

[21] A. P. Monkman, P. N. Adams, P. J. Laughlin and E. R. Holland, *Synthetic Metals*, **1995**, *69*, 183

Chapter 5

Electrospectroscopic Studies on Polyimides and Their Model Compounds Containing Trianiline Segments

X. S. Meng [1], P. Desjardins, and Z. Y. Wang [1]

Department of Chemistry, Carleton University, 1125 Colonel By Drive, Ottawa, Ontario K1S 5B6, Canada

The electrospectroscopic behavior of soluble, film-forming polyimides and their model compound containing amino trianiline, N,N'-bis-(4'-aminophenyl)-1,4-phenylenediamine, was investigated. Similar to polyaniline, two typical redox processes were observed for both the polyimides and the model compound. A reversible electro-polychromic switching behavior of the polyimide films was found and a dynamic absorption change of the film at different wavelength during the redox process was demonstrated.

I. Introduction

Extensive studies on polyaniline (PANI) have been carried out in the last five years. Some unique features related to PANI such as electrical conductivity, highly environmental stability, magnetic and electrochromic properties, and as well as availability have attracted much attention from both academic and industrial sectors (1). Among all the potential applications the most attractive one is the use as electronic devices such as electronic capacitors (2,3), sensors (4), electrochromic displays (2,3,5-9), and photodiodes (10,11). All these applications are based on its unique reduction-oxidation (redox) behavior and the associated spectral properties.

Electrochromic device is a generic form of device applications based on PANI. It may include display and optical memory devices, anti-glare car rearview mirrors, sunglasses, protective eyewear for the military use, and smart windows for car and buildings. Recent reports on flexible electrochromic device composed of PANI represents a new area of applications of PANI (12,13).

It is known that PANI shows multiple colors depending on its oxidation state, transparent yellow ↔ green ↔ dark blue ↔ black (or dark purple) (Scheme 1) (14-16). These colors corresponding to the leucoemeraldine (fully reduced), emeraldine

[1] Corresponding authors.

base and emeraldine salt (polaronic partially oxidized) and pernigraniline (fully oxidized) as shown in Scheme 1.

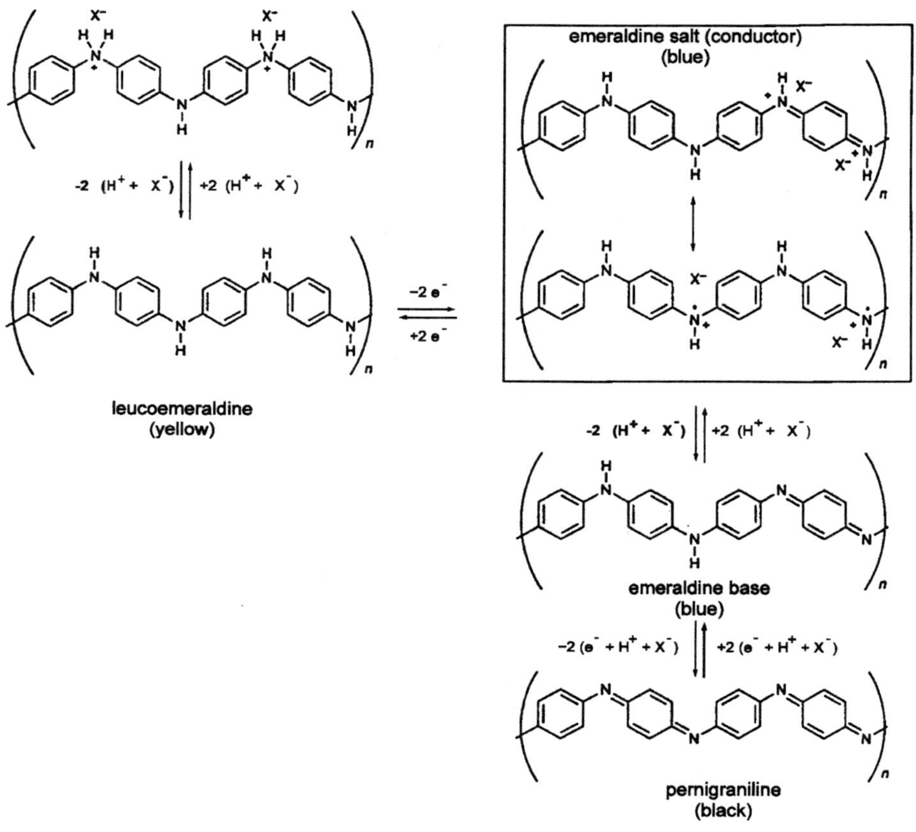

Scheme 1. The redox states of polyaniline (Reproduced with permission from ref. 14. Copyright The Royal Society of Chemistry, 1997)

In the leucoemeraldine form the only electronic transitions present are the benzenoid centered $\pi \rightarrow \pi^*$ and since no long range electronic conjugation exists in the polymer these transitions are in the UV appearing at ~3.8 eV (326 nm), making the material colourless. At the other extreme is the pernigraniline (fully oxidized) form in which exist equal quantities of benzenoid and quinoid rings. Here three bands appear at 4.6 (270 nm), 3.8 (326 nm) and 2.2 (564 nm) eV, and the purple colour of this form arises from the low energy band, which is believed to be due to a charge transfer (17) transition from the benzenoid ring to the quinoid ring. The transitions at 4.6 and 3.8 eV are localized $\pi \rightarrow \pi^*$ transitions of the quinoid and benzenoid units respectively.

The intermediate oxidation state has relatively complicated spectra, dependent upon the pH and dominated by the presence of charge carriers (polarons) in the polymer. Two forms of the polymer exist in this state, each having a proposed 3:1 benzenoid : quinoid composition, the emeraldine base (EB) and the protonated emeraldine salt (ES). The visible spectra of the EB form closely matches that of the pernigraniline form since they both contain similar functional units, the only difference observed in the band at 2.2 eV, which is red shifted to 2.0 eV (620 nm) (18). This red shift accounts for the blue color of this form. The ES form has somewhat more complicated spectra, which is a direct ramification of the presence of polarons in the polymer since the population of polaronic sites is dependent upon the concentration of acid in the solution. Generally, though the ES form possesses the following uv-vis bands, 1.5 (830 nm), 2.75-3.1 (450 nm), and 1.0 (1240 nm). The band at 1.0 eV is assigned to an intrachain free-carrier excitation, while the variable band at 2.75-3.1 eV is due to 'polaron' band and the band at 1.5 eV is the excitonic transition observed for EB (17,19).

Polyaniline shows complex and irreversible electrochemical behavior in the range of –0.2 to 1.0 V (vs. SCE) (15). At potentials lower than 0.5 V a completely reversible, and relatively simple behavior is observed (9). The CV of polyaniline, scanned between 0 and 1 V (vs. SCE) initially shows two redox couples corresponding to the leucoemeraldine - emeraldine couple (base or salt – depending on the pH) at ~0.15 V and the emeraldine – pernigraniline couple at ~0.8 V. After the initial scan, however, a new couple at 0.5 V appears which corresponds to the appearance of a decomposition product, which Genies et al (20-22) has been able to attribute to the cross-linking of the polymer (22). As the material is scanned this 0.5 V couple grows in intensity at the expense of the 0.15 and 0.8 V couples. In non-aqueous solutions, this middle couple reaches a maximum concomitant with complete loss of the other two couples and can be cycled with no apparent loss of intensity. However, under aqueous acid conditions, this middle couple reaches a maximum then also decays over time, which has been attributed to degradation of the polymer to hydroquinone, which has a redox couple at 0.5 V (23). These two proposals seem to be a odds, however it is not unreasonable to imagine that the nature of the decomposition pathway is dependant upon the environment and that both pathways may be overlaid in aqueous acid.

The instability of PANI above 0.8 V aside, the redox behavior of the first couple of polyaniline, leucoemeraldine to emeraldine, is stable and has been exploited in EC systems (9,13,24). Thin layer electrochromic devices have been reported by a number of groups, however despite the seemingly good results derived it would be desirable to address the pernigraniline (blue) form as well. If it were possible to access this state reversibly, then PANI or related materials could find utility in glare reduction applications and in smart windows since the band associated with this form best matches the solar spectrum (25).

In addition to the electrochemical limitation of PANI, poor structural control (due mainly to inadequate polymerization methods) leading to wide variation between the properties of individual preparations of the polymer and generally poor solubility have prevented the commercialization of this polymer in EC devices. Much effort then has been directed at developing synthetic protocols to produce more structurally controlled and soluble PANI (14,26-29). These efforts however, do not remove the underlying electrochemical decomposition problem of PANI above 0.8 V.

As it turns out, similar EC behavior has been observed in oligomeric polyaniline, and thus some effort has been directed toward preparing such systems as both model compounds to enhance the understanding of the behavior of PANI and as EC materials (23,30-33). Although these oligomers do not produce stable films and are subject to similar poor solubility, such structures hold promise since they can be incorporated into a well behaved polymer system. In this way, it may be possible to construct uniform, electrochemically stable, and behaviorally predictable films containing this electrochromic fragment.

Shacklette et al (23) successfully prepared dimeric and tetrameric aniline in order to model the electrochemical behavior of PANI. Electrochemical studies of these molecules were performed and it was shown that they possessed fully reversible electrochemical behavior, unlike PANI. Furthermore, uv-vis spectroscopy with coulombic titration of the tetramer demonstrated that it also possessed similar, but not identical, polychromic behavior to PANI. The discrepancy in the CV was attributed to the longer conjugation length in PANI, which affects the first oxidation couple of the polymer shifting it to less positive potential. As well, the uv-vis spectra of the polymeric material, while possessing similar shape generally, is dominated by a changing background absorption for the ES form due to mobile charge carriers. Due to this changing background, no clear isobestic points were observed. In similar studies conducted by Cao et al (30) and Honzl et al (31), in which, they show that the spectroscopy, though similar between oligomeric and polymeric aniline, do show an overall magnitude dependence. In the case of Cao et al, these model studies, utilizing model oligomers in various degrees of oxidation did confirm the formation of a semiquinone chain upon HCl doping.

The aforementioned promising results notwithstanding, the effect of increasing scale upon the nature of the electrochemical and spectroscopic of these materials seems to indicate that the model compounds do not express identical behavior as that of polymer. In our studies, we wished to investigate whether this difference was due to synergism within the PANI created by essentially 'having a string' of connected oligomeric. It was devised then, to incorporate an aniline trimer (TANI) into polyimide to create a polymer containing many 'isolated' TANI units. In this way the magnitude of the aniline system could be increased without increasing the length of conjugation. As well, we wished to show that the TANI unit could be incorporated into an easily processable polymeric material without loss of polychromic behavior. Beginning with the TANI, which is synthesized by oxidative means from aniline and 4,4'-diaminodiphenylamine, an imide model compound and several polyimides were prepared as described previously (34). In so far as is known, this is the first time such a study has been performed on this type of an aniline oligomer.

Preliminary studies (34,35), which are described in the discussion, of the polymer and model compounds seem to indicate that, when the TANI-polyimide derivatives have nearly identical electrochemical and spectroscopic behavior as that of TANI. This would indicate that, when oligomeric aniline is 'isolated' within a polymer its behavior is unchanged over that of the discrete molecule in solution and bodes well for producing easily processable polymeric materials incorporating aniline like electrochromic behavior.

In this paper, recent results of studies performed on the imide derivatives of the trimeric aniline are presented. Films of TANI-polyimide were also prepared and their electrochemical and electrochromic properties were probed. It is shown from this work that well behaved polymeric materials incorporating aniline trimers can be prepared, these trimers can be addressed electrochemically within such polymeric systems and the behavior of the trimer within the polymer does not differ significantly from that within the model compounds.

II. Experimental

1. Materials. Model **1** was prepared previously and the synthetic aspects reported in (34). The solvent used for CV and uv-vis studies was acetonitrile (Aldrich, spec. grade) and the electrolyte used were tetraethylammoniumperchlorate (TEAP) and tetrabutylammoniumperchlorate (TBAP), which were recrystallized prior to use from methanol/water and dried for 12 hours at 110 °C under vacuum. The solutions were acidified with perchloric acid (Aldrich, double distilled) to a concentration of ~0.01 M.

1

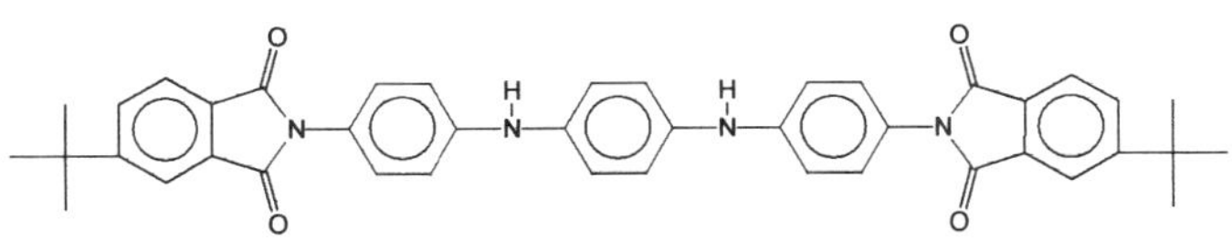

2. Instruments and Characterization. Solution characterization was conducted using a Cary 3 and Cary 5 spectrophotometers, and a custom-made PC controlled voltammeter.

Cyclic voltammetry was performed in a jacketed glass container with an inner volume of ~20 mL. The cell was fitted with a Teflon lid through which holes had been drilled to accommodate the electrodes (BAS 2013 MF working, Pt wire counter, and Ag wire quasi-reference electrode) and a nitrogen or argon gas bubbler. No internal standard was used to give an absolute reference for the $E_{1/2}$ potentials (i.e. NHE) and so all values are versus AgRE.

Spectroelectrochemical spectra were collected using a quartz cell of known design (36) and Cary 5 uv-vis-nir spectrophotometer. During the experiment the potential was varied between 0.2 and 1.1 V while spectra were taken at suitable increments.

3. Film preparation. Polymers (10 mg) containing various amounts of trianiline unit were dissolved in 1 mL of N,N'-dimethylformamide (DMF). To the polymer solution was added a drop of 1 % DMF solution of toluene sulfonic acid. The polymer solution turned to green immediately after acidified. The polymer films were obtained by spin-coating the acidified solution onto the ITO glass followed by drying at 150 °C in vacuum.

The polymer films characterized in a strong acidic electrolyte composed of TEAP and perchloric acid in acetonitrile.

III. Results and Discussions

1. Electrochemical Behavior of the Model Compound. PANI has alternative phenylene-amine units in its main chain and forms conjugated structure upon oxidation. But it is not the case in TANI-polyimide. According to a proposed mechanism (23) for the electrochemical redox reaction of the aniline oligomers the redox process of TANI-diimide unit is presented in Scheme 2. Since amino groups at the ends of TANI segment are capped by the imide the oxidation may only occur at the central part, a structure of phenylenediamine. Thus there are two oxidized states i.e. cationic radical by losing one electron and diimine by losing two electrons. The diamine (reduced form), cationic radical (first oxidation form) and diimie (second oxidation form) in TANI-polyimide are very similar to the leucoemeraldine, emeraldine salt and pernigraniline in PANI. In TANI-polyimide there is no oxidation state like emeraldine base in PANI. Therefore the electrochromic behavior may be simpler than that of PANI.

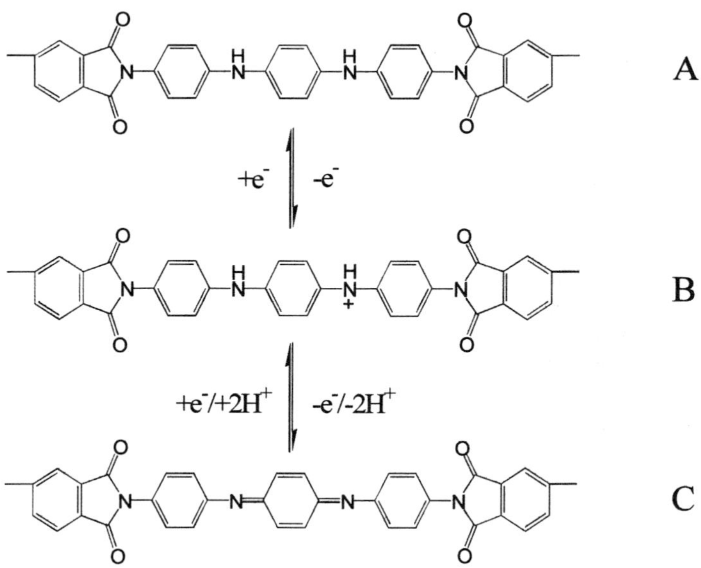

Scheme 2. Proposed redox states of trianiline segment

The CV of **1** (Figure 1) shows two redox couples corresponding to the two redox active nitrogen centers similar to that of the dimer prepared by Shacklette (23).

The reactions believed to occur during the CV experiment are given in Scheme 2. Only quasi-reversibility of the redox couples was demonstrated, as the anodic and cathodic peak to peak separation was 90 mV for couple I and 70 mV for couple II. The slow kinetics of **1** are likely due to the influence of the proton exchange between the amine groups and the solution, which is integral to the redox reaction. The $E_{1/2}$ potentials were 740 mV and 910 mV for the redox couples I and II (respectively). These two potentials were found unchanged with varied scan rate when the scan rate was lower than 100 mV/s.

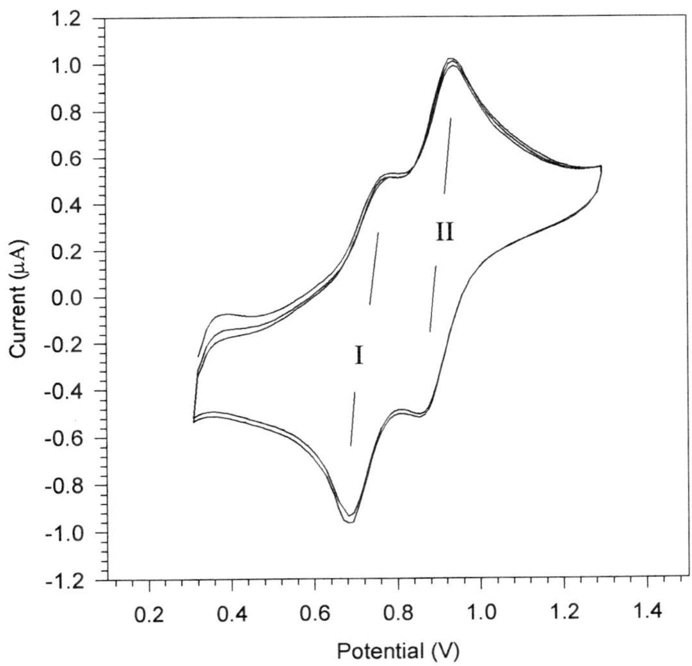

Figure 1. CV of model compound **1** (acetonitrile, ~0.5 M Et_4NClO_4, $HClO_4$)

The spectroelectrochemical spectrum for **1** is shown in Figure 2. The fully oxidized, blue compound (Scheme 2 - C), possesses a strong transition at 580 nm which corresponds to a charge transfer transition between the benzenoid and quinoid units, similar to that of pernigraniline-PANI described by Goff (15). The yellow, fully reduced, compound (Scheme 2 - A) shows two transitions at 300 and 400 nm as well as a very broad transition at 750 nm. Regardless of the oxidation state the compound possesses a high energy phenyl $\pi \rightarrow \pi^*$ transition, likely centered in the terminal phenyl groups, as well as those of the in the trimer in the leucoemeraldine form.

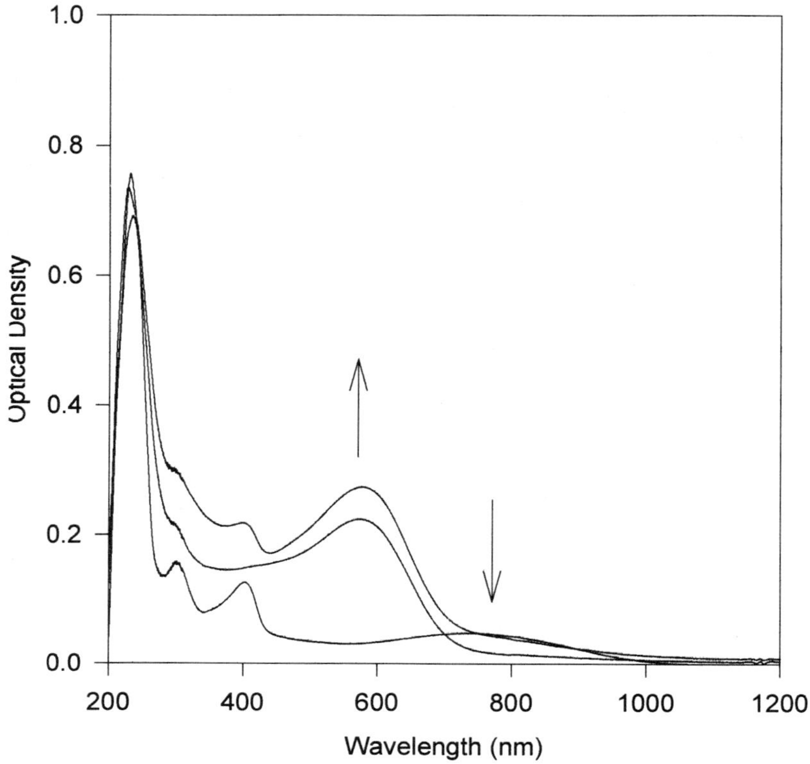

Figure 2. Spectroelectrochemical generation of oxidized model compoud **1** in acetonitrile (1.0 M Bu_4NClO_4, $HClO_4$).

2. Electrochromic Behavior of the Polyimide Containing Trianiline (TANI) Segments. To exam their electrochemical and electrochromic properties all films were made from three copolyimides **2-4** and the homopolyimide **5** respectively. The cyclic voltammograms of copolyimides **2-4** and the homopolyimide **5** displayed two distinct reversible waves and displayed no significant difference in electrochemical behavior. Therefore, only the results for the copolyimide containing 20 % TANI will be discussed in detail. The CV curve from this copolymer is shown in Figure 3. Of significance is the position of the first oxidation wave, which occurred at higher potential (near 800 mV) than those (ca. 300 mV) of reported aniline dimer and tetramer without the terminal amine substituents (23). This can be rationalized by considering the electron-withdrawing imido group, which exerts a stabilizing influence, towards oxidation, over the trimer. Moreover, there is a difference between TANI-polyimide and PANI in the

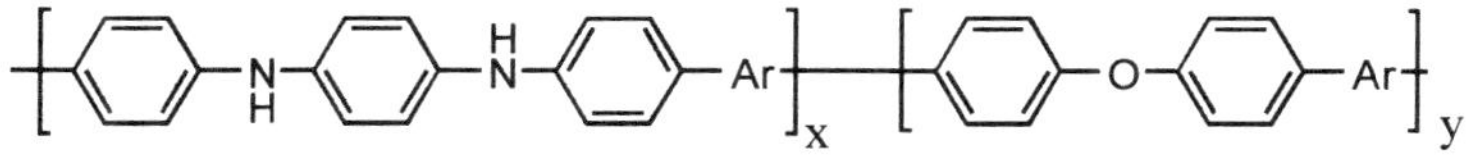

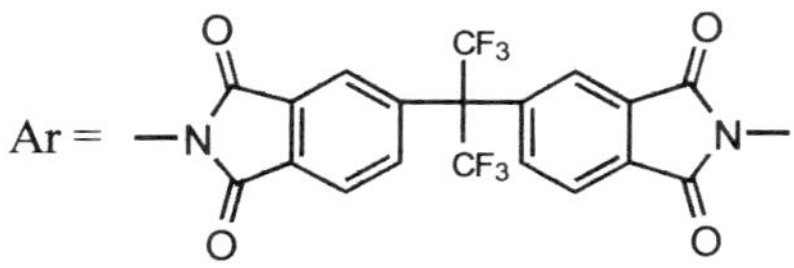

	x : y
2	20 : 80
3	40 : 60
4	80 : 20
5	100 : 0

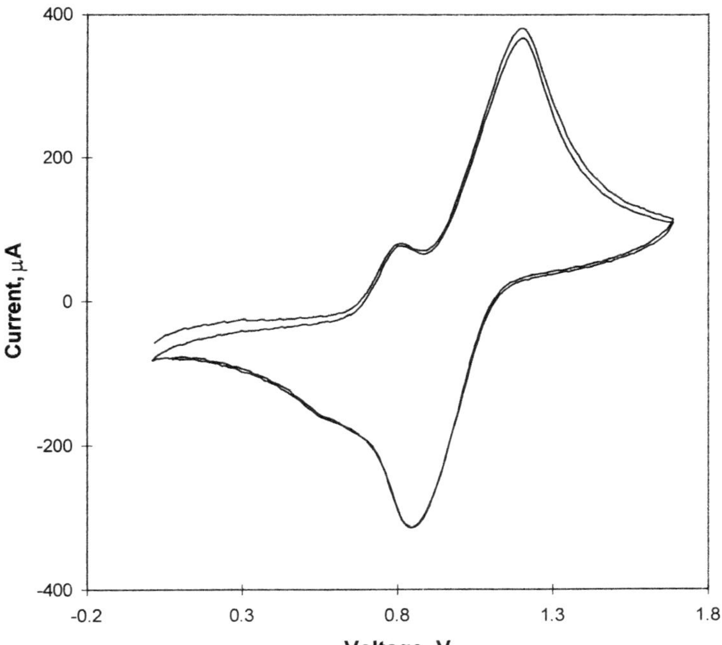

Figure 3. CV of copolyimide **2** film on ITO glass in acetonitrile (1.0 M Et_4NClO_4, $HClO_4$).

oxidation mechanism. Specifically, only one electron is transferred in first oxidation of TANI-polyimide forming a cationic radical while two electrons are transferred in the case of PANI forming emeraldine salt or emeraldine base depending on pH.

The electrolyte effect on the electrochemical behavior of the polyimide film in acetonitrile solution was examined with different perchlorate salts. The difference among the voltammograms obtained in acetonitrile solutions containing $LiClO_4$, $NaClO_4$, Et_4NClO_4, Bu_4NClO_4 respectively was very small. This suggests that these cations are not directly involved in the redox process of the polyimide films.

Comparing with the electrolyte effects the acidity level of the electrolyte solution plays more important rules in the electro-activity of the polyimide film. Similar to that for polyaniline a strong pH dependence in electrochemical behavior of the polyimide containing aniline trimer segments was observed in acetonitrile solution. Lower acidity (concentration less than 1×10^{-4} M) gave only the first oxidation peak indicating the second oxidation is more pH dependent. Both first and second redox pairs appear when the concentration of $HClO_4$ ranges from 1×10^{-3} to 1×10^{-1} M.

The uv-vis spectra of the TANI-polyimide film were recorded in-situ during redox cycling. Spectra for the three oxidation states are presented in Figure 4. The diamine (reduced form) has no absorption in visible region, possessing only the expected phenyl ring centered $\pi \rightarrow \pi^*$ transitions. Two transitions appeared when the first oxidation occurred, one at 390 nm and the other at about 780 nm, arising from the cationic radical (i.e. similar to those of the bipolarons in PANI). The intensity of these two transitions decreased during the second oxidation, accompanied by the appearance of a new transition at 580 nm. At this point the film visibly turned dark blue, attributable to the generation of the diimine form of the trimer. The intensity changes at 390, 580 and 780 nm were monitored when the film was cycled between the boundary potentials of 0.7 and 1.3 V (Figures 5 and 6). In this potential range, the trimer is electrochemically switched between the cationic radical and diimine forms. In Figure 5, A is the triangle wave potential applied on the cell, whereas B, C and D are the visible absorption's as a function of time at the various wavelengths mentioned.

It is important to mention that the polymer film showed apparently reversibility at high potential up to 1.3 V. There is no new peak appear between the two oxidation waves. The CV curves are exactly overlapped and the absorption intensity at all the three wavelength mentioned above were reversibly switched with the potential. The large intensity change observed and reversibility observed in the traces is promising with respect to the EC potential of these materials since these are precisely the properties which lead to high contrast and long cycle life in real world EC applications.

IV. Conclusions

A trianiline (TANI), N,N'-bis(4'-aminophenyl)-1,4-phenylenediamine, was successfully incorporated into a series of high molecular weight, film-forming, electroactive polyimides. Their electrochromic behavior and reversible electrochromic switching behavior was observed. As well, no significant difference in the electrochromic properties between these copolymers was observed.

Comparison to PANI and the aniline oligomers revealed that these electroactive polyimides showed relatively high oxidation potentials. The oxidation resistance of TANI-imide can be attributed to the electron withdrawing terminal imide groups.

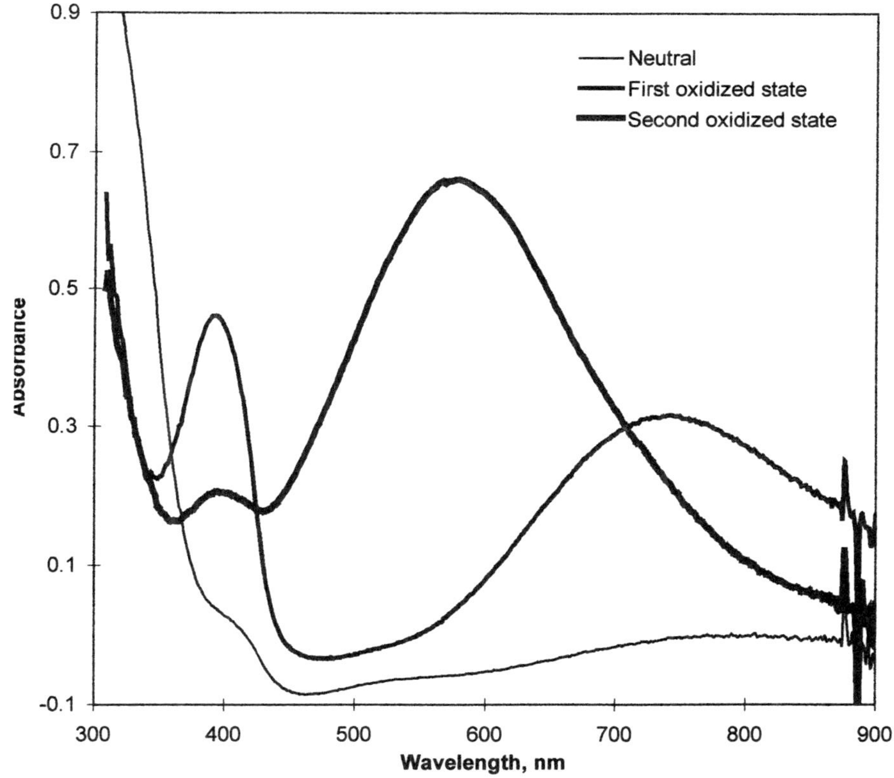

Figure 4. Spectroelectrochemical generation of oxidized copolyimide **2** film on ITO glass in acetonitrile (1.0 M Et_4NClO_4, $HClO_4$) during the cycling.

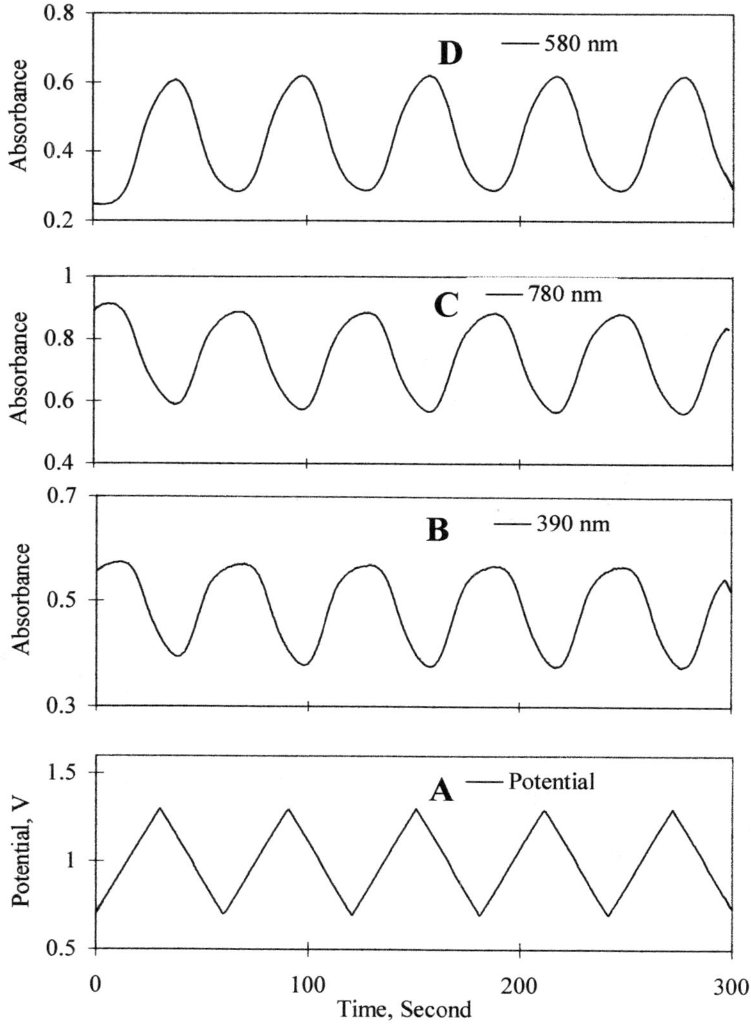

Figure 5. Absorption change of oxidized copolyimide 2 film on ITO glass in acetonitrile (1.0 M Et_4NClO_4, $HClO_4$) within the first 6 cycles at 390 (B), 580 (C) and 780 (D) driven by the potential change over time (A).

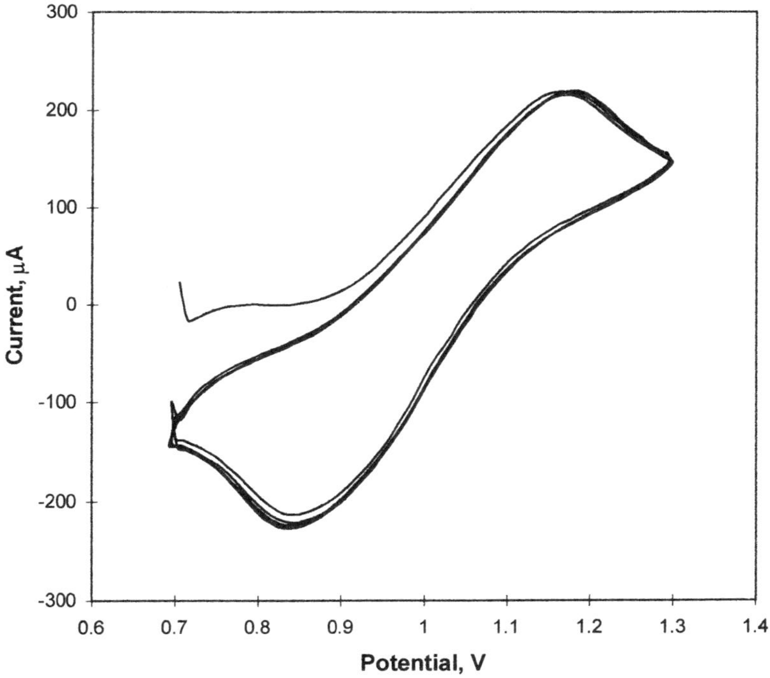

Figure 6. CV of copolyimide **2** film on ITO glass in acetonitrile (1.0 M Et_4NClO_4, $HClO_4$), first 6 cycles from 0.7 V – 1.3 V.

V. Acknowledgement

This work was financially supported by the Natural Sciences and Engineering Research Council of Canada.

References

1. Bodalia, R.; Stem, R.; Batich, C. and Duran, R. *J. Polym. Sci. Polym. Commum.* **1993**, *31*, 2123.
2. Roth, S. and Graupman,W. *Synth. Met.* **1993**, *55-57*, 3623.
3. Zoppi; R. A. and De Paoli; M.-A. *Quím. Nova* **1993**, *16*, 560.
4. Barlett, P. N. and Birkin, P. R. *Synth. Met.* **1993**, *61*, 15.
5. Tassi, E. L. and De Paoli; M.-A. *Electrochimica Acta* **1994**, *39*, 2481.
6. Panero, S.; Scrosati, B; Baret, M; Cecchini, B and Masetti, E. *Solar Energy Materials and Solar Cells* **1995**, *39*, 239.
7. Jelle, B. P.; Hagen, G.; Sunde, S. and Ødegård, R. *Synthetic Metals.* **1993**, *54*, 315.
8. Jelle, B. P. and Hagen, G. *J. Electrochem. Soc.* **1993**, *140*, 3560.
9. Akhtar, M.; Weakleam, H. A.; Paiste, R. M. and Gaughan, K. *Synthetic Metals*, **1988**, *26*, 203.
10. Gustafsson, G.; Treacy, M.; Cao, Y.; Klavetter, F.; Colaneri, N. and Heeger, A. J. *Synth. Met.* **1993**, *55-57*, 4047.
11. Heeger, A. J.; Yang, Y.; Westerweele, E.; Zhang, C.; Cao,Y. and Smith, P. In *Polymeric Materials Encyclopedia* Ed. Salamone, J. C., CRC Press Ltd. New York, USA, 1996, Vol. 7, pp5500-5505.
12. Michalak, F. and Aldebert, P. *Solid State Ionics*, **1996**, *85*, 256.
13. Yamasaki, S.; Terayama, K.; Ohura,H. and Takayanaci, M. *Denki Kagaku* **1995**, *63*, 30-35.
14. Mortimer, R. J. *Chem. Soc. Rev.*, **1997**, *26*, 147.
15. Goff, A. H.-L. In *Handbook of Organic Conductive Molecules and Polymers* Ed. Nalwa, H. S., John Wiley & Sons Ltd. USA. 1997, Vol. 3, pp745.
16. Mattoso, L. H. C. and MacDiarmid, A. G. In *Polymeric Materials Encyclopedia* Ed. Salamone, J. C., CRC Press Ltd. New York, USA, 1996, Vol. 7, pp5505.
17. Cao, Y.; Smith, P. and Heeger, A. *Synth. Met.* **1988**, *32*, 263.
18. McCall, R.; Ginder, J.; Leng, J.; Yeh, H.; Manohar S.; Master, J.; Asturias, G.; 21. McDiarmid, A. and Epstein, A. J. *Phys. Rev.*, **1990**, *41*, 5202.
19. McDiarmid, A.; Chiang, J.; Richter, A.; Epstein, A. J. *Synth. Mater.* **1986**, *18*, 285.
20. Kobayashi, T.; Yoneyama, H. and Tamura, H. *J. Electroanal Chem.* **1984**, *177*, 293.
21. Kobayashi, T.; Yoneyama, H. and Tamura, H. *J. Electroanal Chem.* **1984**, *161*, 419.
22. Genies, E. M.; Lapkowski, M. and Penneau, J. F. *J. Electroanal Chem.* **1988**, *249*, 97.
23. Shacklette, L. W.; Wolf, J. F.; Gould, S. and Baughman, R. H. *J. Chem. Phys.* **1988**, *88*, 3955.
24. Jang, G.-W.; Chen, C.; Gumbs, R. W.; Wei, Y. and Yeh, J.-M. *J. Electrochem. Soc.* **1996**, *143*, 2591.
25. Goff, A. H.-L.; Bernard, M. C. *Proc. Europto.* **1992**, *1728*, 130.

26. Gazotti, W. A.; Faez, Jr., R. and De Paoli, M.-A. *J. Electroanal Chem.* **1996**, *415*, 107.
27. Wei, Y.; Jia, X.; Jin, D.; Mathai, M. W.; Yeh, J.-M.; Narkis, M. and Siegmann, A. *Polym. Prepr.* **1998**, *39(1),* 115.
28. Boone, H. W. and Hall Jr., H. K. *Macromolecules*, **1996**, *29*, 5835.
29. Liao, Y.-H.; Angelopoulos, M. and Levon, K. *J. Polym. Sci. Polym. Chem.*, **1995**, *33*, 2725.
30. Cao, Y.; Li, S.; Xue, Z. and Guo, D. *Synth. Met.* **1968**, *16*, 305.
31. Honzl, J. and Tlustatkova, M. *J. Polym. Sci. Part. C*, **1968**, *22*, 451.
32. Wienk, M. M.; Janssen, R. A. J. *J. Am. Chem. Soc.*, **1996**, *118*, 10626.
33. Sein Jr, L. T.; Kolla, S.; Pasupuleti, P.; Patel, K.; Jansen, S. A. *Polym. Prepr.* **1998**, *39(1),* 117.
34. Wang, Z. Y.; Yang, C.; Gao, J. P.; Lin, J.; Wei, Y. and Li, S. *Polym. Prepr.* **1998**, 47 *39(1),* 119.
35 Wang, Z. Y.; Yang, C.; Gao, J. P.; Lin, J.; Meng, X.; Wei, Y. and Li, S. *Macromolecules*, **1998**, *31*, 2702.
36. Brewer, K. J.; Calvin, M.; Lummplein, R. S.; Otuos, J. W. and Spreer, L. O. *Inorg. Chem.* **1989**, *28*, 4446.

Chapter 6

Conducting, Waterborne Lignosulfonic Acid-Doped Polyaniline

M. Sudhakar [1], A. D. Toland [2,3], and T. Viswanathan [2,3]

[1] Department of Chemistry, University of Southern California, Los Angeles, CA 90089
[2] Department of Chemistry, University of Arkansas at Little Rock, Little Rock, AR 72204

Polyaniline (PANi) was synthesized by using lignosulfonic acid as a template/dopant and ammonium persulfate as the oxidizing agent. The product was water-borne and showed a conductivity of 10^{-1} S/cm. Lignosulfonic acid-polyaniline (Ligno-PANi) was characterized by UV-Vis spectroscopy which showed peaks for both benzenoid and quinoid moieties. After Soxhlet extraction with water and methanol, Ligno-PANi was characterized by UV-Vis and NMR spectra and showed peaks for lignosulfonate in addition to that for polyaniline thereby providing spectroscopic evidence for probable grafting of the polyaniline chain to lignosulfonate. Thermal stability of Ligno-PANi was compared with hydrochloric acid doped Polyaniline (HCl-PANi) and results indicated the Ligno-PANi was slightly less stable than its HCl counterpart.

One of the most desirable characteristics of polyaniline (PANi), especially for technological applications, is its solution processability (*1-3*). Even though the experimental methodology is relatively simple, polymer characterization and usage has been hampered due to the limited solubility of the conducting form of polyaniline in common solvents. Insolubility of polyaniline in common solvents is due to interchain interactions. Hence the synthesis and modification of these polymers in water (*4-6*) will allow for better characterization and enhanced processability. Water solubility has favorable economic and environmental advantages, in terms of reduced toxicity and ease of processing (*7,8*). Several approaches have been used to synthesize water soluble PANi and they are summarized below.

Sulfonic acid ring substituted polyaniline was prepared by reacting the base

[3] Corresponding authors.

form of HCl-doped polyaniline with fuming sulfuric acid at 5 ^{0}C (*9-11*). The presence of one sulfonate group for every two phenyl rings promoted the solubility of the polymer. Chan et al. synthesized a water soluble self doped polyaniline (*12*) using o-aminobenzylphosphonic acid and its hemisodium and monosodium salts.

Phenyl ring substituted aniline monomers like o-toluidine, m-toluidine (*13,14*), o-ethylaniline, o-phenitidine and o-anisidine (*15-19*) were used to prepare substituted polyaniline which were soluble in the dedoped state in solvents like dimethylsulfoxide (DMSO), dimethylformamide (DMF), dichloromethane and chloroform.

Water soluble poly(N-alkylaniline)s were prepared with alkyl (C_1-C_{18}) substituents (*20*). Chen and coworkers reacted the base form of HCl-PANi with 1,3-propane sultone (*21*) which gave the sodium salt of poly(aniline propane sulfonate) which was passed through a protonated cation exchange resin resulting in water soluble poly(aniline propane sulfonic acid).

Copolymerization of aniline with a substituted aniline monomer also results in a water soluble PANi. The comonomers used were butylaniline (*22*), o-toluidine, m-toluidine (*23*), o-phenitidine (5) and sodium diphenylamine-4-sulfonate (*24*).

Template-guided synthesis (*2,4*) is currently being used increasingly to synthesize water soluble polyaniline. A polyacid is used as a template to bind the aniline monomers to form the corresponding anilinium salt. Aniline monomers polymerize to form a complex with the template. The hydrophobic and hydrophilic ends of the template help in solubilizing the hydrophobic polyaniline chains. Some of the polyacids used (*25*) in the synthesis of polyaniline are polystyrenesulfonic acid, poly(acrylic acid) and poly(methacrylic acid).

In our studies, we have used lignosulfonic acid (LSO_3H) as a dopant/template to synthesize water-borne PANi. Lignosulfonates are produced from spent-sulfite liquors which are by-products from the paper industry (*26-28*) and used for various applications. It was chosen due its high water solubility, availability of sulfonic acid groups (for doping), and for its relative abundance and low cost. Lignosulfonic acid can be thermally crosslinked with appropriate curing agents. The crosslinkability should leave the electronic structure of PANi undisturbed which is highly desirable for its utility in various applications. Therefore in this complex, PANi can contribute to the electronic and optical properties and the template can be used for several modifications in order to enhance the processibility.

Experimental

Aniline and ammonium persulfate were purchased from Aldrich Chemical Company. Sodium lignosulfonate REAX 825E was obtained from Westvaco Chemicals. Dowex HCR-W2 cation exchange resin was purchased from Sigma Chemical Company. All other reagents were A.C.S. certified and were used as received.

A sample of 5.0 g of sodium lignosulfonate (REAX 825E) was dissolved in 25 ml water. It was passed through 200 g of a protonated cation exchange resin column (Dowex HCR-W2) which had been washed with deionized water to a neutral pH prior to elution. After completion of exchange, pH of lignosulfonic acid solution eluting from the column was typically around 1.0.

An aqueous solution containing 1g of LSO_3H was added to 1mL (11 mmol) of aniline and the volume was adjusted to 35 ml using deionized water. This solution was cooled to $0^{0}C$ and oxidatively polymerized in the presence of 1.8 g (7.9 mmol) of ammonium persulfate. After three hours of reaction at $0^{0}C$, a green colored water-borne product was formed and centrifuged three times with deionized water to remove the free lignosulfonic acid and ammonium sulfate. The centrifuged sample was dedoped with excess ammonium hydroxide and filtered. The filtered residue was Soxhlet extracted with methanol for 72 hours followed by water for 24 hours, and dried under vacuum for 48 hours.

Hydrochloric acid doped polyaniline (HCl-PANi) was synthesized by a standard method (*29*), dedoped with ammonium hydroxide and Soxhlet extracted using methanol for 72 hours. The extracted sample was dried under vacuum overnight and used for characterization studies.

Conductivities of the pressed pellets (doped form) were measured using a Alessi four-point conductivity probe. UV-Vis spectra were obtained using a Perkin Elmer Lambda 19 spectrometer. Thermal stability of lignosulfonic acid polyaniline (Ligno-PANi) was compared to hydrochloric acid doped PANi (HCl-PANi) by thermogravimetric analysis (TGA) using Mettler-Toledo TC15 TA controller. NMR spectra were obtained using a Bruker 200 MHZ NMR spectrometer using DMSO-d_6 as the solvent.

Results and Discussion

Figure 1 shows the relationship between conductivity of Ligno-PANi vs w/w ratio of lignosulfonic acid : aniline. A weight ratio of 0.125 : 1 shows the maximum conductivity of 0.9 S/cm. Above this concentration, LSO_3H probably acts as an impurity and below this concentration as a dopant. A weight ratio of 1:1 LSO_3H : aniline was determined to give a conductivity of 10^{-1} S/cm in addition to good water compatibility. HCl-PANi synthesized by the standard technique showed a conductivity of 10 S/cm and was completely insoluble in water..

During the synthesis of Ligno-PANi, the color of the reaction mixture changed from dark brown to green after the addition of ammonium persulfate and was monitored by spectral methods. Figure 2 shows the UV-Vis spectrum of the reaction mixture collected at 10 minute time intervals. The peak at 305 nm is due to benzenoid ring absorption in lignin which remains constant with time. In addition we observe a peak developing at 438 nm for the 10 minute sample due to the π- π^* transition of the benzenoid ring. This absorption peak shifts to longer wavelengths as the reaction progresses. Thus, for the 60 minute sample the benzenoid ring absorption is observed at 447 nm. Apparently, as the reaction progresses, the oligomers formed are doped by lignosulfonic acid. The doped chain is conductive, thereby shifting the absorption to a longer wavelength.

NMR spectrum of (doped and soxhlet extracted) Ligno-PANi (Figure 3) reveals some information about the structure of the polymer. Ligno-PANi was washed with NH_4OH to remove all unbound lignosulfonic acid and further soxhlet extracted with water to ensure that the product was free of lignosulfonic acid and NH_4OH. Prior research in our lab on aniline/phenylene diamine oligomers (30) was of help in assigning the peaks in the spectrum. There are two different types of primary amines in the

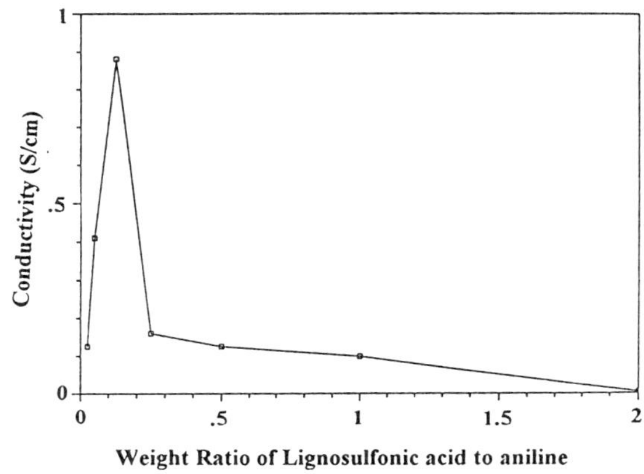

Figure 1. Relation between conductivity of Ligno-PANi for various weight ratios of lignosulfonic acid to aniline

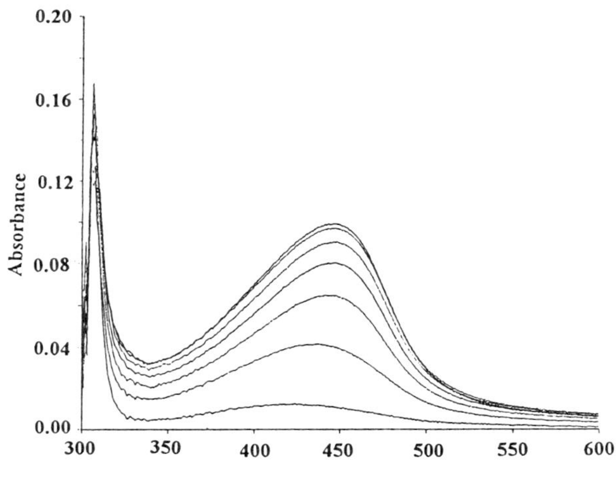

Figure 2. Monitoring the reaction of Ligno-PANi by UV-Vis spectroscopy.

polyaniline chain. Type I primary amine is attached to a phenylimine which is adjacent to a quinoid ring. Type II primary amine is attached to a benzene ring near a secondary amine which is adjacent to a phenylimine. In the NMR spectrum of Ligno-PANi, the peak at 5.7 ppm corresponds to type II primary amine and the peak at 6.3 ppm due to type I primary amine. Since the imine nitrogen is more electron withdrawing, the signal for the type I primary amine is shifted downfield in comparison to the signal from the other primary amine. The signal for the benzenoid aromatic protons are observed at 7.0 ppm and for quinoid protons at 7.4 ppm. The signal at 8.4 ppm is due to the secondary amine flanked by benzene rings. The peak that appears at 1.3ppm is due to alkyl substituents in lignin and are also seen in the NMR spectrum of pure lignosulfonic acid in DMSO.

Elucidation of the structure of the PANi segments in the polymer was possible based on the ratio of peak areas in the NMR spectrum. The quinoid to benzenoid peak area ratio was approximately 1 : 3 implying that the repeat unit structure of the polymer consists of one quinoid ring for every three benzenoid rings. The peak areas for both primary amines were added and was compared to the sum of the peak areas of the quinoid and benzenoid protons. The ratio was found to be approximately 50 : 1.5, which corresponds to an average chain length of 16 units and a M_n value of 1440.

We have also obtained spectroscopic evidence to demonstrate probable grafting of PANi to lignin. Dedoped Ligno-PANi was Soxhlet extracted with water and methanol. Dedoping with excess NH_4OH should have removed all lignosulfonic acid and further extraction with water should NH_4^+ -lignosulfonate formed after dedoping. Therefore if PANi was not grafted to lignin, Soxhlet extraction should remove the lignosulfonate salt and peaks for lignin should be absent. But the UV-Vis and NMR spectra show peaks for lignin thereby confirming the grafting of PANi to lignin. The UV-Vis spectra of Soxhlet extracted sample (Figure 4) shows peaks at 280 nm corresponding to the aromatic ring absorption of lignin in addition to the benzenoid and quinoid ring peaks at 350 and 620 nm respectively. The NMR spectrum (Figure 3) also shows a peak at 1.3 ppm corresponding to alkyl groups in lignin.

Thus we have spectroscopic evidence to demonstrate that polyaniline chains may be grafted to lignin, or that they may form a dopant-polymer complex that are not separable under the conditions employed. If grafting is assumed, Ligno-PANi may be visualized to be composed of hair like growth of polyaniline chains (Figure 5) from the lignin moiety and doped by the sulfonic acid groups present on the lignin molecule. The reaction of phenols in the presence of an oxidizing agent to generate phenoxy radicals is well known (31). These radicals or resonance structures thereof may presumably provide sites for growth of the polyaniline chains.

HCl-PANi was insoluble in all solvents tested and hence its solvatochromic behavior could not be studied. However, doped Ligno-PANi can be dissolved in DMSO and the green colored polyaniline turns blue. This change was followed by UV-Vis spectroscopy (Figure 6). Ligno-PANi dissolved in DMSO shows a peak at 358 nm characteristic of the benzenoid ring absorption and a peak at 590 nm for the quinoid ring in polyaniline. Thus, we can state that dissolution of Ligno-PANi in DMSO leads to dedoping, as the spectrum shows absorptions which are similar to that of dedoped polyaniline sample (Figure 4).

The solvatochromic behavior of Ligno-PANi is exemplified with increasing

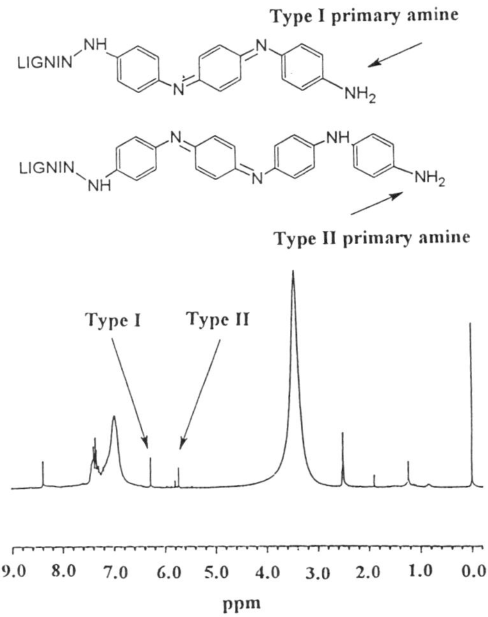

Figure 3. NMR spectrum of Soxhlet extracted Ligno-PANi.

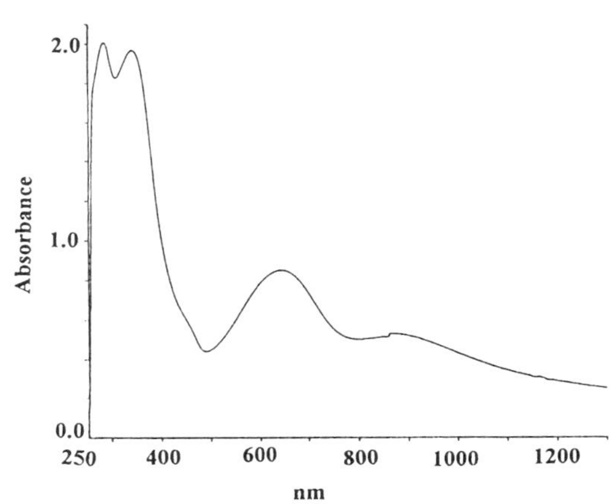

Figure 4. UV-Vis spectrum of Soxhlet extracted Ligno-PANi .

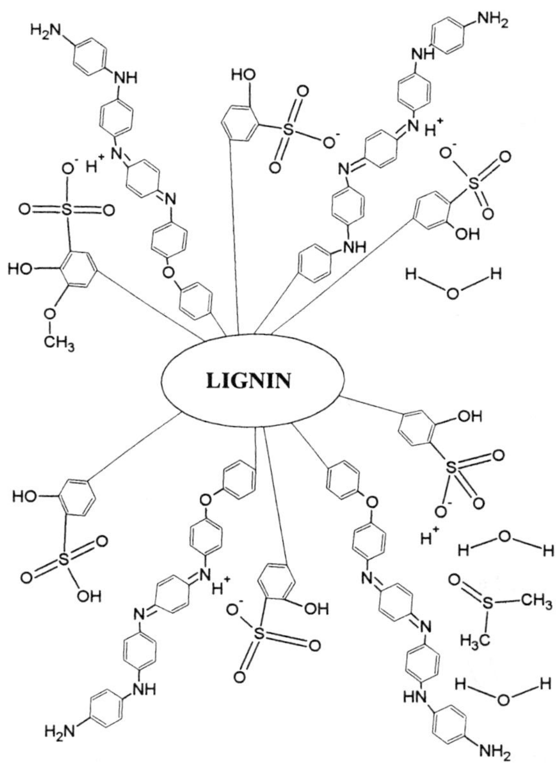

Figure 5. Illustration of probable grafting of polyaniline chains to lignin/ solvatochromic behavior of Ligno-PANi.

amount of water added to the above solution. Intensity of the benzenoid and quinoid absorption peak decreases as the concentration of water increases in the solution. In addition, a free carrier tail which is associated with doping of the polymer also starts forming around 960 nm.

Solvatochromic behavior of Ligno-PANi was also studied by NMR spectroscopy (Figure 7a). The dried Ligno-PANi sample dissolved in DMSO-d_6 was blue in color indicating that the polymer is in a dedoped state. When D_2O is added to this sample, it turns green and the changes associated with this addition was studied by NMR spectroscopy. Since the imine nitrogens are more basic than the amine nitrogens, the former should be protonated in preference to the latter. The protonated polymer gives rise to a chain with alternating radical cations. Protons closer to the electron withdrawing radical cation are shifted further downfield leading to an apparent increase in the peak area of the quinoid protons (Figure 7b).

We also observe a decrease in the intensity of the primary amine peak at 5.7 ppm relative to the primary amine peak at 6.3 ppm (Figure 7a & 7b). Though the lone pair of electrons on both the primary amines are held up in resonance with the aromatic ring, the electrons on type II primary amine is more readily available. The lone pair of electrons on type I primary NH_2 is not as readily available due to the electron withdrawing inductive effect of the imine nitrogens and is therefore protonated to a lesser extent. Hence the peak due to the former disappears whereas the peak for the latter remains relatively unaffected.

Further addition of D_2O increases the peak area (Figure 7c and 7d) for the quinoid protons relative to the benzenoid ring protons but the peak finally broadens out. In addition, type I primary amine is also protonated resulting in a decrease in the intensity of its peak.

Figure 5 is helpful in illustrating the reversible solvatochromic phenomenon in Ligno-PANi. Dissolution of Ligno-PANi in DMSO results in dedoping of polyaniline since the electronegative oxygen in DMSO is hypothesized to form a hydrogen bond with the proton on the imine nitrogen thereby deprotonating the polymer. Thus, addition of DMSO changes the color of the solution from green to blue. But, addition of water to this solution results in redoping of the polymer as evident from the color change from blue to green. The redoping might be due to solvation of DMSO by the water molecules thereby releasing the proton back to the polymer. This solvatochromic behavior is similar to the spectral changes observed in the UV-Vis region for PANi in hexafluoroisopropanol studied by Rasmussen et. al. (*32*).

Thermal stability of Ligno-PANi was compared to HCl-PANi using thermogravimetric analysis (TGA). Figure 8a shows the TGA curve for Ligno-PANi. The weight loss at about 100^0C is probably due to moisture loss. Around 120^0C, the sulfonic acid groups in the dopant might be eliminated. The weight loss after 250^0C may be attributed to the degradation of the polyaniline backbone. The sample loses 50% of its weight at 370^0C and complete thermal degradation is observed at 620^0C.

Figure 8b shows the TGA curve for HCl-PANi. The initial weight loss around 100^0C is probably due to some moisture loss from the sample. The weight loss up to 265^0C may be due to elimination of HCl. Thermal degradation of the polymer begins around 265^0C. About 50% weight loss occurs around 440 C and complete thermal degradation is observed at 680^0C. Thermal studies on polyaniline have also been carried

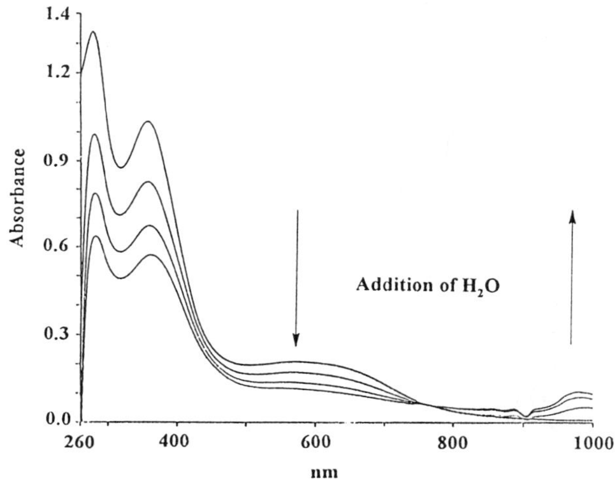

Figure 6. Solvatochromic behavior of Ligno-PANi studied by UV-Vis spectroscopy.

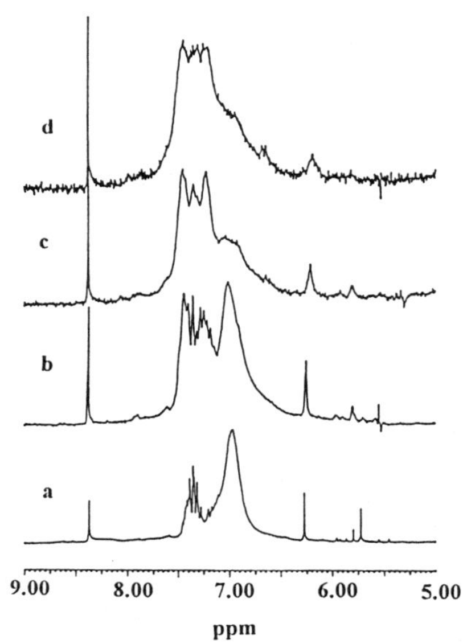

Figure 7. Solvatochromic behavior studied by NMR spectroscopy; a) Ligno-PANi in DMSO-d_6 ; b) 8 drops D_2O; c) 16 drops D_2O; d) 24 drops D_2O.

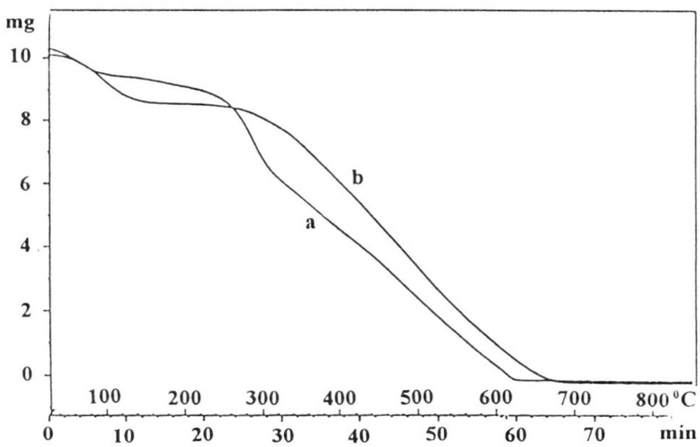

Figure 8. TGA curve of a) Ligno-PANi; b) HCl-PANi.

out by Wei and co-workers (*33,34*). Comparing the TGA of the two polyaniline samples, we can infer that HCl-PANi is slightly more stable than Ligno-PANi.

Conclusion

Polyaniline was synthesized as a water-borne system using lignosulfonic acid as a dopant/template. Characterization of (doped) Ligno-PANi was possible due to its water compatibility and DMSO solubility. The conductivity of a typical (self-doped) ligno-PANi preparation is 10^{-1} S/cm which is adequate for applications like corrosion prevention and anti-static charge dissipation. Even though the template polymerization of aniline was initiated in the presence of lignosulfonic acid the data obtained from NMR and UV-Vis spectroscopy suggests that polyaniline is probably grafted or tightly bound to lignosulfonic acid.

Acknowledgments

The authors wish to acknowledge NASA/KSC (NAG 10-0181G) and NASA/EPSCOR for providing financial support for this work.

Literature Cited

1. Cao, Y.; Qiu, J.; Smith, P. *Synth. Met.* **1995**, 69, 187.
2. Sun, L.; Yang, S.C. *Mat. Res. Soc. Symp. Proc.* **1994**, 328, 209.
3. Chen, S.A.; Hwang, G.W. *J. Am. Chem. Soc.* **1995**, 117, 10055.
4. Sudhakar, M.; Stoecker, P.W.; Viswanathan, T. *Recent Advances in Polymer Science*, Transworld Research Network, in press.
5. Liao, Y.; Angelopoulos, M.; Levon, K. *Proc. ANTEC 1995, SPE,* Brookfield, CT **1995**, 1413.
6. Angelopoulos, M.; Patel, N.; Shaw, J.M. *J. Vac. Sci. Technol.* **1993**, B 11(6), 2794.
7. Viswanathan, T.; Shanks, K.U. *ACS Polymer Preprints* **1996**, 37 (1), 508.
8. Nguyen, M.T.; Leclerc, M.; Diaz, F. *Trends in Polymer Science* **1995**, 3(6), 186.
9. Yue, J.; Epstein, A.J. *J Am. Chem. Soc.* **1990**, 112, 2800.
10. Yue, J.; Wang, Z.H.; Cromack, K.R.; Epstein, A.J.; MacDiarmid, A.G. *J. Am. Chem. Soc.* **1991**, 113, 2665.
11. Yue, J.; Gordon, G.; Epstein, A.J. *Polymer* **1992**, 33 (20), 4410.
12. Chan, H.S.O.; Ho, P.K.H.; Ng, S.C.; Tan, B.T.G.; Tan, K.L. *J. Am. Chem. Soc.* **1995**, 117, 8517.
13. Wei, Y.; Focke, W.W.; Wnek, G.E.; Ray, A. ; MacDiarmid, A.G. *J. Phys. Chem.* **1989**, 93, 495.
14. Leclerc, M.; Guay, J; Dao, L.H. *Macromolecules* **1989**, 22, 649.
15. MaCinnes, D.; Funt, B.L. *Synth. Met.* 1988, 25, 235.
16. Dhawan, S.K.; Trivedi, D.C. *Synth. Met.* 1993, 60, 67.
17. Sathiyanarayanan, S.; Balakrishnan, K. *Electrochimica Acta.* **1994**, 39 (6), 831.
18. Wei, Y. U.S. Patent **1990**, 4,940,517.

19. Kinlen, P.J., Silverman, D.C. and Jeffreys, C.R. *Synth. Met.* **85**, 1327 (1997).
20. Dao, L.H.; Leclerc, M.; Guay, J. and Chevalier, J.W. *Synth. Met.* **1989**, 29, E377.
21. Chen, S.-A.; Hwang, G.-W. U.S. Patent **1997,** 5641859.
22. Bergeron, J.-Y.; Dao, L.H. *Macromolecules* **1992**, 25, 3332.
23. Wei, Y.; Hariharan, R.; Patel, S.A. *Macromolecules* **1990**, 23, 758.
24. Nguyen, M.T.; Kasai, P.; Miller, J.L.; Diaz, A.F. *Macromolecules* **1994**, 27, 3625.
25. Sun, L.; Clark, R.; Yang, S.C. *Synth. Met.* **1997**, 84, 67.
26. Nimz, H.H.; In *Wood adhesives- Chemistry and Technology;* Pizzi, A.,*Ed.;* Marcel Dekker, Inc.; New York, New York, **1983**, pp 247-288.
27. Goel, K. *L' Actualite chimique Canadienne* April **1987**, 9.
28. Glennie, D.W.; In *Lignins- Occurrence, Formation, Structure and Reactions*; Sarkanen, K.V.; Ludwig, C.H.; Ed., Wiley-Interscience, New York, New York, 1971, pp 597-601.
29. Wei, Y.; Hseuh, K.F.; Jang, G-W. *Macromolecules* **1994**, 27, 518.
30. Viswanathan, T.; Helmich, J., Feng, Q.; Toland, A. *J. Elast. Plast.* **1998**, 30, 197.
31. Sarkanen, K.V.; In *Lignins- Occurrence, Formation, Structure and Reactions*; Sarkanen, K.V.; Ludwig, C.H.; Ed., Wiley-Interscience, New York, New York, 1971, pp 116-125.
32. Hopkins, A.R.; Rasmussen, P.G.; Basheer, R.A. *Macromolecules* **1996**, 29, 7838.
33. Wei, Y.; Hsueh, K.F.; *J.Polym.Sci., Part A: Polym. Chem.* **1989**, 27, 4351.
34. Wei, Y., Jang G-W.; Hsueh, K.F.; Scherr, E.M.; MacDiarmid, A.G.; Epstein, A.J.; *Polymer* **1992** 33, 314.

Chapter 7

Direct Evaluation of Injection Efficiency from Metals into Trap-Free Small-Molecule-Based Transport Layers: Probing the Details of Interface Formation

M. Abkowitz [1,2], A. Ioannidis [1,2], and J. S. Facci [1,2]

[1] Wilson Center for Research and Technology, Xerox Corporation, 800 Phillips Road, Webster, NY 14580
[2] NSF Center for Photoinduced Charge Transfer, University of Rochester, Rochester, NY 14625

Studies of interface formation on conventional semiconductor materials are typically carried out under relatively pristine conditions. However, for devices based on the use of electronic polymers there is also compelling interest in exploring the variations in contact behavior that might result under realistic manufacturing conditions like mutilayer device assembly based on solution coating technology. Small molecule doped polymers (MDPs) developed principally as large area coatings for electrophotographic use are now finding wider device applications. These polymers are insulators capable of transporting excess injected charge with a unipolar drift mobility which can be tuned over a wide range by varying the concentration of transport active species. Most significant in the present context, MDPs can be rendered trap free by molecular design. These unique characteristics of MDPs make it possible to analyze the relative injection efficiencies of their interfaces with various contacts simply by a direct comparison of current voltage characteristics with time of flight drift mobility measurements carried out on the same film coatings. In this way, and apart from their intrinsic interest and practical value, MDPs and closely related polymeric media provide the ideal venue for the study of contact phenomena on molecular solids. Measurements were carried out by measuring dark hole injection into the MDP film TPD/polycarbonate and a polymeric analog, PTPB, from various preformed metal substrates as well as evaporated top contacts. For comparison, injection was also studied from various contacts into a small molecule glass of TPD, and an electroluminescent polymer, MEH-PPV. For preformed metal substrates under fully relaxed conditions, it was found that while injection efficiency nominally scaled with the estimated interfacial energy step there was significant variance that in some cases could be clearly associated with the specific details of interfacial chemistry. In the case of metal contacts evaporated onto TPD/polycarbonate and PTPB, a time and temperature dependent phenomenon was delineated in which the hole injection efficiency systematically evolved from emission limited to ohmic. For evaporated Au contacts on TPD/polycarbonate analysis of the kinetics of the entire evolution process from the earliest resolved times to one month suggests that it can be operationally represented as a superposition of two processes: a slow process associated with the details of fabrication of the contacts and assigned to recovery following deposition induced damage, and a rapid forming process which we demonstrate occurs even in the absence of overt damage to the organic surface. This latter process is also observed for TPD/polycarbonate contacts to a preformed Ag substrate and to a liquid Hg top contact. In these cases an interfacial energy mismatch leads to a final contact emission limited steady state. This process which is apparently not associated with surface structural damage could reflect interfacial chemistry or some as yet unspecified surface reorganization. Time dependent contact forming processes sharing the general characteristics described above are also observed for evaporated Au contacts on MEH-PPV and a pure TPD glass.

A key motivation for the application of electronic polymers is based on their potential for simple low cost manufacturing using simple solution coating techniques. Though there is fundamental value in studying interface formation under pristine conditions, there is compelling interest in elucidating the variance in contact behavior that might result under realistic fabrication conditions, namely those attending device assembly using solution coating techniques. In this latter connection it should be noted that the most commercially successful application of organic electronic film coatings is in the mass fabrication of large area electrophotographic receptor devices that are in fact prepared using a variety of solution coating processes.[1] In particular, the transport layers used in these bilayer structures are trap free, unipolar, insulating but charge transporting films, typically solid solutions of a transport active molecule, e.g. a triarylamine in a host polymer like polycarbonate[1] or its polymeric analog.[2] Because certain small molecule transport layers can be rendered trap free, direct comparison of current-voltage data with time of flight drift (TOF) mobility measurements provides an unambiguous injection efficiency figure of merit for any contact under test on such a transport layer.[3,4] The underlying experimental technique is reviewed. Time-dependent phenomena affecting injection efficiency are analyzed and it is demonstrated that a contact formation process occurring on the time-scale of a few hours is inherent to a number of organic interfaces. Furthermore, under fully relaxed conditions, injection efficiency correlates with simple estimates of the interfacial barrier estimated from the difference between published values of the substrate work function and an electrochemical estimate of the small molecule ionization potential in the polymer matrix. The experimental procedure described below provides a means whereby variances in injection efficiency associated with the details of contact structure, composition and formation can be clearly identified. Several of these are delineated and analyzed in the following.

BACKGROUND

Consider a trap-free molecularly doped polymer, MDP, optimized for hole transport. Hole transport involves a field driven chain of redox reactions in which the carrier, under the influence of the electric field, hops from a neutral molecule to its radical cation derivative. The hole transport process is triggered by an initial charge transfer in which a neutral molecule is energetically favored to donate an electron either to an adjacent metal contact or a photoexcited layer. If the neutral transport molecule has an ionization potential lower than any competing species present, then the latter, independent of their concentration, cannot act as hole traps; thus they are invisible to transiting carrier. A complementary situation can be described in terms of relative electron affinities for electron transporting systems. These transport layers are therefore trap-free insulators with finite unipolar mobility which means that while there are no intrinsic free carriers, an extrinsic carrier once injected will, under the influence of an electric field, traverse the sample bulk without being locally neutralized or immobilized.

An ohmic contact is defined as an infinite reservoir of charge that is able to satisfy the demands of the bulk for injected charge up to and including a supply sufficient to sustain steady state space charge limited current.[5] The bulk demands a space charge limited carrier supply when the transit time of any excess injected carrier is less than the bulk relaxation time $\rho\varepsilon$. Under the latter circumstances which must always prevail for an ohmic contact on a finite mobility insulator (J_{TFSCLC}), the trap free space charge limited current (which only an ohmic contact will supply) takes a particularly simple form namely[5] :

$$J_{TFSCLC} = 9/8\ \varepsilon\ \varepsilon_0\ \mu\ E^2 / L \tag{1}$$

Here ε is the relative dielectric constant, μ is the drift mobility, E is the average field defined as V/L where L is the film thickness. At a given field for fixed specimen geometry the current supplied by an ohmic contact is uniquely specified by the drift mobility of the injected carrier at that field. If we measure the drift mobility at a given field and calculate J_{TFSCLC} then compare this calculated current density to the current injected in the dark J_{DC} by a contact under test on the same specimen we can define an injection efficiency figure of merit F, under fully self-consistent conditions namely:

$$F = J_{DC}/J_{TOF} = J_{DC}/J_{TFSCLC} \tag{2}$$

We will specifically use the TOF subscript in some of the figures which follow in order to remind the reader that in each case this is the TFSCL current density computed from TOF drift mobility data. On the other hand J_{TFSCLC} will refer to experimentally measured trap free space charge limited current densities. The associated experimental procedures are fully described in the next section.

EXPERIMENTAL TECHNIQUE

Specimen Preparation. TPD/polycarbonate molecularly doped polymer films were prepared by slow solvent evaporation from a 4-5 wt% solution of the small molecule and polymer. Methylene chloride (Aldrich) used in the preparation of the coating solutions was of the highest purity available. Films were cured in a convection oven at 110 °C for 15-20 min. Films of 100% TPD were evaporated at a rate of 300 Å/s on MystR® (a carbon filled polymer substrate) to thicknesses of 5-20 μm under a vacuum of $1x10^{-6}$ Torr. The polymer MEH-PPV was synthesized according to the general methodology toward soluble PPV derivatives developed by Hsieh et al.[6] and solution coated onto MystR® from a 1 wt% solution in methylene chloride. Films of 6-10 μm were obtained and subsequent to top contact deposition, the samples were maintained and measured in a dry N_2 atmosphere. Carbon filled polymer substrates (MystR®) were commercially obtained. Au substrates were obtained as commercially evaporated 1000 Å or 2000 Å Au films on highly polished 3 inch Si (100) wafers from Polishing Corporation of America. The Au layers were bonded to the wafer with a 5 nm layer of Ti or Cr. AFM characterization of these surfaces showed a continuous smooth pebble like appearance.[7] Au/mica substrates were prepared in-house as previously described by evaporating Au under UHV conditions onto freshly cleaved mica (green, Grade V2 or better).[7] Mica substrates were conditioned by baking in vacuum overnight at 300 °C prior to Au evaporation. Cr, Ag and Pt substrates were evaporated in-house by conventionally accepted processes.[8] Single crystal Pt(111) electrodes were obtained from Aremco. Transport polymer films were coated onto the (111) facet. Glassy carbon (GC) electrodes were obtained from Atomergic Chemmetals.

Time-of-flight secondary ion mass spectrometry (TOFSIMS) was used to analyze the surface and to depth profile the elemental composition of these Au substrates. TOFSIMS is a semi-quantitative surface specific analytical technique which provides information on the nature of atomic species on surfaces. Measurements were done with a Charles Evans & Associates Model TFT-I instrument using Ga as the ion source. Depth profiling was done using Ga as the sputtering ion. Au/Cr and Au/Ti coated Si wafers and Au/mica substrates were analyzed using a 600 picoampere aperture. Evaporated top contacts were analyzed by X-ray diffraction, scanning electron microscopy and transmission electron microscopy and their interfaces to the MDP were analyzed by X-ray photoelectron spectroscopy.

Measurements. Fig. 1a is a schematic drawing of the experimental apparatus for time of flight (TOF) drift mobility and steady state current measurements. Hole transporting (unipolar) specimen films were configured as thin film parallel plate capacitors with the contact under test on one face, and the opposite contact, typically a 300 Å evaporated Al film which was blocking for hole injection and transparent to the laser pulses required for the execution of small signal current mode TOF experiments.[9] The TOF technique was used to make measurements of hole drift mobility. With positive bias applied to the blocking contact the specimen film was photoexcited through the Al contact with a highly attenuated 337 nm pulse from a nitrogen gas laser. Transit times t_{tr} for photoinjected holes at the applied bias could then be determined

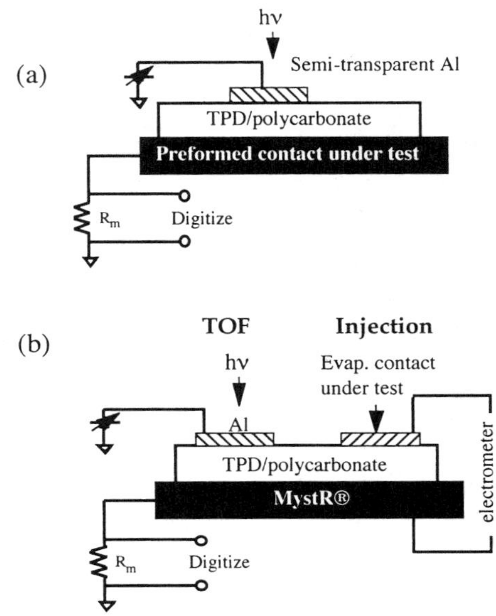

Figure 1. Schematic diagram of experimental apparatus for measuring TOF, transient and dark dc injection in the same trap free hole transporting film. Panel (a) TOF hole transients are observed via small signal excitation through the top Al contact. When negative bias is applied to the contact under test a weak laser light pulse incident on the semi-transparent Al blocking contact induces a positive photogenerated sheet of charge to transit the polymer, enabling the transit time to be measured. In the configuration shown, positive bias is applied to the contact under test for measuring the dc dark injection and steady state current. Panel (b) Experimental configuration for evaluating the injection efficiency of an evaporated metal top contact. TFSCLCs can be computed from hole transit times obtained directly from TOF experiments (at left) or by direct measurement of injection from the ohmic MystR® substrate.

from the current transient and the mobility calculated. In the present case, e.g. evaluation of hole injecting contacts, steady state J-E measurements were done by application of a positive step voltage to the contact under test, capturing the transient for additional analysis, and then waiting for the current to achieve a steady state value. TOF and dark injection experiments were thus carried out systematically under identical conditions on the same specimen bulk and with the same pair of contacts. Fig. 1b is a schematic diagram of the experimental arrangement for evaluating hole injection efficiency from the evaporated metal contact under test. The trap free space charge limited current (TFSCLC) density is computed from TOF hole transit times (at left) and compared with the steady state DC injection current obtained from the top contact on the same specimen. Field dependence of the drift mobility is clearly neglected in the derivation of equation 1. We have in fact previously quantified directly the error introduced when field dependence is ignored for both TPD/polycarbonate and PTPB in the relevant temperature range by using a combined analytical and numerical procedure.[2] On this basis and for purposes of the present paper the correction introduced using a more exact representation of the TFSCLC is of no significance.

RESULTS AND DISCUSSION

1. Polymers Deposited on a Preformed Contact.

Fig. 2 is a plot of current density J versus electric field E for a thin film of a molecularly doped polymer composed of a 40 wt% solid solution of TPD (see structure below) in a bisphenol A polycarbonate binder.

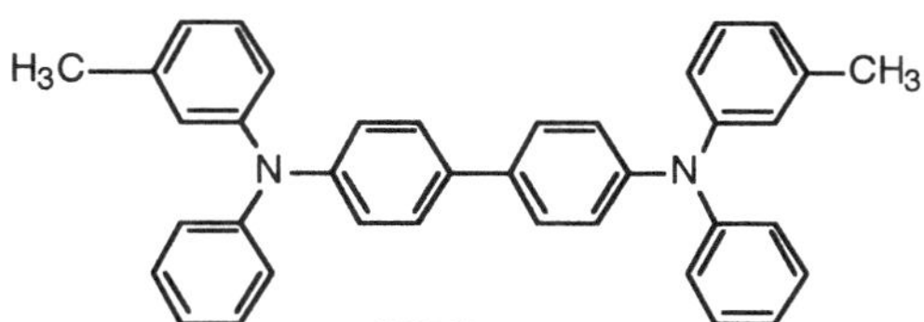

TPD

The polymer film was solution coated onto a carbon loaded polymer film substrate (MystR®) and then overcoated with an evaporated semi-transparent Al top contact. The dark current transient injection experiment is carried out by applying the positive voltage step to the substrate contact and time resolving the transient response of the injected hole current. The early time dark injection current is depicted in the insert and the time at which the maximum current occurs is denoted as t_p. According to the theory of transient space charge limited injection[5], the time t_p can be related to the mobility μ of injected holes from equation 3,

$$\mu = 0.8\ L / (E\ t_p), \tag{3}$$

where L is the specimen thickness and μ is the hole drift mobility. If it is assumed that the dark injection transient is space charge limited, the steady state trap free space charge limiting current

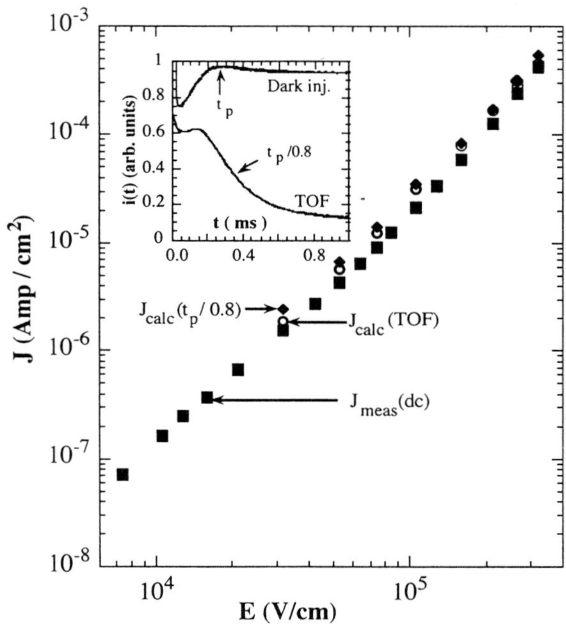

Figure 2. Plots of the measured dc current density: filled squares, steady state current density; open circles, the current density computed from drift mobility measurements and Eq. 1; filled diamonds, the current density computed from the transient dark injection peak values. The contact under test is a carbon filled polymer coated with a transport layer (TPD/polycarbonate) that is known to support trap free hole transport. The insert shows a typical dark injection transient compared to a small signal TOF transient. Conformity of key features of the steady state and transient data with the theory of trap free space charge limited currents provides a self consistent demonstration of contact ohmicity.

(Reproduced with permission from reference 17. Copyright 1998 American Institute of Physics.)

can be computed. Thus substituting the mobility calculated from t_p into equation 1 provides a quantitative estimate of the resulting steady-state trap free space charge limiting current J_{TFSCLC}, as shown by the filled diamonds in the figure. The insert also presents the corresponding time-of-flight transient obtained at the same applied field as the dark injection curve but with positive bias applied to the semi-transparent top contact. The measured transit times t_{tr} provide an independent estimate of J_{TFSCLC} according to the procedure described in the experimental section. Those results are represented by the open circles in the figure. Finally the filled squares represent the measured steady state hole injection current density J_{DC} supplied by the carbon filled polymer substrate contact. The measured values of the steady state dark currents are coincident within experimental error to the two independent calculated estimates of J_{TFSCLC}. Thus taken together these results constitute *prima facie* evidence that the carbon filled contact is ohmic for hole injection. The comparisons presented in Fig. 2 are taken as a benchmark representation of ohmic contact behavior on a molecularly doped polymer and illustrate application of the experimental technique to be used in the following study for evaluation the injection efficiency of each of the contacts under test. The injection behavior illustrated in Fig. 2 is also observed for injection into TPD/polycarbonate from a substrate contact consisting of Au evaporated onto mica. The injection efficiency is defined by the quotient J_{DC} / J_{TOF} where J_{DC} is the measured steady state current density supplied at a field E and J_{TOF} is the trap free space charge limited current calculated from hole transit times measured at the same field in the same specimen as illustrated in Fig. 1. Hole injection efficiencies J_{DC} / J_{TOF} as a function of field for the Au/mica contact (filled squares) are presented in Fig. 3. Note that the latter values are very nearly unity clearly demonstrating the ohmic nature of Au/mica substrate for hole injection into TPD/polycarbonate.

On the other hand, when Au is evaporated onto Si using a thin (5 nm) Cr adhesive layer, the hole injection efficiency is significantly degraded. The latter is also illustrated in Fig. 3. The filled diamonds are the hole injection efficiencies for Au/Cr/Si substrate contacts made to two different specimens. The dashed curve represents an average value of hole injection efficiency. Variability in J_{DC}/J_{TOF} is observed despite the care taken to cast TPD films from the same solution onto different Au/Cr/Si substrates diced from the same wafer. The error bar is presented as an estimated representation of the variability since sufficient replicate runs for a complete statistical analysis were not undertaken.

The severe diminution, with respect to Au/mica, of the hole injection efficiency from the Au/Cr substrates appear to be associated with the known contamination of the Au surface by migration of the "glue" metal Cr to the polymer/metal interface. While this has been discussed previously,[10] we have obtained independent spectroscopic evidence that this occurs in our particular Au/Cr/Si substrates. Comparison of the surface composition and elemental depth profile analyses of the Au/mica and Au/Cr substrates were obtained by the TOFSIMS technique and the salient results are presented in Figs. 4 and 5, respectively. The principle result expressed in Fig. 4 is that relatively few contaminants are present at the Au/mica surface and that only a background level of C is observed. It has been well established that the surface texture of Au/mica is nearly entirely Au(111).[7]

Figure 5 is a depth profile of 1000 Å of evaporated Au on highly polished Si (100) with a thin (5 nm) intermediate layer of Cr used to enhance adhesion. One striking difference, however, is the penetration of Cr atoms into the Au layer all the way to the Au surface. In fact this effect is well known and has been previously described.[10] It is thought that the mechanism of Cr migration to the Au surface involves Cr atom diffusion along the grain

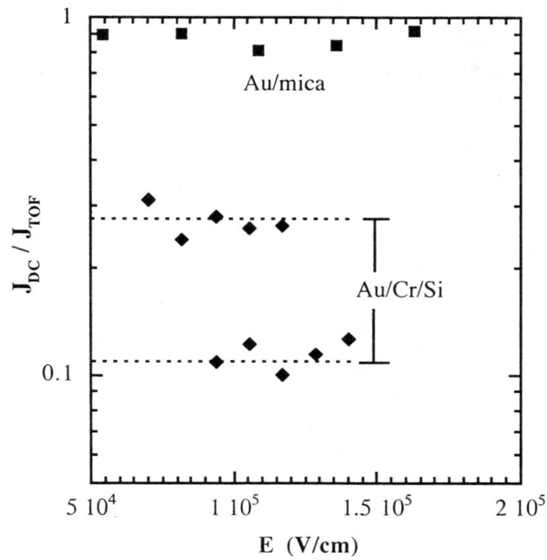

Figure 3. Comparison of injection efficiency figures of merit J_{DC}/J_{TOF} vs. field for hole injection from a Au coated mica substrate (squares) into a film of TPD/polycarbonate with injection from a series of Au/Si substrates (triangles) under analogous conditions. The latter contact uses a Cr adhesion layer between Au and Si. The dashed lines illustrate the nominal variance in behavior observed when a number of such substrate contacts are compared.
(Reproduced with permission from reference 17. Copyright 1998 American Institute of Physics.)

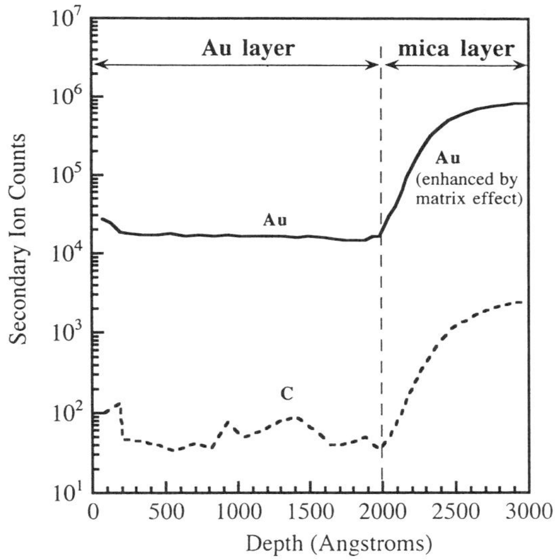

Figure 4. Elemental depth profile of a Au on mica substrate obtained from time of flight secondary ion mass spectroscopy (TOFSIMS). Only elements (background C) found in the Au layer are displayed.
(Reproduced with permission from reference 17. Copyright 1998 American Institute of Physics.)

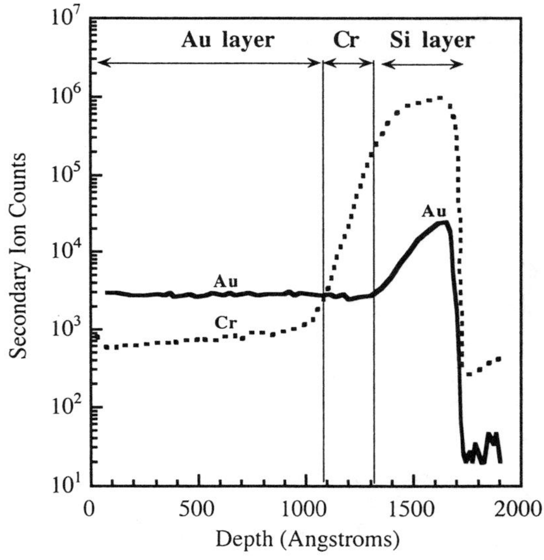

Figure 5. Elemental depth profile of a Au/Cr/Si substrate from TOFSIMS results. Au and Cr atomic profiles are presented. Note that Cr is present within and at the surface of the Au substrate.

(Reproduced with permission from reference 17. Copyright 1998 American Institute of Physics.)

boundaries of polycrystalline Au. Subsequent contamination of the Au surface with Cr suggests that hole injection occurs from a lower work function surface consistent in the present case with an apparent increase in the estimated barrier to hole injection.

Figure 6 presents for various contact systems a compilation of hole injection efficiency data expressed as J_{DC}/J_{TOF} versus estimated interfacial barrier heights. The latter barrier heights are estimated from published values of substrate contact work functions[11] together with our estimate of the transport molecule ionization potential (5.5 eV) which is based on the measured value of the electrochemical half wave potential of neutral TPD and its analogues in solution.[12] Such estimates have an uncertainty associated with them because of the difference in the solvent and MDP polarization energies. A value of -4.7 V was assumed for the electrochemical potential of the SCE (saturated calomel electrode) in the conversion of the electrochemical potential scale to the vacuum scale.[13] While the absolute value of the barrier height are associated with some uncertainty the relative differences in injection barriers are more certain. The vertical bars in Fig. 6 encompass in each case the experimental variation observed in specimens prepared under ostensibly identical conditions. For the cases of Cr and Ag(111) it is not unreasonable to ascribe the large variation observed to uncontrolled air oxidation of the surface of the substrate contact prior to coating with TPD/polycarbonate. In the case of Cr, one of the substrate contacts was solvent coated immediately after evaporation while the other was allowed to air oxidize for several months prior to coating. The injection efficiency of these two Cr contacts with TPD/polycarbonate are, respectively, $8X10^{-3}$ and $1X10^{-3}$.

Glassy carbon (GC) is hard and impervious to gases and liquids. The structure of GC is believed to be graphitized ribbons with a coherence length on the order of tens of angtroms and the electrical conductivity is about one-fourth that of randomly ordered graphite.[14] Commercially obtained glassy carbon electrode disk substrates were extensively pretreated prior to solvent coating. Their surfaces were first brought to a mirror polish with a graded series of abrasives ending in either 0.05 micron aluminum oxide dispersed in highly purified water or 1 micron diamond polish pasted in a high molecular weight hydrocarbon. The aluminum oxide polishing was done against a clean glass substrate. As noted in Fig. 6 the aluminum oxide polishing pretreatment systematically yielded an interface with higher hole injection efficiency than similarly prepared but diamond paste polished GC. It appears that polishing preparation of GC electrode surfaces with diamond paste polishing leads to a surface that contains irreversibly adsorbed organic species likely derived from the pasting vehicle. These sites may be blocking for hole injection. In contrast the aluminum oxide abrasive was a highly purified sample that had no history of contact with organic impurities. Extensive analogous electrochemical studies of interfacial electron transfer rate constants k_s for the oxidation-reduction of aqueous solutions of ferro/ferricyanide at glassy carbon electrodes subject to both aluminum oxide and diamond polish surface pre-treatments have been carried out.[14] The electrochemical studies pointed out a parallel enhancement of k_s in the case of GC electrodes treated with aluminum oxide relative to diamond polished surfaces. It should be noted that the latter rate constants k_s are the electrochemical analog of solid state injection efficiency.

A striking enhancement in hole injection efficiency for another type of carbon surface is illustrated in the upper right hand portion of Fig. 6. In this case the substrate contact consists of highly graphitized carbon particles dispersed in a polymer binder at concentrations exceeding the percolation threshold. Carbon filled polymers are complex percolative systems whose electrical behavior is a function of the filler, the filler interparticle contacts, the host matrix and

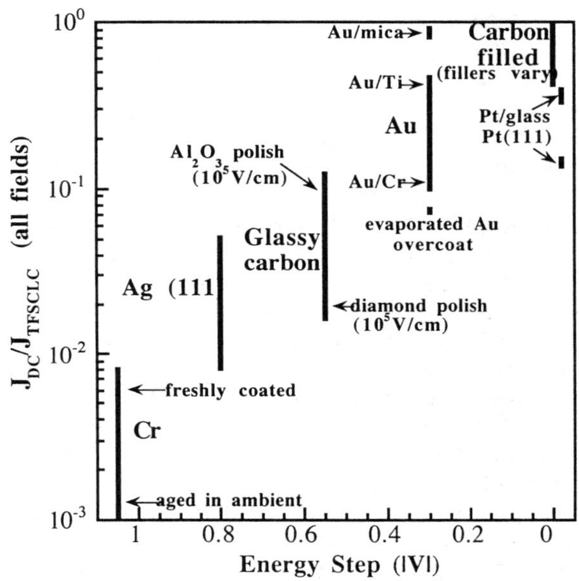

Figure 6. Hole injection efficiency figure of merit for substrate contacts of varying work function vs. energy step across the contact polymer interface estimated from published work function data and electrochemical redox potential data. The height of each bar reflects the variability in injection efficiency due primarily to variation in substrate surface pretreatment and for the particular case of Au, diffusion to the interface of metal atoms from underlying binder layers.

(Reproduced with permission from reference 17. Copyright 1998 American Institute of Physics.)

the resulting internal network topology.[15] Such contact systems were in certain cases found, using the present techniques, to be ohmic for hole injection into TPD/polycarbonate. One such case was illustrated in Fig. 2. Although we generally have no independent information regarding the work function of these carbon filled media *a priori*, it is evident from a rudimentary analysis based on thermionic emission theory that the barrier to hole injection must, in the illustrated cases, be very small.

With the exception of Pt shown on the extreme right hand side of Fig. 6 the average behavior of the injection efficiency nominally scales with the estimated interfacial barrier height. Variations about the average behavior are, however, significant and it is suggested that they reflect significant perturbations of the effective work functions of the contacts. These perturbations involve the influence of factors extrinsic to the two phases involved in the formation of the electrical interface but likely to be encountered in realistic manufacturing environments which exploit the advantages inherent in devices based on solution coatable organic layers.

The behavior of platinum electrodes is known to be especially complex. The two cases described for platinum substrates are, respectively, Pt thermally evaporated onto glass under high vacuum conditions and a Pt(111) single crystal precleaned by a brief immersion in acidic peroxide and rinsed copiously in distilled water. It is well known from electrochemical studies that the electrochemistry of the Pt surfaces and the electron transfer reactions at Pt are highly sensitive to mono- and submonolayer levels of contaminants and adsorbates as well as to the electrode surface texture and morphology.[16] Moreover, it is known from controlled ultrahigh vacuum studies that the work function of platinum varies by approximately 1 eV when a pristine Pt(111) surface is dosed with a single monolayer of water. Thus the sensitivity of the Pt work function and associated interfacial charge transfer reactions to relatively minor variations in surface condition likely underlies the apparent misfit of the Pt data in Fig. 6 which is based on published values of the Pt workfunction[11] obtained by photoemission under high vacuum conditions.

2. Contacts Evaporated onto an Organic Film

In Fig. 7 we augment the hole injection efficiency from various substrates with data obtained when Au is vacuum deposited (10 Å/sec) on the top surface of a PTPB film that is in turn overcoated onto a substrate which is ohmic for hole injection, i.e. when the positive bias is applied to the substrate contact $J_{DC} / J_{TOF} = 1$. PTPB, shown below, is structurally analogous to TPD. Molecularly doped polymers containing TPD have formed the basis of much of this study. In the case of PTPB, however, the functional unit is covalently incorporated into the main chain and offers an additional degree of thermal and solvolytic stability over the molecular solid solution of TPD in the polycarbonate binder. The relaxation behavior in the present figure is very similar to TPD/polycarbonate as demonstrated in elaborate detail in the sections to follow. The evaporated Au data in the figure corresponds to hole injection when positive bias is applied to the top contact.

It is in fact found that the hole injection efficiency from the evaporated Au contact depends on the details of deposition. More interestingly, subsequent detailed analysis on similar systems shows that the injection efficiency also depends on the extent to which the system has been allowed to relax at a given temperature immediately following contact deposition. To our knowledge this is the first observation of time dependent contact forming on an organic film.

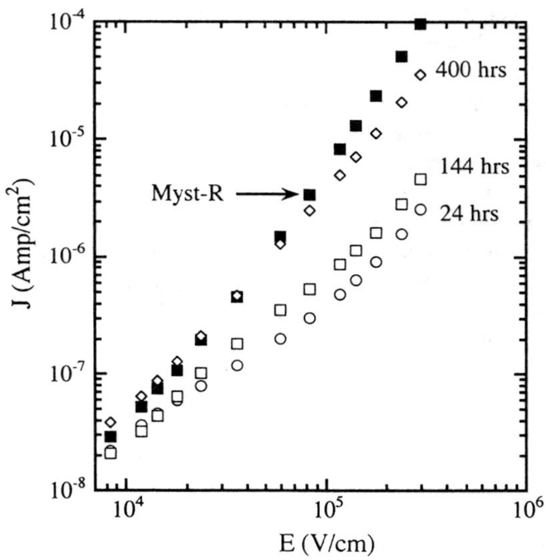

Figure 7. J vs. E plots for hole injection into TPD polycarbonate from a carbon filled polymer substrate contact (MystgR®) known to be ohmic for hole injection (filled squares) and an evaporated Au contact (deposited at 10 Å/sec) as a function of room temperature annealing time following evaporation on the free surface of a specimen film. Open symbols are the corresponding J vs. E data when positive bias is applied to the substrate and depict the evolution of injection from the evaporated Au top contact in time.

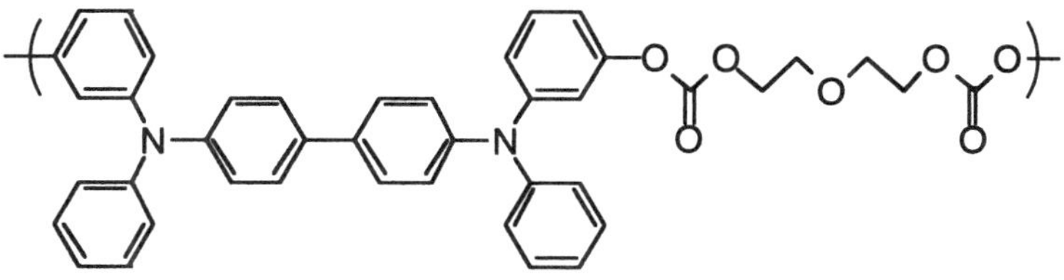

PTPB

The blocking nature of freshly evaporated Au contacts is unexpected since contacts formed by overcoating preformed Au films with either TPD/polycarbonate or PTPB (its polymeric analog) were demonstrated to be ohmic.[17] In addition, estimation of the injection barrier height from relative workfunctions of the interface components is not consistent with a blocking interface. We therefore investigated the injection evolution for the evaporated contact in detail for a series of samples, characterizing the time and temperature dependence of the phenomenon to obtain indications as to what process(es) may be operating.

Rapid sequence data acquisition, beginning typically at 5 minutes after Au deposition, uncovered a rapid evolution within the first few hours after the contact is deposited.[17-20] Representative J_{Au} versus E results are shown in Fig. 8 for a sample aged at 40°C. A limited number of J-E curves are shown for clarity, spanning the time period of the injection evolution. J_{Au} versus E curves were obtained every 10-15 minutes for the first few hours, followed by sampling every few hours and finally every few days as the rate of change in injection current decreased with time. The behavior of all the samples at three temperatures below the glass transition temperature (Tg = 85°C) is qualitatively similar. The initial behavior of the Au contact is blocking, with the injection current J_{Au} being ca. 3 orders of magnitude below the trap-free space charge limiting current, i.e. the maximum sustainable by the bulk under steady state conditions, J_{TFSCLC}. This is especially evident in the mid-to-upper field region 10^4-10^6 V/cm, where the demand of the bulk for charge is greatest.

Furthermore, for an ohmic contact on a semiconductor[5], the slope of a logarithmic plot of current vs. field should be unity at low fields, the ohmic region, where the t_{tr} of any excess injected carrier exceeds the bulk relaxation time $\rho\varepsilon$. Similarly, in the upper field region where t_{tr} becomes less than the bulk relaxation time, corresponding to TFSCLC conditions, the slope should be two[5]. While this behavior is in fact observed for injection from MystR®, neither of these characteristics are displayed by the evolving J_{Au} vs. E curves. However, J_{Au} increases continually over time throughout the field range and the current-field characteristic changes slope such that it approaches the current that would be supplied by an ohmic contact first at the lower fields, where the demand of the bulk for charge is least. After a period of ca. one month, J_{Au} becomes identical with the trap free space charge limited current at the highest fields as well, showing that evaporated Au does in fact become ohmic for injection into the MDP. The latter is the expected behavior of Au towards this MDP, since the ionization potential of the MDP is ~5.5 eV and the Au workfunction between 5.2 and 5.4 eV, depending on Au crystallinity.[11] Furthermore, in the inverse sample geometry, when the MDP is cast onto a preformed Au film, the Au contact behavior has been shown to be ohmic when measured a day after casting.[17] Fabricating the Au/MDP interface by evaporating the Au contact has a

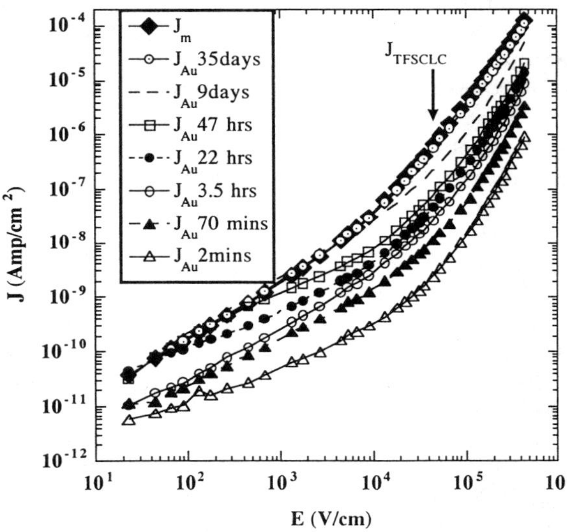

Figure 8. Representative J_{Au} versus E curves illustrating the evolution of the current density from emission limited to ohmic for hole injection from an evaporated Au contact evaporated at 10 Å/sec into a TPD/polycarbonate film. During this entire process the TPD/polycarbonate film was aged in ambient at 40 °C. For comparison the steady state dark injection current J_{TFSCLC} from the ohmic MystR® substrate on the sample is also shown.

consistent degradative effect on the Au injection behavior relative to the preformed Au substrate. Furthermore, a rapid evolution occurs within the first few hours after contact deposition. The following analysis further clarifies the kinetics of the evolution.

The description of the injection evolution from the Au contacts is further quantified by the use of equation 1 to calculate at different temperatures the injection efficiency from the evaporated-Au/MDP interface. A field strength of 10^5 V/cm was selected for the calculation of J_{Au}/J_{TFSCLC}. The injection efficiency is then plotted as a function of time. A typical result for the forming behavior is illustrated in Fig. 9 for 40 wt% TPD/polycarbonate aged in ambient at 23°C. A rapid increase in injection efficiency from blocking to ohmic can be distinguished for this sample in the first few hours, followed by a dramatically slower increase which occurs over a period of weeks. The inset illustrates that the overall behavior can be operationally regarded as a sum of two distinct processes to which two time constants may be assigned.[17,18]

The long-term recovery process shown in Fig. 9 is only weakly temperature dependent. However, this process is strongly dependent on the detailed conditions of Au deposition. The time required for evolution of the injection behavior toward ohmicity is dramatically decreased when the Au deposition is periodically interrupted, i.e. when Au is deposited in a layer by layer manner. For example, deposition of eleven 20 Å thick Au layers at the usual 10 Å/sec leads to contact ohmicity in 12 hours. Other sequential Au deposition conditions lead to a similar result. Details will be presented in the following section. Taken together these results suggest that incremental Au depositions lead to less molecular damage/decomposition at the MDP surface and that the slow kinetics may be related to a repair process involving slow polymer chain motions.[21] Note that this phenomenon was also observed for the polymer PTPB. In the case of the MDP, diffusion of small molecules from the bulk to the surface is also a possibility, if TPD sublimes off the MDP surface during the metal deposition process, creating a TPD-depleted layer near the metal/MDP interface.[22] The inset to Fig. 9 presents in greater detail the short time behavior in order to show clearly the evolution in injection efficiency at early times.[17] The early time behavior is dominated by a rapid time constant for the first 5 hours, depending on temperature. Thus the rate of this early process was found to be thermally activated with an activation energy of 0.3 eV.

In order to investigate the mechanisms underlying the two evolution processes, a sample is regarded as a three-component system, consisting of the MDP, the Au contact and their interface. The role on each of these will be considered.

2.1. Mechanisms related to the Au/MDP interface region. In addition to injection current measurements using the Au/MDP/MystR® sandwich samples, capacitance measurements were obtained periodically as a function of frequency for the same samples during the course of the evolution in injection efficiency. A small frequency dependence (~4 pF peak to plateau) is observable at early times, which is strongest at the lowest frequencies. The frequency dependence decreases with time in parallel with the increase in injection efficiency.[17] In the absence of large scale relaxation processes, a uniform dielectric is expected to exhibit a capacitance independent of frequency, which is observed for these samples only at long times. While not a large effect the consistent trend in capacitance indicates the presence of some surface-to-bulk inhomogeneity which decreases with time. Furthermore, capacitance-frequency measurements were taken over the same time scale on a similar MDP film held at 23°C substituting Al for the Au top contact. Aluminum was chosen because it does

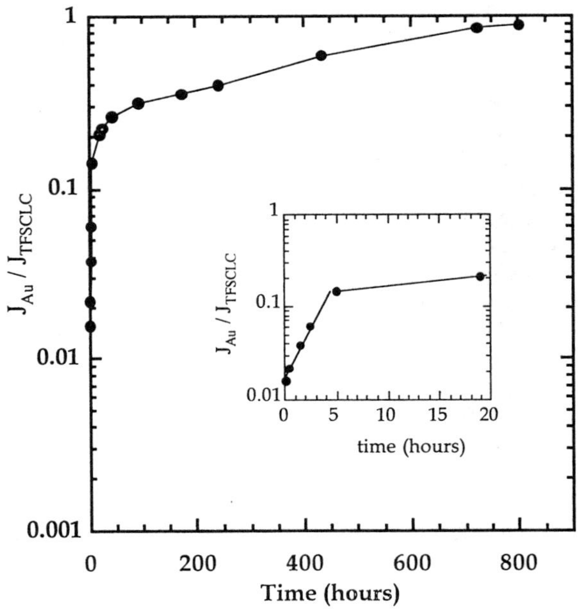

Figure 9. Plots of injection efficiency defined as $J_{au}/J_{kTFSCLC}$ vs. time illustrating the recovery of hole injection efficiency from blocking to ohmic for an evaporated Au contact (10Å/sec) on a 40 wt% TPD/polycarbonate molecularly doped polymer held at room temperature. Inset: Expanded display of the early time behavior illustrating that contact forming can be regarded as the summation of two distinct processes.

not show any evolution in injection behavior. The capacitance shows no frequency dependence or temporal evolution in the latter case, confirming that the inhomogeneity does not arise either from any changes in the bulk of the MDP film or from changes at the MystR® interface.

The capacitance results suggest that the inhomogeneity is constrained to the evaporated Au/MDP interface. In light of the results regarding two processes, the long-term decrease in surface-to-bulk inhomogeneity is consistent with the gradual repair of the MDP surface associated with the slow process. Presumably damage to the MDP surface is induced by the impact of energetic Au atoms. Meanwhile, the capacitance changes more rapidly immediately following Au evaporation, raising the possibility of Au interdiffusion with the MDP. Indeed, a major mechanism known to operate during vapor deposition of various metal contacts on organic thin films is diffusion of metal atoms into the film,[21] as was recently reported for perylenetetracarboxylic acid dianhydride (PTCDA) thin films with a series of metal contacts.[23] In that case, no diffusion was observed for the least reactive metals, Ag and Au and the diffusion for more reactive metals was driven by metal-organic complex formation. Metal-organic complexes have also been reported for thiophene containing thin films with metals.[24] For the MDP, the possibility of Au penetration and diffusion was investigated by transmission electron microscopy (TEM).[18,19] TEM images of the evaporated Au/MDP interface were obtained with evaporated Au films prepared under the same 10 Å/s Au deposition rate and the interface features were monitored over a period of 2 months in each sample. Time-lapsed results obtained at 1 hr, 4 hrs, 24 hrs and 2 months are shown in Figs. 10a-d, respectively, at the highest obtainable resolution of 1 nm. The interface is seen to be abrupt to within 1 nm or ~2-3 monolayers of Au over the entire course of the evolution of the electrical behavior of the interface.

Note that to within the resolution of the technique the roughness of the interface is invariant over the entire period of injection evolution. Careful handling allowed injection measurements to be obtained periodically on the same sample from which sections were being cut for the TEM measurements. This verified that the usual evolution in injection efficiency is occurring in the corresponding TEM samples. Au/MDP interpenetration is not observed on this scale and seems unlikely to account for the evolution in the injection from the Au contact. Accordingly, Au penetration would not be able to account for the surface-to-bulk inhomogeneity in the dielectric, which evolves in tandem with the injection efficiency. Moreover, no change in the topography or surface texture of the evaporated Au contact was detectable as a result of X-ray diffraction (XRD) and Scanning Electron Microscopic (SEM) studies of the Au surface over time, as detailed in the following section .[19]

Therefore no time dependent characteristics of the Au film are detected that would account for the recovery of the injection efficiency and the associated effects on capacitance. The remaining possibility involves changes in the MDP side of the interface. It is neither the bulk properties of the MDP that are changing, as no such change appeared in any of the control measurements, nor the uncontacted MDP surface that evolves with time. Aging and control experiments were run on each MDP film. Evaporated Au contacts were deposited and the recovery of injection efficiency measured over the course of several weeks. A control section of the same MDP film was allowed to age in ambient without a Au contact over the same period of time. When the first Au contact became ohmic, the control section was fitted with a Au top contact and found to be blocking. Thus the recovery of injection efficiency is unrelated to separate processes which may be occurring in the bulk, but rather must be due to physical or

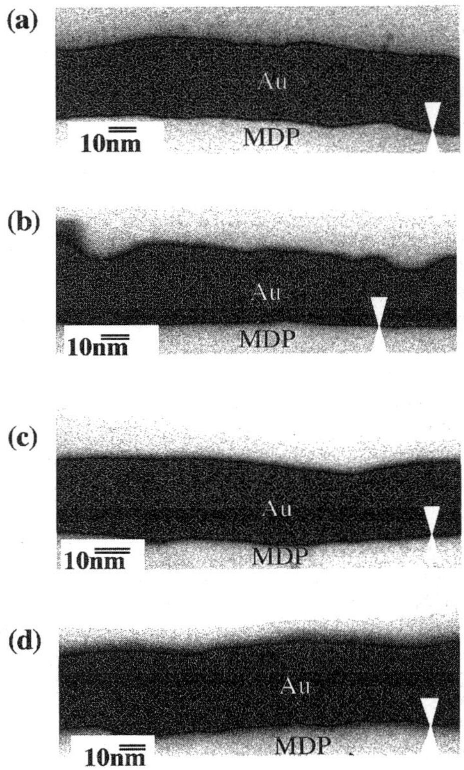

Figure 10. Transmission electron microscopy of the interface of evaporated Au with TPD/polycarbonate analyzed (a) 1 hour, (b) 4 hours, (c) 24 hours and (d) 2 months after Au evaporation, respectively.

chemical changes which are restricted to the MDP side of the interface subsequent to Au deposition.

2.2. Contact fabrication effects on injection evolution. SEM and XRD measurements were obtained as a function of time for 220 Å thick Au films deposited at 10 Å/s on the MDP and exposed to ambient conditions in-between measurements. These measurements were performed in order to investigate the possibility that the structure of the metal film itself is evolving structure with time, thus affecting its workfunction or actual contact area. The XRD results show two peaks of equal amplitude at (111) and (222) and neither their amplitude nor their relative ratio changes from the earliest resolvable time (1 hr) to 2 weeks after Au deposition, a time span involving both evolution processes. SEM results obtained at resolutions of 300 nm, 600 nm and 1 μm show a cracked Au film morphology which is invariant over the same 1 hr to 2 week time span. Therefore, the evolution in injection efficiency can not be readily assigned to changes in metal film surface texture or morphology.

The possibility still remains, however, that during metal evaporation, energetic Au atoms or the accumulation of a hot Au layer on the MDP may result in molecular damage to the TPD transport sites and/or the polymer binder at the surface. Similarly, surface TPD molecules may sublime from the MDP surface as it is heated during the metal evaporation, decreasing their concentration at the interface. Accordingly, a systematic variation of metal evaporation conditions was performed.[19] A comparison of four Au evaporation conditions is shown in Figs. 11a-d. The metal evaporations in Fig. 11a-c are each performed in stages, i.e, the metal is deposited in increments separated by 1-2 minutes during which the source is not heated and a shutter covers the MDP surface. All layered depositions that begin with a thin ca. 20 Å Au layer resulted in a significant reduction in the time scale of the long-term evolution so that ohmic injection is achieved within 20 hours. This compares very favorably to the ~800 hours obtained for a continuous uninterrupted deposition of Au as shown above in Fig. 8. However, when the deposition rate is changed, from 10 Å/s to the 2 Å/s as shown Fig. 11d, no diminution in the time required for the long term relaxation process is noted. The results of Fig. 11 taken together therefore strongly suggest that the MDP surface is thermally damaged by the accumulating hot Au layer during a typical uninterrupted evaporation and this damage can be virtually eliminated if Au is deposited in multiple steps separated by short cool-down periods. Molecular repair processes may plausibly involve slow polymer chain motions which would act to replace damaged surface segments as well as diffusion of TPD through the film which would act to increase its surface concentration over time. Indeed, the time scale of the slow evolution is not inconsistent with such polymer motions or diffusion mechanisms.[21] It should be noted in Fig. 11 that variations in the Au fabrication conditions do not materially affect the initial blocking nature of the interface nor the rapid initial rise in injection efficiency associated with short term relaxation. Thus consistent with the prior kinetic studies, these results suggest that two main processes appear to govern the evolution in contact behavior.

Contact forming behavior analogous to that observed with evaporated Au contacts has also been identified for the case of an evaporated Ag contact on TPD/polycarbonate. In Fig. 12 we illustrate the data for the case of a layer by layer evaporation of Ag thereby eliminating the longer term damage induced component of relaxation. Injection efficiency versus time is shown in Fig. 12 (open squares) for a sample of evaporated-Ag/MDP/ MystR®. However, in this case an interfacial energy mismatch with TPD/polycarbonate leads to an injection current that remains emission limited in the steady state, i.e., $J_{Ag}/J_{TFSCLC} = 0.6$. For comparison, the

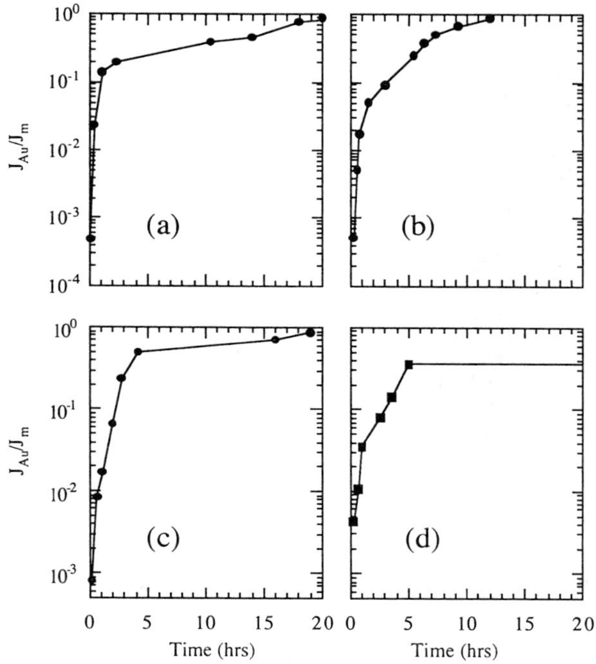

Figure 11. Temporal evolution of the injection efficiency at 1.0 X 10^5 V/cm of evaporated Au contacts on 40 wt% TPD/polycarbonate as a function of Au deposition conditions. All Au contacts are 220 Å. Panel A: Au is deposited in two steps, 50 Å and 170 Å, at 10 Å/sec. Panel B: Au is deposited in 11 layers of 20 Å each at 10 Å/sec. Panel C: Au is deposited in 11 layers at 2 Å/sec. Panel D: Au is deposited in a single continuous step at 2 Å/sec. In all cases the injection efficiency is initially blocking. The overall relaxation process occurs in a much shorter time when Au deposition is carried out in several stages.

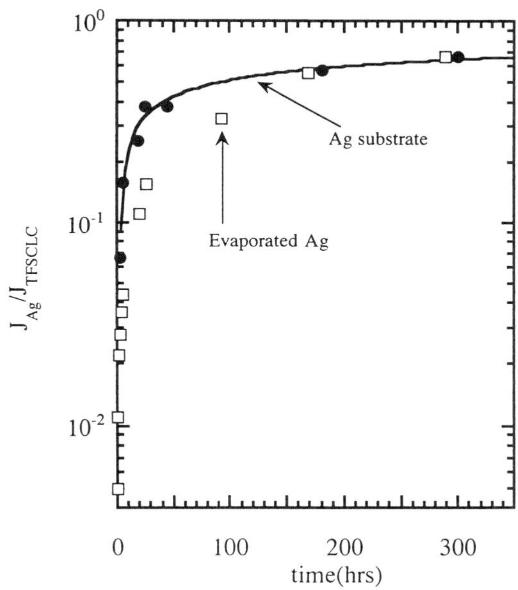

Figure 12. Comparison of the temporal evolution of the injection efficiency J_{Ag}/J_{TOF} for a freshly deposited Ag substrate with the injection efficiency J_{Ag}/J_{TFSCLC} of an evaporated Ag top contact.

injection efficiency for a Ag-substrate/MDP/Al sample is monitored beginning 3 hours after contact is made, i.e., after the MDP is cast onto the substrate and cured. Nearly identical characteristics are observed for both the MDP contact made to the Ag substrate and to the Ag top contact evaporated in stages as shown in Fig. 12. Thus we emphasize that even in the absence of overt thermal damage associated with evaporated metal contacts, a contact forming process is still observed for hole injection from a preformed Ag substrate into TPD/polycarbonate. Interestingly, the overall relaxation process, dominated in this case by the short term relaxation component, is slower than the corresponding behavior observed in the case of Au contacts.

Fig. 13 depicts injection efficiency versus time for a Hg/MDP/MystR® sample, showing that an intimate contact using a liquid Hg droplet (0.316 cm^2) also produces an evolution in injection behavior which can be resolved on a sufficiently short time scale. Note that the manner in which such liquid contacts are fabricated preclude any possibility of thermal degradation of the interface and this is manifested in the absence of a process which can be identified with interfacial damage. Thus the key point is the persistence of a short term relaxation process not easily identified with structural damage to the interface.

Fig. 14 shows a comparison of the early time evolution of hole injection efficiency from evaporated Au top contacts (staged evaporation) into two specimens of 40 wt% TPD/polycarbonate. Note that sequentially evaporated Au contacts were employed in order to minimize the effect of interfacial damage. Fig. 14 shows a comparison of the temporal evolution of the injection efficiency from Au into a MDP specimen as prepared (curve a) with a separate specimen of the same thickness that was vapor doped for 30 minutes in a saturated atmosphere of methylene chloride vapor just prior to analysis (curve b). Fig. 14 clearly shows that prior vapor doping increases the initially observed injection efficiency J_{Au}/J_{TFSCLC} as well as the overall rate of recovery of the injection efficiency to ohmic behavior. Control transport measurements made in parallel with those of Fig 14b also showed that hole drift mobility and injection from MystR® remain unchanged by vapor doping with methylene chloride. Previous vapor doping studies[25] of the analogous main chain polymer PTPB showed the persistent presence of plasticizing solvent vapor in the substrate after the vapor doped sample is exposed to ambient conditions. It was also noted in those studies that transport remains invariant with the exposure to vapor. The persistence of plasticizing methylene chloride in the TPD/polycarbonate matrix after exposure to a local atmosphere of its vapor is thus to be expected on the time scale of the results presented in Fig. 14b. The present result suggests that the rate of contact forming is associated with molecular scale relaxation processes that are enhanced upon plasticization, i.e. when imbibed solvent enhances molecular mobility.

2.3 Contact formation on various organic films. Fig. 15 illustrates that "forming," i.e. the enhancement in time of the injection efficiency from evaporated Au (10 Å/sec) into a 6 μm thick film of MEH-PPV (curve a) occurs in a manner analogous to TPD containing films but on a much shorter time scale. The figure compares the MEH-PPV results with the early time results on 40 wt% TPD/polycarbonate (curve b) presented above. Assuming that MystR® is an ohmic or near ohmic contact for hole injection into MEH-PPV and taking the measured relative dielectric constant to be 2.57 yields a mobility value of $\mu=1.4\times10^{-6}$ cm^2/Vs at fields of 10^4-10^5 V/cm, consistent with mobilities reported for PPV.[26,27]

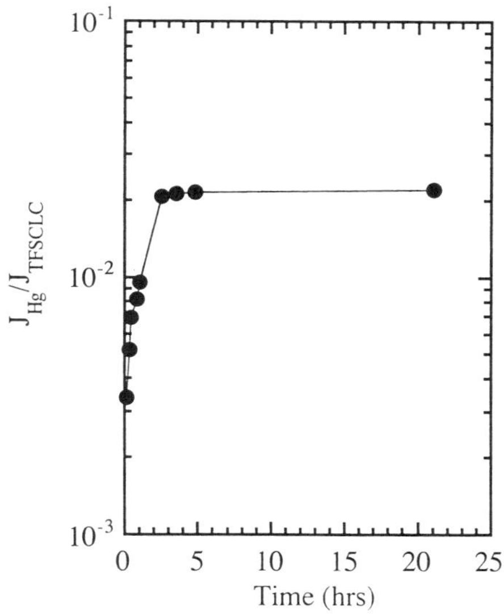

Figure 13. Plot of time dependence of the hole injection efficiency J_{Hg}/J_{TFSCLC} from a Hg droplet with a contact area of 0.316 cm^2 on the top surface of 40 wt% TPD/polycarbonate. Steady state injection efficiency is emission limited at 0.02.

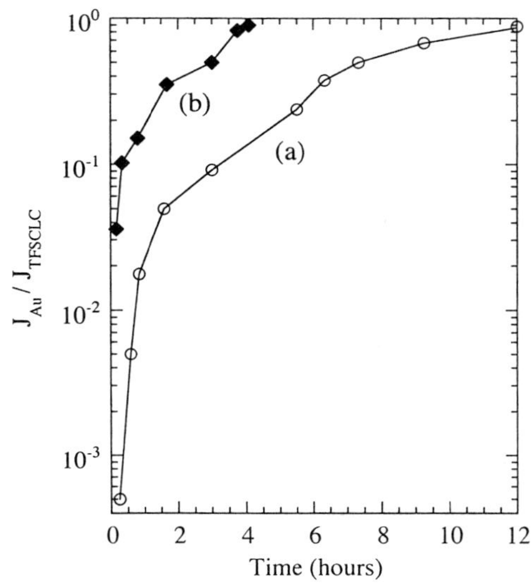

Figure 14. Comparison of the initial contact forming response of sequentially deposited Au on 40 wt % TPD/polycarbonate as prepared (a) and after exposure to a saturated atmosphere methylene chloride (b).

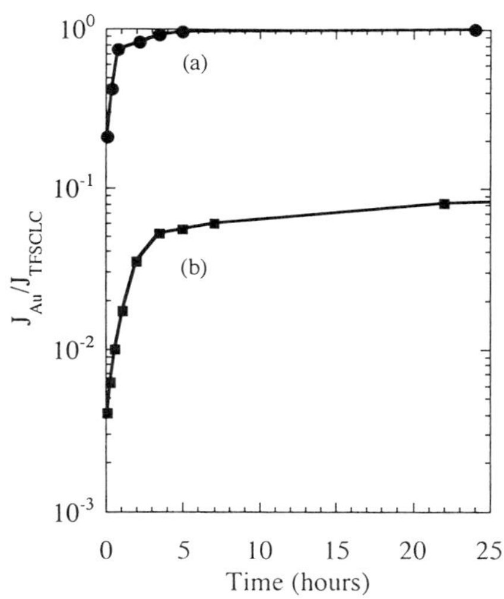

Figure 15. Contact forming behavior of a continuously evaporated (10 Å/sec) Au contact on (a) MEH-PPV and (b) 40 wt% TPD/polycarbonate thin glassy films. Note that as displayed in Fig. 9 the Au/TPD contact in curve (b) achieves contact ohmicity at a much longer time (ca. 800 hrs).

SUMMARY OF KEY RESULTS

1. An ohmic contact is defined as one that is able to act as an infinite source of charge and thus able to supply the current demands of the bulk for injected charge up to and including a supply sufficient to fully sustain space charge limited (SCL) conditions. An ohmic contact made to a unipolar transport medium will by definition supply the space charge limited current (SCLC) at applied fields which make the transit time of any excess injected carrier less than the bulk dielectric relaxation time. This condition is assured over any practical experimental field range when the transport medium has no population of thermal free carriers at equilibrium. Polymer transport layers initially developed for electrophotographic application are insulators with finite unipolar mobilities which are trap free by molecular design. An ohmic contact on such a transport layer will inject trap free space charge limited current which under fixed experimental conditions is uniquely determined by the injected carrier drift mobility. The latter invites the use of such trap free media as a venue for the study of interfacial electron transfer. Thus an unambiguous yet simple test for determining relative contact injection efficiency consists of measuring the steady state current density J_{DC} supplied by the contact under test to a trap free transport layer and comparing the current density at a given field to the trap-free space charge limited current density, J_{TFSCLC}, calculated from independent measurements of drift mobility carried out in the same specimen at the same field. The figure of merit for classifying injection efficiency is therefore $J_{DC} / J_{TOF} = J_{DC} / J_{TFSCLC}$.

2. Using this experimental technique the relative injection efficiencies of a series of contact media interfaced to a trap free hole transport polymer have been compared to the respective interfacial barriers that are expected to result assuming equilibrium is achieved. The magnitude of the barriers are estimated from the substrate work function and electrochemical estimates of the TPD redox potential. In all cases but one the transport polymer was coated onto a preformed substrate under conditions that would realistically prevail in a clean coating facility. Measurements were repeated over many specimen coating runs so as to clearly discriminate experimental artifacts. For selected specimens measurements were repeated over a period of several weeks under ambient conditions to elucidate any aging phenomena.

3. Although it would be plausible to find that injection is controlled by extrinsic factors (for instance metal-compound formation, a well established feature during the metalization of compound semiconductor surfaces) when an interface is formed under non-pristine conditions, it is instead found that on average injection efficiency tracks the estimated barrier height. However, deviations about the average behavior are significant. It is suggested that for Ag and Cr substrates random variation about the norm may reflect differences in the growth of surface oxides which are then present prior to polymer-interface formation. However, the systematic sensitivity of Au substrate injection efficiency to the thin Cr adhesive layers used to bind Au to a supporting Si substrate is revealed by TOFSIMS to be correlated with diffusion of Cr to the Au/polymer interface.

4. The Au on mica substrate is ohmic for hole injection into coated layers of TPD/Polycarbonate when measured a day after coating. However, when Au is <u>evaporated</u> on the same molecularly doped polymer (MDP) transport layer it is found that the injection efficiency is a strong function of the deposition conditions. For Au evaporated at normal deposition rates (2-10 Å/ sec) the contact injection behavior evolves from emission limited to ohmic as a systematic function of time and temperature.

5. For the evaporated (10 Å/sec) Au contact on TPD/polycarbonate analysis of the kinetics of the entire evolution process from the earliest resolved times to one month suggests that it can practically represented as a superposition of two processes. The early time process is thermally activated (0.3 eV) while the slower process exhibits only a very weak temperature dependence.

6. The evaporated Au-organic interface was extensively studied by spectroscopic techniques as a function of time, over the same time frame as the contact forming process., The interface was found to be abrupt and invariant over time as demonstrated by transmission electron microscopic studies of the interface at a resolution of 10Å. In addition, changes in metal surface texture (crystallinity) and morphology over time were not detectable by X-ray diffraction and scanning electron microscopy. Finally, Au and Ag chemical compound formation with TPD could not be resolved by X-ray photoelectron spectroscopy.

7. A systematic variation in the conditions of metal evaporation shows that the slow, long-term component of the evolution is clearly associated with the method of evaporation and can be suppressed by performing an interrupted layer-by-layer metal deposition. This result supports the proposition that the molecularly doped polymer surface is thermally damaged in some way during a typical metal deposition (e.g. 220 Å of hot Au accumulated continually on the surface for approximately 20 sec). The long-term evolution process then appears to reflect healing of the damaged MDP surface (and its main chain analog PTPB) following Au deposition. This recovery may be due to polymer chain motions that act to replace damaged segments at the surface or in the case of the solid solution diffusion of the molecular dopant TPD that could act to restore a surface concentration of TPD depleted by sublimation during the heating of the MDP surface.

8. The phenomenon of evolving injection efficiency has been confirmed to apply for various metals under different fabrication conditions. Differently prepared Au, Ag and Hg contacts all show an initially severely limited hole injection efficiency into the MDP which however improves (increases) with time. Once again two processes governing the evolution in efficiency can be operationally distinguished for evaporated Au and Ag contacts. In the case of Au, whose workfunction is in close agreement with that of the MDP (~5.5 eV) the injection current evolves to J_{TFSCLC}. In the cases of Ag and Hg where an interfacial energy mismatch is to be expected at equilibrium the steady state injection efficiency is in fact emission limited.

9. The injection evolution has been confirmed in organic materials other than molecularly doped polymers, i.e. a small molecule glass of TPD, and in an undoped polymer film of MEH-PPV.

10. We emphasize that in certain key cases (preformed Ag substrates, Hg droplet top contacts) relaxation occurs even in the absence of overt evidence for structural damage. These processes could involve interfacial chemistry immediately adjacent to the interface and not easily detected by spectroscopic means, or some as yet unspecified surface reorganization process.

Acknowledgments. The authors gratefully acknowledge a Science and Technology Center Grant (CHE-9120001). The assistance of Dominic Salamida (TOFSIMS), Harry Freitas and Jack Czerniawski (TEM measurements), Witold Niedzalkowski (SEM), Thomas Debies and Roger Laforce (XPS measurements) and Scott Ingham (X-ray diffraction measurements) in obtaining analytical measurements is also gratefully acknowledged. MEHPPV was kindly supplied by Dr. Bing Hsieh.

REFERENCES

1) Borsenberger, P. M.; Weiss, D. S. *Organic Photoreceptors For Imaging Systems*; Marcel Dekker: New York, 1993.

2) Abkowitz, M. A.; Facci, J. S.; Stolka, M. *Chem. Phys.* **1993**, *177*, 783-92.

3) Abkowitz, M. A.; Pai, D. M. *Phil. Mag. B* **1986**, *53*, 193.

4) Abkowitz, M. A.; Facci, J. S.; Stolka, M. *Applied Physics Letters* **1993**, *63*, 1892.

5) Lampert, M. A.; Mark, P. *Current Injection in Solids*; Academic Press: New York, 1970.

6) Hsieh, B. R.; Yu, Y.; VanLaeken, A. C.; Lee, H. *Macromolecules* **1997**, *30*, 8094.

7) Guo, L.-H.; Facci, J. S.; McLendon, G. L.; Mosher, R. A. *Langmuir* **1995**, *11*, 4588.

8) Holland, L. *Vacuum Deposition of Thin Films*; Chapman and Hall: London, 1963.

9) Dolezalek, F. K. ; Dolezalek, F. K., Ed.; Elsevier: New York, 1976, pp 71.

10) Josowicz, M.; Janata, J.; Levy, M. *J. Electrochem. Soc.* **1988**, *135*, 112.

11) Michaelson, H. B. *Journal of Applied Physics* **1977**, *48*, 4729.

12) Facci, J. S.; Abkowitz, M. A.; Limburg, W. W.; Knier, F.; Yanus, J. *J. Phys. Chem.* **1991**, *95*, 7908.

13) Khan, S. U. M.; Kainthla, R. C.; Bockris, J. O. M. *J. Phys. Chem.* **1987**, *91*, 5974.

14) McCreery, R. L. *Carbon Electrodes: Structural Effects on Electron Transfer Kinetics*; Marcel-Dekker: New York, 1991; Vol. 17.

15) Heaney, M. B. *Appl. Phys. Lett.* **1996**, *69*, 171.

16) Angerstein-Kozlowska, H. *Surfaces, Cells, and Solutions for Kinetic Studies*; Plenum Press: New York, 1984.

17) Abkowitz, M. A.; Facci, J. S.; Rehm, J. *J. Appl. Phys.* **1998**, *83*, 2670.

18) Ioannidis, A.; Facci, J. S.; Abkowitz, M. A. *J. Appl. Phys.* **1998**, in press.

19) Ioannidis, A.; Facci, J. S.; Abkowitz, M. A. *J. Appl. Phys.* **1998**, manuscript in preparation.

20) Ioannidis, A.; Facci, J. S.; Abkowitz, M. A. *Phil. Mag. Lett.* **1998**, manuscript in preparation.

21) Wool, R. P. *Polymer Interfaces*; Hanser-Verlag: Munich, 1995.

22) Borsenberger, P. M., private communication.

23) Hirose, Y.; Kahn, A.; Aristov, V.; Soukiassian, P.; Bulovic, V.; Forrest, S. R. *Phys. Rev. B* **1996**, *54*, 13748.

24) Lachkar, A.; Selmani, A.; Sacher, E.; Leclerc, M.; Mohkliss, R. *Synthetic Metals* **1994**, *66*, 209.

25) Abkowitz, M.; Facci, J. S.; Stolka, M. *Appl. Phys. Lett.* **1994**, *65*, 1127.

26) Blom, P. W. M.; de Jong, M. J. M.; van Munster, M. G. *Phys. Rev. B* **1997**, *55*, R656.

27) Lebedev, E.; Dittrich, T.; Petrova-Koch, W.; Karg, S.; Brutting, W. *Appl. Phys. Lett* **1998**, *in press.*

Chapter 8

Interface Control of Polymer- and Oligomer-Based Light-Emitting Devices

A. J. Epstein and Y. Z. Wang

Department of Physics, The Ohio State University, Columbus, OH 43210–1106

Bilayer and multilayer polymer structures provide opportunities for new photophysics and new types of light emitting devices. Photoluminescent and electroluminescent studies of bilayer heterojunctions formed from a poly(pyridyl vinylene phenylene vinylene) (PPyVPV) derivative and poly(vinyl carbazole) (PVK) show an emission peak which cannot be ascribed to either the PPyVPV derivative or PVK layer. Through studies of absorption and photoluminescence excitation (PLE) spectra we demonstrated that the additional feature results from an exciplex at the bilayer interface. We also discuss here the fabrication of color variable dc and bipolar/ac light-emitting devices based on conjugated polymers and also use of oligomers. The bipolar/ac devices consist of blends of pyridine-phenylene and thiophene-phenylene based copolymers sandwiched between the emeraldine base form and the sulfonated form of polyaniline. ITO and Al are used as electrodes. The devices operate under either polarity of driving voltage with different colors of light being emitted, red under forward bias, and green under reverse bias. The voltage-controlled dc color variable devices include layers of para-sexiphenyl (6P), PVK, or derivatives of polyacetylene together with pyridine based polymers in their structures. Voltage dependent multicolr emission was observed in both bilayer and trilayer devices. The emission colors of single devices cover a wide range of visible spectra whose CIE color coordinates vary from blue to white to green with increasing voltages.

Conjugated polymer based light-emitting devices[1,2,3,4,5,6,7,8] have become a topic of great interest since the report of electroluminescent (EL) properties in poly(phenylene vinylene) (PPV).[1] A large variety of polymers, copolymers, and their derivatives have been shown to exhibit EL properties, including polypyridines,[5,6] and poly(pyridyl vinylene)s.[6,9] The configurations of these devices may consist of a simple single layer,[1,2] bilayers,[2] or blends[7] used to enhance efficiency and tune the emission wavelength, or multilayers that may allow the device to operate under an ac or dc applied voltage and change color of emission.[6]

In single layer devices the low efficiency frequently is due to the imbalance of electrons and holes. Inserting a hole-transport (electron-blocking) or electron transport (hole-blocking) layer provides a means to enhance minority carriers and block the majority carriers and confine them to the emitter layer, which increases the probability of recombination.[2] Poly(N-vinylcarbazole) (PVK) has been used as a hole transport layer[8] and occasionally in blends with the emitter polymer.[7,8] PVK is a well-studied photo-conductive polymer which often forms exciplexes with organic molecules, e.g., dimethyl terephthalate.[10] An exciplex is a transient donor-acceptor complex between the excited state of the donor and the ground state of the acceptor.[10]

For most devices the color of the emitted light is fixed once the device is fabricated. Recently there has been great interest in developing color variable light-emitting devices, i.e., individual devices that can generate two or more colors of light. In color variable devices based on blends of polythiophene derivatives, different components in the blend emit different colors of light simultaneously with the intensity of each component varying with the applied voltage.[11] Though such devices can emit multiple colors of light, they have very limited control of the brightness at a desired color. Color variable LECs,[12] which emit two independent colors of light, also have been developed. The two color LECs offer an improved control of the color and brightness: the color is controlled by the polarity and the brightness is controlled by the magnitude of the driving voltage. However, due to the involvement of ionic species in the device operation, the response of the devices is intrinsically slow and not suitable for applications that require rapid switching of colors. More recently, multilayer light emitting devices which generate two independent colors were achieved at liquid nitrogen temperature by inserting a blocking layer in between two different emitting polymer layers.[13] The two colors also can be controlled by the polarity of the driving voltage. Such an approach improves the device response time. However, it raises the device operating voltage due to the introduction of the charge blocking layer and retains the stability concerns of conventional polymer LEDs.

We discuss here bilayer devices with PVK as the hole transport layer and a derivative of the copolymer poly(pyridyl vinylene phenylene vinylene) (PPyVPV) as the emitter layer. Absorption, photoluminescence (PL) and electroluminescence (EL) results demonstrate emission due to exciplex formation at the interface between the PVK and copolymer. The PL and EL of bilayer films are dramatically different from that of a single layer film.

We also discuss an approach to color variable light-emitting devices which generates two independent colors of light at room temperatures. The devices consist of a layer of active electroluminescent polymers sandwiched between two different redox polymer layers. The redox polymer layers modify the charge injection and transport properties such that the device can be operated under both forward and reverse bias. Also, at least one of the redox polymers is capable of modifying the emission properties of the emitting polymers at the interface so that the interface emits different colors of light than the bulk does. In this approach, the colors of light are controlled by selecting the desired emission locations which in turn are controlled by the polarity of driving voltage and the charge injection and transport properties of the emitting polymers. Since motion of ionic species is not required for device operation a relatively fast time response is expected, allowing the colors to be switched rapidly.

Figure 1 presents the chemical structures of PPyVP($COOC_{12}H_{25}$)$_2$V and PVK. Figure 2 shows the PL of a single layer of the copolymer, a single layer of PVK and a bilayer of PVK and the copolymer. The PL of single PVK layers excited at 3.6 eV has a peak emission energy at 3.05 eV, similar to previous reports of the PL of PVK.[7,8] The PL for single layer copolymer films excited at 3.1 eV shows an emission peak at 2.05 eV. The bilayer when excited at an energy less than the absorption edge of the PVK, but greater than the absorption edge of the copolymer shows PL peaked at the same energy as for the copolymer along with a low intensity tail to the blue side. When the bilayer was excited at energy equivalent to the excitation energy for the single PVK layer (3.6 eV), the PL emission spectrum contains contributions from both single layers (3.05 and 2.05 eV), as well as from a completely new species, which we identify with an exciplex. To the low energy side of the exciplex PL is a weak shoulder near the PL energy for the single layer of the copolymer. Figure 3 shows the PL intensity as a function of both the *excitation* energy and the *emission* energy. At excitation energies above 3.6 eV the PL due to the exciplex and PVK are apparent, but if the excitation energy is lowered below 3.4 eV these peaks have essentially disappeared. As the excitation energy is further lowered into the peak absorption of the copolymer, PL from the copolymer strongly predominates (excitation energy 2.6-3.0 eV and principal emission energy 1.8-2.2 eV).

We fabricated bilayer devices using ITO as the anode and aluminum as the cathode. The inset of Fig. 4 shows the EL spectrum of a typical device with the PL spectrum from the same device. The devices can easily be seen in a brightly lit room, appear bright green to the eye, and have internal quantum efficiencies of ~0.1%-0.5% when operated in air (without encapsulation) under constant dc drive. Although the PL efficiencies are comparable, the EL efficiency of the bilayer configuration, ~0.1%-0.5%, is much greater than for a single layer device which has an EL efficiency of less than 0.0001%. The similarity between the PL and EL of the bilayer device demonstrates that the exciplex is responsible for the EL emission. Figure 4 shows the current-density voltage and brightness-voltage characteristics for a typical bilayer device. The turn-on voltage of the bilayer devices depends on the thickness of the polymer layers and in this case is ~18 V, with the brightness following the current. The generality of this concept has been demonstrated using several other pyridine-based copolymers.[14] Through the use of polyaniline network electrodes[15] we have lowered the threshold voltage to below 10 V while maintaining the same efficiency.[16]

Thus heterojunctions of PVK and PPyVP($COOC_{12}H_{25}$)$_2$V show a strong photoluminescence and electroluminescence feature due to exciplex emission at the interface. The absorption and PLE spectra have shown that the exciplex is not directly accessible from the ground state. The exciplex is also the primary species of electroluminescence emission in the bilayer devices. The efficiency of the bilayer

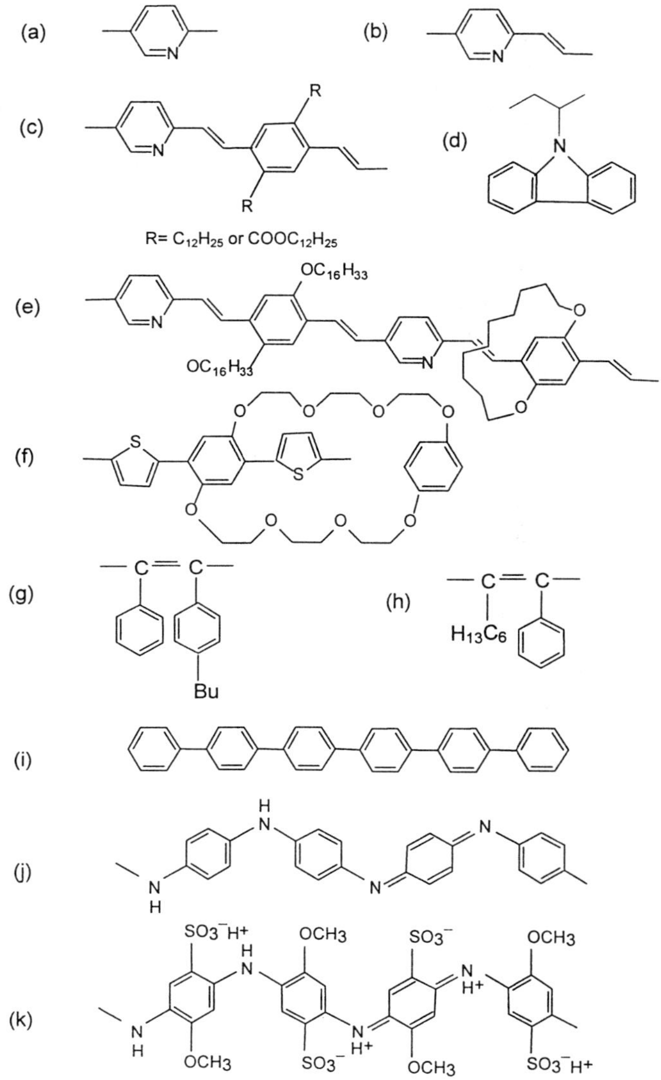

Figure 1. Schematic units of the pyridine-containing polymers and other materials reported here.
(a) poly (p-pyridine) (PPy); (b) poly(p-pyridyl vinylene) (PPyV); (c) copolymer of PPV and PPyV (PPyVP(R)$_2$V); (d) poly(vinyl) carbazole) (PVK); (e) wrapped copolymer of pyridyl vinylene and phenylene vinylene (PPyVPV*); (f) wrapped copolymer of dithienylene and phenylene (PTP*); (g) poly (diphenyl butyl acetylene) (PDPA-nBu); (h) poly(hexyl phenyl acetylene) (PhPA); (i) para-sexiphenyl (6P); (j) emeraldine base (EB) form of polyaniline; (k) sulfonated polyaniline (SPAN).

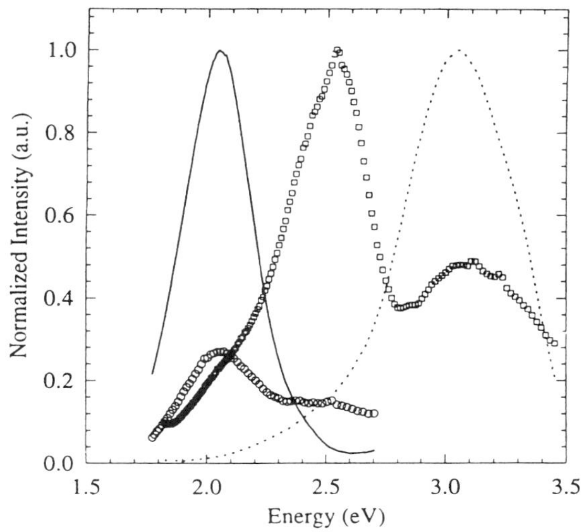

Figure 2. PL of $PPyVP(COOC_{12}H_{25})_2V$ at 2.8 eV excitation energy (—), a bilayer of PVK and $PPyVP(COOC_{12}H_{25})_2V$ at 3.6 eV excitation energy (square) and 2.8 eV excitation energy (O), and PVK at 3.6 eV excitation energy (···), all on quartz substrates. Reproduced with permission from reference 23.

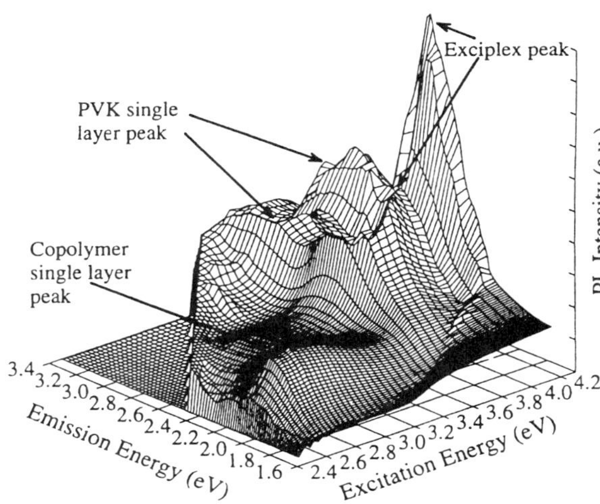

Figure 3. PL of a bilayer of PVK and $PPyVP(COOC_{12}H_{25})_2V$ as a function of both emission energy and excitation energy. The 3D plot shows three prominent features: a peak due to the PVK (excitation energy from 3.6 to 4.2 eV, emission energy from 2.8 to 3.4 eV), a peak due to the copolymer (excitation energy from 2.4 to 3.0 eV, emission energy 1.8 to 2.2 eV), and the exciplex peak (excitation energy from 3.6 to 4.2 eV, emission energy 2.2 to 2.8 eV). Reproduced with permission from reference 23.

devices is greatly enhanced over single layer devices due to charge confinement and exciplex formation and emission at the interface.

For two-color devices presented here, a copolymer of poly(pyridyl vinylene) and poly(phenylene vinylene) derivative, PPyVPV*, and a copolymer of polythiophene and polyphenylene derivative, PTP*, were used as the emitting materials; sulfonated polyaniline (SPAN)[17] and the emeraldine base (EB) form of polyaniline were used as the redox materials; ITO and Al were used as electrodes. Figures 5 shows the schematic device structure. The EB layer was first spin coated at ~3000 rpm from N-methyl pyrrolidinone (NMP) solution (concentration of ~5mg/ml) onto precleaned patterned ITO substrates (with a sheet resistance of 15 Ω/square). The emitting layer was then spin coated over the EB layer from a blend of PPyVPV* and PTP* (3:2 weight ration) in xylenes or trichloroethylene (total concentration ~10 mg/ml). The SPAN layer was subsequently spin coated over the emitting layer from an aqueous solution. This device differs from the prior SCALE device[18] in having different polyaniline forms on the two sides of the light emitting polymer. Figure 6 shows the typical I-V and luminance-voltage characteristics of the devices configured as in Figure 5. The devices have typical turn-on voltages of ~4-8 V depending upon film thickness and work equally well under both polarities of driving voltage as reported earlier for similar symmetrically configured ac light-emitting (SCALE) devices, with different colors of light being emitted. The light appeared red and green to the eye under forward and reverse bias, respectively, and was clearly visible under normal indoor lighting. Internal device efficiencies of up to 0.1% photons/electron has been achieved for the initial devices when operated in air under constant dc voltage. The EL spectra under forward and reverse bias are shown in the inset of Fig. 6. The CIE chromaticity x,y coordinates of the two spectra are calculated to be (0.654,0.345) and (0.471,0.519), respectively, showing both colors to be relatively pure or saturated. The colors of the devices have been switched rapidly, up to ~20kHz, depending upon device impedance and geometry.

The EL spectra under forward bias is substantially different from that of the single layer devices of either PPyVPV* or PTP*, suggesting that the light is generated from the interface between the emitter blend and either EB or SPAN under forward bias. The red light is generated from the PPyVPV*/SPAN interface on the cathode side under forward bias. Studies suggest the formation of complex species due to the quarternization of the pyridyl units by SPAN.

When the PTP* is replaced by derivatives of polyacetylene, poly(diphenyl butyl acetylene) (PDPA-nBu) or poly(hexyl phenyl acetylene) (PhPA), voltage dependent emission colors were observed under reverse bias. Fig. 7 shows the EL spectra of an ITO/ PPyVPV*:PhPA/SPAN/Al device. The weight ratio of PPyVPV*:PhPA is 1:1. Under forward bias, the device emits red light similar to the PTP* device. Under reverse bias, however, the emission colors change from orange to green as the voltage increases. Further studies of this type of device are underway.

Here we summarize the fabrication and study of multicolor light-emitting devices combining the pyridine-based polymers with the 6P oligomer in separate layer configurations. Voltage dependent multicolor emission was observed in both bilayer and trilayer devices. In the bilayer devices with the configuration of ITO/6P/PPy/Al, blue light is emitted at low voltages. As the driving voltage increases, a component of green light grows, and eventually becomes dominant. When PVK or PPyVPV* is introduced in between ITO and 6P, the trilayer device

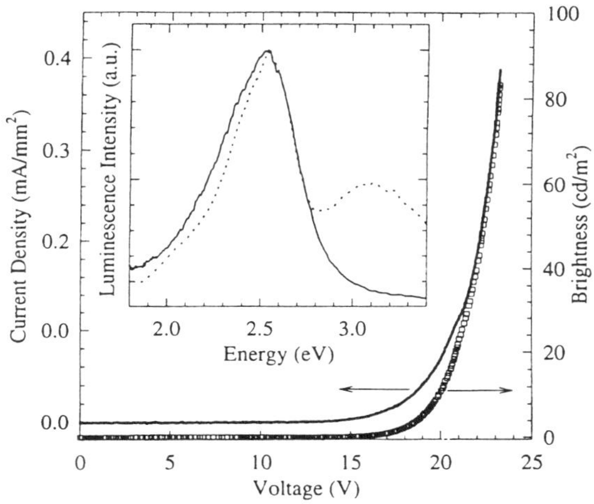

Figure 4. Current-voltage (—) and brightness-voltage (square) characteristics of a typical bilayer light-emitting device. Inset: PL(···) and EL(—) of a bilayer light-emitting device. Reproduced with permission from reference 23.

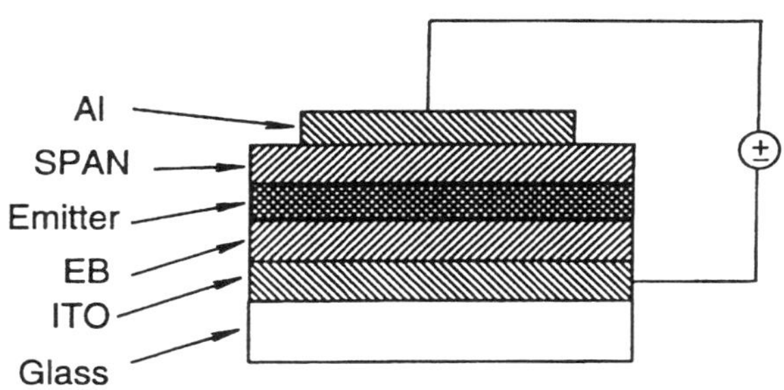

Figure 5. Schematic diagram of the color variable light-emitting device. Reproduced with permission from reference 24.

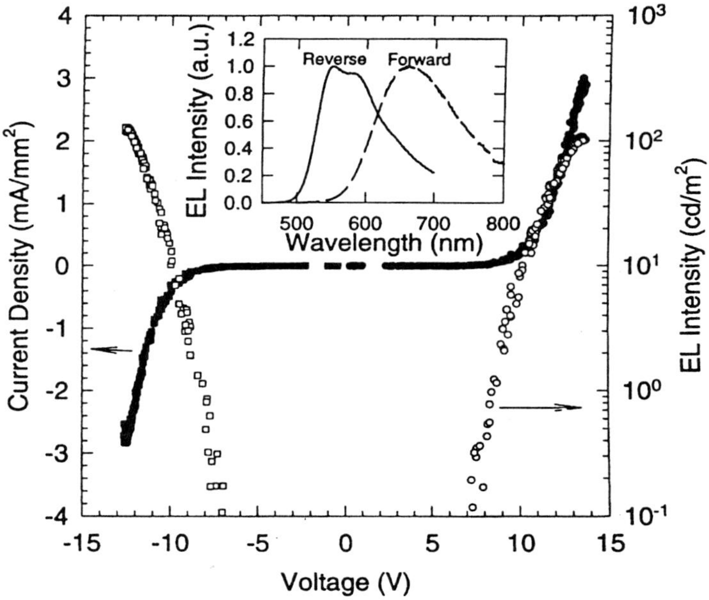

Figure 6. Current-voltage and light-voltage characteristics of color variable light-emitting devices under forward and reverse bias conditions. The device emits red light under forward bias and green light under reverse bias. Inset shows the electroluminescent spectra of such a device under forward and reverse bias conditions. Reproduced with permission from reference 24.

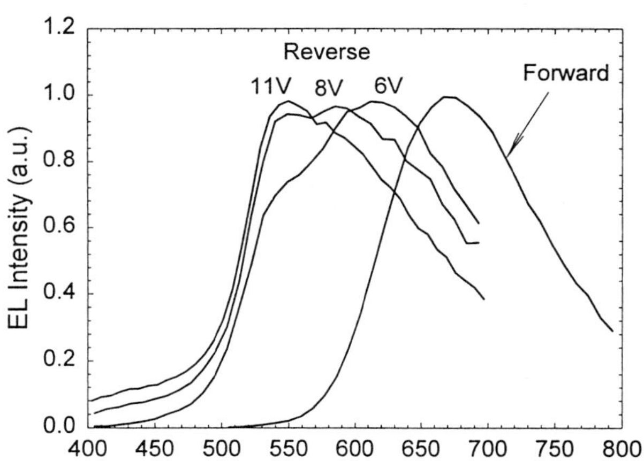

Figure 7. EL spectra of the ITO/PPyVPV*:PhPA/SPAN/Al device under forward and reverse bias conditions. Under forward bias, the EL spectrum is voltage independent. Under reverse bias, the EL spectrum blue shifts as the voltage increases.

generates light from blue at low voltages to nearly white at high voltages whose color coordinate traverses along a straight line in the CIE chromaticity diagram.

Figure 8 shows the EL spectra of a bilayer device ITO/6P/PPy/Al operated under different applied voltages. Under low voltages, the EL spectra show two peaks at 425 nm and 450 nm, in addition to shoulders at 400 nm and 480 nm. The bilayer EL spectrum can be roughly represented by a superposition of the two individual 6P[19] and PPy EL spectra, suggesting that the light is generated near the interface between 6P and PPy. The EL appears light blue to the eye. As the voltage increases, a peak at 565 nm appears and eventually becomes dominant.

Figure 9 (a) and 9 (b) show the PL and EL of a three layer device ITO/PPyVPV*/6P/PPy/Al. When excited at 350 nm, the PL spectra of the three-layer device ITO/PPyVPV*/6P/PPy/Al are dominated by the PL spectrum of 6P. As the excitation wavelength increases, the contribution from PPy increases, and becomes dominant at excitation wavelength of 410 nm. As the excitation wavelength further increases, the contribution from the PPyVPV* layer grows and becomes dominant at 470 nm excitation. The EL spectra are dominated by the 6P layer at low applied voltages, appearing blue to the eye. As the voltage increases, a component at ~ 700 nm and a component at ~ 530 nm appear and grow. At 17-21 V, the strength of the blue, green, and red components are almost equal, making the light appearing white to the eye. As the voltage further increases, the green component keep growing while the blue one is decreasing. The color in the CIE chromaticity diagram of this device is shown in Fig. 10.

The PL of a three-layer ITO/PVK/6P/PPy/Al device is similar to those of the two-layer one, except that there is some emission from PVK at 350 nm excitation. At low applied voltages, the EL spectrum is very similar to that of the bilayer device, which is light blue in color. The CIE chromaticity x,y coordinates of the EL spectra at 17 V is calculated to be(0.237,0.224). As voltage increases, the peak at 450 nm grows. At the same time, a new peak at ~ 605 nm appears and grows and drives the location of the EL spectra in the CIE diagram towards white along a straight line, as shown in Fig. 11 inset. At 27 V, the emitted light appears bright white to the eye. The CIE chromaticity x,y coordinates of the spectra at 27 V is calculated to be (0.322,0.315). Below 27 V, the voltage dependent EL spectra are more or less reversible. Above 27 V, the devices go through an irreversible change. The features in the blue region disappear and the EL spectrum is dominated by a broad peak at 605 nm. The light appears greenish white to the eye and its CIE chromaticity x,y coordinates is calculated to be (0.400,0.411). It is noted that the EL spectra do not change with applied voltages above 27 V although its brightness increases with increasing applied voltages.

A modest quantum efficiency and brightness were achieved for the bilayer devices. The incorporation of hole transporting layer PVK in the three-layer device significantly improved the device performance. Figure 11 shows the typical current-voltage (I-V) and luminance-voltage characteristics of the ITO/PVK/6P/PPy/Al three layer devices. The devices have a typical turn on voltage of ~ 15-17 V. After turn on, the light intensity roughly followed the current density. The discrepancy between them is probably due to the spectral change with the applied voltages. The maximum brightness is well above 100 cd/m^2 when the EL spectrum at 27 V is used to calculate the brightness. The external quantum efficiency of greater than 0.1% has been achieved for the initial unoptimized three layer devices.

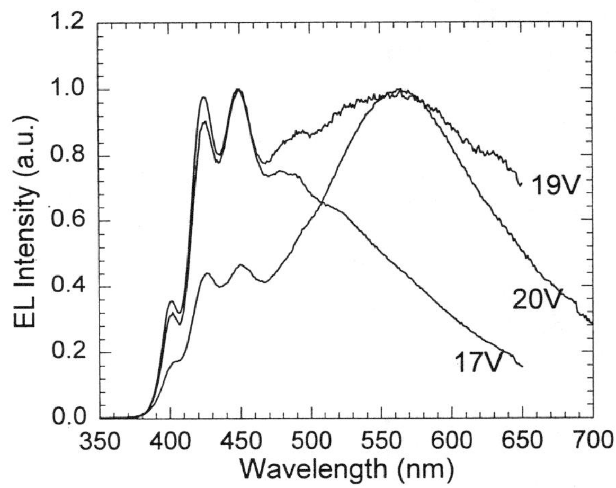

Figure 8. Normalized EL spectra of an ITO/6P/PPy/Al device under different applied voltages.

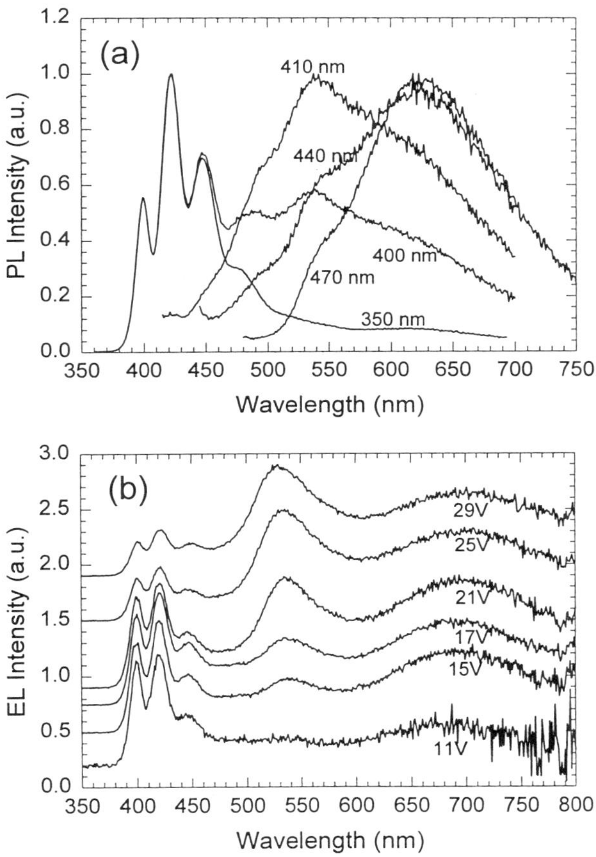

Figure 9. (a) Normalized PL spectra of PPyVPV*/6P/PPy with different excitation wavelengths; (b) Normalized EL spectra of ITO/PPyVPV*/6P/PPy/Al device under different applied voltages. The spectra are offset for clarity.

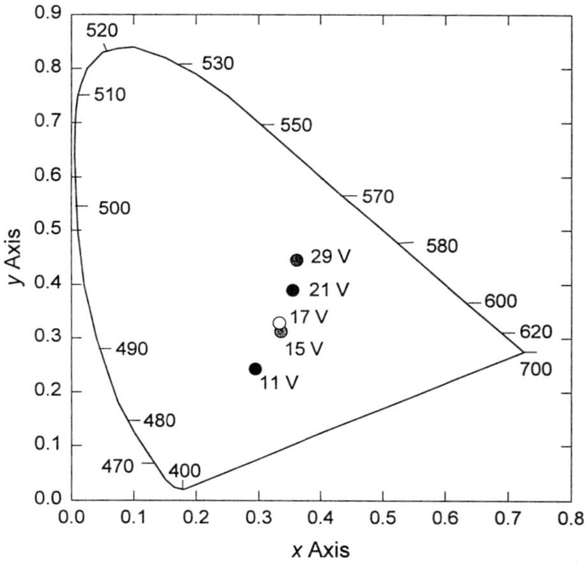

Figure 10. The CIE color coordinates of the EL spectra of the ITO/PPyVPV/6P/PPy/Al device under different applied voltages.

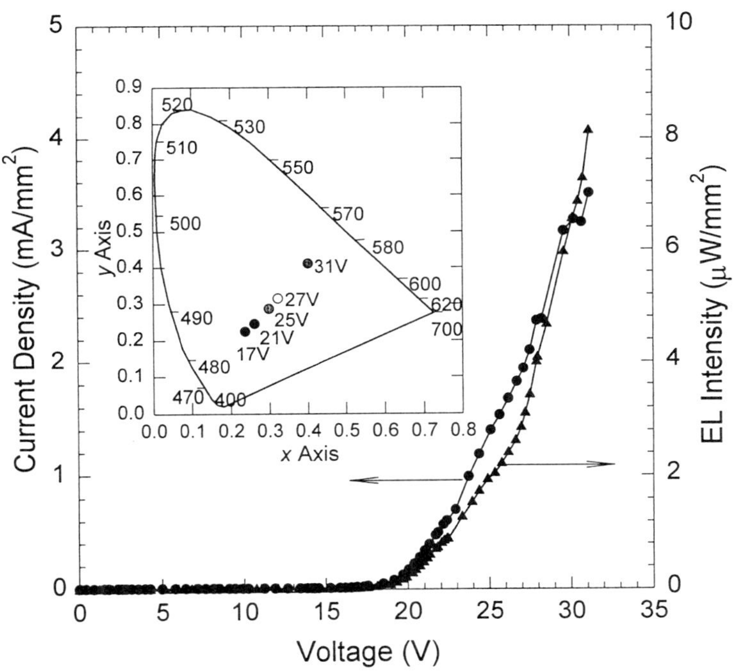

Figure 11. Current-voltage and luminance-voltage characteristics for the ITO/PVK/6P/PPy/Al device. Inset: The CIE color coordinates of the EL spectra of the ITO/PVK/6P/PPy/Al device under different applied voltages.

The phenomena of voltage dependent EL previously have been observed most often in blend systems.[20,21] In the blend systems, the change of the EL spectra with applied voltage is usually due to different charge injection thresholds for each component. The emission typically starts from low band gap component and gradually shifts to high band gap ones (towards blue) as the applied voltage increases. This mechanism clearly can not explain the devices presented here. Recently there is one report[22] on voltage controlled color-variable multilayer LEDs in separate layer configurations. In that system, the emitted light is composed of a green component from Alq_3 and a red component from a perylene bisimide pigment (PBP). For relative thin PBP layer, as the applied voltage increases, the red component decreases while the green one keeps constant. For thick PBP layer, on the other hand, the red component increases with applied voltage. This was explained in terms of electric field-induced quenching of excited states and voltage evolution of the recombination zone in the red emitting material. Field induced quenching is unlikely the mechanism for the devices reported here because the changing component in the EL spectra *grows* with increasing applied voltage. The fact that the new component in the EL spectra does not belong to any of the individual material suggests that the new component comes from the interface instead of bulk. The exact mechanism is under further study. We believe that the voltage dependent emission location due to different field dependent mobility of the materials may play an important role.

In summary, high electron affinity of pyridine-based polymers enables the use of relatively stable metals such as Al as efficient electron injecting contacts. Taking advantages of the better electron transport properties of the pyridine-containing polymers, we have fabricated bilayer devices utilizing PVK as hole transporting/electron blocking polymer. The bilayer device structure improves the device quantum efficiency and brightness significantly due to the charge confinement and exciplex emission at the PVK/emitting polymer interface. The introduction of emeraldine base (EB) form of polyaniline (PAN) on both sides of the emitting layer enables the device to work under both forward and reverse bias, as well as in AC modes. When the EB is replaced by sulfonated polyaniline (SPAN) on the cathode side and the emitting layer is properly modified to balance electron and hole transport, the device generates different colors of light, red under forward bias and green or other colors under reverse bias depending on the materials used. Use of PhPA and PPyVPV* in the blend of the emitting layer, under reverse bias yields voltage dependent emission colors. Voltage (magnitude) controlled multicolor devices were fabricated by combining the pyridine-based polymers with the 6P oligomer. Voltage dependent multicolor emission was observed in both bilayer and trilayer devices. The emission color can be tuned from blue to white to green along almost a straight line in the CIE chromaticity diagram.

This work was supported in part by Office of Naval Research. We acknowledge Dr. D. K. Fu, Professor T. M. Swager, Professor G. Leising and Professor A. G. MacDiarmid for providing samples used in this work. We thank Dr. D. D. Gebler, Dr. R. G. Sun and F. Meghdadi for discussions and experimental assistance.

Dedication: This paper is dedicated to Alan G. MacDiarmid who has shown unfailing leadership and creativity in the field of electronic polymers. He has continuously challenged the electronic polymer community to produce high quality science. His insights into synthesis, processing and properties of electronic polymers have had a major impact on us and the electronic polymer community as a whole. We look forward to his continued leadership and success in his noble pursuit of science and technology. We wish good health and happiness to Alan and his family.

References

[1] Burroughes, J. H.; Bradley, D. D. C.; Brown, A. R.; Marks, R. N.; Mackay, K.; Friend, R. H.; Burns, P. L.; Holmes, A. B. *Nature (London)* **1990,** *347* , 529.

[2] Bradley, D. D. C. *Synth. Met.* **1993,** *54*, 401 and references therein.

[3] Brown, A. R.; Bradley, D. D. C.; Burroughes, J. H.; Friend, R. H.; Greenham, N. C.; Burn, P. L.; Holmes, A. B.; Kraft, A. *Appl. Phys. Lett.* **1992,** *61*, 2793.

[4] Yang, Y; Pei, Q; Heeger, A. J. *J. Appl. Phys.* **1996,** *79*, 934.

[5] Gebler, D. D.; Wang, Y. Z.; Lin, L. B.; Blatchford, J. W.; Jessen, S. W.; Gustafson, T. L.; Swager, T. M.; Wang, H. L.; MacDiarmid, A. G.; Epstein, A.J. *J. Appl. Phys.* **1995,** *78,* 4264.

[6] Wang, Y. .Z.; Gebler, D. D.; Lin, L. B.; Blatchford, J. W.; Jessen, S. W.; Wang, H. L.; Epstein, A. J. *Appl. Phys. Lett,* **1996,** *68,* 894.

[7] Hu, B.; Yang, Z.; Karasz, F.E. *J. Appl. Phys.* **1994,** *76,* 2419.

[8] Zhang, C.; von Seggern, H.; Pakbaz, K.; Kraabel, B.; Schmidt, H-W.; Heeger, A. J. *Synth. Met.* **1994,** *62,* 35.

[9] Blatchford, J. W; Jessen, S. W; Lin, L. B.; Gustafson, T. L.; Wang, H. L.; Swager, T. M.; MacDiarmid, A. G.; Epstein, A. J. *Phys. Rev. B* **1996,** *54,* 9180.

[10] Pope, M.; Swenberg, C. E. *Electronic Processes in Organic Crystals*, Oxford University Press: New York, 1982, p. 739.

[11] Berggren, M.; Inganaas, O; Gustafsson, G.; Rusmusson, J.; Ansersson, M. R.; Hjertberg, T.; Wennerstrom, O. *Nature (London)* **1994,** *372,* 444.

[12] Yang Y.; Pei, Q. *Appl. Phys. Lett.* **1996,** *68,* 2708.

[13] Hamaguchi, M.; Yoshino, K. *Appl. Phys. Lett.* **1996,** *69,* 143.

[14] Gebler, D. D.; Wang, W. Z.; Fu, D.-K.; Swager, T. M.; Epstein, A. J. *J. Chem. Phys.* **1998,** *108,* 7842.

[15] Yang, Y.; Heeger, A. J.; *Appl. Phys. Lett.* **1994,** *64,* 1245.

[16] Wang, Y. Z.; Gebler, D. D.; Fu, D.-K.; MacDiarmid,A. G.; Swager, T. M.; Epstein, A. J. *Synth. Met.* **1997,** *85,* 1179.

[17] Lee, W.; Du, G.; Long, S. M.; Epstein, A. J.; Shimizu, S.; Saitoh, T.; Uzawa, M. *Synth. Met.* 1997, 84, 807..

[18] Wang, Y. Z.; Gebler, D. D.; Lin, L. B.; Blatchford, J. W.; Jessen, S. W.; Wang, H. L.; Epstein, A. J. *Appl. Phys. Lette*r **1996,** *68,* 894.

[19] Meghdadi, F.; Tasch, S.; Winkler, B.; Fischer, W.; Stelzer, F.; Leising, G.; *Synth. Met.* **1997,** *85,* 1141.

[20] Berggren, M.; Inganas, O.; Gustafsson, G.; Rusmusson, J.; Andersson, M. R.; Hjertberg, T.; Wennerstrom, O. *Nature* **1994,** *372,* 444.

[21] Uchida, M.; Ohmori, Y.; Noguchi, T.; Ohnishi, T.; Yoshino, K. *Jpn. J. Appl. Phys.* **1993,** *32,* L921.

[22] Kalinowski, J.; Di Marco, P.; Cocchi, M.; Fattori, V.; Camaioni, *Appl. Phys. Lett.* **1996,** *68,* 2317.

[23] Gebler, D. D.; Wang, Y. Z.; Blatchford, J. W.; Jessen, S. W. *Appl. Phys. Lette*r **1997,***70,* 1644.

[24] Wang, Y. Z.; Gebler, D. D.; Fu, D. K.; Swager, T. M.; Epstein, A. J. *Appl. Phys. Lette*r **1997,***70,* 3215.

Chapter 9

Polymer–Metal Interfaces and the P-I-N Structure of the Polymer Light-Emitting Diode

Yang Yang and Jayesh M. Bharathan

Department of Materials Science and Engineering, University of California at Los Angeles, 6531 Boelter Hall, Los Angeles, CA 90095–1595

Conjugated polymers are often treated as semiconductors with low doping concentrations. The nature of the metal/polymer interface is quite sensitive to the work function of the contact metal. In this manuscript, we present evidence to show that the pinning of the surface Fermi level effect commonly observed at the silicon/metal interface can also be observed at the metal/polymer interface. It is achieved by controlling the doping level at the metal/polymer interface. For the cathode side, the heavily n-doped region pins the surface energy level, hence the polymer interface at the cathode side is no longer sensitive to the work function of the overcoated metal. On the other hand, a p-doped region at the anode side pins the surface energy level and makes the contact insensitive to the work function of the anode. Therefore, an efficient polymer LED with the p-i-n structure has been demonstrated without the matching of the work function of the metal electrodes.

Due to their unique mechanical and optoelectronic properties, conjugated polymers have been identified as a novel class of semiconductors. (1,2) Various electronic and optoelectronic devices, such as light-emitting diodes, transistors, and photodiodes, have been demonstrated. Conjugated polymers, such as poly[2-methoxy-5-(2'-ethyl-hexyloxy)-1,4-phenylene vinylene] or MEH-PPV, are semiconductors with very low dopant concentrations. Therefore a rigid band energy structure at the polymer/metal interface is often adopted. In this model, the nature of the polymer/metal interface is very sensitive to the metal work function and the charge injection efficiency is low when the barrier height is large. These phenomena are very different from silicon wherein the mid-gap surface energy states, caused by the dangling bonds, pin the surface Fermi level. Thus the silicon/metal interface weakly reflects the difference of the metal work functions. Thus, for the polymer light-emitting diodes, an efficient charge injection requires the use of high work function conductors, such as gold or indium-tin oxide, to inject holes, and low work function metals, such as calcium, to inject electrons. Hence Au/MEH-PPV/Ca devices exhibit much better performances than Au/MEH-PPV/aluminum devices.

The calcium/polymer interface has been extensively studied by several groups using XPS and UPS (3), and the evidence suggests that Ca dopes the first few monolayers of the conjugated polymer and mid-gap bipolaron energy states are subsequently created. The photoluminescence is thus quenched at the polymer/metal interface. (4, 5) However, the role of calcium in the device performance is still unclear. It is uncertain whether Ca provides the low work function which is necessary for the electron injection, or it simply n-dopes the first few monolayers of polymer and changes the nature of the interface such that the interface is no longer sensitive to the metal work function.

In this manuscript, we demonstrate that the pinning effect observed in the inorganic semiconductors (i.e. the nature of the metal/semiconductor contact is insensitive to the work functions of the contact metals (6)) can also be produced in polymer/metal interfaces. We summarize some of the recent investigations on MEH-PPV LEDs in our group, which indicate that a 2 Å layer of Ca is sufficient to n-dope the surface of MEH-PPV, and this doped layer subsequently screens the MEH-PPV from sensing the work function of the overcoated aluminum electrode. Alternatively, it is perhaps convenient to state that this n-doped layer pins the surface energy level of MEH-PPV. (7). Similarly, a thin layer of p-doped MEH-PPV, obtained by depositing a thin layer of polystyrene sulfonic acid (PSSA) on the indium-tin oxide (ITO) anode, significantly enhances the hole injection at the MEH-PPV/ITO interface. Therefore, we demonstrated an efficient MEH-PPV LED in a p-i-n structure, i.e. ITO/(p-doped MEH-PPV)/MEH-PPV/(n-doped MEH-PPV)/Al.

Experiment

The LEDs discussed in this manuscript are consisting of a polymer active layer sandwiched between two electrodes, the cathode and the anode. For this study, aluminum (Al) was used as the cathode, and ITO was used as the anode. In order to achieve the doping effect at the metal/polymer interface, two thin interfacial layers were inserted between the metal electrodes and the active polymer. For the anode side, this thin layer was made up of polystyrene sulfonic acid (PSSA) of different thickness. For the cathode side, the thin layer comprised of a calcium layer of varying thickness. The ITO substrates underwent a routine cleaning procedure, which included sonication in a detergent followed by rinsing in deionized water and isopropanol. The polymer acid used in this study was 30% aqueous PSSA solution, purchased from Polyscience. Four different concentrations of PSSA were used such that varying thickness of PSSA could be obtained. In addition, a device with a PEDOT/ITO bilayer anode was also fabricated as a standard device. (PEDOT is an abbreviation for the conducting polymer 3,4-polyethylenedioxythiophene-polystyrenesulfonate.) Both PSSA and PEDOT were spin-coated onto the pre-cleaned ITO surfaces and subsequently baked at 100°C for a few hours. The active luminescent layer was obtained by spin-casting a solution of MEH-PPV in xylene at 2500 rpm. The MEH-PPV films were typically 1000Å thick, as obtained by using an alpha-step profilometer. Thermal evaporation of the cathode metal from a tungsten boat was carried out at pressures of around 1×10^{-6} Torr. Different calcium thickness ranging from 0.2, 2, 120, to 1000 Å, determined by a Sloan STM-100 quartz crystal thickness monitor, were deposited followed by an overcoating of a 1500 Å thick aluminum layer. The diode areas were 12 mm^2. Solution preparation and the thin film spin-casting process was carried out in a nitrogen gas environment. The current-brightness-voltage measurements (I-V and L-V curves) were carried out using a HP 4145A semiconductor parameter analyzer.

Results

n-doping at the cathode/polymer interface. Figure 1 (a) shows the I-V curves and (b) the light-voltage (L-V) curves of MEH-PPV LEDs with different thickness of calcium layers of 2, 120 and 1000 Å. The device performance, which includes the injection current, brightness and quantum efficiency, are almost identical, independent of the Ca thickness.

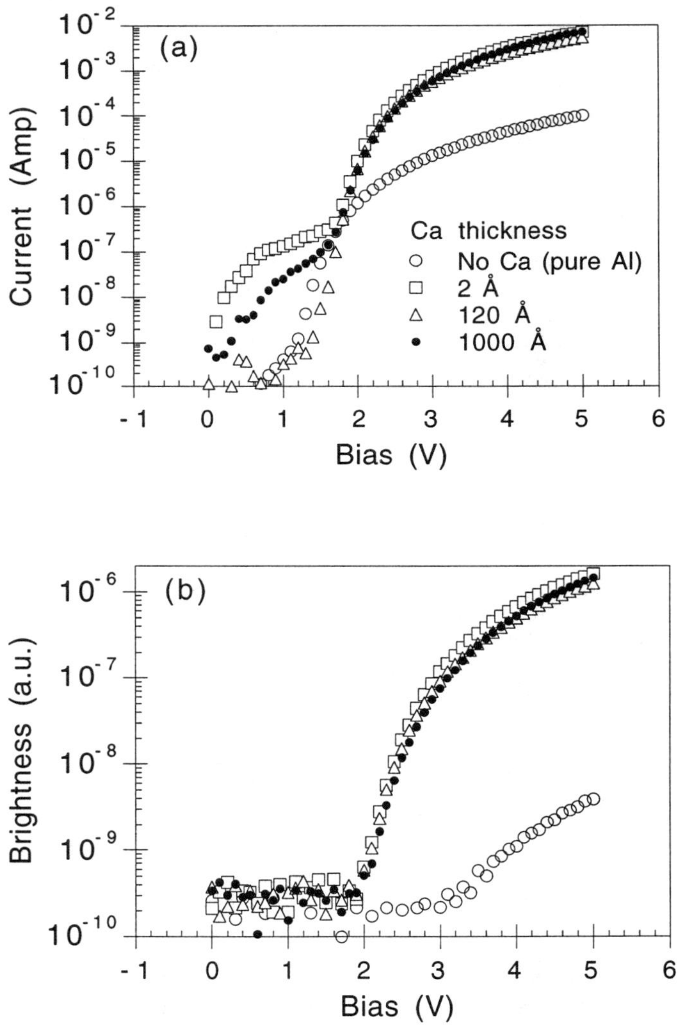

Figure 1 The I-V (a) and the L-V (b) curves of the ITO/PEDOT/MEH-PPV/Ca/Al LEDs. The thickness of calcium films ranged from 2 Å to 1000 Å, and the device performance is almost identical. The data of the device without Ca layer also included for comparison.

For the purpose of comparison, LEDs without Ca layer have also been fabricated and the data is also plotted in the same figures. This discovery indicates that the calcium film modifies the surface properties of MEH-PPV polymer such that the nature of the interface is no longer sensitive to the work function of the over coated metal layer. The n-doping of conjugated polymer by the alkali metals is an efficient process and previous studies shows that one alkali atom can effectively dope at least 6 to 7 repeat units of PPV polymer. (8) This probably explains why a 2 Å layer of Ca is sufficient to achieve the doping effect of the surface of MEH-PPV. We also made devices with 0.2 Å thickness of Ca between the MEH-PPV/Al interface. The injection current is not as strong as the device with the 2 Å thick Ca layer. This observation is not surprising since 0.2 Å of Ca is probably insufficient to effectively dope the whole surface of MEH-PPV. However, if the MEH-PPV film with 2 Å of Ca is exposed to oxygen (dry air) before the evaporation of the Al layer, the device performance is similar to the devices using only Al as the cathode material. Once the oxygen reacts with calcium to form CaO, it negates the doping effect or "de-dopes" the n-type MEH-PPV, thereby again making the interface sensitive to the work function of the metal electrode. The origin of this screening (or pinning) effect is still unclear. It is still uncertain whether this effect is due to the bipolaron energy states at the MEH-PPV interface, or it is simply due to the sharp band bending at the interface.

These observations are consistent with the photoluminescence (PL) study done by Gao et al. which reveal that a 0.1 Å calcium thin film quenched 50% of the PL efficiency at the organic/metal interface. (4) The quenching of PL efficiency is due to the n-doping (or charge transfer) effect wherein the electrons are transferred from the Ca to the polymer. This quenched PL could be recovered by introducing oxygen into the sample, which is similar to the de-doping process discussed above. We believe that this thin doped layer screens the polymer from seeing the real work function of the metal, thereby permitting the use of metals of higher work function such as Al for electron injection. However, this n-doped layer is not ideal for hole injection, hence, the device still shows the rectification behavior (not shown in the Figure 1). This doping effect is indeed widely used in the inorganic semiconductor for the fabrication of ohmic contacts. Usually, a heavily doped n-type silicon is introduced in between the n-type silicon and the metal layer for the formation of a good ohmic contact.

p-doping at the anode/polymer interface At the anode side, we suspected that conducting polymers such as polyaniline and PEDOT have similar function, that is they dope the interface layer of MEH-PPV into a p-type semiconductor. This doping process is evident by the decrease of photoluminescence of MEH-PPV at the PANI/MEH-PPV interface. The high conductivity of the conducting polymer can be attributed to the doping effect, which is obtained by the use of a strong acid. It is suspected that the same acid also changes the first few layers of MEH-PPV into a heavily p-doped region.

In order to test this interface doping hypothesis, we designed an experiment to dope the interface of MEH-PPV without the use of the conducting polymer. Polystyrene sulfonic acid, the dopant for the PEDOT conducting polymer, was spin cast onto the clean ITO/glass substrates at different concentrations, ranging from 1.5% to $3x10^{-3}$% such that a wide range of thickness of PSSA could be obtained. MEH-PPV LEDs, with the structure of ITO/PSSA/MEH-PPV/Ca were then fabricated to prove our point. The I-V and L-V curves of these devices are shown in Figure 2 (a) and (b) respectively. It can be clearly seen that the operating voltage, which is defined as the voltage at which devices reach equal brightness, progressively increases for the devices treated with 0.075%, 0.03%, and 0.003% PSSA and is the highest for the device with the blank ITO.

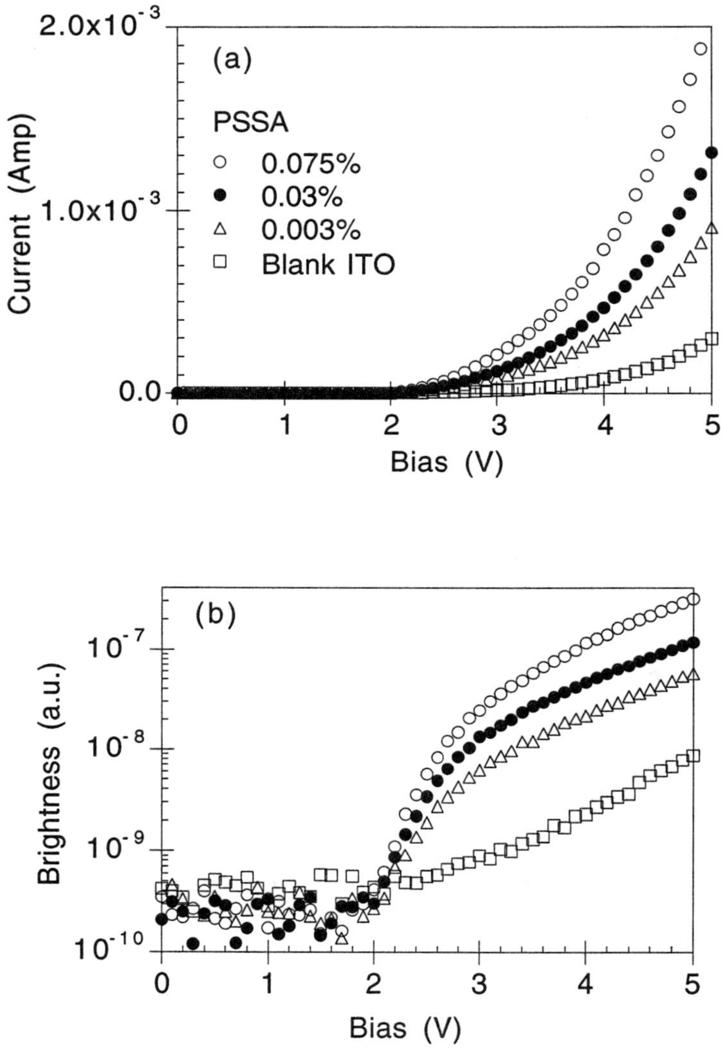

Figure 2 The I-V (a) and the L-V (b) of the ITO/PSSA/MEH-PPV/Ca/Al LEDs. For comparison, the device with pure ITO substrate is also shown.

However, for the device processed by using 1.0% PSSA, the operating voltage significantly increased. This could be due to the very low conductivity of polystyrene layer beyond a certain thickness of which it behaves as an insulating material. LEDs made of the ITO/polythiophene bilayer anode have also been made for comparison, and these LEDs still show the lowest operating voltage. However, the device quantum efficiency of the ITO/PSSA(0.075%) is similar to that of the device using ITO/PEDOT as the anode.

Using a base vapor, the de-doping process which has been described for the cathode side has also been achieved on the ITO/PSSA and ITO/PEDOT anodes. The de-doped anodes have completely changed the devices' characteristics. This de-doping process was performed by exposing the ITO/PSSA and ITO/PEDOT substrates to the vapor of triethylamine for a few minutes. ITO/d-PEDOT/MEH-PPV/Ca devices and ITO/d-PSSA/MEH-PPV/Ca devices were fabricated, and the device performance was compared to similar devices without the de-doping process. The d-PEDOT and d-PSSA represents the de-doped PEDOT and PSSA respectively. Furthermore, to confirm the de-doping effect, the conductivity of the PEDOT was measured before and after the base vapor treatment. The conductivity was found to decrease by more than one order of magnitude after the base vapor treatment. Figure 3 shows the I-V curves of the devices using PEDOT and d-PEDOT as the anode interfacial layer. Figure 4 shows the I-V curves of the devices using PSSA and d-PSSA as the anode interfacial layer. Both the figures reveal poor device performances after the base vapor treatment, similar to the devices with blank ITO as the anode wherein there is no doping effect at MEH-PPV/ITO interface. The device quantum efficiency shown in the inset of the same figures also drops significantly. From these figures, we conclude that the interfacial doping effect is more profound when using the conducting polymer as the doping layer, since devices made using PEDOT as the anode interfacial layer exhibited the best performance. This can be due to the fact that the high concentration of the acid used to dope the PEDOT also facilitates effective p-doping of the MEH-PPV interfacial layer. On the other hand, simply using a higher concentration of PSSA onto the substrates results in a thicker (insulating) interfacial layer thereby significantly reducing the charge injection, even though this layer might have achieved a better doping effect than the PSSA layer of lower concentration (thinner layer).

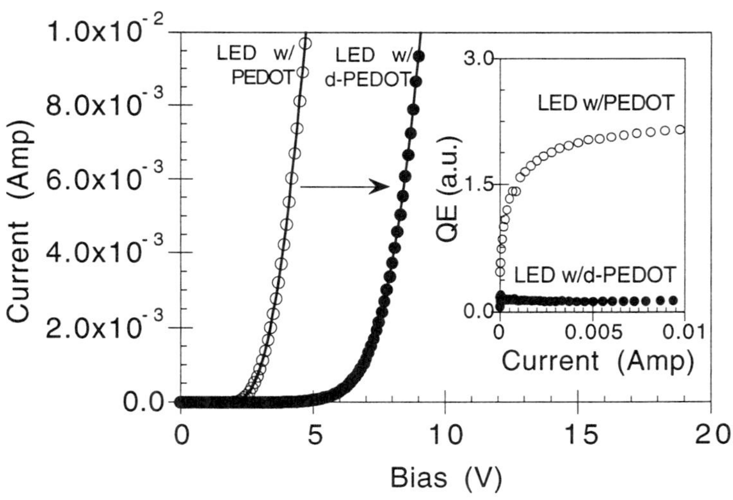

Figure 3 The I-V curves of LEDs with PEDOT and d-PEDOT as the interfacial doping layer. The device performance of LEDs with d-PEDOT has a similar characteristics as that of the LED with pure ITO anode. The QE vs. I curves are shown in the inset.

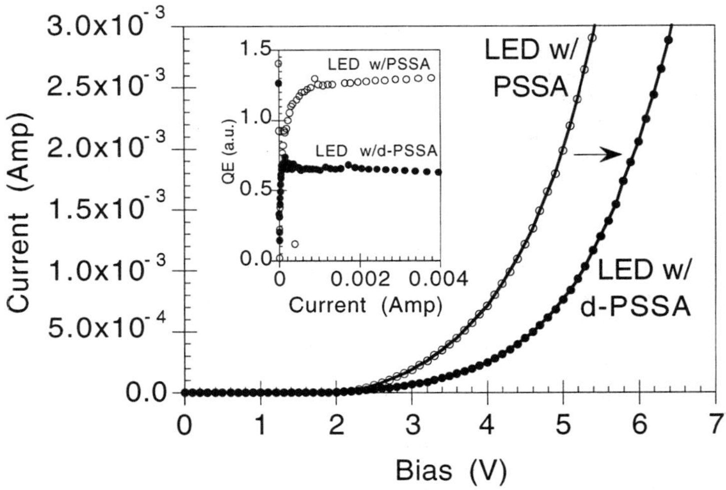

Figure 4 The I-V curves of LEDs with PSSA and d-PSSA as the interfacial doping layer. The device performance of LED with d-PSSA has a similar characteristics as the LED with pure ITO anode. The QE vs. I curves are shown in the inset.

Previously, the enhancement of the device performance in ITO/polyaniline/MEH-PPV/Ca devices has also been observed (9). The metallic emeraldine salt form of PANI is prepared by doping and complexation with functionalized sulfonic acids (e.g. camphor sulphonic acid, CSA), yielding a conducting PANI-complex soluble in common organic solvents. Hence, based on the results obtained from our research, we conclude that the enhancement of the device performance in ITO/PANI/MEH-PPV/Ca devices is probably due to the fact that the CSA dopes the surface of MEH-PPV and subsequently enhances the charge injection and lowers the device operating voltage.

Polymer LED in the p-i-n structure From the previous descriptions, we conclude that the structure ITO/PSSA (or PEDOT)/MEH-PPV/Ca/Al indeed comprises of a p-i-n structure made up of three regions: a p-doped region, an intrinsic region, and a n-doped region. This observation will be useful in designing device structures that yield the best device performance. We describe the device structure as the following:

1. p-doped region: this region should be doped to a high level such that a high charge injection level can be achieved without the matching of the anode work function. In addition, the hole mobility also significantly increases.
2. n-doped region: Similarly, the n-doped region should also be doped to a high level so as to increase the injection and mobility of electrons.
3. intrinsic region: this region is the region where the radiative recombination occurs and the thickness of this layer may play an important role in determining the actual device performance. It should be thick enough such that electrons and holes can have a good probability of encountering each other, and it should also be narrow

enough to achieve optimum quantum efficiency without losses through scattering mechanisms. Figure 5 illustrates the quantum efficiency vs. MEH-PPV thickness data for devices in the ITO/PSSA/MEH-PPV/Ca configuration. The best device performance is obtained when the thickness of MEH-PPV is around 800 Å.

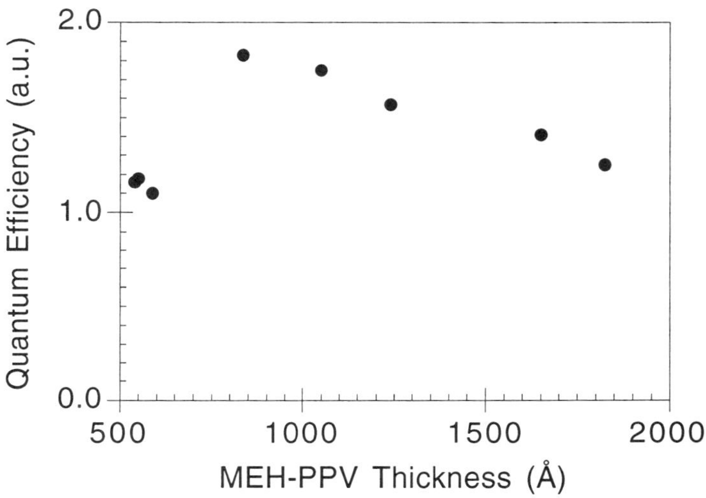

Figure 5 The QE vs. the thickness of MEH-PPV layer. The highest QE is obtained at a film thickness around 800è . The slightly increase of QE is probably due to the interference effect.

Finally, two existing examples can be used to further illustrate this concept of doped interfaces and the p-i-n structure of the organic electroluminescent devices:

1. Polymer light-emitting electrochemical cells (LECs) (10): A LEC is a device in which the p- and n- doped regions are introduced through the electrochemical doping process at the metal/polymer interface. However, the difference between an LEC and the p-i-n structure discussed above is that the p and n regions in the p-i-n structure are limited to the first few mono layers at the metal/polymer interface. Therefore, for an LED with a 1000 Å thick polymer layer, there is a 800 - 900 Å thick intrinsic region for carriers to travel through. For an LEC, the p- and n- doped regions nearly occupy 80% of the total thickness (10, 11), and the intrinsic region becomes a narrow region between two heavily doped regions. Therefore, an LEC can be turned-on at near the bandgap voltage due to the extremely thin intrinsic layer.

2. Another example has been demonstrated by Kido et al.: A bilayer organic LED was fabricated by co-evaporating Alq_3 and Li as the cathode (12). In this case the Li atoms dope the Alq_3 and create a n-doped region not limited to the interface but which extends into the bulk. This device shows rather strong charge injection efficiency and high brightness, and it is believed that the high current injection is due to the fact that the n-doped region is closer to the radiative recombination zone, which is defined at the heterojunction of the bilayer structure.

Conclusions

A p-i-n structure of MEH-PPV LED using ITO and Al as the anode and cathode respectively has been demonstrated. To achieve the doping effect at the interface, a thin layer of PSSA and Ca were deposited between the ITO and MEH-PPV and between the Al and MEH-PPV respectively. For the cathode side, the thickness of calcium film ranged from 2 Å to 1000 Å with no change in the device performance. On the other hand, the doping effect was enhanced at the anode side when a higher concentration of acid was used at the interface. The best p-doping effect is achieved when a doped conducting polymer was used as the interfacial layer. Poor device performance was observed when these interfaces were "de-doped" by using a base vapor or oxygen at the anode and cathode interfaces, respectively. From these observations coupled with the results obtained from previous XPS, UPS (3), and PL measurements (4), we conclude that the thin n-doped layer of MEH-PPV significantly changes the nature of the metal/polymer interface. It serves to pin the surface energy level, thereby making the contact insensitive to the work function of the metal electrode. Hence, an efficient electron injection can then be achieved from high work function metals, such as Al. Similarly, based on the result of PSSA/MEH-PPV interface and the quenching of photoluminescence at the interface, it is also reasonable to conclude that an ultra-thin layer of PSSA or PEDOT at the anode interface plays a similar role to modify the anode interface. This layer serves to p-dope the interfacial MEH-PPV. Efficient hole injection is therefore achieved regardless of the work function of the anode. De-doping processes performed at the anode and cathode interfaces resulting in poor device performances have been observed, further confirming the theory of the p-i-n structure.

Acknowledgments

The authors are indebted to the important discussions with Prof. Y. Gao on the cathode/polymer interface, Prof. B. Schwartz for suggesting the de-doping experiments at the anode interface, Prof. F. Wudl and Prof. R. Menon on the doping effect on conjugated polymers and for supplying the MEH-PPV polymer. This research is partially supported by a grant from the School of Engineering, University of California/Los Angeles.

References:

1. (a) *Handbook of Conducting Polymers 2nd Edition*, Skotheim, T.A., Elsenbaumer, R.L., and Reynolds, J.R., Eds., Marcel Dekker, New York, 1998. (b) *Conjugated polymers and related materials: the interconnection of chemical and eletronic structure*: Proceedings of the Eighty-first Nobel Symposium, Salaneck, R., I, Lunstrom., and Ranby, B., Oxford University Press, Oxford, 1993.

2. Parker, I. D., *J. Appl. Phys.*, 1994, *75, 1656-1666.*

3. Salaneck, W.R. and Bredas, J.L., *MRS Bulletin*, 1997, *22, 46-51.*; Park, Y., Ettedgui, E., Choong, V.E., Gao, Y., Hsieh, B.R., Wehrmeister, T., and Mullen, K., *Appl. Phys. Lett.*, 1996, *69, 1080-1082.*

4. Choong, V.E., Park, Y., Shivaparan, N., Tang, C.W., and Gao, Y., *Appl. Phys. Lett.*, 1997, *71, 1005-1007.*

5. There are other evidences indicating that the quenching of PL is dominated by the dipole interactions due to the presence of the metal electrode and that the doping phenomenon only affects the first few mono layers of the polymer at the interface

(not 50 Å as claimed in ref. 4). However, a few mono layers of doped polymer at the interface is sufficient to change the nature of the charge injection.

6. Physics of Semiconductor Devices, S.M. Sze (John Wiley & Sons, 1981), p. 276.

7. The "screening effect" in the polymer/metal interface is referred as "the nature of the polymer/metal interface of MEH-PPV LED is not sensitive to of the work function of the contact metal electrode due to the presence of a thin doping layer between the metal electrode and the polymer". This effect is analogous to the pinning of surface Fermi level of the inorganic semiconductor, hence we also use the same terminology in this manuscript.

8. Mao, G., Winokur, M.J., Karasz, F. E., *Phys. Rev. B*, 1996, *53, R463-467.*

9. Y. Yang and A.J. Heeger, *Appl. Phys. Lett.* 1994, *64, 1245-1247.*

10. Pei, Q., Yu, G., Zhang, C., Yang, Y., Heeger, A.J., *Science,* 1995, *269, 1086-1088.*

11. Yang, Y. and Pei, Q., *Appl. Phys. Lett.*, 1996, *68, 2708-2710.*

12. Kido, J. and Matsumoto, T., Digest of Technical Papers, Society for Information Display, International Symposium, , 1997, Volume XXVIII, *775-777.*

Chapter 10

Ion Migration Effects in the Degradation of Polymeric Electroluminescent Devices

James R. Sheats, Ying-Lan Chang, and Daniel Roitman

Hewlett Packard Laboratories, 3500 Deer Creek Road, Palo Alto, CA 94304

We have analyzed degradation in polymer electroluminescent devices using time-of-flight secondary ion mass spectrometry (TOF-SIMS). The devices consisted of polyfluorene-based polymers cast on indium tin oxide (ITO) coated glass substrates, with evaporated calcium cathodes. The as-cast polymer films were found to have a significant number of pinholes. Indium, concentrated in spots, was found on the polymer surface in increasing amounts with increasing electrical stress. Similarly, the amount of residual calcium on the polymer surface (left after washing) increased with stress; in this case the metal ions were more uniformly distributed (but still inhomogeneous). The anode was found to be physically roughened. We propose that the shorts commonly observed during initial operation of devices are related to cathode metal filling in pinholes, and the shorts that develop as a function of stress are caused by migration of indium ions toward the cathode and their reduction to a metallic pathway.

Electroluminescence (EL) in organic materials has evolved from the studies which laid the scientific foundations in the early 1960's, to a burgeoning field of R&D today*(1)* which now has led to at least one commercial product (from Pioneer Electric Corp.). Lifetimes are commonly being reported in the several thousand hour and above range,*(2-4)* with steady increases over the last few years suggesting that more progress is to come. This greatly reduces one of the most serious concerns that organic materials initially faced, since the useful lifetimes at desired operating luminances were far less than 1000 hours. This progress has been reviewed in several places,*(1,5,6)* which also provide background on device structure and principles of operation.

There are several reasons for these improvements that are relatively well understood. It was recognized rather quickly that all organic EL materials are to varying extents susceptible to photooxidation,*(1, 7,8)* and so must be hermetically sealed from the atmosphere. Similarly, protection of the reactive cathode metals from water*(2, 9)* (and to a lesser extent oxygen) is vital. Pinholes in the metal, often caused by particles, are especially deleterious. Purification of materials is essential.*(10)* Finally, the introduction of various ultrathin "buffer layers" next to either the anode*(11, 12)* or cathode*(13, 14)* have often been found to prolong lifetime.

Lifetime remains a very important realm of research, however, since from the commercial point of view more is always better, and not all materials exhibit suitably long life; temperature dependence is also a significant issue. We have accordingly initiated studies using TOF-SIMS to try to better understand the chemical changes that might be responsible for device degradation. TOF-SIMS is advantageous for this purpose because of its remarkable sensitivity (ppm or less) as well as resolution:*(15)* for example one can easily distinguish In (mass 114.904) from $C_7H_{15}Si$ (mass 115.094), or O (15.994) from CH_4 (16.031). This resolution is great enough that it is usually possible to unambiguously identify fragments of several hundred amu with little additional chemical information. The TOF-SIMS experiments done here are so-called "static SIMS": small pulses of incident ions are used, at a surface density of ~$10^{12}/cm^2$; this results in essentially no sputtering, so that what is seen is truly a pristine surface (1-2 atomic layers). The chief disadvantage is the difficulty in quantifying concentrations, but this can often be addressed by comparison to Auger or XPS analysis at higher concentrations. Even without absolute concentrations, the *relative* ion yields for a given sample type provide reliable information about relative concentrations.

Experimental

Devices were prepared from polymers (see below) supplied by Dow Chemical Co. They were spin-cast, usually from toluene, in total thicknesses of ~1800Å (measured by AlphaStep profilometer) onto 15 x 15 x 1 mm glass substrates with ~20Ω/sq. ITO (Donnelly Corp.), in a room with yellow lights (designed for photolithography). Ca cathodes (nominally 2000Å) were evaporated in a chamber (base pressure in the 10^{-6} torr range) connected to a nitrogen glovebox (typically about 0.1 - 0.2 ppm O_2 and H_2O, monitored by bare W light bulbs and by an Illinois Instruments oxygen monitor). The devices, kept inside the glovebox, were stressed at constant current using HP power supplies, and luminance (measured by a photodiode calibrated by a photometer) was recorded along with bias by a PC with an 8-bit digitizer card. Each substrate contains four 9.0 mm^2 "fingers" (2.0 x 4.45 mm) separated by 0.50 mm, and next to them is a ground finger, connected to the ITO by scratching away some polymer before cathode deposition. All metal fingers extend onto portions of the substrate with no ITO, to allow easy external contact using a spring-loaded clip.

The light emitting layer consisted of a poly(9,9′-disubstituted-fluorene), of which there were two varieties. PFH is 9,9′-dihexylfluorene, and PFSE (polyfluorene silyl ether) is a 10:1 copolymer of 9,9′-dioctylfluorene and 9-octyl-9′-t-butyl-dimethylsiloxyphenylfluorene. Hole transport materials are abbreviated as TPATA (a 4:1 copolymer of tolyldiphenylamine and anisolyldiphenylamine) and 3AS: p,p′,p″-(diphenylamino)-triphenylamine. When a multilayer structure with 3AS under TPATA was attempted, the 3AS was almost entirely washed out (not more than 100 Å remained). However, casting PFSE or PFH on TPATA was possible with some loss of thickness (planned thickness of 2000 Å went to 1800 Å) This is the structure of the four samples (denoted FO9, FP3, FP4 and GP2) for which data are shown in the figures.

After stress, the calcium electrodes were washed off with deionized water and blown dry in the room with yellow lights, and packaged under Al foil for transportation to the TOF-SIMS system (Charles Evans Associates), where they were inserted with minimal (but not zero) exposure to room light into the vacuum system. A Ga ion beam at 7.5 keV generated the signal. The images were acquired from an 80 μm x 80 μm raster scan; the beam size is about 1 μm for the spatial maps. The Tof-Pak software from Phi was used for analysis. To gain access to the ITO surface, the device was immersed in a beaker of toluene for ~30-60 sec.; the films came off readily.

Results

Figure 1 shows the behavior of one sample of PFSE/TPATA/3AS under stress. The shape of the decay curve is typical of these devices: there is a rapid initial decay followed by a slower long time behavior; near the end there appears to be very little if any decay (the device did not fail; the experiment was stopped to do the SIMS analysis). Note the respectable initial efficiency of around 1-2 cd/A (although the power efficiency is worse, since the bias required to reach 100 cd/m^2 is always at least 15-16 V for these materials). After some time, the devices typically cease luminescing and the voltage drops to a low value (a few V at these currents). By simply turning the constant current bias off and back on, the devices are usually resuscitated; the luminance initially tends to be higher than it was when it went off, but decays very rapidly to a curve that looks like a smooth continuation of the previous one. This "recovery effect", which is often seen in polymer LEDs,*(16)* becomes less pronounced as it is repeated. The curves shown in Figure 1 contain several such shorts; the data recorded during the shorted condition has simply been removed from the time series.

In the case of the device in Figure 1, a partial short occurred around 60 hrs: V and L dropped, but L did not go to zero. This short was healed gradually over the next 25 hrs. Other such partial shorts can be seen; these are actually less common than complete ones and this sample was atypical in that respect.

Two samples (FP4 and GP2) were used for TOF-SIMS analysis of the polymer surface (same polymer as Figure 1). Three fingers of FP4 were stressed as follows: (#2) 20 hrs at 22 mA/cm^2 (1.58 kC/cm^2 total charge passed); (#3) a few L-I-V (luminance-current-voltage) scans only; (#4): 200 hrs at 44 mA/cm^2, followed by 100 hrs at 44 mA/cm^2 and 60°C (47.5 kC/cm^2). GP2, #4 was treated with 35.4 hr at 11 mA/cm^2 (1.4 kC/cm^2), and #2 with L-I-V only.

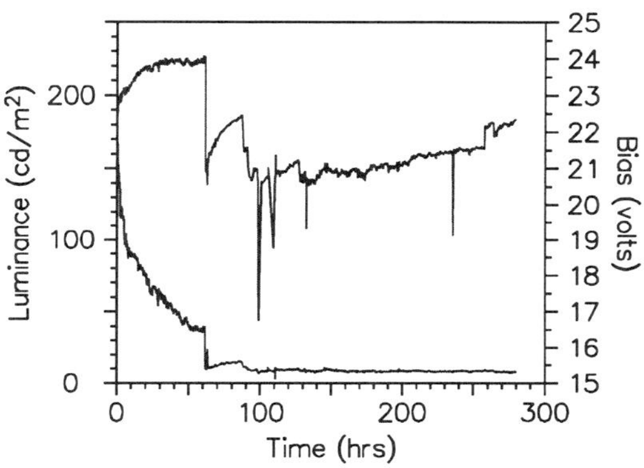

Figure 1. Luminance and bias *vs.* time at 22 mA/cm^2 constant current stress (FP3). Initial luminance 225 cd/m^2.

Figure 2 shows TOF-SIMS images of these two devices with no stress (the images are acquired from the region of the ground finger, so there was no electric field across the polymer). A small but detectable In signal is found. (There is essentially no "noise" in these spectra: a single count always represents at least one ion.) Figure 3 shows moderate stress, and Figure 4 corresponds to heavy stress; note that the same gray scale intensity is used for all the figures, to allow easy comparison. (Figure 4 shows only one finger of one sample - FP4, #4 - but looking at two different places; Figure 4b is a damaged region, as described below.) There are two unambiguous results: there is a clear increase in the amount of In as a function of stress (amount of current passed), and even after no stress there is definitely In visible, which is distributed very inhomogeneously. In a separate experiment, films of PFH were cast on silicon wafers, and a Si signal was seen (Si in positive ion spectra, amounting to 10 – 30% of total; SiO_2 and other oxide fragments in negative ions). This appears to leave no explanation other than the presence of pinholes, through which the incident ion beam can penetrate to the substrate.

The increase (described quantitatively in Table I), however, cannot be accounted for by pinholes, since the lateral density increases. This data clearly

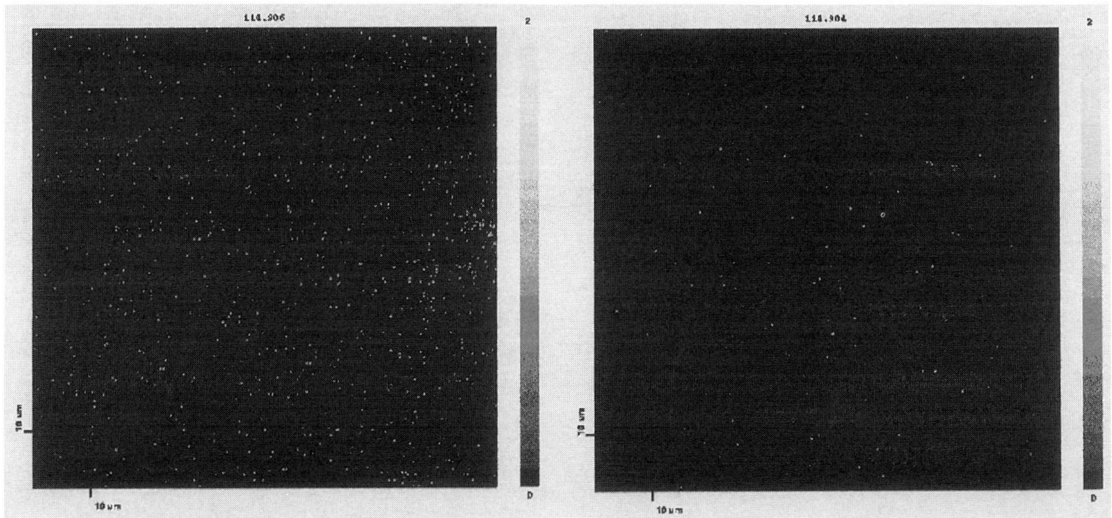

Figure 2. Indium images for samples with no stress. All images are 80 μm wide.

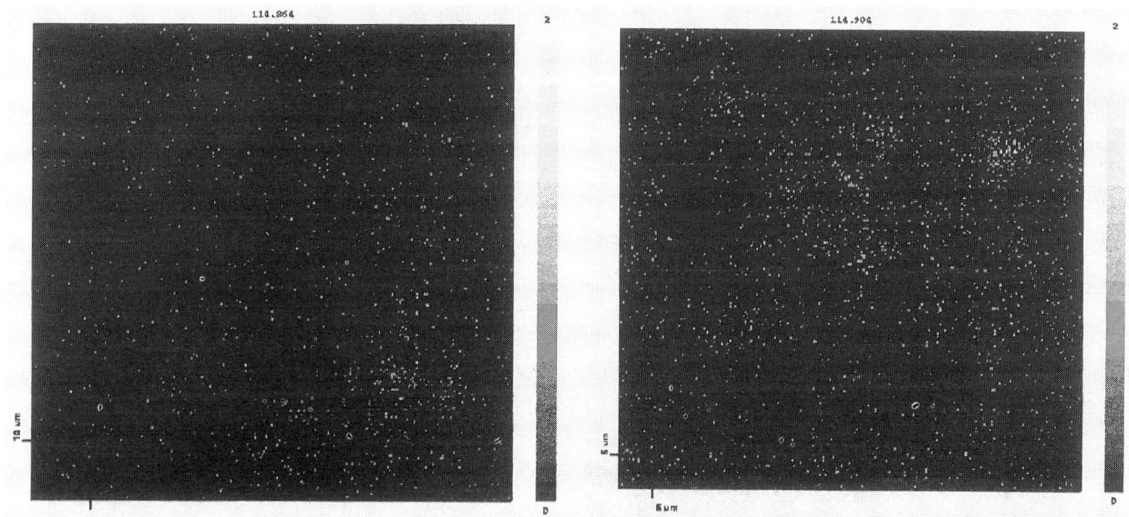

Figure 3. Indium images for samples with moderate stress.

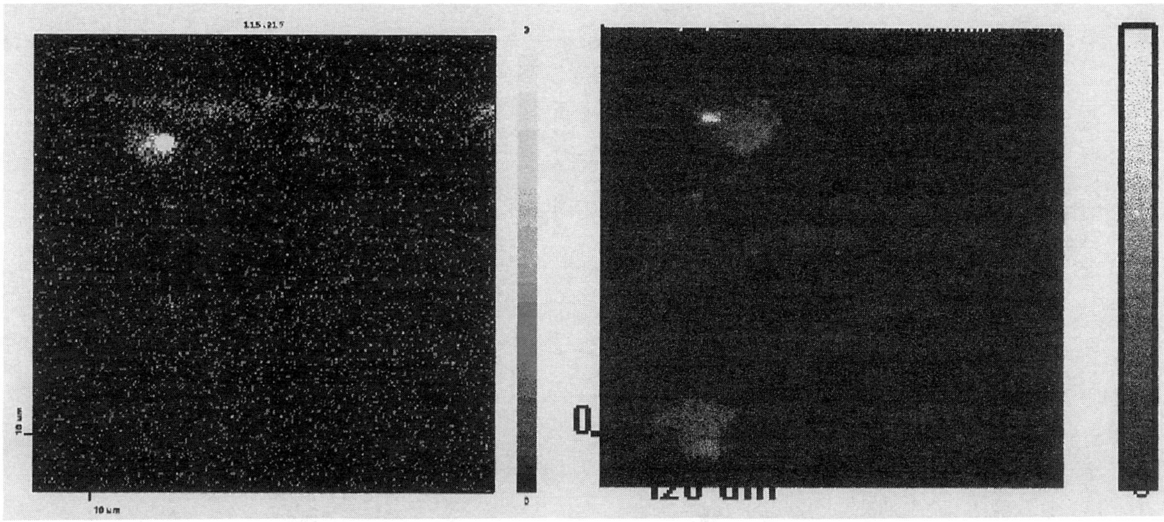

Figure 4. Indium images for samples with heavy stress.

shows ion migration as a consequence of electrical stress. The detailed mechanism of this effect must of course be a matter of conjecture; however it is at least in the right direction for electric field-induced migration. It is somewhat surprising that In ions can be removed from the ITO so readily; however an indication of the instability of ITO during operation of an OLED was already given by Scott, et al.*(17)* in their measurements of oxidation of 2-methoxy,5-(2′-ethylhexoxy)-1,4-phenylene vinylene (MEH-PPV). That the ITO is indeed disrupted is graphically demonstrated in Figure 5, which shows both SIMS data and profilometry of a heavily stressed finger after removal of the polymer: the pitting is in fact visible by microscope. Figure 4b also is of such a damaged region; atomic force microscope images (not shown) indicated maximum pit depths up to ~150 nm. These images show clearly the preferential extraction of In from the ITO: the ITO in Figure 5 is thicker than the profiled pit; so a dark region must still be "ITO", but depleted of In.

Table I. Fractional positive ion counts for In and Ca in samples with various amounts of stress.

Unstressed			Moderate Stress			Heavy Stress		
In (%)	Ca (%)	Stress (kC/cm^2)	In (%)	Ca (%)	Stress (kC/cm^2)	In (%)	Ca (%)	Stress (kC/cm2)
0.123	8.97	–	0.23	5.08	L-I-V	0.481	24.8	47.5[b]
0.066[a]	4.58	–	0.12	13.3	1.6	1.70	14.7	47.5[b,c]
0.050[a]	6.33	–	0.29	20.4	L-I-V			
			0.16	14.3	1.4			
0.08	6.6	avg.	0.20	13	avg.			avg.[c]

[a,b]Two different spots on the same sample. [b]This sample received 1/3 of its stress at 60°C. Each datum is from a different finger except as noted. [c]This spot is most likely derived from an area of highly damaged ITO surface, with little organic material left. Therefore an average is not meaningful. Sample identification (in order top to bottom): unstressed: FP4, GP2; moderate: GP2, FP4, FP4, GP2; heavy: FP4.

Although Table I clearly shows an average trend toward higher In concentration with stress (charge passed), the concentrations in Figs. 2a and 3a are quite similar. We note that the finger used for Figure 3a, though passing a similar amount of current as in Figure 2a, did not exhibit any shorts during its operation. The significance of this observation is discussed later.

The Ca signals at the same spatial positions (Figures 6-8) also reveal interesting trends. If there were no interaction of Ca with the polymer, it should be completely removed by washing. Certainly there is ample evidence that Ca does react with typical EL polymers, forming doubly charged Ca ions and

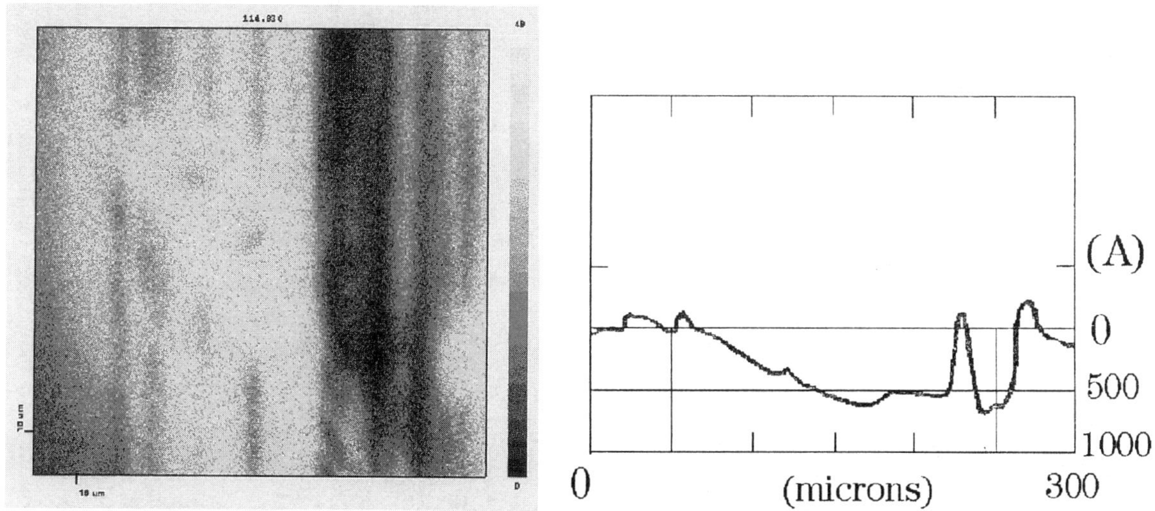

Figure 5. Indium image and profilometry of the ITO electrode of a heavily stressed sample (FO9; image and profile are not necessarily at the same spot).

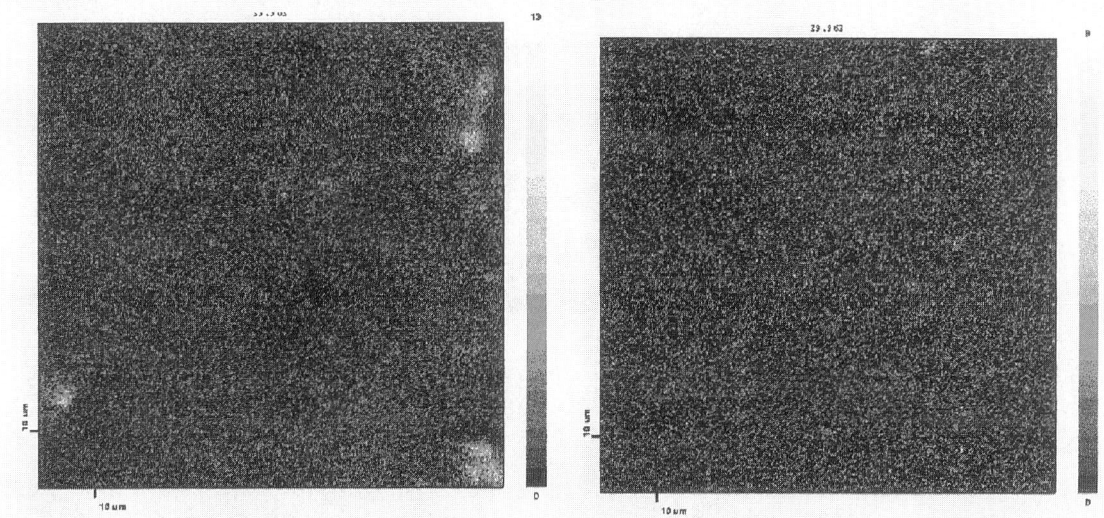

Figure 6. Calcium images with no stress (same locations as Figure 2). The gray scales for the Ca images differ by up to 50% between images.

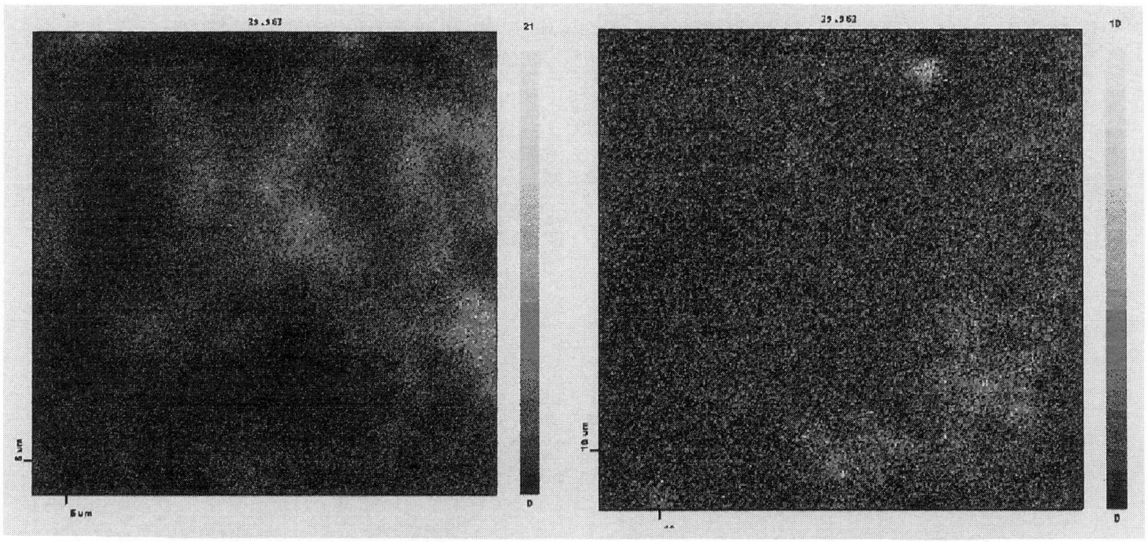

Figure 7. Calcium images after moderate stress (same locations as Figure 3).

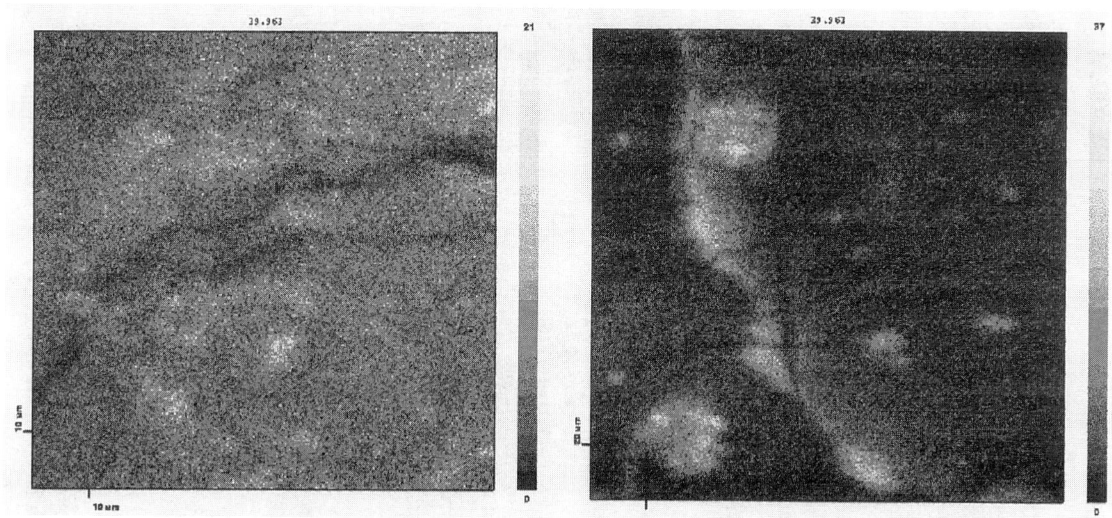

Figure 8. Calcium images after heavy stress (same locations as Figure 4). Note the right-hand image is 160 μm wide.

corresponding polymer anions.*(18)* In the presence of water (used to wash off the Ca), one would expect these anions to become protonated. The SIMS data show that some Ca nevertheless remains bound tightly enough to the surface that it does not dissolve in copious amounts of water. It is possible that this Ca is bound to oxygen-containing sites on the polymer, since the spectra do show evidence of oxidation. Since the polymer is hydrophobic, water should not penetrate the surface readily, and the quenching of the proposed reactive hydrocarbon ions would then be by oxygen (from the first air exposure). Alternatively, the oxygen may be that which was already present at the time of cathode evaporation. At any rate, the statistics showing an increase with stress are robust, and strongly suggest that these binding sites must be formed as a consequence of electrical operation.

The Ca is distributed relatively homogeneously, as would be expected if it is bound to reacted sites on the polymer. The spots in the unstressed sample may be places where the dissolution process was incomplete, or they might be the locations of pinholes where the Ca would be removed more slowly. It is important to realize that this Ca cannot simply be patches of undissolved metal, or it would be impossible to see other elements given the surface selectivity of TOF-SIMS. Ca does indeed constitute the largest positive ion signal in most cases, but it does not mask other ions, and large negative ion signals (in which of course Ca is essentially totally absent, and which are not predominantly oxygen) are also obtained.

We also looked at the Ca distribution on devices from which both cathode and polymer layers had been removed (as for example in Figure 5). Ideally there should be no Ca on the ITO surface, unless it is deposited in pinholes or can penetrate the film during deposition (which Salaneck, et al., have found not to happen*(18)*). Figure 9 shows some examples, again as a function of stress. The lightly stressed case (Figure 9a) has a bright spot which may be the location of a pinhole; immediately on application of bias one should expect that Ca ions will be repelled and dispersed over the surface, giving rise to the observed distribution. Eventually they should migrate to the cathode, and indeed longer periods of bias result in loss of Ca, as shown in Figs. 9b and especially 9c. (These figures are good evidence that the observed Ca is not an artifact of the sample preparation, where it might be supposed that Ca from the polymer surface could make its way down to the ITO surface during dissolution of the polymer in toluene. If that happened, one would not see the Ca-depleted images corresponding to long stress.)

Discussion

The primary results of this study are: 1) pinholes can be observed by TOF-SIMS analysis in films that appear to be of high optical quality; 2) indium penetration into the EL polymer (which has been observed by others, but not as a consequence of stress*(19,20)*) is driven by electrical operation; and 3) the amount of calcium strongly bound to the polymer surface increases with electrical stress. Each of these observations has relevance to one or more modes of device degradation.

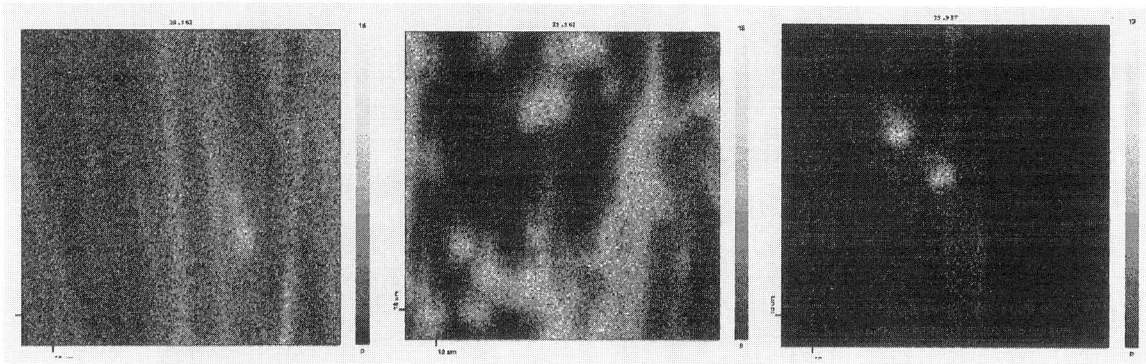

Figure 9. Ca images on ITO surface after polymer removal (FO9). The first is lightly stressed (32 C/cm^2); the next two heavily (69 and 32 kC/cm^2, respectively). The last is the same area for which the In image is shown in Figure 5. The fraction of positive ion counts is 6.1%, 5.2%, and 0.78%, respectively.

1) *Initial shorts.* As often noted, OLED devices (especially single layer ones) frequently show noisy and unstable I-V curves (especially in that bias regime before light emission is visible) on initial testing; after one or more scans the instability "burns out" and the curve becomes reproducible and smooth. Only the current shows this effect: the luminance curve is generally smooth. The extent of this problem varies with individual samples and does not correlate strongly to any particular material type. It is most effectively eliminated by using a layer of conducting polymer such as polyaniline (PANI) or poly(ethylenedioxythiophene) (PEDOT) between the ITO and EL polymer.*(11)* Other hole transport layers can help, but do not in general eliminate the phenomenon: we made many devices with poly(fluorene) emitters and various polymeric hole transporters, and shorts were seen with all of them, with various frequencies that did not appear to correlate with any particular material choice.

The lateral density of pinholes seen with the TOF-SIMS analysis suggests that pinholes are a major contributor to these observations. Even if one has two layers, and pinholes would presumably not be in register in the two layers, cathode metal would penetrate to the hole transport layer, and a relatively high hole current would result. This would be exacerbated by the fact that the small metal tip in the pinhole enhances the electric field.

The greater efficacy of conducting polymers in eliminating shorts is sufficiently unambiguous as to require comment. We suspect that this may result from the temperature sensitivity of these polymers, which lose conductivity at about 200°C or less (depending on the dopant). The basis of "burning out" shorts is presumably a rise of the local temperature due to conduction through a small cross-section; this would result in burning out the short by changes in the conducting polymer, which may happen at a lower temperature (and hence lower bias) than is required to burn out the metal.

Previous work on poly(phenylene vinylene), or PPV, has shown that during curing of the precursor, acid is released which reacts with ITO to cause In doping of the polymer.*(20)* Schlatmann, *et al.(19)* have reported indium contamination of soluble polymer LEDs, using Rutherford backscattering and x-ray photoelectron spectroscopies. They conclude that their results cannot be explained by pinholes, although there is no known precedent for In reacting with the EL polymer or otherwise possessing significant mobility in it. Without knowing more about the specific experimental conditions, it is difficult to make a comparison with our results. Our observation of Si through the poly(hexylfluorene) layer on Si wafers is unambiguous, and field-free mobility of this element through a typical EL polymer is even less probable than for In. Casting conditions could vary significantly from one sample to another, since these solutions were from relatively small, "laboratory scale" stocks, and no serious effort had been made to perfect film quality. Also, the surface energies for different polymers cast on different surfaces such as oxidized Si, ITO, and polyer may well affect film quality. Nevertheless, the concordance of results on the different samples in this study is strongly supportive of a pinhole interpretation. We are also able to observe peaks that are consistent with Sn at roughly the same relative intensity as on bare ITO, although the signal strength (for the case where high resolution spectra are available) is very weak (Sn has seven isotopes with 5% or more of the total abundance), and there are possible interferences. Thus we cannot at this time completely rule out the possibility that In migrates through the polymer before operation, but the sum of the evidence argues against it.

2) *Developing shorts.* As shown clearly in the degradation curves in this paper and in ref. 5, devices that either are initially stable or have become so by burn-in eventually develop shorts; in most cases these shorts can also be burnt out by an electrical pulse of higher than normal current (as happens when a constant current power supply is switched on), although eventually this fails to revive the device.

We suggest that the applied field is sufficiently strong to occasionally pull In ions out of the ITO surface, after which they migrate through the polymer. If they come into electron transfer distance of a negative charge carrier, however, they will be immediately reduced to In. In this way whiskers of In can be built up which bridge the electrodes and cause shorts. They would be relatively fragile, however, and an intense current pulse can typically burn them out.

The remedy for shorts, according to this postulate, is to introduce a layer of conducting material between the ITO and the polymer; then the field across the ITO interface is reduced to zero (or a small value, if the material is not a true metallic conductor but a very good hole transporter). Empirically, this is found to be the case: all long-lived devices (either polymer or small molecule based) use such a layer, and these devices lose luminescence steadily without shorting.*(2-4,21)* For polymer devices this is usually a conducting polymer such as PANI or PEDOT (though these also work for small molecules,*(22)* the most common structure there is a high-mobility hole transport molecule that is known to have an "ohmic" interface with ITO*(23)*). Whether the prevention of In ion migration is the mechanism or not remains to be definitively proven. The analytical experiments reported here should be carried out on devices with and without buffer layers to provide greater confidence.

3) *Chemical degradation.* There has long been speculation about the role of reactions at the cathode interface,*(24,25)* where metals known to be oxidatively reactive are used to get good electron injection. Some concern about ion migration from this electrode has been expressed; however, positive ions should not migrate away from a negative electrode. In fact, it appears that reaction of cathode metal (calcium in particular) is in fact essential for good charge injection and long-lived devices.*(26)*

Nevertheless, a buildup of high concentrations of charged reaction products (dianions or bipolarons, or their further decay products) is clearly detrimental for two reasons: they are luminescence quenchers,*(27)* and they are immobile, so that they constitute a fixed space charge which raises the voltage required for electron injection. In previous publications*(5, 16)* we have suggested that such reactions are involved in EL degradation, based in part on the fact that relatively inefficient, "hole-only" devices appear to decay far more slowly (if at all) even after the same amount of charge is passed.

The present data on Ca binding supports these conclusions. While more analysis is required to determine exactly what species are present at the surface, it is clear that *a*) they bind Ca strongly enough to prevent its removal by water, and *b*) their concentration is increased with electrical operation. During this stress, the bias for constant current operation rises steadily, and the efficiency (luminance at constant current) decreases, as predicted by the hypothesis.

There are several other effects that could contribute to the observations. For example, impurities might diffuse to the surface. However, if they are neutral, then the rate of accumulation should depend only on time (and not on current passed or even on whether the device is operated or not). Our data show clearly that degradation is a function of electrical stress, since the luminance does not decay when the device is off. If the impurities are charged, then they must be positive to be attracted to the cathode, and then they would not bind Ca. On the other hand, the impurities might be there from the beginning, and could account for reactions whose plausibility is not obvious when examining the structure of the idealized polymer. These "impurities" could also be a part of the polymer chain.

Scott, et al. showed*(17)* that MEH-PPV became oxidized by device operation, even when run in a nitrogen glovebox, and postulated that the ITO (which was shown by FTIR to be degraded) was the source of oxygen. Since we also observe disruption of the ITO (indeed, if In is extracted, the associated O must go somewhere), it is possible that oxygen molecules are produced, which then photooxidize the polymer. Such species would be capable of binding Ca as well as producing the deleterious effects discussed above. Indium might be transported as a positive InO ion, and is reduced near the cathode (where the electron concentration is highest) to In metal and oxygen. The chemistry remains to be clarified, but the increase of Ca binding represents clear evidence of some degree of reaction to produce negatively charged products. It should be noted that even though the bound Ca shows up strongly in TOF-SIMS, data on the organic species (not shown here) show only minor changes in the polymer structure. Thus it may be very difficult or impossible to detect these changes by, for example, FTIR.

Elsewhere we have suggested*(16)* that the "recovery effect" (as described in the beginning of the Results section) arises because deeply trapped electrons are released while the bias is off, and then it takes some time for this trapped charge to build back up again and reduce the luminesence. The relationship between such traps and the Ca-binding species identified above is not clear: the latter would seem to be an irreversible chemical change. However, in the language of traps this may simply represent a continuum of trap depths: at some point a "deep trap" becomes a permanently stable species.

Conclusions

We have presented TOF-SIMS data showing a number of degradation processes in polymer LEDs. According to our interpretation, they point out the occurrence of significant morphological defects in routinely prepared spin-cast films, the need for low electric fields at the ITO interface (or a more stable substitute for ITO), and the development of relatively small concentrations of negatively charged stable chemical species at the cathode interface. The first of these problems can be addressed by sufficiently careful engineering practice; the second appears to be satisfactorily solved by the use of conductive polymer buffer layers. The third is one that is fundamentally in the realm of materials design and preparation. It would be useful to determine if localized electron species are in fact responsible for device degradation as postulated, and if so, whether this reaction is intrinsic to the polymer or occurs primarily at chain defect sites. We suspect that the latter may well be the case, and that meticulous attention to quality in synthesis and purification will lead to constantly improving lifetimes. Since no polymer can be "perfect" in the sense of a perfect crystal, however, it is of some importance to know more about the nature of the defects and which ones are harmful.

Acknowledgments

We thank Liz Carr for the AFM studies, Homer Antoniadis and Mark Hueschen for many useful discussions, Ron Moon for support, and Ed Woo and collegues at Dow Chemical Co. for the polymers.

Literature Cited

1. Sheats, J.R., Antoniadis, H., Hueschen, M., Leonard, W., Miller, J., Moon, R., Roitman, D. and Stocking, A. *Science* **1996**, *273*, 884.

2. Berntsen, A., van de Weijer, P., Croonen, Y., Liedenbaum, C. and Vleggaar, J. (1997) *Soc. Inform. Display Conf. Proc.* (Internat. Display Res. Conf., Sept. 15-19, Toronto, Canada); Soc. Inf. Display, 1997, p. F28.

3. Carter, J.C., Grizzi, I., Heeks, S.K., Lacey, D.J., Latham, S.G., May, P.G., Ruiz de los Paños, Pichler, K., Towns, C.R. and Wittman, H.F. *Appl. Phys. Lett.* **1997**, *71*, 34.

4. Cao, Y., Yu; G., Zhang, C.; Menon, R.; Heeger, A.J. *Syn. Metals* **1997,** *87,* 171.

5. Sheats, J.R.; Roitman, D.B. *Syn. Metals* **1998,** *95*, 79.

6. Kraft, A.; Grimsdale, A.C.; Holmes, A.C. *Angew. Chem. Int. Ed.* **1998,** *37*, 403.

7. Cumpston, B.H; Jensen, K.F. *Trends Poly. Sci.* **1996,** *4*, 151.

8. Scurlock, R.D.; Wang, B.J.; Ogilby, P.R.; Sheats, J.R.; Clough, R.L. *J. Am. Chem. Soc.* **1995**, *117*, 10194.

9. McElvain, J.; Antoniadis, H.; Hueschen, M.A.; Miller, J.N.; Moon, R.L.; Roitman D.M.; Sheats, J.R. *J. Appl. Phys.* **1996,** *80*, 6002.

10. Antoniadis, H.; Hueschen, M.R.; McElvain, J.; Miller, J.N.; Moon, R.L.; Roitman, D.B.; Sheats, J.R.; *Macromol. Symp.* **1997,** *125*, 59.

11. Roitman, D.B.; Sheats, J.; Antoniadis, H.; Hueschen, M.; Yang, Y.; 27th SAMPE Intern. Conf.,1995, *27*, 681.
12. Karg, S.; Scott, J.C.; Salem, J.R.; Angelopoulos, M. *Synth. Metals* **1996,** *80*, 111.
13. Namiki, T.; Sato, H.; Nagayama, K.; Watanabe, T. *U.S.P. #5,457,565* **1995.**
14. Hung, L.-S.; Tang, C. W. *U.S.P. #5,677,572* **1997.**
15. Wien, K. *Nucl. Instrum. Methods in Phys. Res. B* **1997,** *131*, 38.
16. Sheats, J.R.; Mackie, W.A.; Anz, S.; Xie, T. *Proc. SPIE* **1997,** *1348*, 219.
17. Scott, J.C.; Kaufman, J.H.; Brock, P.J.; DiPietro, R.; Salem, J.; Goitia, J.A. *J. Appl. Phys.* **1995**, *79*, 2745.
18. Salaneck, W.R; Brédas, J.L. *MRS Bulletin* **1997,** *22* (no. 6, June), 46.
19. Schlatmann, A.R.; Floet, D.W.; Hilberer, A.; Garten, F.; Smulders, P.J.M.; Klapwijk, T.M.; Hadziioannou, G. *Appl. Phys. Lett* **1996,** *69*, 1764.
20. Sauer, G.; Kilo, M.; Hund, M.; Wokaun, A.; Karg, S.; Meier, M.; Riess, W.; Schwoerer, M.; Suzuki, H.; Simmerer, J.; Meyer, H.; Haarer, D. *Fresenius Z. Anal. Chem.* **1996,** *353*, 642.
21. Carter, S.A.; Angelopoulos, M.; Karg, S.; Brock, P.J.; Scott, J.C. *Appl. Phys. Lett.* **1997,** *70*, 2067.
22. Antoniadias, H.; Roitman, D.B.; Miller, J.N. *U.S.P. #5,719,467* **1998.**
23. Giebeler, C.; Antoniadias, H.; Bradley, D.D.C.; Shirota, Y. *Appl. Phys. Lett.* **1998,** *72*, 2448.
24. Cacialli, F., Friend, R.H., Moratti, S.C., Holmes, A.B. (1994) *Syn. Met. 67*, 157.
25. Wei, X.; Jeglinski, S.A.; Vardeny, Z.V. *Syn. Metals* **1997,** *85*, 1215.
26. Bröms, P.; Birgersson, J.; Johansson, N.; Lögdlund, M.; Salaneck, W.R. *Syn. Metals* **1995,** *74*, 179.
27. Greenham, N.C.; Shinar, J.; Partee, J.; Lane, P.A.; Amir, O.; Lu, F.; Friend, R.H. *Phys. Rev. B.* **1996**, *53*, 13528.

Chapter 11

Conjugated Polymer Blends as Emitting Layer for White Light LED

Show-An Chen and En-Chung Chang

Department of Chemical Engineering, National Tsing Hua University, Hsinchu 30043, Taiwan, Republic of China

Studies on white light organic light emitting diodes with organic molecules and polymers as emitting materials are reviewed. The white light can be obtained in the following ways: (1) single-emitting materials, such as metal complex of 2-(2-hydroxyphenyl) benzothiazole with zinc ($Zn(BTZ)_2$) and emeraldine base poly-aniline; (2) poly(vinyl carbazole) (PVK) doped with red, green and blue emitting organic materials; (3) three emitting organic layers each emitting red, green and blue lights; (4) polymer blends, in which the major and minor (or trace) components are blue and red light emitting materials, respectively; (5) bilayer with good mixing at the interface, in which exciton and exciplex emissions occur.

1. Introduction

Organic light emitting diodes (OLED) have drawn great attention for they process several advantages over the conventional LEDs, such as lower power consumption, large area and easy to fabricate and over the liquid crystal display (LCD) such as wide view angle, lower cost, and no need of backlight. OLEDs involve the two categories in accordance with their emitting materials: OMLED with organic molecular dye as emitting materials and polymer LED (PLED) with conjugated polymers as emitting materials. Extensive studies on OLED began from the discovery of the thin device: ITO/diamine/Alq_3/Mg:Ag prepared by thermal evaporation by Tang and Vanslyhe (*1*) in 1987; this device has the brightness 1000 cd/m^2 at the bias 10 V, and external quantum efficiency 1 %. Later the Cambridge group (*2*) in 1990 found that poly (phenylene vinylene) (PPV) can be used as light-emitting layer for LEDs, which can emit light as bright as 500 cd/m^2 at 8 V (film thickness about 1000 Å) when using calcium as the cathode material. Introducing side groups at the phenylene ring and/or vinylene unit allow a control of the color of the emitted light by providing a blue (or red) shift as electron withdrawing (or donating) groups are introduced, in addition to an increase in solubility. A side chain that imparts a steric hindrance to the conjugation could also lead to a blue shift. Considerable efforts on the molecular design of conjugated polymers for tuning the color of emitted light through adjustment of the bandgap has been attempted. Other conjugated polymers that have been found to impart the luminescent property are poly (*p*-phenylene)s and polythiophenes (*3-7*).

OLED can be fabricated to emit light with various colors from red to blue covering the entire visible range. It can also emit white light by use of a proper single emitting material or by use of a polymer blend or multi-emitting layers each emitting different colors. The chemical structures of the materials involved in this article are listed on Figure 1. Such OLEDs have wide range of applications as displays and backlight sources for LCD, flat television screens, personal computer displays and wristwatch. This paper reports a survey on the research works in this area.

2. White light from a single emitting material

Hamada et. al. (1996) (*8*) have synthesized a metal complex of 2-(2-hydroxyphenyl) benzothiazole (BTZ) with zinc (Zn) to yield $Zn(BTZ)_2$ which can emit greenish-white light from the device ITO/TPD/ $Zn(BTZ)_2$/MgIn (Figure 2), in which TPD behaves as a hole transport layer (HTL). Its chromaticity coordinates are x=0.246 and y=0.364, and brightness over 10,000 cd/m^2 at the bias 8 V.

Our laboratory (1996) (9) has found that the emeraldine base polyaniline (PAn) can emit nearly white light covering the full range of visible light (380-750 nm) (Figure 3) from the device with ITO coated glass as the hole ejector and deposited aluminum (or magnesium) thin film as the electron ejector. It is found that the phase with reduced repeat units (amine form) can emit white light, while the phase with oxidized repeat units (quinoid form) has no contribution to the emitting light. The turn-on voltages for eye-observable light intensity are 13 V and 6 V for the LEDs with aluminum and magnesium electrodes, respectively at the thickness of the emitting layer 800 Å. The EL spectrum covering the range 300-750 nm is much broader than that of its PL spectrum, 350-510 nm.

3. White light from multilayer OMLED

OMLEDs have been fabricated to emit white light in two ways, one is by dispersing organic dyes (that can emit various colors of light) in polymer as the emitting layer, and the other is by depositing several emitting layers, each emitting different colors of light.

Kido and coworkers (1994) (10) have constructed a white light emitting device using PVK doped with various dyes TPB (5 mole %, blue light), coumarin 6 (0.3 mole %, green light) and DCM1 (0.2 mole %, orange light) as the emitting layer and using TAZ and Alq_3 as ETL. The structure of the device so constructed is ITO/doped PVK/TAZ/Alq_3/Mg:Ag; its EL spectrum is broad and covers the active visible range having the characteristic wavelengths: 450 (from TPB), 510 (from coumarin 6) and 550 (from DCM1) as shown in

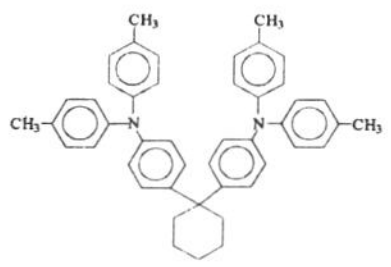
Diamine

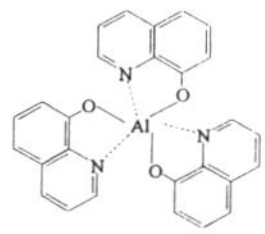
Alq_3

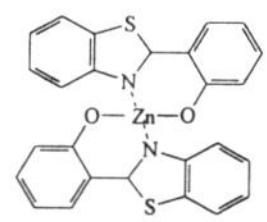
$Zn(BTZ)_2$

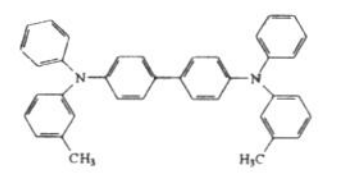
TPD

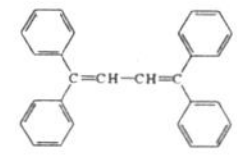
TPB

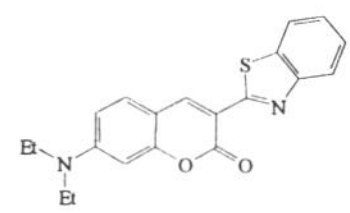
Coumarin 6

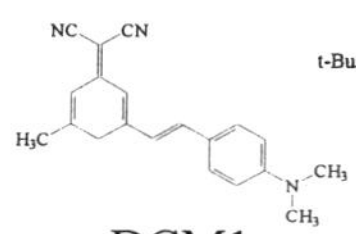
DCM1

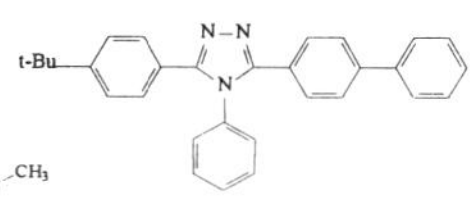
TAZ

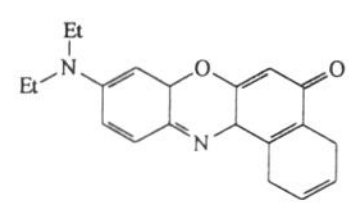
Nile Red

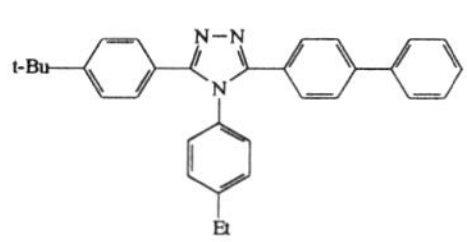
p-EtTAZ

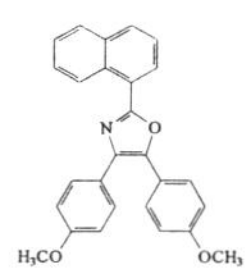
NAPOXA

Figure 1(a). The chemical structures of the organic molecular materials involved in this article.
Diamine: diamine derivative
Alq_3: tris(8-hydroxyquinoline)aluminum
$Zn(BTZ)_2$: bis(2-(2-hydroxyphenyl)benzothiazolate)zinc
TPD:N,N'-diphenyl-N,N'-bis(3-mathlphenyl)-1,1'-biphenyl-4,4'-diamine
TPB: 1,1,4,4,-tetraphenyl-1,3-butadiene
Coumarin 6: 3-(2-benzothiazolyl)-7-(diethylamino)coumarin
DCM 1: 4-(dicyanomethylene)-2-methyl-6-(4-dimethylamino-styryl)-4*H*-pyran
TAZ: 3-(4'-tert-butylphenyl)-4-phenyl-5-(4''-biphenyl-1,2,4-triazole
Nile Red: Nile Blue A oxazone
p-EtTAZ: 1,2,4-triazole derivative
NAPOXA: 2-naphthyl-4,5-bis(4-methoxy-phenyl)-1,3-oxzole

Continued on next page.

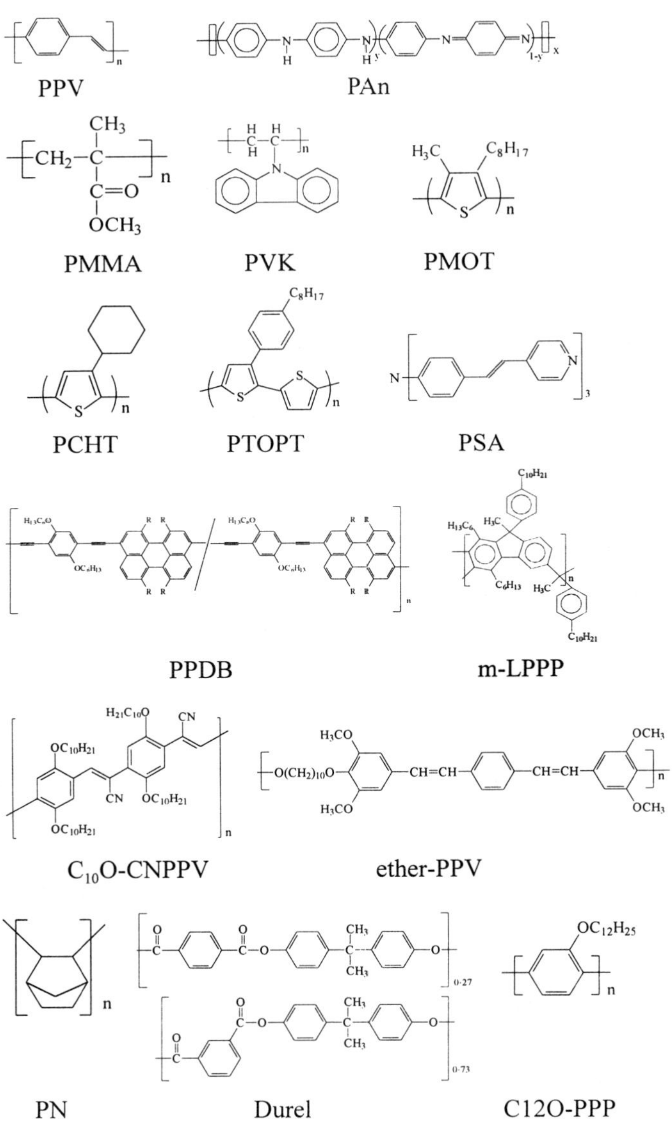

Figure 1(b). The chemical structures of the polymers involved in this article.

PPV: poly(phenylene vinylene)
PAn: polyaniline
PMMA: poly(methyl mathacrylatte)
PVK: poly(*N*-vinylcarbazole)
PMOT: poly(3-methyl-4-octylthiophene)
PCHT: poly(3-cyclohexylthiophene)
PTOPT: poly[3-(4-octylphenyl)-2,2'-bithiophene]
PSA: tristilbene amne
PPDB: poly(perylene-co-diethylbenxene)
m-LPPP: methyl substituted laddertype poly(paraphenylene)
C_{10}O-CNPPV: cyano-substituted poly(2,5-didecyloxy-*p*-phenylene vinylene)
Ether-PPV: ether-type phenylene vinylene based copolymer
PN: polynobornene
Durel: polyarlate Durel
C12)-PPP: poly(2-dodecyloxy-*p*-phenylene

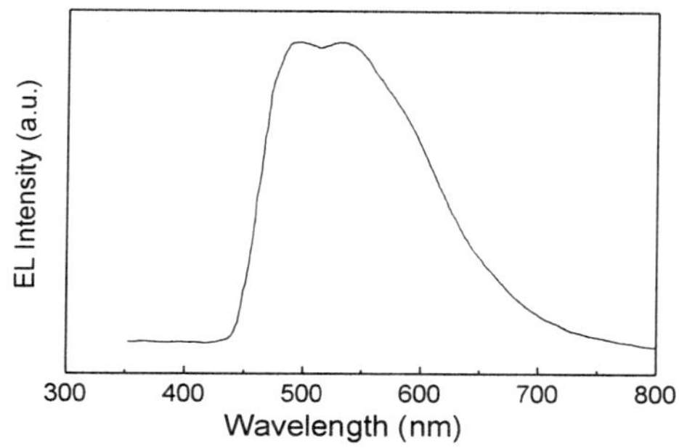

Figure 2. EL spectrum for ITO/TPD/$Zn(BTZ)_2$/Mg:In.

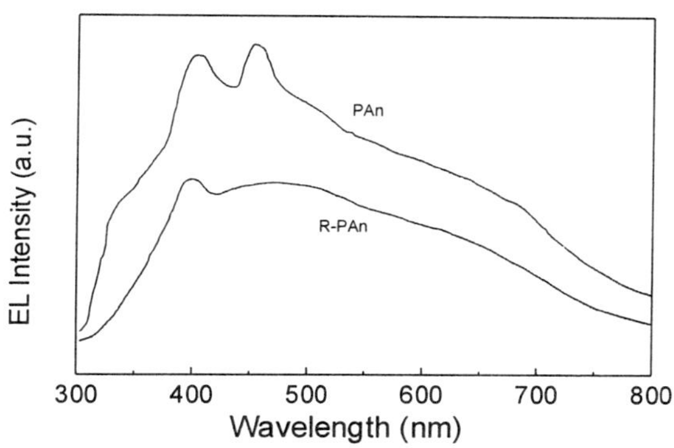

Figure 3. EL spectra for ITO/PAn/Mg/Ag and ITO/R-PAn/Mg/Ag.

Figure 4 (curve a). The turn-on voltage of the device is 4V, and has the brightness 50 cd/m^2 at 10 V and 3400 cd/m^2 at 14 V. Later (1995), they have also prepared a white light emitting device in the same way, the dopants used are TPB, coumarin 6, DCM1 and Nile red (Figure4 curve b) (11). Their device has the brightness 4100 cd/m^2 at 20 V. However, effects of the applied voltage on the color of emitted light have not been reported.

Kido and co-workers (1995) (12) have prepared a multi-emitting-layer white light OMLED having the structure ITO/TPD(400 Å)/p-EtTAZ(30 Å)/Alq_3 (50 Å)/Nile Red-doped (1 mol %)Alq_3(50 Å)/Alq_3(400 Å)/Mg:Ag, in which red, green and blue (RGB) emitting materials are included (Figure4 curve c) and this device has the brightness 2200 cd/m^2 at the bias 16 V.

Strukelj et al (*13,14*) in 1996 have added the blue light emitting materials (2-naphthyl-4,5-bis(4-methoxy-phenyl)-1,3-oxzole, NAPOXA) to the HTL, bis(triphenyl)diamein (TAD), and to the ETL tris(8-hydroxy quinoline) aluminum (Alq3) and in addition, the red light emitting material DCM1 to the latter to yield the device, ITO/TPD/NAPOXA(blue)/Alq_3/DCM1(0.5%,red) doped Alq/AIQ/Li:Al, which emits white light (Figure 5). This device has the brightness, 4000 cd/m^2 at 15 V and external quantum efficiency greater than 0.5 %.

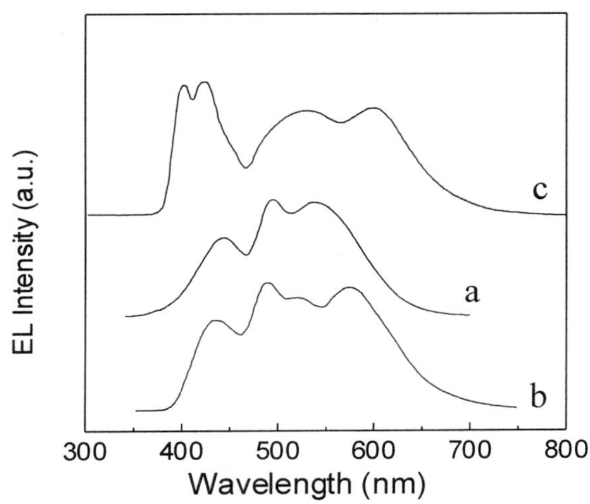

Figure 4. EL spectra for (spectrum a) ITO/dye-doped PVK/TAZ/Alq/Mg:Ag (PVK is doped with 5 mol % TPB, 0.3 mol % Coumarin 6, and 0.2 mol % DCM 1), (spectrum b) ITO/dye-doped PVK/Mg:Ag (PVK is doped with 30 wt % PBD, 3 mol % TPB, 0.04 mol % Coumarin 6, 0.02 mol % DCM 1, and 0.015 mol % Nile Red), and (spectrum c) ITO/TPD/p-EtTAz/Nile Red-doped Alq/Alq/Mg:Ag.

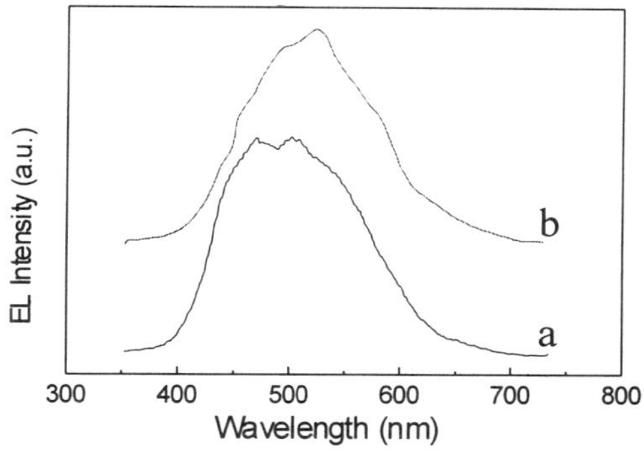

Figure 5. EL spectra for (spectrum a) ITO/TAD/NAPOXA/Alq/Al and (spectrum b) ITO/TAD/NAPOXA/Alq_3/Alq_3+DCM 1/Alq/Al.

4. Polymer blend as the emitting material

Inganas and coworkers (1995) (*15*) have used the blend of three polythiophene derivatives, which can emit red, green, and blue light respectively, as the emitting layer. This device so constructed allows a turning of the color of the emitting light by varying the applied voltage. At the higher voltage 20 V, as all components are activated, the device can emit white light. The variation in the color of emitting light is resulted from the occurrence of phase separation in the blend, in which the domain size of each component is greater than the thickness of the emitting layer causing that each component has its own passways for the charge carriers. They then introduced poly(methyl methacrylate) (PMMA) in the blends of the polythiophenes at the weight ratio, PMOT (blue): PCHT (green): PTOPT (red): PMMA (10:4:1:1), and obtained white light at 20 V having a quantum yield of 0.4-0.6% (*16*). However, at the lower voltages 5 V and 12 V, the emission from the red light polymer (PTOPT) is dominant.

Leising, Müllen and coworks (1997) (*17,18*) found that even by introduction only a very small amount about 1% of red-light-emitting polymer, poly(perylene-co-diethylbenzene) (PPDB), into the blue-light-emitting ladder type polymer, ladder type poly(paraphenylene) (*m*-LPPP), the PL and EL spectra appear as those from PPDB. This is due to the excitation energy transfer from *m*-LPPP to PPDB. The color of the emitted light can be varied and, in addition, the quantum yield can be increased by adjusting the amount of PPDB. As the amount of PPDB is reduced to 0.05%, the device emits white light (as shown in Figure 6), which does not vary with the applied voltage.

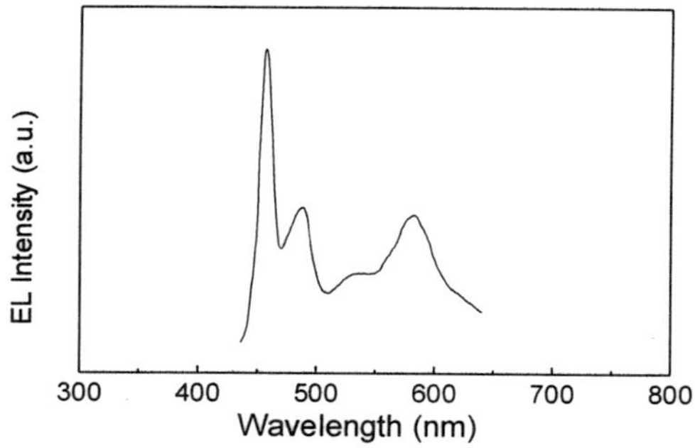

Figure 6. EL spectrum for ITO/m-LPPP:PPDB(PPDB 0.05 %)/Al.

We have studied EL spectra from the blends of PC10 with C_{10}O-CNPPV in 1998 (*19*). The devices with the blends (the weight ratio of PC10/C_{10}O-CNPPV, 1/1 and 4/1) emit light with various colors from red to yellow-orange light, depending on the applied voltage. When the ratio is 150/1, it can emit the light with broad spectrum near white light covering the entire visible region at all applied voltages (as shown in Figure 7). Similar results are also obtained from the blend of PC10/RO-PPV at the weight ratio, 14/1.

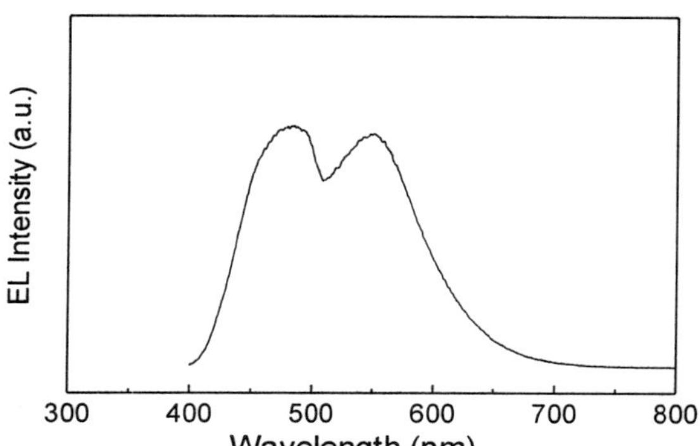

Figure 7. EL spectrum for ITO/ether-PPV:C_{10}O-CNPPV(150:1)/Al.

The optical properties of the blend systems, PC10 with C_{10}O-CNPPV, were further investigated using PL, photoluminescent excitation (PLE) and near field scanning optical microscopy (NSOM). In these blends, a new emission in the PL and EL spectra appears, which can be attributed to the emission from an exciplex formation due to the energy transfer between the two polymers. The NSOM shows that the two components in the blend system have good miscibility. As the blend ratio 150/1 for the former blend, the minor component C_{10}O-CNPPV is uniformly dispersed in the major component PC10 having the domain size smaller than the film thickness of the emitting layer.

As a forward bias is applied, the charges can not pass through the domains of C_{10}O-CNPPV directly, and the light emitting from the component should be originated from the energy transfer. Thus, the color of the emitting light from the device does not change with the applied voltage.

Wendroff (1997) (*20*) found that as the emitting molecular material PSA in the blend of polynobornene (PN) with polyarylate Durel at the weight ratio 1:2, its EL spectrum is broad and nearly white as shown in Figure 8. This is a guest-host system with a chromophore as the guest molecule and a polymer blend as the host. PSA can form radical cations and is a hole transport material. Their PL spectra vary with host polymer, thus the host polymer blend can provide an emission of white light only at a particular composition.

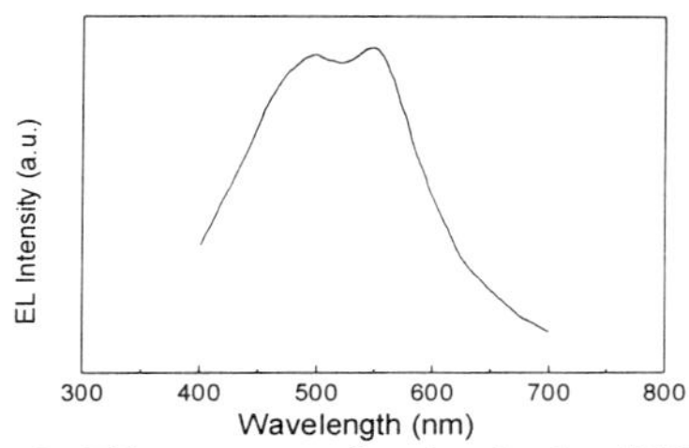

Figure 8. EL spectrum for the device ITO/PSA:Durel:PN/Al

We recently (1998) found that bilayer polymer light-emitting diodes with two blue light-emitting materials, poly(N-vinylcarbazole) (PVK) and poly(2-dodecyl-p-phenylene) (C12O-PPP), can emit blue or white light, depending on the solvent used in the fabrication of the second layer, C12O-PPP (*21*). If hexane (the nonsolvent for PVK) is used, the device emits blue light as the single layer device with C12O-PPP. However, if toluene (the cosolvent for the two polymers) is used, the device emits white light originating from an exciplex emission at the bilayer interface in addition to the exciton emission from the C12O-PPP (as shown in Figure 9). At low temperatures, the intensity of the exciplex emission drops and that of the exciton emission becomes dominant.

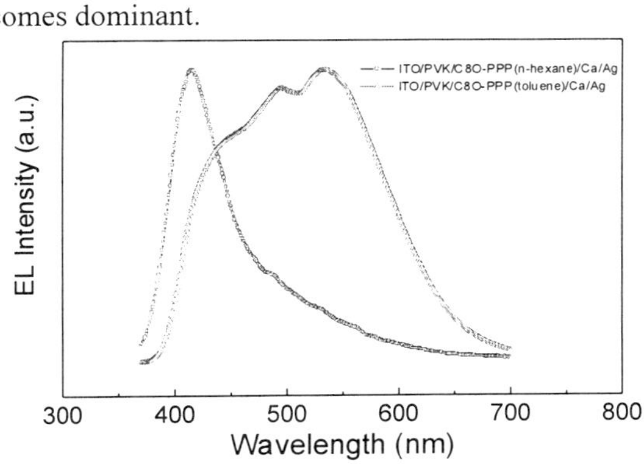

Figure 9. EL spectra for (spectrum a) ITO/PVK/C8O-PPP(n-hexane)/Ca/Ag and (spectrum b) ITO/PVK/C8O-PPP(toluene)/Ca/Ag.

6. References

1. C. W. Tang and S. A. Vanslyke, ***Appl. Phys. Lett.*** **1987, *51*,** 914.
2. J. H. Burroughes, D. D. C. Bradley, A. R. Brown, R. N. Marks, K. Mackay, R. H. Friend, P. L. Burns, A. B. Holmes, ***Nature,*** **1990, *347*,** 539.
3. D. Braun, A. J. Heeger, Kroemer-H, ***J. Elec. Mater.,*** **1991, *20*,** 945.
4. G. Grem, G. Leditzky, B. Ullrich, G. Leising, ***Adv. Mater.***, **1992, *4*,** 36.
5. Y. Ohmori, K. Yoshino, M. Uchida, ***Jpan. J. Appl. Phys***., **1991, *11B*,** 1941.
6. D. Braun, A. J. Heeger, ***Appl. Phys. Lett.***, **1991, *58*,** 1982.
7. J. Gruner, P. J. Hamer, R. H. Friend, H. -J. Huber, U. Scherf, and A. B. Holmes, ***Adv. Mater.***, **1994, *6*,** 748.
8. Y. Hamada, T. Sano, H. Fujii, Y. Nishio, H. Takahashi, K. Shibata, ***Jpan. J. Appl. Phys***., 1996, *35*, L1339.
9. S. A. Chen, K. R. Chuang, C. I. Chao, H. T. Lee, ***Synth. Met.,*** **1996, *82*,** 207.
10. J. Kido, K. Hongawa, K. Okuyama, K. Nagai, ***Appl. Phys. Lett.***, **1994, *64*,** 815.
11. J. Kido, H. Shionoya, K. Nagai, ***Appl. Phys. Lett.***, **1995, *67*,** 2281.
12. J. Kido, M. Kimura, and K. Nagai, ***Science***, **1995. *267*,** 1332.
13. M. Strukelj, R. H. Jordan, and A. Dodabalapur, ***J. Am. Chem. Soc***., **1996,** *118,* 1213.
14. R. H. Jordan, A. Dodabalapur, M. Strukelj, and T. M. Miller, ***Appl. Phys. Lett.***, **1996**, *68*, 192.
15. M. Berggren, O. Inganas, G. Gustasson, J. Rasmusson, M. R. Andersson, T. Hjertberg, and O. Wennerstorm, ***Nature***, **1994, *372*,** 444.
16. M. Granstrom, and O. Inganas, ***Appl. Phys. Lett.***, **1996, *68*,** 147.
17. S. Tasch, E. J. W. List, C. Hochfilzer, G. Leising, P. Schlichting, U. Rohr, Y. Geerts, U. Scherf, K. Mullen, ***Phys. Rev. B***, **1997, *56*,** 4479.
18. S. Tasch, E. J. W. List, O. Ekstrom, W. Graupner, G. Leising, P. Schlichting, U. Rohr, Y. Geerts, U. Scherf, K. Mullen, ***Appl. Phys. Lett.***, **1997, *71*,** 2883.
19. S. A. Chen, E. C. Chang, K. R. Chuang, C. I. Chao, J. H. Hsu, P. K. Wei, W. S. Fann, ***Polym. Prepr.***, 1998, *39*, 105; paper submitted to ***Macromolecule***.
20. T. Christ, A. Greiner, R. Snder, V. Stumpflen, J. H. Wendorff, ***Adv. Mater.***, **1997, *9*,** 219.
21. C. I. Chao, S. A. Chen, ***Appl. Phys. Lett.***, **1998, *73*,** No. 4.

APPLICATIONS

Chapter 12

Polyanilines: Progress in Processing and Applications

Vaman G. Kulkarni

Americhem Inc., 723 Commerce Drive, Concord, NC 28025

Despite its ease of synthesis, facile chemistry and novel doping by simple treatment with protonic acids, industrial, real-world applications of polyaniline and other conducting polymers are limited by processability and electrical decay during formulation and use. Over the last several years significant advances have been made in processing of Intrinsically Conductive Polymers (ICP's). ICP coatings and blends have been tested for application in electrostatic discharge (ESD) applications and corrosion protection and are ready for commercial applications. This article will discuss progress in processing and applications of polyaniline coatings and melt processable blends with special emphasis on processing by dispersion techniques.

Background

Doped polyaniline (conductive) is a green powdery material with a bulk conductivity of 1-10 S/cm. In this form it is insoluble in most solvents and infusible, therefore, it has been traditionally categorized as intractable polymer. The neutral or emeraldine base form of polyaniline is a coppery bronze powder with a conductivity of less than 10^{-10} S/cm. Polyaniline is an unique ICP in that its conductivity can be reversibly controlled chemically and electrochemically. The emeraldine base is soluble in a wide variety of amine solvents such as N-methylpyrrolidone (NMP), N, N' dimethylacetamide, dimethylformamide and N,N'-dimethylpropylene urea (DMPU). However, even in these solvents the solubility is limited and solutions have limited stability.

Polyanilines have been processed from solutions of neutral polyaniline in NMP, DMPU and others for some time[1,2]. Significant strides have been made in making fibers form these solutions. Nonetheless, the technique suffers from the disadvantage that processed articles are non-conductive and need to be doped in a secondary step. The technique is not suited for preparation of coatings on a commercial scale. Processability of polyaniline in the doped form is more attractive as it removes the subsequent doping step. Functionalized protonic acids such as camphor sulfonic acid, preferably in the presence of m-cresol and dodecyl benzene sulfonic acid which dope polyaniline and simultaneously render it soluble in common organic solvents [3,4] provide such an alternative. Highly transparent and conductive coatings have been reported by this technique. However, application of these coatings is limited to very few substrates that are resistant to attack by m-cresol, such as PET and glass. A further disadvantage may be seen in the fact that some of the m-cresol remains in the film and potential toxicological problems arise both during process and later use. Polyaniline doped with dodecyl benzene sulfonic acid offers processability from xylene and may be better suited for industrial applications. However, the maximum solubility is low (0.5% w/w[5]). Dodecyl benzene sulfonic acid has also been shown to plasticize polyaniline and form melt processable complexes[6]. More recently, an emulsion polymerization process for preparation of an organically soluble polyaniline has been reported [7]. This process uses dinonylnaphthalene sulfonic acid as dopant and the reaction is carried out in 2-butoxyethanol. The polyaniline/2-butoxy ethanol solution is readily soluble in xylene and other selective organic solvents for further processing. The technique represents a major advancement in solution processing of polyaniline, however limitations can be seen in its solubility in select organic solvents and environmental friendly formulations.

Processing of Polyanilines Using Dispersion Techniques

Processing of polyaniline or other intrinsically conductive polymers (ICPs) by dispersing the doped conductive polymer in a processable non-conductive matrix offers a direct and practical route. Polyaniline is prepared in the conductive form by oxidative polymerization of aniline in the presence of a protonic acid. Conversion to the neutral form for processing requires dedoping of the conductive polymer using an alkaline solution - a secondary step. While other processing techniques rely on a specific dopant, solvent or a particular form of polyaniline, dispersion techniques offer the freedom to choose the matrix into which the conductive polymer is dispersed. This key feature provides the ability to tailor performance properties of the resulting coating or blend. No other processing technique offers this flexibility to prepare conductive materials based on polyaniline or other ICPs. The technique is suitable for preparing coatings and melt processable blends. A detailed discussion on understanding of processing of intrinsically conductive polymers using dispersion techniques has been reported earlier[8,9].

In this technique, the intractable doped polyaniline is dispersed in the solution or melt of a non-conductive processable matrix. The resulting processable dispersion is ready for use without further modification. Key advantages of dispersion techniques for processing polyaniline are:

- Direct and practical technique
- Processable in the doped form
- Independent of the chemistry
- Solution and melt processable
- Tuned conductive coatings

- Control over performance properties of the coating/blend
- Enhanced environmental stability especially in thin films

Conductive Coatings/Solution Processing. Conductive coatings can be prepared by dispersing doped polyaniline in a solution of a non-conductive film forming resin in common organic solvents [10-12]. Coated substrates and films are prepared by applying the dispersion and flashing off the solvent. Coatings can also be prepared by dispersing polyaniline in polymerizable monomers, oligomers and polymer precursors. Such formulations are environmentally friendly, since there is no volatile organic content (VOC) and 100% of the formulation is converted to film. Film formation is brought about by heat or UV/EB radiation curing. The powdery conductive polymer, is broken down to primary particles of the conductive polymer, in the range of 200 to 300 nm and encased in the polymer matrix during the dispersion process. Coatings thus prepared exhibit good transparency and a characteristic green color. Due to the excellent dispersion quality, the coatings are often indistinguishable from true solutions by visual observation. The conductivity of the coating composition depends on processing, concentration of polyaniline and the percolation behavior of polyaniline in the film forming matrix. While polyaniline offers the electrical properties, other properties such as transparency, abrasion resistance and flexibility can be tailored by proper choice of the film forming matrix. Figure 1 shows typical transmission spectra of polyaniline coating prepared by dispersion of polyaniline.

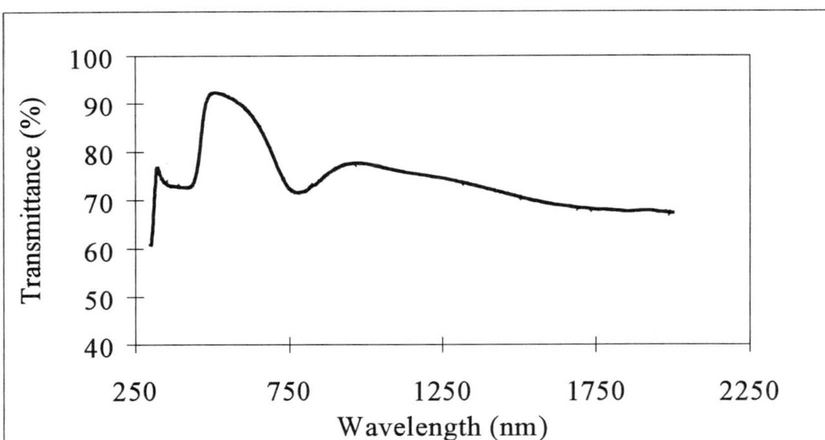

Figure 1. Spectra of polyaniline coating prepared by dispersion of polyaniline in polymethyl methacrylate

Figure 2 shows a typical percolation curve for polyaniline in a film forming matrix. The resistivity of the system remains unchanged until a critical volume fraction of the polyaniline is reached, at which point there is a sudden very large decrease in the resistivity of the system. With further addition of the polyaniline, there is only a marginal change in the resistivity of the system and the resistivity of the system is saturated. The critical volume fraction of the conductive filler for the sudden onset of conductivity is referred to as the percolation threshold. Typically this occurs with a change of conductivity of 10^{-12} to 10^{-5} S/cm within a change of volume fraction of conductive phase of 0.5 to 3 volume percent.

By controlling the composition and the amount of polyaniline, the electrical properties could be tailored over a range of 100 to 10^9 ohms/square.

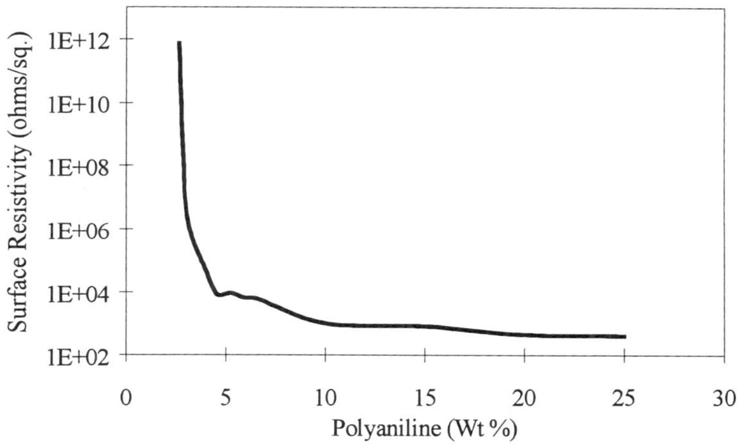

Figure 2. Percolation curve for polyaniline in polymethyl methacrylate coating system.

The insulating film forming matrix, in addition to providing processability, enhances the environmental stability of polyaniline by encasing it. Coatings prepared with dispersion technology withstand contact with water for an extended period of time and even show resistance to slightly alkaline atmospheres [12]. Neat polyaniline coatings, on the contrary, lose their conductivity on exposure to environments with pH greater than about 4. Other desired properties of industrial coatings such as abrasion resistance, chemical resistance, flexibility etc. are controlled by judicious choice of the polymer matrix into which polyaniline is dispersed. UV/EB cured coatings provide exceptional abrasion resistance. Coatings with pencil hardness of greater than 3 have been prepared using UV/EB formulations.

Tuned Conductive Coatings. One of the major uses of conductive polymer coatings is in electrostatic discharge protection. For effective electrostatic discharge protection, the ideal situation is one wherein the electrical charges are dissipated nearly as rapidly as they are generated. In order to do this effectively, the surface resistance of the packaging material should be less than 10^{11}

ohms/square and preferably between 10^5 to 10^{10} ohms/square, even under extreme climatic conditions. With a very narrow percolation threshold shown in Figure 2, it has been very difficult to control the surface resistivity of polyaniline coatings in the 10^5 to 10^{10} ohms/square range.

Recently, we have observed that use of amine, hydroxy functional and other select acrylate film forming matrices exhibit relatively broad percolation behavior[13]. By prudent choice of functionalized film forming matrices, it is now possible to make polyaniline coatings with surface resistivity in the 10^5 to 10^{10} ohms/square range with tight control. Figure 3 compares the percolation curves for polyaniline in polymethyl methacrylate and hydroxy functional and other film forming matrices which clearly demonstrate the ability to tune the surface resistivity of polyaniline coatings.

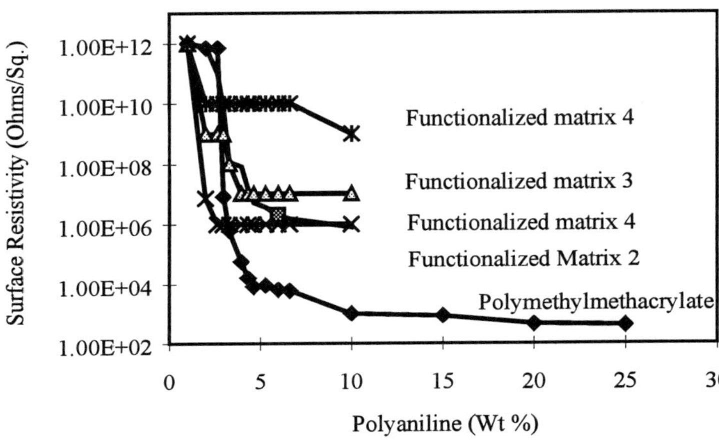

Figure 3. Percolation curves for polyaniline in various functionalized film forming matrices compared with polymethylmethacrylate.

Although not wishing to be bound by any theory, it is believed that these conductivity controlling materials, disrupt the normal percolation behavior of the dispersed conductive polymers in the insulating film forming polymer matrices, thereby causing only gradual or continuous increase in the conductivity of the matrix as a function of the loading of the conductive phase in the mix. It is also possible that the polymer matrix interacts with polyaniline causing controlled undoping of polyaniline. Compositions show surprisingly broad percolation thresholds and in some cases the disappearance of the percolation threshold over a wide range of concentration of the conductive polymer. This gradual increase in conductivity as opposed to abrupt rapid increase in conductivity provides for reproducible control and tunability of conductivity in coatings and blends in the range of 10^5 to 10^{10} ohms/square.

Melt Processing. Melt processability has been demonstrated for sometime now [8,14]. In this technique, polyaniline is dispersed in the melt of a thermoplastic matrix. The resulting blends can be processed by a variety of melt processing techniques such as extrusion or injection molding. The thermal stability of polyaniline is the limiting factor in the selection of the matrix thermoplastic polymer. Polyaniline should be stable at the processing temperature of the thermoplastic polymer it is blended in. The thermal stability of polyaniline is strongly dependent on the choice of the dopant [15-17]. Toluene sulfonic acid doped polyaniline, with a stability of greater than 230°C, still falls short of processing temperature for many engineering plastics. Some thermoplastic matrices such as polycaprolactone, polyvinyl chloride, chlorinated polyethylene and PETG show very low percolation thresholds. Polar additives, especially, lactones have been shown to enhance the conductivity of these blends by more than an order of magnitude [18]. Figure 4 shows the effect of butyrolactone and N-methyl pyrrolidone on the percolation behavior of polyaniline. Conductivities in the range of 25-80 S/cm have been demonstrated with blends containing 40% polyaniline.

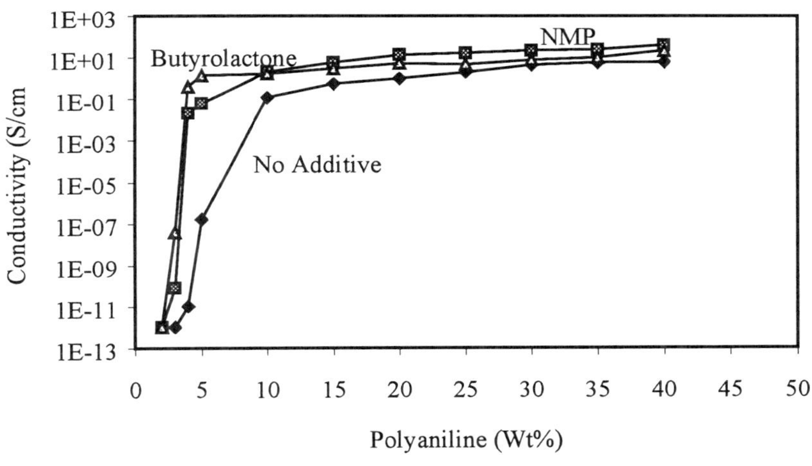

Figure 4: Effect of N-methyl pyrrolidone (NMP) and Butyrolactone on percolation and conductivity of polyaniline-polymethacrylate blends.

Colored Conductive Compounds and Coatings. Due to the dark color of polyaniline, formulations derived from it are deep green to black in color. Blends/coatings containing polyaniline in combination with titanium dioxide and other pigments with high tinting strength, exhibit light green color with little or no loss in conductivity. In some cases, the blends show higher conductivities than compositions containing no additive [19]. A possible explanation is the increase in the volume fraction of the conductive polymer due to the high specific gravity of inorganic pigments. The color of these compounds/coatings can be further enhanced by use of other colorants. Figure 5 shows the reflectance of two coatings, one containing polyaniline and the other containing a mixture of polayniline and titanium dioxide with comparable surface resistivities of around 750 ohms/square. The coating containing titanium dioxide is considerably lighter and displays a characteristic spectra of a green color. The spectra of coating containing only polyaniline displays very low and flat reflection characteristic of black color. Thus in addition to enhancing the appearance, the technique provides for cost advantage in coatings and blends of polyaniline where transparency is not needed.

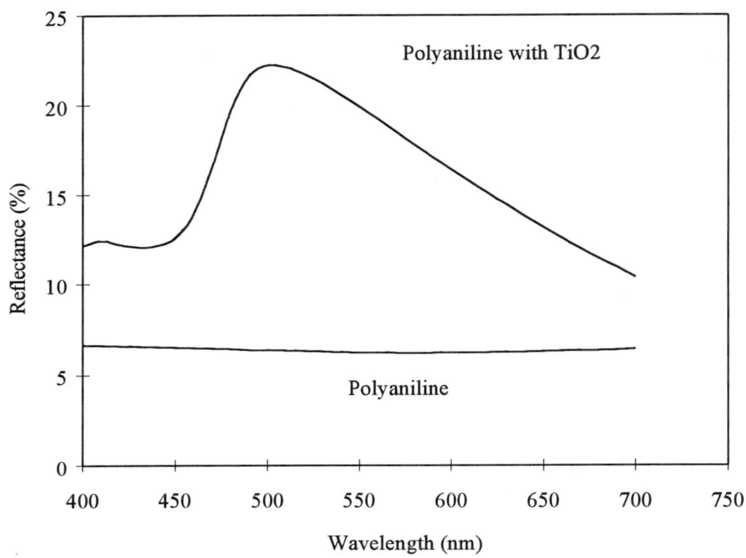

Figure 5. Reflectance spectra of polyaniline and polyaniline/TiO_2 coatings

In transparent thin films, polyaniline coatings exhibit the characteristic light green color as shown in Figure 1. Other transparent colored coatings can be prepared by combining polyaniline with conventional dyes and pigments. Figure 6 shows the optical spectra of a variety of transparent colored conductive coatings in comparison with the characteristic green color. In most cases the conductivity of the coating is unaffected. The neutral gray colored coating is particularly appealing since it offers relatively high degree of transmission over the entire visible range.

Applications

Polyaniline compounds and coatings find application in electrostatic dissipation, anti-corrosion coatings, EMI Shielding, Sensors, Microelectronics, and the like.

Electrostatic Dissipation and EMI Shielding. Coatings and films in particular find application in a variety of electrostatic dissipative (ESD) applications ranging from electronic component packaging to static safe labels to clean benches and partitions in clean room applications. When used in ESD packaging, excellent transparency of these coatings allows identification of components without opening the package. Coatings can be applied to plastic and other substrates using conventional coating techniques such as rotogravure, spray and spin coating. Using the rotogravure technique, extremely thin coatings can be applied as a highly transparent green colored film with surface resistivity in the 10^2 to 10^7 ohms/square range. Coated sheets can be thermoformed, with only marginal loss in conductive properties.

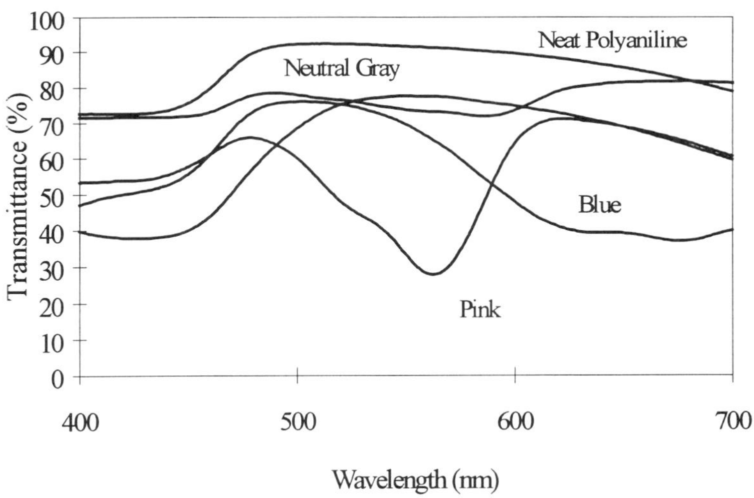

Figure 6. Optical spectra of transparent colored conductive coatings with polyaniline

Static Dissipative Lables. Static dissipative labels find application in electronic and PC board industry. Static discharge to the tune of 10,000 V have been reported with conventional labels at the removal of adhesive liner prior to label application and removal from circuit board. Polyaniline coatings have been shown to provide effective protection with reducing the discharge to 50 to 500 volts.

EMI Shielding and Other Applications. Polyaniline compounds can offer higher conductivity and improved processing over other filled conductive compounds. While excellent EMI shielding has been reported for polyaniline compounds, at the present time the conductivity of polyaniline falls short of commercial interest. With advances in synthesis of conductive polymers with higher conductivity, this may be an attractive area of application for conductive polymers. Polyaniline compounds also find application in industrial clean rooms where sloughing is a problem. Prototype antistatic footwear have been manufactured and tested with excellent ESD properties[20].

Corrosion Prevention. Polyaniline has been extensively investigated for its corrosion inhibition properties[21,22]. It has been shown to provide passivation to steel and other substrates by forming a thick impervious layer of oxide on the surface of the metal. In addition to providing passivation, the coating must offer good adhesion and blister resistance needed for long term protection. The dispersion technique is especially suited for preparing coatings for corrosion protection where choice of conventional organic coatings such as epoxies and urethanes provide double benefit. In addition to the passivation provided by polyaniline, the matrix resin provides the toughness of conventional coatings that the coating industry relies on.

Figure 7 shows the corrosion effectiveness of several polyaniline coatings compared with a two-component epoxy and a 16% zinc chromate coating. The coatings were sprayed on to cold rolled steel and top coated with 2-component epoxy. The testing was carried as per ASTM-B117 accelerated corrosion test and evaluated using ASTM-D1654, D610 and D714.

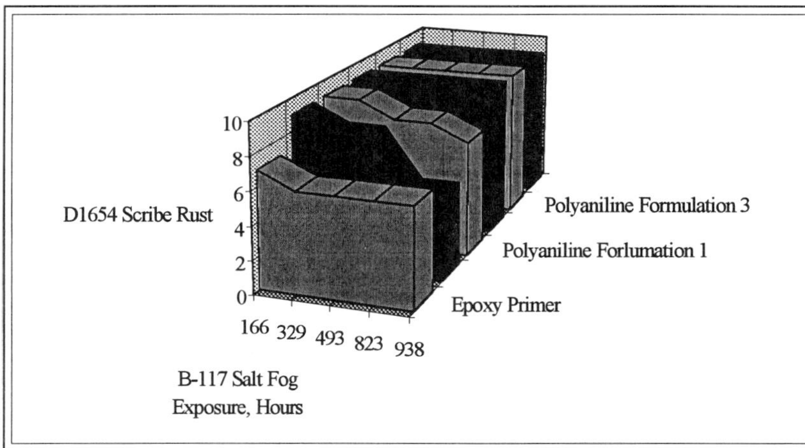

Figure 7. Corrosion Protection of polyaniline coatings compared with 2-component epoxy and zinc chromate coatings

Acknowledgments

The author wishes to acknowledge Perry Durbin, Tim Chen, John Campbell and Eric Jackson for their assistance in part of the study and analytical testing.

References

1. Angelopoulos, G.E. Asturias, S.P. Ermer, A.Ray, E.M. Scherr, A.G. MacDiarmid, M. Akther, Z.Kiss and A.J. Epstein, Mol. Cryst. Liq. Cryst., 160:151-163 (1988)
2. R.V.Gregory, "Solutions Processing of Conductive Polymers" in Handbook of Conducting Polymers, 2nd edition Ed. Terje A. Skotheim, Ronald L.Elsenbaumer, John R. Reynolds, Marcel Dekker Inc., 1998
3. Y. Cao, P.Smith and A.J. Heeger, Synth. Met. 48:91-97 (1992)
4. A.J. Heeger, Synth. Met., 55-57:3471-3482 (1993)
5. Y.Cao, P.Smith and A.J. Heeger, US Patent 5,232,631 (1993)
6. Toivo Karna, Jukka Laakso, Timo Niemi, Heikki Ruohonen, Esko Savolainen, Helge Lindstrom, Esa Virtanen, Olli Ikkala, Alejandro Andreatta, US patent 5,340,499
7. P.J. Kinlen, J. Liu, Y.Ding, C.R. Graham and E.E. Remsen, paper presented at "International Seminar on the Technology of inherently Conductive Polymers. March 2-4, 1998, Hilton Head, SC.
8. B.W. Wessling, H. Volk, W.R. Mathew and V.G. Kulkarni, Mol. Cryst. Liq. Cryst., 160:205-220 (1988)
9. B.Wessling, "Dispersion as the Key to Processing Conductive Polymers" in Handbook of Conducting Polymers, 2nd edition Ed. Terje A. Skotheim, Ronald L.Elsenbaumer, John R. Reynolds, Marcel Dekker Inc., 1998
10. V.G. Kulkarni, W.R.Mathew, L.W. Shacklette and J.C. Campbell, US patent 5,494,609
11. V.G. Kulkarni, " Transparent Conductive Coatings" in Handbook of Conducting Polymers, 2nd Edition, Ed. Terje A. Skotheim, Ronald L. Elsenbaumer and John R. Reynolds, Marcel Dekker Inc., 1998
12. V.G. Kulkarni, J.C. Campbell and W.R. Mathew, Synth. Met. 55-57:3780-3785 (1993).
13. V.G. Kulkarni, to be published
14. V.G. Kulkarni, B.W. Wessling, US patent 5,217,649
15. V.G. Kulkarni, L.D. Campbell and W.R. Mathew, Synth. Met., 30(3):321-325 (1989)
16. V.G. Kulkarni, Thermochimica Acta, 188(2):265-272 (1991)
17. V.G. Kulkarni, B.W. Wessling, S. Blaettner, Synth. Met., 41(3):1009-12, (1993)
18. V.G. Kulkarni, J.C. Campbell, US patent 5,595,689
19. V.G. Kulkarni,, W.R. Mathew, ANTEC'89, 390
20. C.McCall, J.A. Dalton, G. Billis and V.G. Kulkarni, Proceedings of SPE's Annual Technical Conference, ANTEC 1995, Boston, MA.
21. B.W. Wessling, Adv. Mater. 6(3):226-118 (1994)
22. W.K. Lu and R.L. Elsenbaumer in Handbook of Conducting Polymers, Second Edition, Ed. By Terje A. Skotheim, Ronald L. Elsenbaumer and John R. Reynolds, Marcel Dekker Inc., 1998

Chapter 13

Doped Polymer Light-Emitting Devices and Sensors for Volatile Organic Compounds

Alan G. MacDiarmid, Feng Huang, and Jing Feng

Department of Chemistry, University of Pennsylvania, Philadelphia, PA 19104

The presence of even traces of doped (ionic) species in MEH-PPV light-emitting devices, either in the light-emitting layer or in a polyaniline layer adjacent to an electrode, can greatly reduce electron and/or hole injection energies and mask or render irrelevant the conventional work functions of the electrode materials. This effect can improve certain important parameters of the device including light emission in a reverse bias mode in addition to the conventional forward bias mode. Reversible gas sensors using octaaniline or tetraaniline for detection of volatile organic compounds (VOCs) in the atmosphere readily detect toluene vapor admixed with air down to ~35 ppm by measuring resistance changes with a simple, inexpensive ohmmeter. They exhibit excellent sensitivity, reversibility, environmental stability and have little response to water vapor, a common source of interference.

During recent years thin films of organic conjugated polymers, "electronic polymers", have been increasingly used in a variety of devices including light emitting devices (LEDs) (*1-28*) and sensors for volatile organic compounds (*29-50*).

We have been particularly interested in light emitting devices involving polyanilines (*1-3*), more recently those which contain ionic species (*4-12*) and in reversible sensors for volatile organic compounds (*40-45*). Application in both of these areas have in general involved the use of spun thin films of the polymers. More recently, however, "self-assembled" multilayer light-emitting devices fabricated by a dip-coating technique from solution have gained increasing attention (*13*).

Light Emitting Devices

As is well known, since their discovery by Richard Friend's group at Cambridge University in 1990 (*14*), LEDs involving conjugated electronic polymers (*15, 16*) and also those involving films of organic molecules (*17*) as the light-emitting species, basically involve the injection of an electron into the light-emitting film from one side of the film and the injection of a positive hole from the other side of the film. The electrons migrate through the film towards the positive (anode) electrode from a D.C. power source while at the same time, the holes migrate towards the negative (cathode) electrode.Under ideal circumstances, the electrons and holes would have the same mobility and would recombine in the center of the film with the emission of light having an energy approximately equal to that of the bandgap in the case of an emissive polymer. The effect is phenomenologically the reverse of a photovoltaic cell in which light is converted to an electric current. The effect is illustrated diagramatically in Figure 1.

Electrons are injected most readily from electropositive (low work function) electrodes such as Al or Ca while holes are injected most readily from electronegative (high work function) electrodes such as gold or transparent glass electrodes coated with a thin layer of conducting indium tin oxide, "ITO electrodes".

Obviously if the electric current were carried entirely by holes or entirely by electrons, no light emission would occur. Ideal light emission is promoted if the current is carried by an equal number of electrons and holes, which itself is determined by the relative mobility of the electrons and holes and by the relative number of holes and electrons injected from their respective electrodes. Unfortunately the mobility of electrons and holes in a given material usually differ significantly as do also the ease with which electrons or holes are injected from a given electrode material. Hence, much effort has been expended on interspersing layers of other materials in LEDs which increase the mobility of electrons or holes and or their ease of injection from an electrode. Multilayer structures can also be useful, for example, in blocking the transport of electrons at the site where electron and hole transport layers interface. Electrons and holes therefore concentrate at the interface promoting increased light emission (*15, 51*).

One way of increasing the ease of electron or hole injection has involved the fabrication of electrochemical LEDs (*4-7, 18*) in which an electrochemical reaction occurs during operation by which the polymer adjacent to the electron injection electrode becomes n-doped (reduced) while the interface adjacent to the hole injection electrode becomes p-doped (oxidized). The doped-polymer/electrode interface is believed to possibly form an ohmic junction which greatly reduces charge injection barriers. *It would be stressed that these types of cells produce a chemical change in composition of the polymer i.e. there is a change in the number of positive or negative charges on the polymer backbone.* The number of attendant counter anions or cations respectively, "dopant" ions, also change. Before a potential is applied, the polymer is "non-doped" and has no charge and is therefore associated with no counter dopant ions. Upon application of a potential, ionic species are immediately formed of the type (polaron/bipolaron)$^+$A$^-$ and (polaron/bipolaron)$^-$M$^+$ where A$^-$ and M$^+$ are attendant dopant anions and cations,respectively.

Very recent elegant studies (*19*) involving the chemical doping of the polymer surfaces in contact with an electrode have confirmed conclusively the postulated electrochemical doping described above. They have also shown that the effects of the work functions of the electrodes which inhibit both injection of electrons or holes can be essentially eliminated, the thin doped layers screening the emissive polymer from seeing the real work function of the electrodes.

With one exception (*20*) all polymer LEDs studied have used non-doped polymer as the emissive layer. However, it has been shown recently that dramatic improvement of the performance of a polymer LED is obtained upon *light* doping of polystyrene-poly(3-hexylthiophene) (PS-P3HT) copolymer by $FeCl_3$ (*20*). The cause of the unusual phenomena is not yet clear, although it has been suggested very recently, that ionic species improve the properties of organic LEDs when purposefully incorporated as an attachment to the emissive polymer backbone (*21,22*) or even when introduced unintentionally during the synthesis of the emissive polymer (*23*). Indeed it has only recently been shown (*24,25*) that poly(phenylene vinylene) (PPV) becomes partly p-doped during its synthesis from its precursor in an LED because of chemical reaction of the precursor polymer with the indium tin oxide coating on the ITO glass, resulting in a small contact resistance at the ITO/PPV interface.

As a continuation and extension of our studies on light-emitting devices containing a polyaniline layer in their fabrication(*1-6*), we have become especially interested in the use of a a *doped* polymer as the light emitting layer (*8-12*) or as a layer adjacent to an electrode (*10-12*) resulting in the following experimental observations:

(i) Even unsuspected trace amounts of residual doping may impart a new set of properties to a "conventional" light-emitting polymer.

(ii) The doped polymer may also be a non-emissive polymer adjacent to an electrode.

(iii) No added ionic salt, (e.g. $Li^+(CF_3SO_3)^-$ is required as in light-emitting electrochemical cells (LECs).

These observations are consistent with a model in which:

(i) The above effects result from (polaron/bipolaron)$^+$ A^- p-doped ionic species initially present in the polymer, the number of which does not change during operation.

(ii) Potential-induced diffusion of A^- ions controls the properties of the systems.

We have selected poly(2-methoxy-5-(2'-ethyl-hexyloxy)-1,4-phenylene-vinylene), MEH-PPV, a typical light-emitting conjugated polymer, for our investigation.

It should be noted that all our devices have been fabricated and tested in air.

The effect of intentionally introducing (ionic) inorganic salts into *molecular* LEDs has recently been reported (*26*); some of the resulting observations may be related to those reported for electrochemical cells involving polymer light emitting materials.

Results

Lightly Doped Emissive Polymer (*8-12*). It has previously been reported (*52*) that MEH-PPV is doped by iodine vapor and that it becomes spontaneously de-doped at room temperature presumably due to the loss of iodine vapor. Recent UPS studies (*53*) are consistent with doping of the type

$$(\text{MEH-PPV})_x + 1.5\,(xy)\,I_2 \rightleftarrows (\text{MEH-PPV})_x^{+y} + (xy)\,(I_3)^-$$

We find that the UV/Vis spectra of the nondoped and iodine doped MEH-PPV film after pumping in vacuum (*9*) are also consistent with the above reaction although we noted that the doping level was significantly reduced by the vacuum treatment. Also, it is not known whether the iodine might have reacted chemically in whole or in part with the aluminum during its deposition during fabrication of the device described below. It was concluded that the iodine doping level of the MEH-PPV beneath the aluminum is certainly very small.

We find(*8-12*) that the performance of polymer light-emitting diodes (LEDs) having the configuration Al/MEH-PPV/ITO where MEH-PPV=poly(2-methoxy-5-(2'-ethyl-hexyloxy)-1,4-phenylene-vinylene) are drastically altered by light iodine doping of the emissive MEH-PPV layer. For nondoped Al/MEH-PPV/ITO devices, light emission has been reported only in the forward bias mode (*27*) (λmax=~580 nm) as confirmed previously in our earlier studies (*5*) (λmax=~575 nm). The reported turn-on voltage was ~10 V (*6,27*). However, for our lightly iodine doped MEH-PPV device, current was observed in both forward *and* reverse bias modes (See Figure 2). Light emission was also observed in forward and reverse bias modes. The electroluminescence (EL) spectra of the doped LED device are shown in Figure 3. The λmax of light emission was ~580 nm for both forward and reverse bias modes, experimentally identical to that observed for the nondoped MEH-PPV, thus

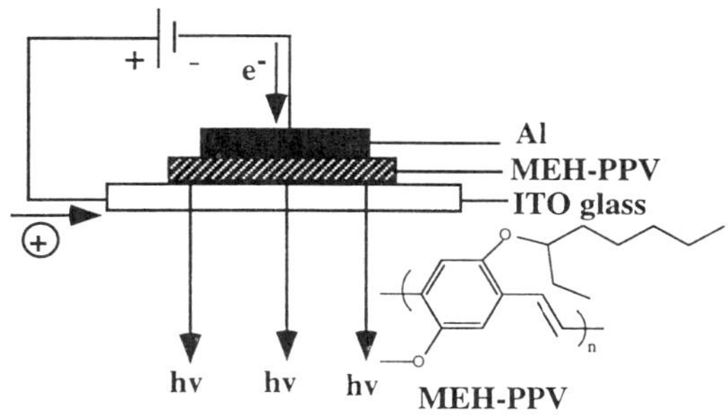

Figure 1. Schematic diagram of an Al/MEH-PPV/ITO LED device.

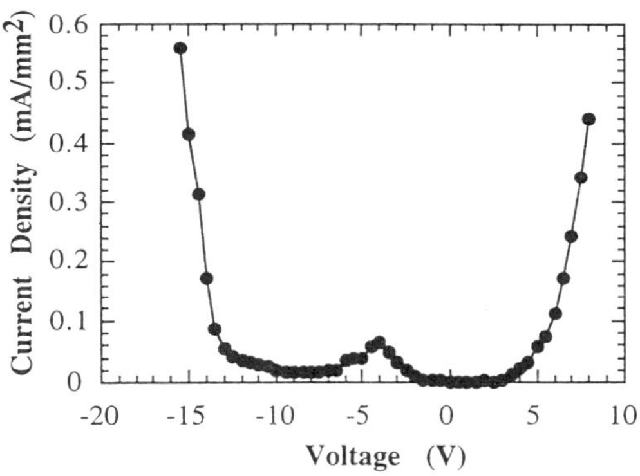

Figure 2. Current-Voltage characteristics of an Al/I_2*-doped* MEH-PPV/ITO device. Reproduced with permission from reference 9. Copyright 1997 American Insitute of Physics.

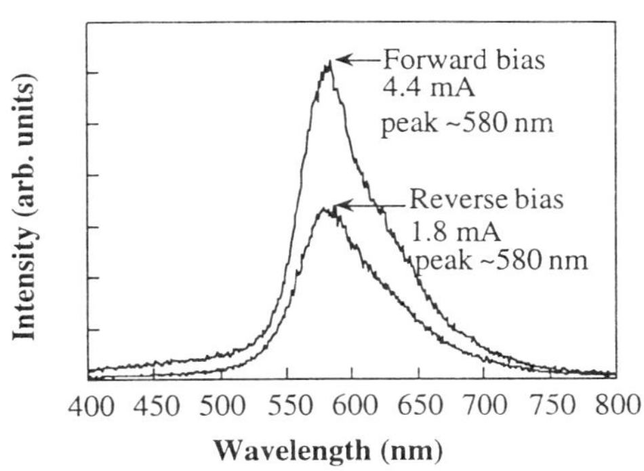

Figure 3. Electroluminescence spectra of an Al/I_2*-doped* MEH-PPV/ITO device in forward and reverse bias modes. Reproduced with permission from reference 9. Copyright 1997 American Insitute of Physics.

showing that the same light-emitting species and/or phenomenon were present both in the nondoped and in the *iodine doped* MEH-PPV. The turn-on voltage at forward bias is ~5 V and ~12 V in the reverse bias mode. It is therefore evident that the Al/I_2-*doped* MEH-PPV/ITO device behaves very differently from the nondoped conventional MEH-PPV LED device. The external quantum efficiency of the I_2-*doped* LED device was ~8x10^{-3}% (measured at 17 V), an increase of approximately one order of magnitude compared to a similarly fabricated nondoped LED device which had an external quantum efficiency (*5*) of ~4x10^{-4}% measured at a higher voltage (~25 V). An integrating sphere was not used in the efficiency studies. The power conversion efficiency (*28*) of the above I_2*-doped* MEH-PPV device is ~30 times larger (~9x10^{-4}%) than that of the device (~3x10^{-5}%) using non-doped MEH-PPV, reflecting the lower voltage used when measuring the external quantum efficiencies. The above observations demonstrate that upon *light* iodine doping the MEH-PPV emissive polymer layer, the performance of the LED is improved in terms of turn-on voltage and external quantum and power conversion efficiencies and light-emission is also observed in the reverse bias mode.

An unusual phenomenon is observed in this LED system in that when a potential is applied such that light is emitted, the intensity of light emission slowly increases with time during several minutes followed by a slow decrease in intensity as illustrated in Figure 4. Of particular interest is the fact that the current *increases* with time as the light intensity *decreases*, exactly the opposite of what might be intuitively expected! Furthermore, the rate at which the intensity of emitted light increases is highly dependent on the applied potential as shown in Figure 5 strongly suggestive of a potential-induced ionic diffusion process. The 16 V study was recorded after the 12 V study had been completed at *exactly* the same "spot" on the LED device. The maximum intensity at 16 V was in excess of an order of magnitude greater than observed at 12 V.

An unusual effect is apparent in Figure 6. The LED was operated at 14 V for 30 seconds (line 1) and the applied potential was turned off for a "rest" period of 5 minutes at room temperature. The potential was again applied and as can be seen, the initial light intensity of the second line was experimentally identical to the initial light intensity of line 1. After 30 seconds, the applied potential was again turned off for a "rest" period of 5 minutes. Re-application of the potential, again resulted in the initial light intensity of line 3, being experimentally identical to the initial light intensities of lines 1 and 2. The similarity of lines 1, 2 and 3 show that no significant degradation had occurred during the study which was performed in laboratory air.

Doped Polyaniline Adjacent to an ITO Electrode (*10-12*). Since it is difficult to determine a doping level in iodine-doped MEH-PPV, we directed our further investigations to polyaniline in which we have had considerable experience in actually determining doping levels. It should be stressed that these studies involve doping of a *non*-emissive polymer layer; however, many of our earlier studies, as mentioned briefly below, have resulted in observations which have certain resemblance to those described above.

We have recently shown (*1,2*) that the incorporation of spun films of polyaniline (emeraldine base, "EB") adjacent to the ITO and/or the metal electrodes in LED devices incorporating the conjugated light-emitting polymer poly(2,5-dihexadecanoxy phenylene vinylene pyridyl vinylene), (PPV.PPyV) in configurations such as metal/EB/PPV.PPyV/ITO, metal/PPV.PPyV/EB/ITO, or metal/EB/PPV.PPyV/EB/ITO lead to remarkable improvement in performance. In the case of metal/EB/PPV.PPyV/EB/ITO (where metal=Al, Cu or Au) light emission in both forward and reverse bias modes is observed (*1,2*)! Thus electrons are readily injected from the high work function Cu and Au electrodes and holes are injected

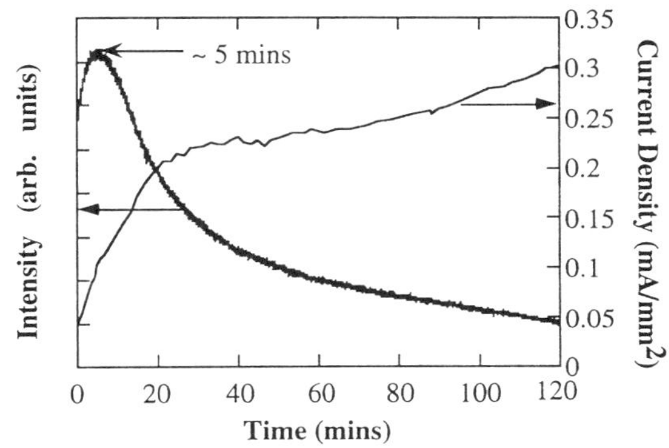

Figure 4. Electroluminescence intensity and current density vs. time for an Al/I_2*-doped* MEH-PPV)/ITO device at 7V in the forward bias mode. Reproduced with permission from Synthetic Metals (in press). Copyright 1998 Elsevier Science Ltd.

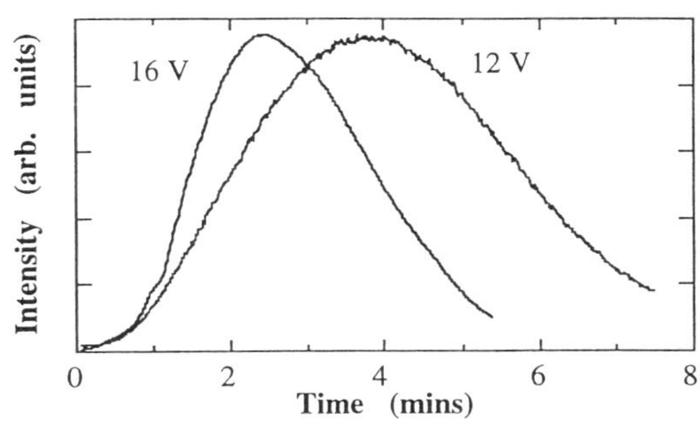

Figure 5. Normalized electroluminescence intensities of light emission (at ~580nm) vs. time for Al/I_2*-doped* MEH-PPV/ITO device operated at 12V and 16V in the forward bias mode respectively. Reproduced with permission from reference 12. Copyright 1997 Society of Photo-Optical Instrumentation Engineers.

readily from the low work function Al electrode. Possible causes for this effect will be discussed in a later section.

We have also noted (*5,6*) a somewhat similar effect when MEH-PPV is used as the emissive layer. Thus in Al/MEH-PPV/ITO and Al/MEH-PPV/EB/ITO, light emission is observed in forward bias only for the first device but in forward and reverse bias modes for the second device analogous to the I_2*-doped* MEH-PPV device.

When we reported the first synthesis of the emeraldine base form of polyaniline 13 years ago (*54*) by the deprontonation (removal of HCl) from the emeraldine hydrochloride form of polyaniline by aqueous ammonium hydroxide, we were content with elemental C,H and N analyses which added up to ~99%. In some cases, we reported the Cl content (as HCl) as <0.4%. Few research groups investigating conducting or non-conducting forms of electronic polymers report elemental analyses and fewer still pay attention to small amount of impurities. It has generally been believed that the properties of interest in such polymers are not particularly sensitive to traces of impurities. However, we have recently found as shown below, that selected properties of interest in the LED field are *very* sensitive to traces of certain impurities.

In our previous studies using EB in LEDs, we had assumed, like all others, that the EB was “pure”. We now find this was not the case! Even very small,apparently unimportant differences (*55*) involving “over” and “under”-oxidized forms of polyaniline produced in the synthesis of doped or non-doped polyaniline can often lead to important changes in certain properties. On re-examining some of our previous studies (*1-7*) on light-emitting devices and in view of the results described below, we began to suspect that our apparently “pure” EB might contain traces of the doped polymer, which had not been completely undoped by our deprotonation of the initially-synthesized polyaniline salt according to the reversible process:

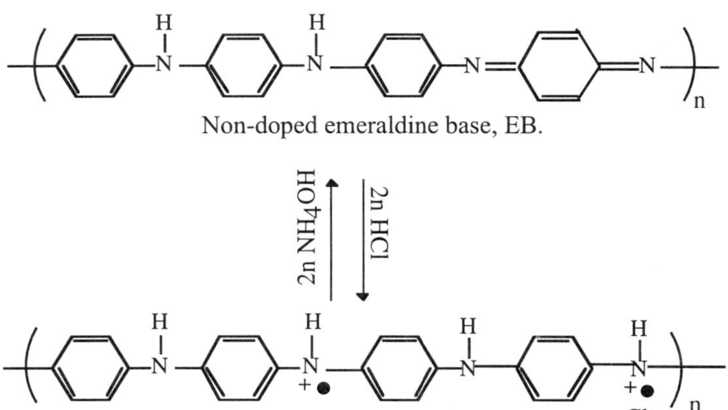

Non-doped emeraldine base, EB.

100% protonated (doped), "EB".

If all the imine nitrogen atoms in the EB are protonated, the resulting salt is said to be 100% prontonated. Protonation of the imine nitrogen atoms is immediately followed by an internal redox reaction to give the product in its polaronic form shown above.

Representative elemental analyses for chlorine (*56*) are given below. Deprotonation of the initial 100% doped salt with 0.1M aqueous NH_4OH:

(i) for 15 hours: 0.35 ± 0.05 wt% Cl corresponding to 1.76% doping.
(ii)for an *additional* 48 hours: 0.075 ± 0.015 wt% corresponding to 0.36% doping.
(iii) for an *additional* 48 hours (i.e. total deprotonation time of 4.6 days): 0.019 ± 0.003 wt% corresponding to 0.09% doping.

In this report, "EB" refers to emeraldine base which is partly (<~2%) doped with HCl. Where known, the level of doping (protonation) is specifically stated in each case. In examples employing 100% doped EB, d,l -camphorsulfonic acid was used as the dopant acid for convenience in spinning films.

In view of the observations below, we now believe that the effect of the "EB" layer in changing the characteristics of the PPV.PPyV and MEH-PPV devices mentioned above may be caused by traces of ionic material in the "EB" resulting from traces of HCl remaining from incomplete deprotonation during its synthesis.

The devices used in the present study had the configurations, Al/MEH-PPV/ITO, Al/MEH-PPV/"EB"/ITO and Al//MEH-PPV/*100% doped* EB/ITO. From Figure 7, it can be seen clearly that the intensity of emitted light is time dependent, taking ~4 seconds to reach maximum intensity when no polyaniline layer was employed, ~4.2 minutes when "EB" was employed and ~7.2 minutes when 100% *doped* EB was used. *The data given in Figure 8 are extraordinarily important since they highlight the large effect of a trace and the change in a trace of doping level on certain specific properties of the LEDs, all other variables being constant.* It can be seen that increasing the Cl^- content from 0.09% doping to 1.76% doping increases the time to reach maximum light intensity from 1.0 minute to 4.2 minutes! This is consistent with the less highly doped polyaniline having a smaller conductivity which promotes a greater field gradient in the polyaniline film which promotes a greater rate of diffusion of Cl^- ions which promotes a greater rate of increase of light intensity than the corresponding more highly doped film.

The information given in Figure 9 is qualitatively very similar to that for the I_2-*doped* MEH-PPV device (Figure 6) and is again consistent with re-establishment of ionic equilibrium at room temperature. The Al/MEH-PPV/*100% doped* EB/ITO device was first operated at 18 V for 60 seconds (1st line) and the applied potential was turned off for a "rest" period of 5 minutes at room temperature. The potential was again applied and as can be seen, the initial light intensity of the second line was approximately identical to the initial light intensity of line 1. After 60 seconds, the applied potential was again turned off for a "rest" period of 5 minutes. Re-application of the potential again resulted in an initial light intensity approximately the same as that of lines 1 and 2. As in the analogous case with I_2-*doped* MEH-PPV device (Figure 6), the similarity of lines 1, 2 and 3 show the absence of any significant degradation during the study.

A completely different result (Figure 10) however, was observed when an analogous study was performed with an Al/MEH-PPV/"EB"/ITO device in which the "EB" was slightly doped (1.76% doping level). In this case, the emitted light intensity did not decrease during the 5 minute "rest" period but continued on at the same intensity as the final intensity of the preceding curve, showing that re-establishment of ionic equilibrium at this low doping level was slow at room temperature. No apparent observable degradation occurred during the study.

A most important result is given in Figure 11 involving an Al/MEH-PPV/"EB"/ITO device in which the "EB" was 0.09% doped. The effect of applied potential on time to reach maximum light intensity was recorded *at exactly* the same "spot" on the device, i.e., all variables except the applied potential were constant. As can be seen, a change from 14.0 V to 11.0 V resulted in an increase in time from 0.17 minute to 1.0 minute to reach maximum intensity strongly suggesting a potential-induced ion diffusion process. Re-application of a potential of 14 V gave an identical curve.

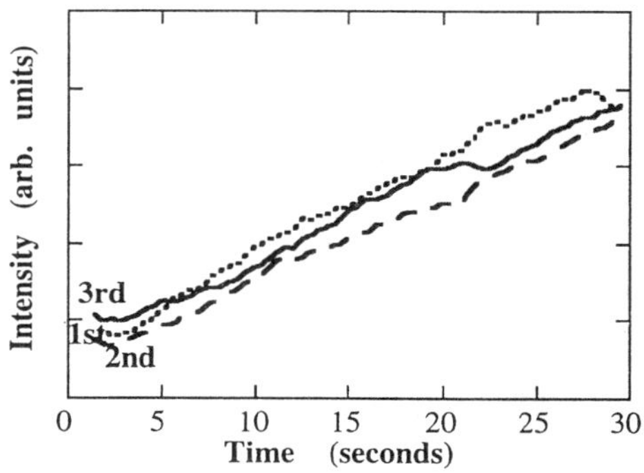

Figure 6. Al/I_2-*doped* MEH-PPV/ITO device: (1) "......" operated at 14V for 30 seconds; then turned off for 5 minutes. (2) "- - -" operated again at 14V for 30 seconds; then turned off for 5 minutes. (3) "——" operated again at 14V for 30 seconds. Reproduced with permission from reference 12. Copyright 1997 Society of Photo-Optical Instrumentation Engineers.

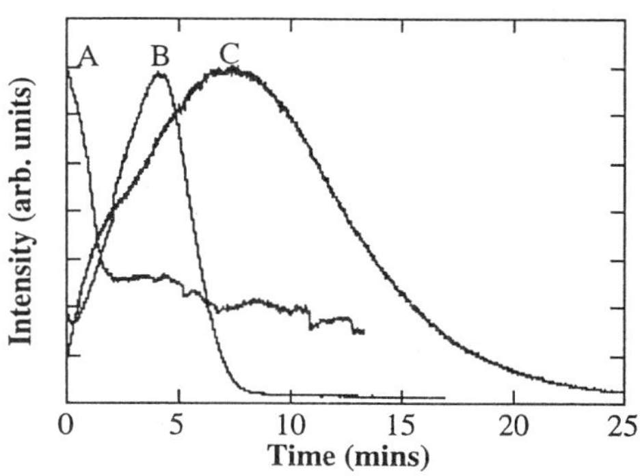

Figure 7. Normalized electroluminescence intensities of light emission (at ~580nm) vs. time for three types of devices operated at 11V in the forward bias mode, A: MEH-PPV; B: 1.76% HCl-doped "EB"; C: 100% *doped* EB. Reproduced with permission from reference 12. Copyright 1997 Society of Photo-Optical Instrumentation Engineers.

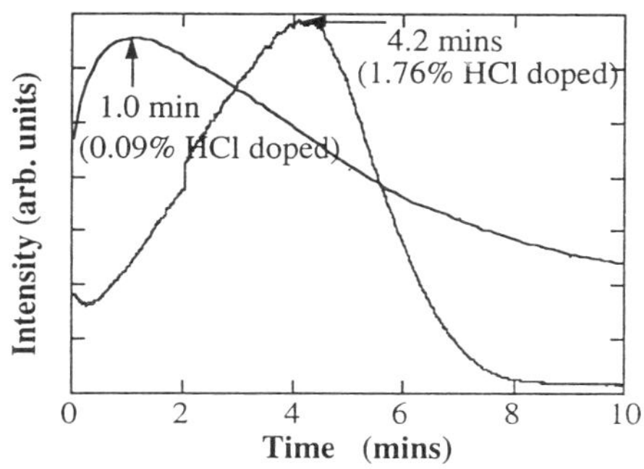

Figure 8. Time taken to reach maximum intensity of emitted light from Al/MEH-PPV/"EB"/ITO device at: the same applied potential (11.0 V); the same thickness of EB film (610±15 A); the same total thickness of (EB+MEH-PPV) film (1500 A). (Thickness measured by Tencor Profilometer). Reproduced with permission from Synthetic Metals (in press). Copyright 1998 Elsevier Science Ltd.

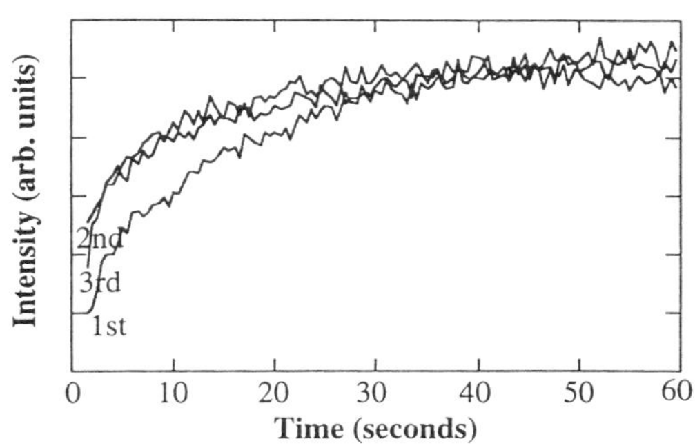

Figure 9. Al/MEH-PPV/*100% doped* EB/ITO device: (1) operated at 18V for 60 secs.; then turned off for 5 mins. (2) operated again at 18V for 60 secs., then turned off for 5 mins. (3) operated again at 18V for 60 secs. Reproduced with permission from reference 12. Copyright 1997 Society of Photo-Optical Instrumentation Engineers.

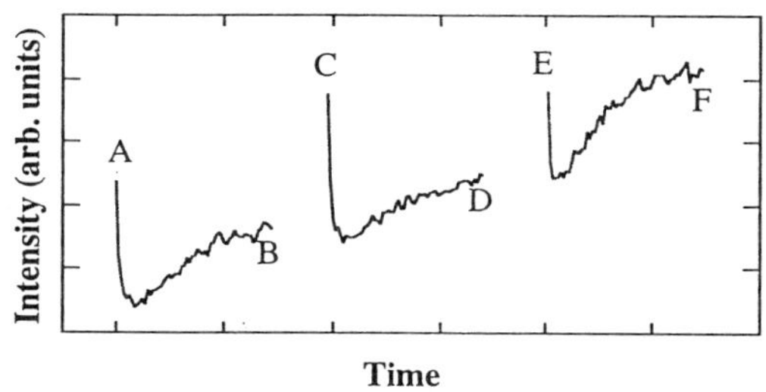

Figure 10. Al/MEH-PPV/*1.76% HCl-doped* "EB"/ITO device:(1) operated at 28V for 30 seconds (A-B;C-D;E-F). (2) turned off for 5 minutes (B-C; D-E). Reproduced with permission from reference 12. Copyright 1997 Society of Photo-Optical Instrumentation Engineers.

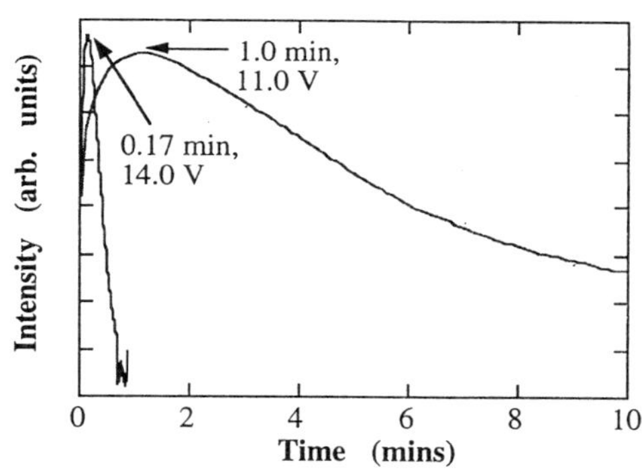

Figure 11. Normalized electroluminescence intensities of light emission (at ~580nm) vs. time for an Al/MEH-PPV/*0.09% HCl-doped* "EB"/ITO device operated at 11V and 14V in the forward bias mode respectively. Reproduced with permission from Synthetic Metals (in press). Copyright 1998 Elsevier Science Ltd.

An unusual effect, similar to that observed in the Al/*I_2-doped* MEH-PPV/ITO device (Figure 4) was also found in the devices Al/MEH-PPV/"EB"/ITO ("EB" was 0.09% doped), (Figure 12) and Al/MEH-PPV/*100% doped* EB/ITO (Figure 13), viz., as the light intensity decreased as a function of time, the current increased, again, exactly the opposite of what would be intuitively expected!

Particularly important information is given in Figure 14. An Al/MEH-PPV/*100% doped* EB/ITO device was operated at forward bias at 10 V for 30 minutes, the potential was then disconnected for a 30 second "rest" period and was then re-applied for a 30 seconds and then disconnected for a 5 minute "rest" period. The potential was then re-applied for a 30 second period. This procedure was repeated two more times. As can be seen, the light intensity increased significantly after each "rest" period highly suggestive of a polarization effect commonly observed in flashlights if the batteries are nearing discharge; a "rest" period results in an increased light intensity. It should be noted that a 26% loss in light intensity was observed during the 27 minutes following attainment of maximum light intensity (~3 minutes); however, 53% of the lost intensity was recovered during the "rest" periods. This clearly demonstrates that at least *a very significant fraction of the lost intensity was not due to an irreversible degradation process* even though the study was performed in air.

It might be noted that a recent detailed study of MEH-PPV LEDs involves measurements in an inert atmosphere (*26*). The life-time of devices in the present studies are comparable even though carried out in air.

Interpretation of Experimental Results

All the experimental results in the preceding section - those involving *I_2-doped* MEH-PPV and also those involving films of polyaniline, are consistent with the properties of both types of systems being governed by the potential-induced diffusion of dopant, A^- anions towards the positive electrode. Charge compensation of electrons and holes injected from the corresponding electrodes by mobile ions of appropriate charge occurs adjacent to the respective electrode.

Lightly Doped Emissive Polymer: Al/*I_2-doped* MEH-PPV/ITO Device (*8-12*). According to our model (Figure 15) which applies regardless of the chemical nature of the electrodes, injected electrons will find themselves in the vicinity of a positive charge on the polymer backbone, thus making it easier for the electron to be accommodated in the polymer backbone as compared to the case when an electron is entering a neutral polymer. Simultaneously with electron injection, the anion, A^- migrates towards the positive electrode. Hence there is a net potential-induced *diffusion* of A^- away from the negative electrode towards the positive electrode. Also simultaneous with electron injection, hole injection occurs at the positive electrode. The injection of a hole is facilitated by the presence of near-by A^- ions; *diffusion* of A^- ions towards the positive electrode provides the A^- ions necessary for compensation of the newly-injected hole. These effects lead to both reduced electron and hole injection energies, consistent with the lower turn-on voltage in the *I_2-doped* MEH-PPV LEDs (Figure 2) and the concomitant increased quantum efficiency previously described and observed reverse bias light emission (Figure 3), masking the effects of the work functions of the electrodes. It is also proposed that the increased concentration of A^- ions (high doping) at the positive electrode during forward bias operation induces metallic conductivity in the MEH-PPV promoting an ohmic or near ohmic junction at the electrode. Hence an even lower hole injection energy results, increasing still further the ease of hole injection at the positive electrode. The high doping level at the positive electrode is *not* caused by an increase in the number of (polaron/bipolaron)$^+A^-$ species since there is no source from which

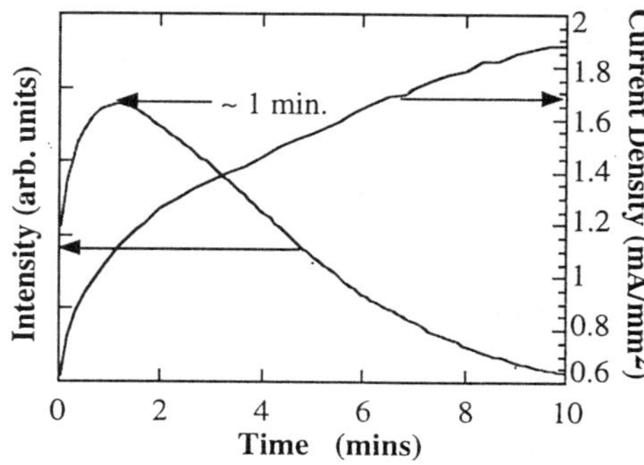

Figure 12. Electroluminescence intensity and current density vs. time for an Al/MEH-PPV/*0.09% HCl-doped* "EB"/ITO device operated at 11V in the forward bias mode. Reproduced with permission from Synthetic Metals (in press). Copyright 1998 Elsevier Science Ltd.

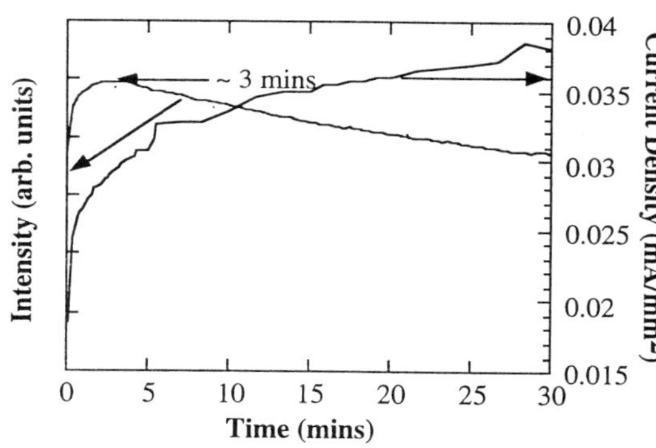

Figure 13. Electroluminescence intensity and current density vs. time for an Al/MEH-PPV/*100% doped* EB/ITO device operated at 10V in the forward bias mode.

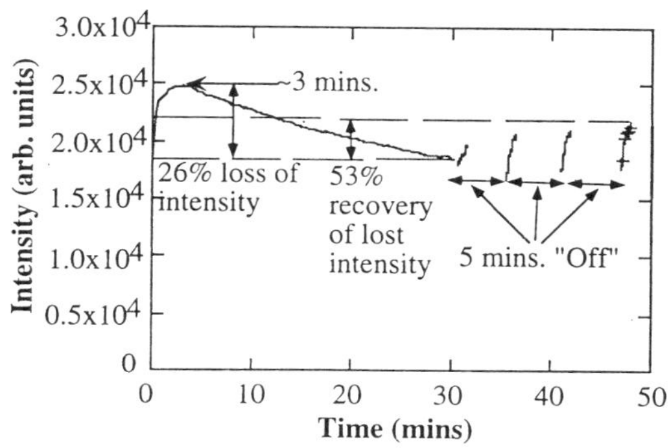

Figure 14. Electroluminescence intensities of light emission (at ~580nm) vs. time for an Al/MEH-PPV/*100% doped* EB/ITO device operated at 10V in the forward bias mode. Reproduced with permission from reference 12. Copyright 1997 Society of Photo-Optical Instrumentation Engineers.

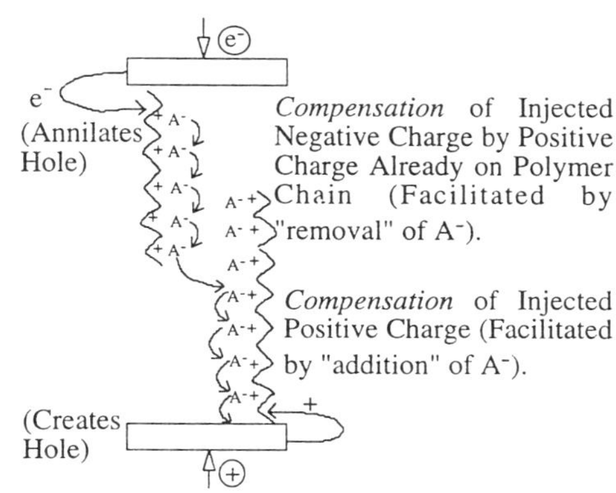

Figure 15. Schematic representation of ion-promoted injection potential reduction for electrons and holes, (A^- : anion). Reproduced with permission from reference 12. Copyright 1997 Society of Photo-Optical Instrumentation Engineers.

the additional A^- ions could come. The increased doping level, in the vicinity of the positive electrode (concomitant with decreased doping level in the vicinity of the negative electrode) is due to the potential-induced diffusion of A^- ions away from the negative electrode towards the positive electrode.

The important role of diffusion is demonstrated by the data in Figure 5 in which the time taken to reach maximum light intensity is reduced by application of a higher potential, resulting in a higher field gradient and hence increased rate of ion diffusion. This concept is also consistent with results given in Figure 4 in which the current *increases* as the light intensity *decreases.* As the steady state concentration of A^- ions near the positive electrode is approached, the resulting increase in concentration of A^- ions will promote an increased rate of hole injection. Simultaneously, the decreased concentration of positive charges on the polymer due to the decreased concentration of A^- ions at the negative electrode will decrease electron injection at the negative electrode. Essentially all the current will be carried by holes and hence no light emission can occur. If the applied potential is disconnected (Figure 6), thermal re-equilibration of all ionic species occurs resulting in similar light intensity-time effects when the potential is again applied.

Since it is likely that the aluminum electrode will react with the *I_2-doped* MEH-PPV with which it is in contact to give AlI_3, it is entirely possible that the Al^{+3} ions adjacent to the aluminum will also promote electron injection from the aluminum (forward bias mode) by charge compensation. The expected overall diffusional properties of such ions are not entirely clear.

Doped Polyaniline Adjacent to an ITO Electrode (*10-12*). Analogous effects to those described above are found in an entirely different system in which films of polyaniline(100%, 1.76% and 0.09% doped) are present between the ITO electrode and the MEH-PPV. The dopant anions in the polyaniline and the positive charges on the polymer chain reduce hole and electron injection energies, respectively, from the electrodes in a similar manner to that described above for *I_2-doped* MEH-PPV. The energy barrier for passage of electrons and holes between the polyaniline and the MEH-PPV is assumed to be small compared to the injection energies at the electrodes. Hence the properties of the LEDs are governed by the same factors as those described in the *I_2-doped* MEH-PPV systems, i.e., by the presence and corresponding behavior of $(\text{polyaniline})^+A^-$ p-doped species. We believe a similar effect may be involved, as described earlier, in our observance of light emission in PPV.PPyV in both forward and reverse bias modes using aluminum, copper and gold electrodes (*2*).

The data given in Figure 8 highlight the effect of even traces of residual dopant in the "EB". The small doping level results in a lower conductivity of the polyaniline resulting in a greater potential gradient in the polyaniline film which itself gives rise to a greater potential-induced rate of diffusion. The information given in Figure 7 shows analogous data for pure MEH-PPV (curve A), 1.76% doped polyaniline (curve B) and 100% doped polyaniline (curve C). A reduced hole injection energy for a device of this type involving 100% doped polyaniline has been reported previously (*57*). This is again consistent with an increased concentration of negative ions in the polyaniline adjacent to the ITO electrode analogous to that described *I_2-doped* MEH-PPV. We believe the very small (~4 seconds) time to reach maximum light intensity (curve A) in the MEH-PPV may be related to presence of traces of residual doping in the form of $(\text{polaron/bipolaron})^+A^-$ species in the MEH-PPV where A^- is an unkown dopant anion in the MEH-PPV (*58*).

The effect of "rest" periods on light intensity induced by disconnecting the applied potential for intervals on an Al/MEH-PPV/*100% doped* EB/ITO device (Figure 9) are qualitatively identical to those observed in an *I_2-doped* MEH-PPV device (Figure 6). A similar explanation applies. However, when lightly-doped (1.76

% doping) "EB" was studied in a similar way (Figure 10), the light intensity was identical to the final light intensity of the immediately previous curve, implying that at the concentrations and potentials employed, little thermal re-equilibration of ions occurred during the "rest" periods. A very short, intense peak is observed when the potential is first applied. This may be caused by the formation of an electrical double layer as is commonly observed when a potential is applied to an electrode in an electrolyte containing ions; a transitory sharp peak resulting from a sharp increase in current which would result in a momentary sharp increase in intensity of emitted light. However, if a reverse bias of 18 V is applied (Figure 16) instead of the 5 minute "rest" period, the initial light intensity of the second curve approaches that of the first curve. The application of reverse bias increases the rate of diffusion of ions in the reverse direction to that promoted by the application of a forward bias.

Figure 11 shows the effect of different applied potentials on the time taken to reach maximum light intensity. They are analogous to those given for I_2*-doped* MEH-PPV LED (Figure 5); a similar explanation applies. Figures 12 and 13 show the analogous dependency of both light intensity and current density on time for two different doping levels of polyaniline("EB") in devices of the type Al/MEH-PPV/"EB"/ITO. The effects are qualitatively identical to those found with the I_2*-doped* MEH-PPV film given in Figure 4. A similar explanation applies. From the foregoing discussion, it is immediately apparent why "lost" light intensity (Figure 14) is recovered during a "rest" period. It is due to thermal re-equilibration of ionic species.

Sensors for Volatile Organic Compounds

There has recently been considerable interest in the detection and identification of air-borne volatile compounds in such diverse areas as quality control of perfume to detection of toxic gases. Miniaturization, the potential low cost of sensors and the variety of applications promise an enormous market (*29*, *30*). Chemical sensors (*29*) for volatile compounds operate on varied principles and can be classified according to the method of functioning into basic groups such as *electrical* (field-effect transistors, metal oxide semiconductors and organic semiconductors), *optical* (spectrophotometric, luminescence, optothermal) and *sensors that are sensitive to a change of mass* (piezoelectric and acoustosurface).

A novel, promising, recent approach using electrical measurements (*31*) involves a dispersion of carbon black particles in conventional polymers coated on to an interdigitated electrode. Adsorption of an analyte vapor in the polymer causes a swelling of the polymer which increases the distance between carbon particles resulting in an overall increase in resistance. Desorption of the analyte causes a corresponding decrease in resistance. Considerable attention has been paid to sensors which contain an electroactive polymer, "conducting polymer" as the sensing material (*32*); "electronic noses" containing an array of different electroactive sensing polymer elements are commercially available (e.g. Aromascan, Bloodhound, Alphamos Fox, Neotronics eNose) (*33*). Conducting polymers are attractive because many different variants of a basic polymer are possible (for example by choice of dopant or by chemical modification of the polymer backbone) which exhibit different sensitivity patterns to a range of gases or vapors.

Sensors involving polymer sensor material have usually involved a thin film of polymer on an interdigitated gold electrode whose resistance changes are measured in various environments. A novel approach has been reported in which the interdigitated electrode is replaced by fabric whose fibers are coated with a conducting polymer such as polypyrrole (*34*). This method, involving integration of sensor and electroactive material deserves further attention.

We are especially interested in *reversible* sensors for volatile organic compounds (VOCs) in the atmosphere using conducting polymers as the sensing material. The most extensively used conducting polymer to date is polypyrrole (*34-38*) which is applied in certain commercial sensing instruments (*39*). Polyaniline, either in its protonated (doped) or non-doped forms, has been studied less extensively (*40-45*) and polythiophene to an even smaller extent (*46-48*).

We are investigating VOCs which are apparently "chemically inert" to the sensor material and the scientific reasons for their reversible response in resistance to a given analyte vapor (*8, 10, 11, 49, 50*). To better understand the processes involved, detailed studies of *pure* conjugated polymers and oligomers with selected dopant anions rather than their *blends* in conventional polymers have been undertaken. In general, a sensor can either react reversibly or irreversibly with a given VOC. For example, the base form of polyaniline will react chemically and irreversibly with an acidic vapor, e.g., HCl with a decrease in resistance. The "doped" polyaniline hydrochloride salt will likewise react irreversibly with an alkaline vapor such as ammonia which combines with the HCl dopant resulting in an increase in resistance. These may be regarded as "one-use" only sensors. In principle they may often be more sensitive than a reversible sensor since the final change in resistance will be strongly dependent on "dosage", i.e., product of analyte concentration and exposure time. This is particularly so if the chemical reaction goes essentially to completion at very low analyte concentrations.

Our special interest involves the use of sensors which can be addressed by *simple*, *inexpensive*, *portable* electronic devices including *low-cost*, *portable* electronic accessories for use in detecting toxic gases such as those, for example, which might be released by a terrorist group in subways such that thousands of sensors could be distributed in a given subway system. Applicability of such sensors to detection of toxic gases by military personnel in the field and for monitoring of toxic gases released in old military or industrial dumps, etc. is immediately obvious. The nature of the response properties (e.g. change in resistance with time) of a given sensor to a series of individual, different analytes of interest is recorded. Data are also obtained with a different sensor material for the same analyte. Each sensor will give a different response to a given VOC. A sensor "array", comprised of a number of different sensors is interfaced with a computer integration system such that when the array is exposed to a *mixture* of VOCs, the most probable composition of the mixture is reported. Individual arrays can be connected by radio to near-by relay stations which then re-transmit signals to more powerful relay stations.

We report studies on polypyrrole sensors (*8, 49, 50*), the prime objective of which was to attain reversibility of response and to demonstrate that selectivity to "chemically inert" VOCs could be realized. Less extensive studies were performed on polythiophene (*8, 50*). The unexpectedly promising properties of sensors containing oligomers of aniline are also reported.

Results

Polypyrrole Sensors (*8, 49, 50*). The soluble polypyrrole (*59*) used in the sensors in Examples 1-7 inclusive in Figure 17 was spun onto interdigitated gold electrode arrays (15μm spacing between electrodes) using a solution of soluble polypyrrole (dodecylbenzene sulfonate anion) and polystyrene dissolved in chloroform. The change in resistance was measured after a 5-second exposure of the sensor to static air saturated with the analyte at room temperature in the consecutive numerical order given in Figure 17. For comparison, our results using polypyrrole synthesized in the presence of polystyrene using a molybate oxidizing agent (*37*) are given in Examples 8 and 9 in Figure 17. The percent change in resistance, sometimes refered to as

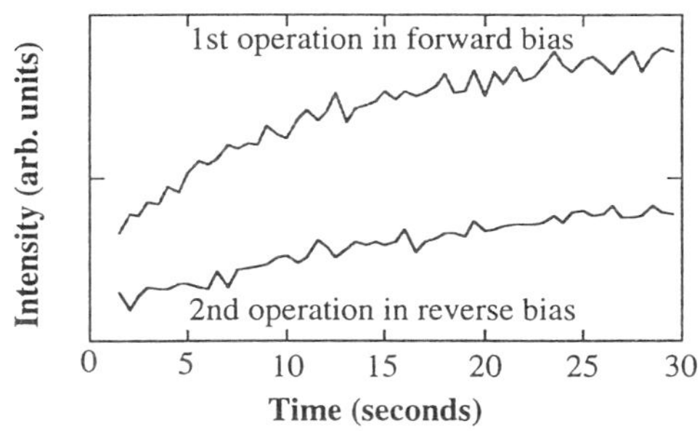

Figure 16. Al/MEH-PPV/*1.76% HCl-doped* "EB"/ITO device: (1) operated at 18V in forward bias for 30 seconds, then operated at 18V in reverse bias for 5 minutes, (2) operated again at 18V in forward bias for 30 seconds. Reproduced with permission from reference 12. Copyright 1997 Society of Photo-Optical Instrumentation Engineers.

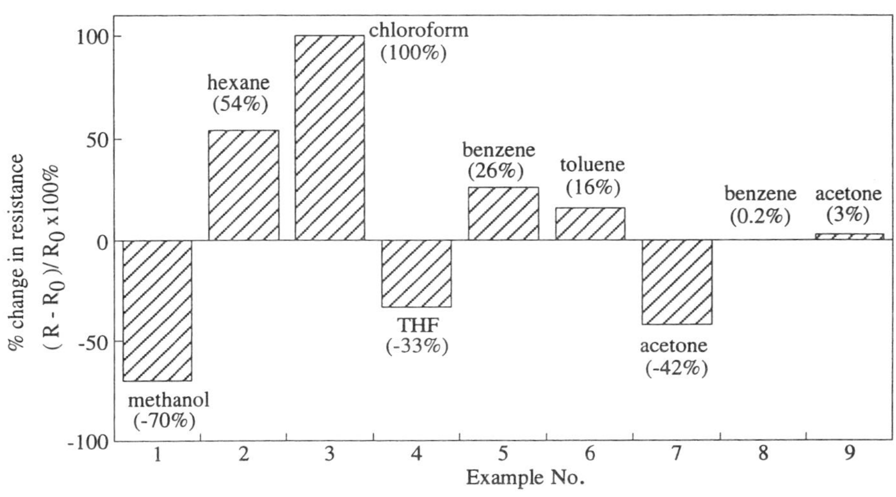

Figure 17. Percentage change in resistance of a sensor coated with doped polypyrrole on exposure to air saturated with analyte vapor at room temperature, R_0=initial resistance of the sensor; R=resistance of sensor after 5 seconds exposure to the analyte/air mixture. Reproduced with permission from reference 49. Copyright 1997 Elsevier Science Ltd.

"sensitivity" is given by the relationship: %change in resistance ("sensitivity", $\Delta R\%$) = $(R-R_0)/R_0 \times 100\%$, where R_0 is the initial resistance of the sensor device and R is the resistance after exposure to gas or gaseous mixture under a given set of conditions for a given time. *The fact that the same sensor can give a large increase or decrease in resistance with different analytes is promising for possible technological use for this type of device.*

Preliminary, non-optimized results using a soluble polypyrrole sensor and toluene vapor were obtained as follows. A solution of polypyrrole (dodecylbenzenesulfonate anion) in chloroform was spun-cast on to a gold interdigitated array attached to an ohmmeter. The thickness of the film was ~200-300 Å. A stream of nitrogen (1500ml/min) which had been saturated with "chemically inert" toluene vapor at 0°C (partial pressure of toluene, 6.8 torr; ~8,900ppm) was passed over the sensor for 10 minute intervals followed by a 1500ml/min stream of pure nitrogen for 10 minute intervals. The results are given in Figure 18 for the first complete ten 20 minute cycles, after which the cycle time was increased to 60 minutes. Exposure to the toluene/nitrogen gas mixture resulted in an increase in resistance; exposure to pure nitrogen resulted in a decrease in resistance.

An approximately 5% change in resistance was observed in each cycle. As can be seen in Figure 18, excellent reversibility is obtained. It should be noted that the initial responses to a mixture of toluene vapor and nitrogen are very rapid, large changes (~70-75% of the maximum change) were obtained during the first ~8 seconds. Similar rapid changes were observed on exposure of the sensor to a pure nitrogen stream.

Polythiophene Sensors (*8, 49, 50*). In order to gain information concerning the type of interaction between polymer sensing material and analyte vapor, we have performed preliminary studies with a non-doped regioregular poly(3-hexylthiophene) (Rieke Metals Inc.) sensor, since the conformation of polythiophene chains is known to be potentially sensitive to external stimuli which are frequently accompanied by measurable changes in their Vis/UV spectra (*60*).

In these studies with poly(3-hexylthiophene), we have used an experimentally easier but less rigorous technique than that employed for the toluene/nitrogen flow system studies. For the resistance measurements, the sensor was simply held for 3~4 minutes above the surface of toluene (vapor pressure ~ 26.8 torr; ~35,000ppm) at room temperature in a container previously filled with laboratory air. The sensor was then removed from the toluene/air mixture and was suspended in laboratory air for a similar time period. This procedure was used in all cycles given in Figure 19 except for the last cycle. As can be seen, the reversibility, percentage change in resistance (~200%) and rapidity of response effected by the analyte and then by subsequent exposure to air in these relatively crude studies are most promising.

It is important to note the effect of toluene vapor on the Vis/UV spectrum of a thin film of poly(3-hexylthiophene) spun from a chloroform solution on to a microscope slide (*8, 49*). A small piece of the slide together with attached film was placed in a cuvette in the spectrometer with air in the reference cuvette and its spectrum was recorded. The film was removed, a few drops of toluene were placed in both cuvettes and the film was carefully suspended in a cuvette and its spectrum was again measured after five minutes. There was no change in the spectrum after an additional 20 minutes in the cuvette. The film was removed and permitted to remain in the air for five minutes. As can be seen from Figure 20, toluene vapor results in a partly reversible effect on the Vis/UV spectrum of the poly(3-hexylthiophene). We have found that gas desorption is frequently slower than gas adsorption in systems of this type; we have not yet determined whether a longer period in air would result in greater reversibility.

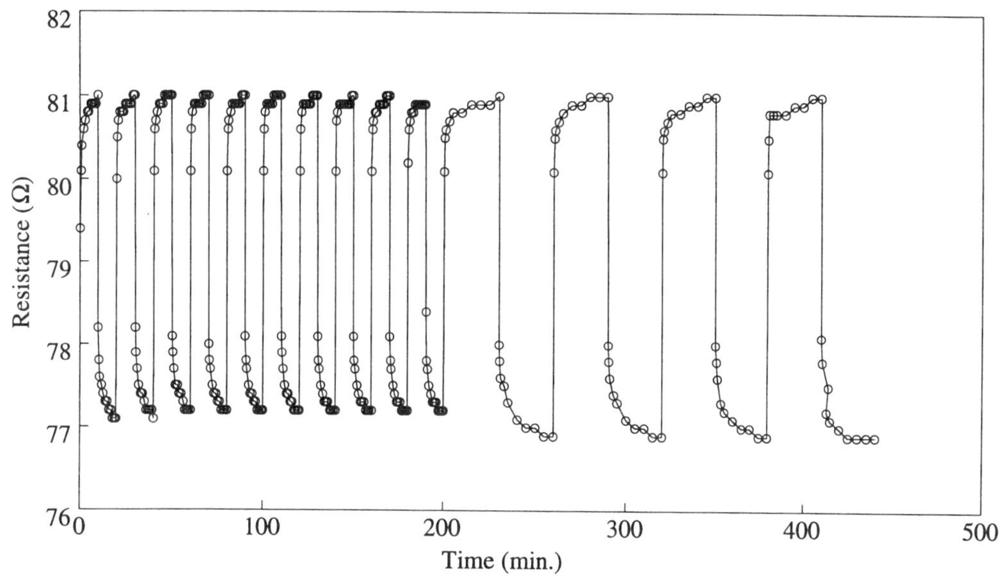

Figure 18. Response of a soluble polypyrrole-based sensor to a stream of nitrogen containing toluene vapor. The higher resistance is obtained with the toluene/nitrogen gas mixture (see text). Reproduced with permission from reference 49. Copyright 1997 Elsevier Science Ltd.

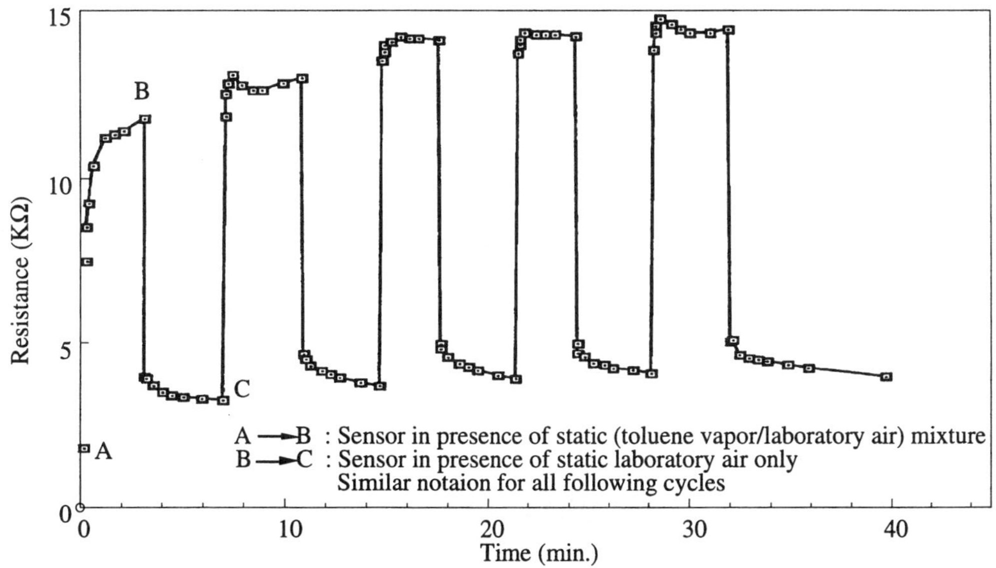

Figure 19. Response of a poly(3-hexylthiophene) sensor to toluene vapor. Reproduced with permission from reference 50. Copyright 1997 Society of Plastics Engineers.

Octaaniline Sensor (*10, 11*). Octaaniline, 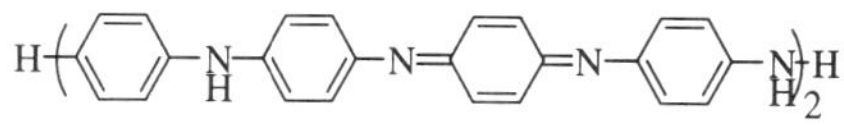

synthesized in the emeraldine oxidation state as we have previously described (*61, 62*) was "doped" with dodecylbenzenesulfonic acid. Its ~1.0wt% solution in chloroform was spun onto interdigitated gold electrode arrays. The resulting sensor unit (film thickness 300~ 500Å) was first held in normal laboratory air during which time its resistance was recorded. It was then suspended for a convenient arbitary 120 second period in static laboratory air saturated with toluene vapor at room temperature (25°C, ~26.8 torr, ~35,000ppm) during which time its increase in resistance was again recorded. It was then held in static laboratory air at room temperature for 120 seconds during which time its decrease in resistance was recorded.

The reversibility (~99.7%) of the octaaniline sensor is excellent as shown in Figure 21. The environmental stability of the sensor is most satisfying. As can be seen from Figure 22, sensitivity, on exposure to a toluene/air mixture as discussed above, is reduced fairly rapidly during the first day of exposure to laboratory air, but then essentially stabilizes after a few days.

As can be seen from Figure 21, the sensitivity of the octaaniline sensor is remarkably high, especially upon noting that the maximum increase in resistance had not been reached during the time interval (120 seconds) at which data collection was discontinued. By comparison, exposure of our polypyrrole sensor to an identical toluene/air mixture resulted in a ~250% change in resistance. However, up to 80,000% change in resistance has been obtained using an octaaniline sensor material under identical experimental conditions and using an identical dopant! This highlights the *enormous* effect of sensor material on sensitivity. Moreover, as shown in Figure 23, the sensor displays very little sensitivity to water vapor. This is most fortunate since most sensors encounter serious interference problems with water vapor.

For ppm concentrations of toluene vapor in air, standard solutions of toluene in paraffin oil of different concentrations were prepared. After equilibration with air, the concentrations of toluene in the air were determined by gas chromatography. As shown in Figure 24 , we can easily detect 35ppm of toluene vapor in air using an inexpensive simple, portable, undergraduate laboratory-type ohmmeter. The data in Figure 24 were corrected for traces of an unidentified volatile component in the paraffin oil (m/e=135 by mass spectrometry).

Tetraaniline sensors. Tetraaniline, 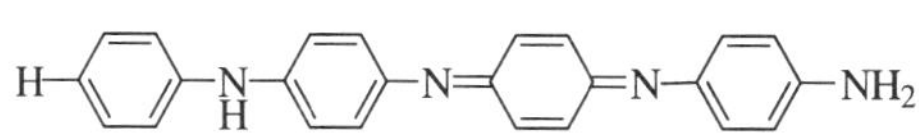

synthesized in the emeraldine oxidation state as we have previously described (*61, 62*) was "doped" with dodecylbenzenesulfonic acid. The procedure for making the sensor was similar to that used for the octaaniline sensor. The response of the tetraaniline sensor is similar to that of octaaniline as shown in Figure 25. By exclusion of the paraffin oil vapor background, the tetraaniline sensor can easily detect down to 100ppm toluene vapor in air.

Discussion

It is evident from the large number of publications and patents involving polypyrrole or polyaniline cited earlier that these polymers or their derivatives or composites

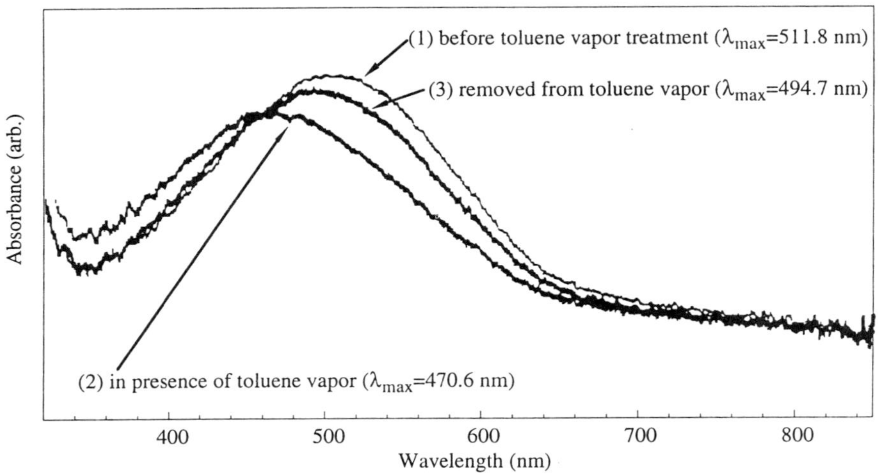

Figure 20. Effect of toluene vapor (~25°C; ~26 torr; 35,000ppm) on Vis/UV spectrum of a thin regioregular poly(3-hexylthiophene) film spun from chloroform solution. Reproduced with permission from reference 50. Copyright 1997 Society of Plastics Engineers.

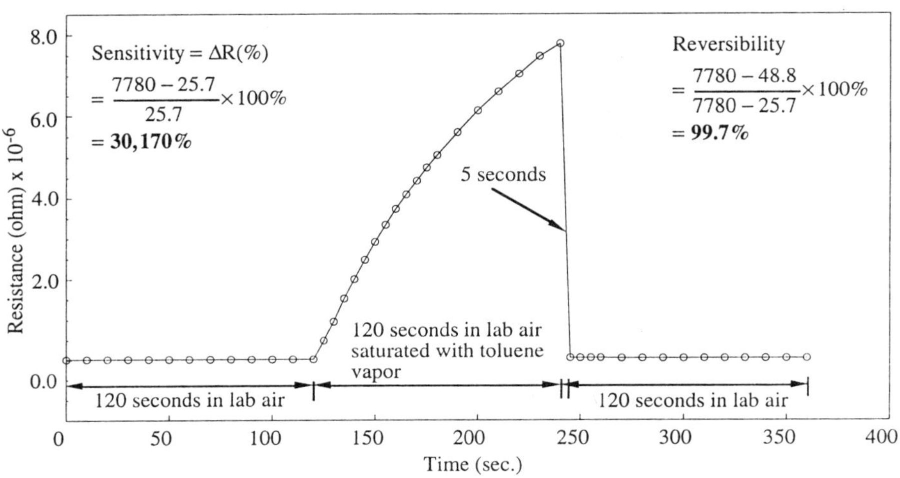

Figure 21. Typical 120 seconds resistance responses of an octaaniline sensor to normal laboratory air, laboratory air saturated with toluene vapor at room temperature (~26 torr; ~35,000ppm) and normal laboratory air. (R_0=25.7x10^3 Ω; R_{max}=7.78x10^6 Ω; R_{final}=48.8x10^3 Ω). Reproduced from reference 10.

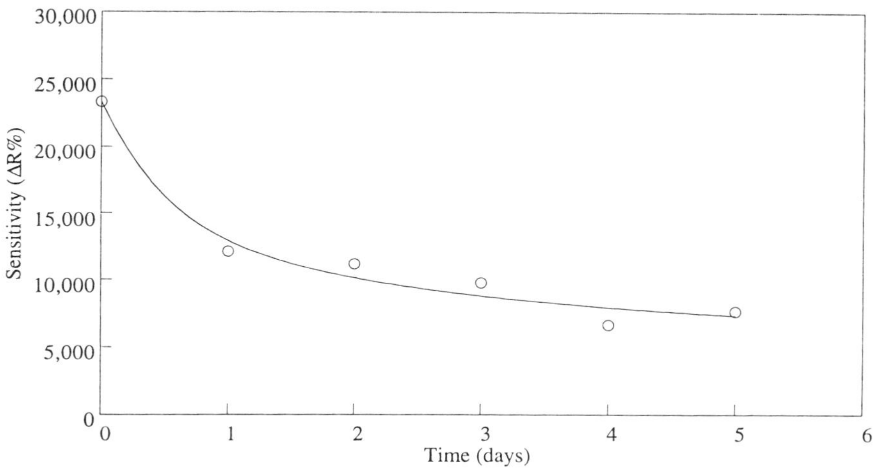

Figure 22. Change in sensitivity (ΔR%) of an octaaniline sensor to saturated toluene vapor (25°C, ~26.8 torr, ~35,000ppm) as a function of number of days exposure of the sensor to ambient laboratory air at room temperature. Reproduced from reference 10.

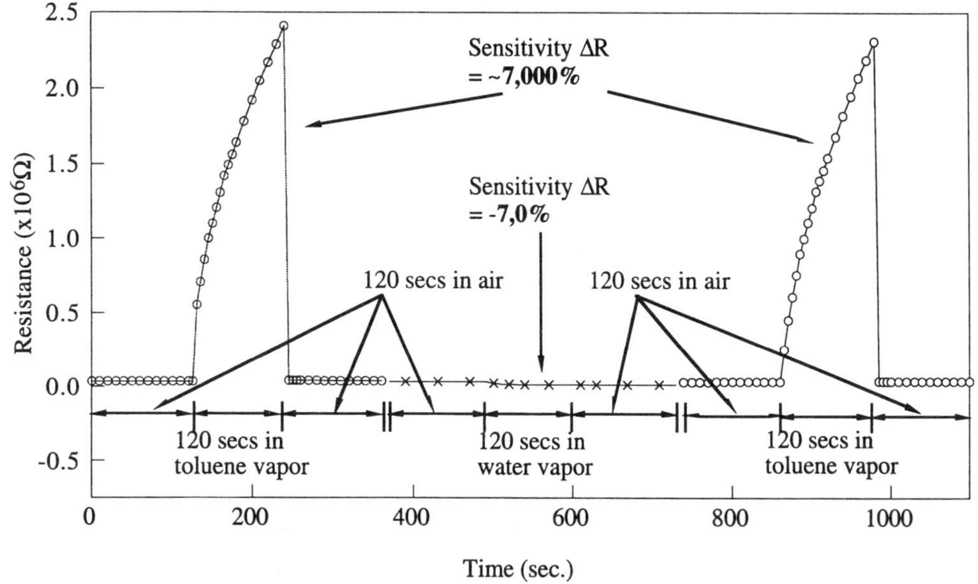

Figure 23. Effect of water vapor in air at 25°C (23.7 torr; 31,300ppm) on an octaaniline sensor. Reproduced with permission from Synthetic Metals (in press). Copyright 1998 Elsevier Science Ltd.

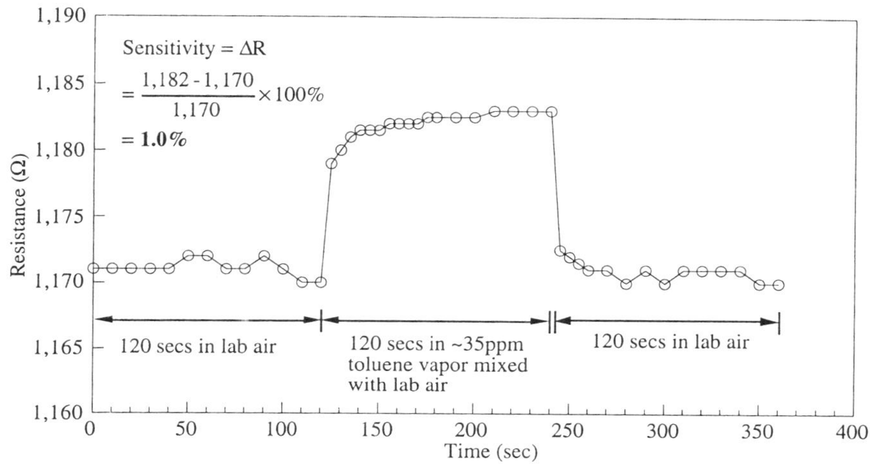

Figure 24. Reversible octaaniline sensor detecting 35ppm of toluene vapor admixed with laboratory air. Reproduced with permission from Synthetic Metals (in press). Copyright 1998 Elsevier Science Ltd.

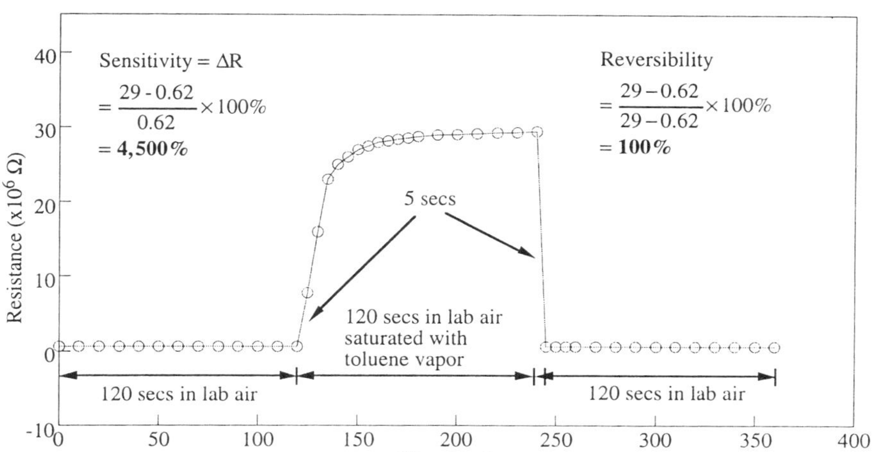

Figure 25. Typical 120 seconds resistance responses of a tetraaniline sensor to normal laboratory air, laboratory air saturated with toluene vapor at room temperature (~26 torr; 35,000ppm) and normal laboratory air. (R_0=620x10^3 Ω; R_{max}=29x10^6 Ω; R_{final}=620x10^3 Ω).

with conventional polymers hold considerable promise as sensors for volatile organic compounds. A problem frequently encountered is the absence of good reversibility i.e., a sloping baseline is encountered on successive exposures to analyte vapor followed by air containing no analyte vapor. We have demonstrated that these problems are essentially eliminated, or at least greatly reduced, in our polypyrrole, octaaniline and tetraaniline sensors discussed above. An additional problem in most sensors is the sensitivity to water vapor -- an almost universal contaminant in all real-life detection of VOCs in the atmosphere -- which is greatly reduced in the octaaniline sensor.

We had initially limited our studies to polypyrrole based sensors with several different "chemically inert" VOCs and to polythiophene sensors with toluene analyte. However, because of the relatively low sensitivity of the polypyrrole sensors and because of photosensitivity of the polythiophene sensors we have recently directed our attention to a new type of oligomer of aniline which we have recently synthesized and characterized (*61*, *62*).

Polypyrrole and Polythiophene Sensors. As noted above, excellent reversibility was obtained with toluene vapor using polypyrrole sensors. Even after approximately one year storage in laboratory air, their sensor response properties to toluene/air mixtures were essentially unchanged. We believe their reversible change in resistance on adsorption of analyte vapor is related to at least two possible simultaneous factors: (1) when an analyte is absorbed, the volume of the mass in polymer increases so that there is less conducting material per unit volume, thus resulting in an increase in resistance; (2) the adsorbed analyte promotes a change in the molecular conformation of the polymer from e.g., "expanded coil" to "tight-coil" or vice verse. We have previously shown (*63*) for polyaniline that such a change can result in up to a thousand fold change in resistance. A change of this type results in a change in the π conjugation of the polymer backbone, with concomitant change in its conductivity. Such changes are also greatly dependent on the nature of the dopant anion. In this respect it should be noted that in general, response to a given analyte vapor is often greatly dependent on the nature of the dopant anion.

The Vis/UV spectrum of polythiophene is known to be very sensitive to changes in its molecular conformation (*60*). Hence polythiophene was an excellent material on which to search for conformational changes accompanying resistance changes. Its observed blue shift in the absorption maximum in the presence of toluene vapor is consistent with a reduction in π conjugation along the polymer chain (more "coil-like" conformation) (*63*) which is also consistent with the observed increase in resistance of the polymer in the presence of toluene vapor. No analogous spectral changes on exposure of the polypyrrole to toluene vapor have yet been observed. The polythiophene sensors appeared promising but we found they displayed a significant photoconductivity effect even in normal laboratory lighting, which decayed only slowly in the dark (*64*). Because of this complicating factor their study was discontinued.

Oligomeric Aniline Sensors. Promising results have been reported (*44*, *45*) using polyaniline sensors on gold interdigitated electrodes for the detection of "chemically inert" VOCs. We have briefly evaluated polyaniline sensors with toluene vapor analyte and found that their sensitivity was greatly dependent on the type of dopant anion used. Since recent x-ray studies (*65*) show that various forms of polyaniline undergo spontaneous changes which might complicate interpretation of sensor data we were more interested in testing our more simple, highly characterized sensor materials involving oligomers of aniline as discussed above (*10*, *11*).

In all our studies, as is clearly evident, for example, from Figures 21and 23, that at least two different kinetic processes are involved in the resistance changes. In each case a slow increase in resistance is observed in the presence of toluene vapor followed by a very rapid (< 5 seconds) decrease in resistance to the original level after re-exposure to air. This rapid decrease in resistance must obviously be due to some type of surface desorption process. A rapid increase in resistance can be seen in Figures 24 and 25 to give a "plateau" region followed by a rapid decrease in resistance after re-exposure to air. We do not as yet understand these phenomena although we believe the adsorbed analyte vapor may promote a change in molecular conformation or configuration and also act as a "plasticizer" permitting the change to occur. In this respect, it should be noted that the sensitivity appears to be related to film thickness. In a possibly related phenomenon we, and others, have noted that a static mixture of air and analyte vapor frequently gives a larger sensitivity response than a flow system for identical analyte concentrations. The reasons for this are still unclear.

The conductivities of HCl-doped (compressed powder pellets) of polyaniline, octaaniline and tetraaniline show a progressive decrease from ~3 S/cm to $2x10^{-2}$ S/cm to $3x10^{-3}$ S/cm (*61, 62*). It seems likely that the sensitivity of a given sensor material may be related to the absolute conductivity of the material. In the conducting regime of $\sim 10^{-2}$ to 10^{-5} S/cm, the conduction process occurs primarily by a variety of hopping processes of charge carriers between different segments in the same or different molecules. The ease with which the hopping process occurs falls off exponentially with distance between hopping sites. Thus, if a sensor film expands on adsorption of analyte vapor (or contracts on desorption of the analyte) the conductivity will either decrease (or increase), respectively. In the more highly doped metallic regime, conduction occurs by a completely different process, "band conduction", which is far less sensitive to interchain spacing due to the fact that weak bands are formed between molecules (as well as strong bands within a molecule). The presence of weak bands is contingent on having close packing (crystallinity) between chains.

It is highly probable that the degree of crystallinity/amorphicity in the polymer film may play a key role in sensitivity -- this will also be significantly affected by the nature of the dopant ion and by the rate of evaporation of the solvent from its solution of the polymer while spinning a film on the gold electrodes. It is well known that amorphous regions are more reactive than crystalline regions -- hence the preferential adsorption of the analyte in the amorphous regions is to be expected.

Since "tight coil/expanded coil" conformational changes of the type discussed for polypyrrole and polythiophene are less likely to occur with octaaniline and are impossible with tetraaniline, we believe that the resistance *changes* in the oligomers may be caused by a reversible change in isomeric composition induced by the adsorption/desorption of analyte vapor not only in the aniline tetramer but also in the octamer and possibly also in polyaniline. An unexpectedly large number of isomers can in principle exist for oligomeric and polymeric anilines (*66*) in the non-doped emeraldine oxidation state. The octamer, for example, can in principle exist in the fifteen different positional isomeric forms shown in Figure 26. Each of these can in principle exist as distinct, different isomeric forms associated with the imine/quinoid double bonds (*cis-trans* isomers) (*67*) and each of these in turn could exist in different isomeric forms (ring-flip isomers) (*68*)! Each of the three positional isomers of the tetramer (*69*) are shown in Figure 27 could in principle also exist in cis and trans forms on protonation.The number of possible isomers in the protonated ("doped") bipolaronic form is smaller but still significant. It is of interest to note that even doped (protonated) species can exist as a distinct geometrical isomer as shown by the *trans* protonated isomer of the phenyl-end capped tetramer obtained from x-ray structural studies (*70*).

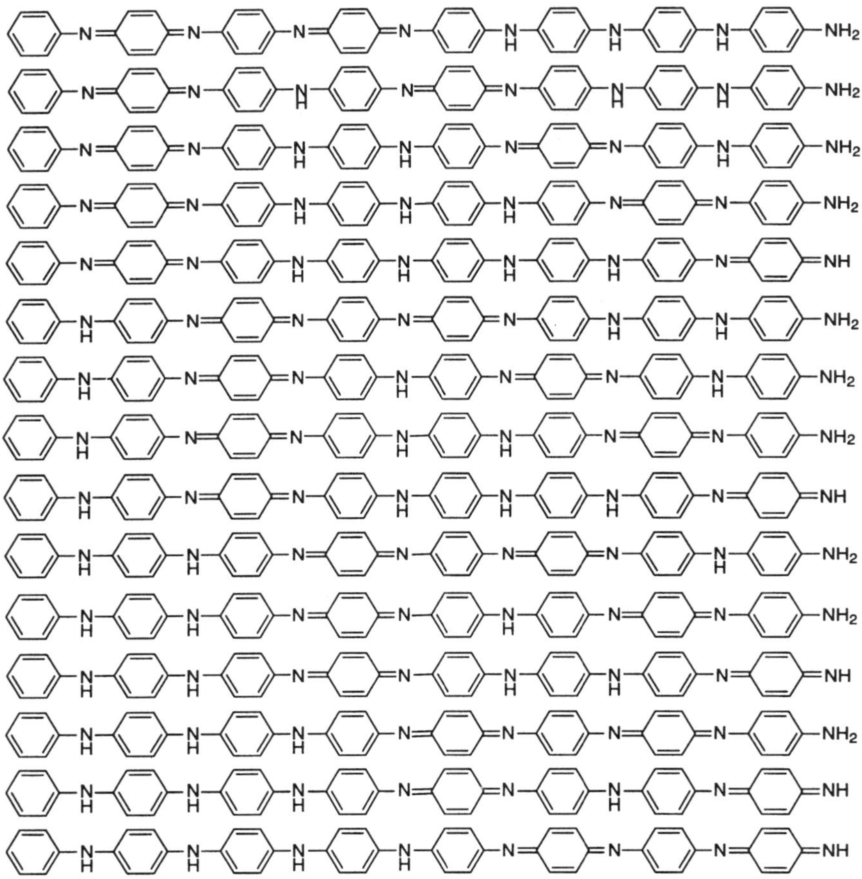

(1) (2) (3) (4) (5) (6) (7) (8) (9) (10) (11) (12) (13) (14) (15)

Figure 26. The fifteen different positional isomeric forms of octaaniline in the non-doped emeraldine oxidation state. Reproduced with permission from Synthetic Metals (in press). Copyright 1998 Elsevier Science Ltd.

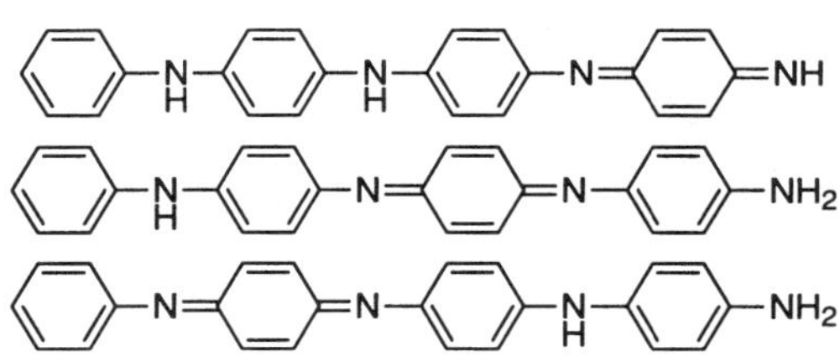

Figure 27. Three possible positional isomeric forms of tetaaniline in the non-doped emeraldine oxidation state.

Conclusions

Even unsuspected trace amounts of residual doping either in the light-emitting polymer layer or in a polymer layer adjacent to an electrode in a light-emitting device, can greatly reduce electron and/or hole injection energies and mask the conventional work functions of the electrode materials. This effect can improve certain important parameters of a device including light emission in a reverse bias mode in addition to the conventional forward bias mode. The effects are consistent with the presence of $(polaron/bipolaron)^{+} A^{-}$ p-doped ionic species initially present in the polymer, the number of which does not change during operation. Potential-induced diffusion of A^{-} ions controls the properties of the systems. Removal of the applied potential promotes room-temperature re-establishment of ionic equilibrium with resulting recovery in emitted light intensity showing that at least a significant amount of the decrease in light intensity during operation of a LED device is not necessarily due to irreversible degradation.

The use of electroactive polymers as reversible sensing materials for the detection of volatile organic compounds holds very considerable promise. In particular, our preliminary findings on oligomers of aniline in which a few ppm of "chemically inert" toluene in air can be detected gives new directions for investigation. Elucidation of the fundamental science involved promises to lead to new conductivity/structure relationships in the anilines as well as to new possible technologies.

Acknowledgements

The LED studies were performed by F. Huang. Sensor studies were performed by J. Feng (aniline oligomers), W.J. Zhang (polyaniline and polythiophene) and D. Adebimpe (polythiophene). The authors wish to thank Dr. B. R. Hsieh (Xerox Corp.) for samples of MEH-PPV. The authors gratefully acknowledge the University of Pennsylvania Materials Research Laboratory (MRSEC NSF Award No. DMR-96-32598) for financial support of studies on light-emitting devices and a dual use Technology Reinvestment program, Contract No. N00014-95-2-0008 subcontract from Hughes Research Laboratory Co. and Abtech Scientific Inc. to the Univ. of Penn. together with the Office of Naval Research N00014-92-1-1369 for the sensor studies.

References

1. Wang, H. L.; MacDiarmid, A.G.; Wang, Y. Z.; Gebler, D. D. and Epstein, A. J. *Synth. Met.*, **1996**, *78,* 33.
2. Wang, H. L.; Huang, F.; MacDiarmid, A.G.; Wang, Y. Z.; Gebler, D. D. and Epstein, A. J. *Synth. Met.*, **1996**, *80*, 97.
3. Wang, Y. Z.; Gebler, D.D.; Lin, L.B.; Blatchford, J. W.; Jessen, S.W.; Wang, H. L. and Epstein, A. J. *Appl. Phys. Lett.,* **1996**, *68,* 894.
4. MacDiarmid, A. G.; Wang, H. L.; Huang, F.; Avlyanov, J. K.; Wang, P. -C.; Swager, T. M.; Huang, Z.; Epstein, A. J.; Wang, Y. Z.; Gebler, D. D.; Ranganathan, S.; Calvert, J. M.; Crawford, R. J.; Vargo, T. G.; Wynne, K. J.; Whitesides, G. M.; Xia, Y.and Hsieh, B. R. in *Materials Research Society, Symposium Proceedings* , Eds. Jen, A.K.-Y.; Lee, C.Y-C.; Dalton, L.R.; Rubner, M.F.; Wnek, G.E.; Chiang, L.Y.; *43*, **1996**, 3.
5. Huang, F.; Wang, H. L.; Feldstein, M.; MacDiarmid, A.G. and Epstein, A. J. *Synth. Met.*, **1997**, *85*, 1283.

6. Huang, F.; Xie, S.J.; MacDiarmid, A.G. and Hsieh, B.R. *Proc. Soc. of Plast. Engr. Annu. Tech. Conf.* (ANTEC), 1997, p. 1261.
7. MacDiarmid, A.G.; Avlyanov, J.K.; Huang, F.; Huang, Z.; Wang, P-C and Epstein, A. J. *Macromol. Symp.,* **1997**,*118*, 445.
8. MacDiarmid, A.G.; Feng, J.; Zhang, W.J.; Huang, Z.; Wang, P-C and Huang, F. *Polym. Prep.* **1997**, *38(2),* 98.
9. Huang, F.; MacDiarmid, A.G. and Hsieh, B.R. *Appl. Phys. Lett.,* **1997**, *71(17),* 2415.
10. MacDiarmid, A.G.; Zhang, W.J.; Feng, J.; Huang, F. and Hsieh, B.R. *Polym. Prep.* **1998**, *39(1),* 80.
11. MacDiarmid, A.G.; Zhang, W.J.; Feng, J.; Huang, F. and Hsieh, B.R. *Proc. Soc. of Plast. Engr. Annu. Tech. Conf.* (ANTEC), 1998, p. 1330.
12. Huang, F.; MacDiarmid, A.G. and Hsieh, B.R. *Proc. SPIE,* **1998**, *3281,* 156.
13. Fou, A.C.; Onitsuka, O.; Ferreira, M.; Rubner, M.F. and Hseih, B.R.*J. Appl. Phys.* , **1995** , *79*, 1316.
14. Burroughes, J. H.; Bradly, D.D.; Brown, A.R.; Marks, R.N.; Mackay, K.; Friend, R. H.; Burns, P.L. and Holmes, A.B. *Nature*, **1990**, *347*, 539.
15. Kraft A.; Grimsdale A. C.; Holmes, A.B. *Angew. Chem. Int . Edit.,* **1998**, *37(4),* 402 and references therein.
16. Yang, Y. *Mater. Res. Soc. Bull.*, **1997**, *22 (6)*, 31 and references therein.
17. Tsutsui, T. *Mater. Res. Soc. Bull.*, **1997**, *22 (6)*, 39 and references therein.
18. Pei, Q.; Yu, G.; Zhang, C.; Yang, Y. and Heeger, A. J. *Science*, **1995**, *269*, 1086.
19. Yang, Y. and Bharathan J. *Polym. Prep.* **1998**, *39(1),* 98.
20. Romero, D.B.; Schaer, M.; Zuppiroli, L.; Cesar, B. and Francois, B. *Appl. Phys. Lett.,* **1995**, *67,* 1659.
21. Schoo, H.F.M.; Demandt, R.C.J.E.; Vleggaar, J.J.M. and Liedenbaum, C.T.H. *Polym. Prep.* **1997**, *38,* 337.
22. De Jong, M.J.M. and Blom, P.W.M. *Proceedings of the 23rd Int. Conf. on the Physics of Semiconductors*, Ed. Scheffler M. and Zimmerman R. (World Scientific), 1996, p.3351.
23. Aziz, H. and Xu, G. *J. Phys. Chem. B* , **1997** ,*101*, 4009.
24. Karg, S.; Meier M. and Riess, W. *J. Appl. Phys.* , **1997** , *82(4)*, 1951.
25. Meier, M.; Karg, S. and Riess, W. *J. Appl. Phys.* , **1997** , *82(4)*, 1961.
26. Zou, D.; Yahiro, M. and Tsutsui, T. *Appl. Phys. Lett.* **1998**,*72(19)*, 2484.
27. Braun, D. and Heeger, A.J. *Appl. Phys. Lett.* **1991**,*58*, 1982.
28. Karg, S.; Scott, J.C.; Salem, J.R. and Angelopoulos, M. *Synth. Met.*, **1996**, *80*, 111.
29. Myasoedov, B. F. *Bull. Russ. Acad. Sci.--Div. Chem.Sci., Part 1*, **1992**, *41 (3)*, 383
30. Gopel, W. *Sensors and Actuators*, **1995**, *B24/25*, 17
31. Lonergan, M. C.; Severin, E. J.; Doleman, B. J.; Beaber, S. A.; Grubbs, R. H.; Lewis, N. S. *Chem. Mater.*, **1996**, *8*, 2298
32. Gardner, J. W.; Bartlett, P. N. *Sensors and Actuators*, **1995**, *A51*, 57
33. Harsis, P. D.; Arnold, W. M.; Andrews, M. K.; Partridge, A. C. *Sensors and Actuators*, **1997**, *B42*, 177
34. Collins, C. E.; Buckley, L. J. *Synth. Met.*, **1996**, *78,* 93
35. Slater, J. M.; Watt, E. J.; Freeman, N. J.; May, I. P.; Weir, D. J. *Analyst*, **1992**, *117*, 1265
36. Partridge, A. C.; Harris, P. D.; Andrews, M. K. *Analyst*, **1996**, *121*, 1349
37. Barisci, J. N.; Conn, C.; Wallace, G. G. *TRIP*, **1996**, *4 (9)*, 307
38. Freund, M. S.; Lewis, N. S. *Proc. Natl. Acad. Sci. USA*, **1995**, *92*, 2652
39. Gardiner, J. W.; Bartlett, P. N. *EP 0,596,973 B1*, **1995**
40. Unde, S.; Ganu, J.; Radhakrishnan, S. *Adv. Mat. Opt. Elect.*, **1996**, *6*, 151

41. Kukla, A. L.; Shirshov, M.; Piletsky, S. A. *Sensors and Actuators*, **1996**, *B37*, 135
42. Agbor, N. E.; Petty, M. C.; Monkman, A. P. *Sensors and Actuators*, **1995**, *B28*, 173
43. Monkman, A. P.; Petty, M. C.; Agbor, N. E.; Scully, M. T. *US 5,536,473*, **1996**
44. Miller, L. J.; van Ast, C. I.; Yamagishi, F. G. *US 5,4147,100*, **1995**
45. Yamagishi, F. G.; Stanford, T. B.; van Ast, C. I.; Braatz, P. O.; Miller, L.; Gilbert, H. C. *Proc. Soc. Plast. Engr. Annu. Tech. Conf.* , **1998**, *2*, 1335.
46. Hanawa, T.; Kuwabata, S.; Hashimoto, H.; Yoneyama, H. *Synth. Met.*, **1989**, *30 (2)*, 173
47. Ohmori, Y.; Takahashi, H.; Muro, K.; Uchida, M.; Kawai, T.; Yoshino, K. *Jap. J. Appl. Phys.*, **1991**, *30 (7B)*, L1247
48. Kim, S. R.; Choi, S. A.; Kim, J. D.; Kim, K. J.; Lee, C.; Rhee. S.B. *Synth. Met.*, **1995**, *71 (1-3)*, 2027
49. MacDiarmid, A. G. *Synth. Met.*, **1997**, *84 (1)*, 27
50. MacDiarmid, A. G.; Zhang, W. J.; Adebimpe, D. E.; Wang, P.-C.; Huang, Z. *Proc. Soc. Plast. Engr. Annu. Tech. Conf.*, **1997**, *2*, 251
51. Tang, C. W. and Vanslyke, S. A. *Appl. Phys. Lett.* **1987**,*51(12)*, 913.
52. Wudl, F. and Srdanov, G. U.S. Patent 5,189,136 (1993).
53. Salaneck, W., Linköping University, Personal communication, Oct. 12, 1997.
54. MacDiarmid, A.G.; Chiang, J.-C.; Richter, A.F.; Somasiri, N.L.D. and Epstein, A.J. in *Conducting Polymers* ; Alcacer, L., Ed.; Eidel Publ: Dordrecht, Holland,1987;p.105.
55. Asturias, G.E. and MacDiarmid, A.G. *Synth. Met.*, **1996**, *29*, E157.
56. Schwarzkopf Microanlytical Lab, Inc., Woodside, New York 11377.
57. Yang, Y. and Heeger, A.J. *Appl. Phys. Lett.* **1991**,*64*, 1245.
58. Hsieh, B.R.; Yu, Y.; VanLaeken, A.C. and Lee, H. *Macromolecules*, **1997**, *30*, 8094.
59. Lee, J. Y.; Kim, D. Y.; Kim, C. Y. *Synth. Met.*, **1995**, *74*, 103
60. Anderson, M. R.; Selse, D.; Berggren, M.; Jarvinen, H.; Hjertberg, T.; Inganas, O.; Wennerstrom, O.; Osterholm, J.-E. *Macromolecules*, **1994**, *27*, 6503
61. Feng, J.; Zhang, W. J.; MacDiarmid, A. G.; Epstein, A. J. *Proc. Soc. Plast. Engr. Annu. Tech. Conf.* , **1997**, *2*, 1373
62. Zhang, W. J.; Feng, J.; MacDiarmid, A. G.; Epstein, A. J. *Synth. Met.*, **1997**, *84 (1)*, 119
63. Avlyanov, J. K.; Min, Y.; MacDiarmid, A. G.; Epstein, A. J. *Synth. Met.*, **1995**, *72*, 65 and references therein
64. MacDiarmid, A. G.; Zhang, W. J. *Bull. Amer. Phys. Soc.*, **1997**, *42 (1)*, 212
65. Winokur, M. J.; Mattes, B. R. *Phys. Rev. B*, **1996**, *54 (18)*, 637 and references therein
66. Richter, A. F.; Ray, A.; Ramanathan, K. V.; Manohar, S. K., Furst, G. T.; Opella, S. J.; MacDiarmid, A. G. *Synth. Met.*, **1989**, *29*, E243
67. Sandberg, M.; Hjertberg, T. *Synth. Met.*, **1989**, *29*, E257
68. Kenwright, A. M.; Feast, W. J.; Adams, P.; Milton, A. J.; Monkman, A. P.; Say, B. J. *Polymer*, **1992**, *33*, 4292
69. Cao, Y.; Li, S.; Xue, Z.; Guo, D. *Synth. Met.*, **1986**, *16*, 305
70. Shacklette, L. W.; Wolf, J. F.; Gould, J. F.; Baughman, R. H. *J. Chem. Phys.*, **1988**, *88 (6)*, 3955.

Chapter 14

Electroconducting Fibers

Joseph A. Swift and Lewis O. Jones

Joseph C. Wilson Center for Technology, Xerox Corporation, MS 0114-22D, 800 Phillips Road, Webster, NY 14580

In this chapter, we examine recent advancements in high technology fibers and their use in a variety of electric and electrostatic applications. Conducting fibers are characterized by a framework that stems from their unique range of contact and bulk properties. In this light, a wide variety of conducting and semi-conducting fibers have emerged as the materials of choice for important components. We will discuss the properties and uses for controlled conductivity fibers, which span the conductivity range from metallic, through semi-conducting, to near insulating. The future of conducting fibers is extremely optimistic. Polymer fibers with engineered conductivity will play an expanding role in the lighter, more reliable products of the next millenium.

Conducting fibers have rapidly emerged over the past two decades to the point where they are now in widespread use as electrostatic contacts, electrical contacts and interconnects. Fiber containing devices that apply, remove, or modify charge, such as; static eliminator brushes, voltage-biased cleaner brushes, and photoreceptor charging brushes, are examples of electrostatic contacts. Fiber-based slip rings, sliding contacts, and connectors are typical of electric contacts and interconnects. Depending upon the engineering requirements, these devices may use "metallic", or "semi-conductive", or "semi-resistive" fibrous materials. We define an electroconducting fiber in the simplest of terms as any slender, elongated structure that exhibits conductivity. Thus, other areas of application for conducting fibers include their use in EMI/RFI shielding, precision electrodes, biological sensors, fillers for advanced composites, and substitutes for traditional wire. This chapter details the properties and characteristics of modern electroconductive fibers which enable their use in a wide and growing range of applications. Products, developed for the xerographic industry, are cited to illustrate the underlying principles of fiber conductors and contacts. We will begin with a discussion of the electromechanical contact which is fundamental to many of the applications cited.

Figures 1 and 2 illustrate the features of a pair of traditional solid metal contacts in comparison to those of fiber contacts, identified in Figure 2 as Distributed Filament Contacts (DFCs). Holm(*1*) is attributed as the first to discover the existence of highly localized, microscopic regions of current conduction within the macroscopic structure of metal contacts and to describe these as contact asperities, or, a-spots. Holm's a-spot theory posits that energy is transferred across a contact interface via only a small number of active sites that establish the most intimate electromechanical contact. As shown in Figure 1, these sites represent only a small percentage of the available contact surface. The phenomenon of constriction resistance arises from the idea that the interface regions in clean, mated-metal contacts, having the lowest electrical resistance are also the areas in closest mechanical contact. Since these occupy only a very small percentage of the available contact surfaces regardless of how smooth the surfaces may be, or how much pressure is applied, much of the available surface area in paired-metal contacts does not participate in the conduction process. Therefore, current is constricted though small regions of metal contacts.

Other researchers including; Greenwood(*2*), Dorsey and Hayes(*3*), Williamson and Hunt(*4*) etc. (*5,15*) have examined the dependency of constriction resistance upon apparent area. Many have concluded that the real electrical contact area is independent of the apparent geometric area of contact. It is important to note that a distinction must be made between constriction resistance, which again is the measure of conduction across the a-spots of theoretically clean, albeit impractical(*14*), surfaces, and, film resistance(*16*) which is representative of current flow across in-situ contacts where thin layers of inorganic and/or organic film(s) are present. Thus, the so-called "contact resistance" is the sum of these two and is nearly always dominated by the film resistance component. Understandably then, "contact resistance" has been the subject of many research studies that have focused upon such factors as: metal composition(*5-8*), metal hardness(*5*), shape(*16*), pressure(*5*), roughness(*4-9,13*), environmental degradation(*10*), contact overcoatings(*5,11*), contaminants(*13,14,16*), and lubricants(*5,12*).

As shown in Figure 2, the area of contact between mated fiber contacts is strikingly different from conventional, paired-solid metals. Owing to the fact that many individual fibers can independently contact, the real areas of contact are a function of many factors, such as; the number, size, distribution, surface roughness, hardness, load and distribution, and the like. Given that the length of the fibers within a contact pair is long with respect to the diameter, adjoining fibers can establish contact along their sides as well as at the tips. Thus, the contact area can actually be of a different type and potentially much more complex. For very small diameter (i.e. less than about 10 - 15 micron), non-interacting fibers of the contacts in Figure 2, the number of individual contacts can fall in the range of 1,000 to 2,000 contacts per square millimeter. Clearly, this number suggests that the potential number of current carrying sites may be several orders of magnitude higher than the a-spot regions within the metal contacts of Figure 1. In this case, the potential for high contact redundancy and for randomness of the individual contact points can lead to very high contact reliability(*18*).

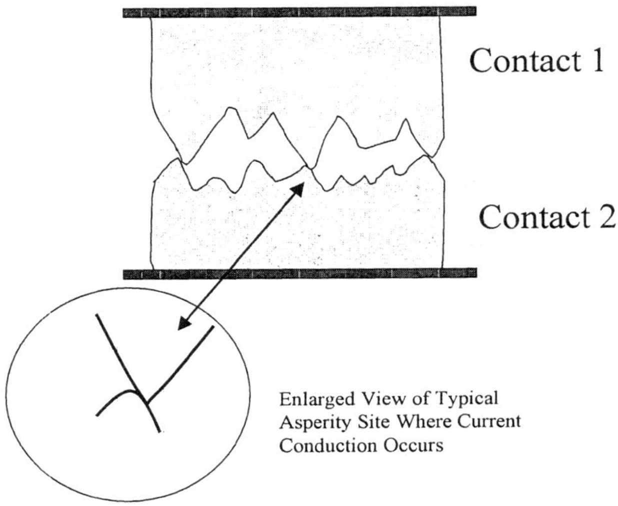

Figure 1. Illustration of the contacting surfaces of a pair of typical metal Contacts.

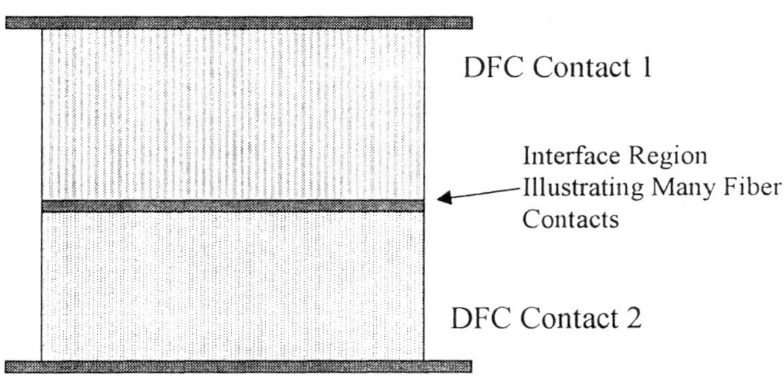

Figure 2. Illustration of the contacting surfaces of a Distributed Filament Contact (DFC) mated pair.

Further, a contact pair of very small diameter fibers, under certain conditions may favor annular quantum mechanical tunneling(*17*), in addition to the conventional conduction mode of current flow. The mechanical wedging effect of dust or other particulate debris which can act to separate solid contacts as detailed by Harper(*13*) is clearly less of a failure concern with the fiber contact. There are two reasons for this. First, the void areas surrounding the fibers can act as a receptacle for debris and regulate its presence at the contact interface. Second, the flexibility of the fibers permits inter-fiber movement which theoretically can dislodge debris from the regions of highest contact pressure. High flexibility and resiliency of fiber-contacts accounts for other desirable physical characteristics, such as softness and suppleness, which can lead to resistance saturation at low contact pressure(*18*), low-bounce, -wear, and -abrasion. Thus, it is clear that fiber contacts, especially those comprising a large number of independently acting contacts, can represent a significant departure from the traditional, monolithic solid contact.

With this framework in mind, we will now turn attention to the various types of fibers. Using constituent materials and configurational features, modern conductive fibers can be grouped into four major categories. Table I identifies the various types of conductive fibers delineated by these variables (i.e. materials: metal vs polymer and configuration: single vs multiple component structures). The table is arranged in decreasing order with respect to conductivity. Most of the values originate from the literature cited and are believed to be typical of the materials shown. The reader is cautioned that there are large variations amongst individual samples of these fibers. For our work, a simple, two-electrode technique using well dried, silver print contacts coupled with a Keithley Model 617 programmable electrometer-ohmmeter was used. It is important to note that sources of measurement error may exist, particularly with irregular, multiple layer, and/or complexly shaped fibers. The two probe-method may itself contribute to the variation noted. Since determination of the conducting layer's cross section in these cases is often imprecise, we calculate an "effective" conductivity for the layered configurations by using estimates of the average area available for conduction from electron micrographs. This area is the basis for the conductivity calculation. Therefore, conductivity is calculated for the effective conducting area recognizing that the fiber's entire cross sectional area may be larger. Wherever effective conductivity is reported, it is so designated.

Type 1: Metal and Metallized Fibers

The modern use of Type 1, solid metal fibers as high performance contact elements can be traced, at least, to the mid 1960s where, for example, Evanicsko and Deibel(*19*) worked with refractory metal fibers (i.e. tungsten, molybdenum, and their alloys) imbedded in metal matrix to produce a fiber-composite contact. While the fibers of their studies were comparatively large, typically 125 -250 microns in diameter, these researchers were able to demonstrate improvements in the operational life of electric contacts used in high current arcing environments. Similar studies were reported in the early 1980s by Reichner(*31,32*) who showed that it was possible under specific conditions to use copper fibers as sliding contacts at extremely high currents (12 MA/m^2, 8,000 A/in^2). More recently, Wilsdorf, et. al.(*20, 21*) is attributed with making significant advancements in metal fiber contacts based upon much smaller sized fibers

Table I
Classification of Conducting Fibers

Type	Category	Description	Typical Range of Conductivity (S/cm)
1	Metal	Fine wire drawn to diameters of 50μ or less. Typical materials are the transition metals and their alloys.	$10^3 - 10^6$
1a	Metallized metal	Thin metal overcoating on a metal wire fiber.	$10^3 - 10^6$ * (* based upon most conductive layer)
1b	Metallized polymer	Thin metal overcoating (e.g. 1-3μ) on an insulating polymer.	$10^3 - 10^6$ *
1c	Metallized carbon	Thin metal overcoating on a carbon fiber.	$10^3 - 10^6$ *
2	Carbon	Carbonized or partially carbonized polymer.	$10^3 - 10^{-10}$
3	Conducting Polymer	Polyacetylene (PAc), polyaniline(PAn), polypyrole(PPy) fiber.	$10^2 - 10^{-3}$
3a	Conducting composite polymer	Conducting polymer overcoating on an insulating polymer core or blends of conducting and insulating polymers.	$10^2 - 10^{-4}$*
4	Composite Polymer	Fiber forming host fiber containing conductive carbon black, metal particles, or conducting polymer powder.	10^3-10^{-16}
4a	Bicomponent, composite polymer	Conductive particulate filled polymer segmented within, or, on an insulating host polymer.	10^3-10^{-16}
4b	Ionic composite polymer	Ionic salt filled composite polymer.	$10^2 - 10^{-4}$

(i.e. 1 – 100 microns in diameter) along with the underlying science and technologies thereof. The 1995 work by Kuhlmann-Wilsdorf(*17*) is particularly authoritative and comprehensive. Highly conductive, metal fibers have found contemporary industrial use as sliding contacts in, for example, slip rings that are made and marketed by numerous commercial sources. These applications indicate that Type 1 contacts are particularly suited for low to high current devices employing sliding motion(*18*).

During the 1960s and 1970s, metal fibers were a research subject within the xerographic industry where a variety of applications centered upon use in brush-like devices for transferring developed images(*22*), for charging an electrically insulating surface(*23,24*), for grounding and for controlling static(*24*). The latter is perhaps one of the most successful and longstanding uses of metal fibers in xerography. Here, the use of fine wire-fiber contacts, such as, stainless steel at approximately 12μ diameter, see Figure 3, having metallic conductivity emerged on a large commercial scale as the passive type, static eliminator brush similar to that illustrated in Figure 4.

This device is a long, thin, strip-type brush made from bundles of individual, continuous length fibers approximately 1/2 inch long. It functions to remove excessive levels of static charge from moving copy sheet by contact, or near contact, with the charge bearing sheet and by providing multiple conductive pathways to ground for the charge to flow. Removal, or, at least reduction of residual charge from the sheet is a requirement for reliable paper handling in high speed machines. Otherwise, the movement of the paper is retarded by electrostatic drag forces in the manner noted by Crowley(*25*) and can result in improper process timing and machine jams, particularly in dry environments. This type of static eliminator is characterized as "passive" because, other than a connection to a suitable ground, no external power source is required for its function. Thus, design simplicity, compact size, and low cost have been determinants of their enduring popularity. In fact, a very large number remain today in many xerographic products unchanged from their early design. In spite of this success and popularity, the metal (Type 1) and most metal plated (Types 1a-c) fibers have been limited in this industry to more, or less, this type of electrostatic discharge application. The main reason for this relates to their characteristic, metallic conductivity which has caused undesirable arcing observed during attempts to use them in other applications such as, in charge placement, for example. Metal wire fibers also exhibit poor resiliency and can have a tendency to work harden which has led to problematic fiber set and fiber breakage. As discussed later, we will see a trend towards use of the Type 2, carbon fibers with semi-conducting properties as the replacement for metal and metal containing fibers in the more recent static brushes and other contemporary devices.

Metallized metal, polymer and carbon, Types 1a, 1b, and 1c, are variants of the solid metal fibers and are distinguished therefrom by a metal layer upon the base fiber's periphery. They are fabricated by electrochemical deposition or grafting of a suitable metal, such as nickel, copper, aluminum, and their alloys, as a thin layer upon the fiber's surface. In general, these variants evolved in attempts to improve upon one, or more, properties of the Type 1 fibers. Applications of metal-on-metal, Type 1a, are typified by the structures described in 1972 by McNab(*26*) who used refractory, non conducting, base fibers for example, aluminum oxide and boron nitride, upon which were deposited films of noble metals. McNab's objective was to improve upon the strength and flexibility of Type 1 fibers by selecting a base fiber for its mechanical

Figure 3. Photomicrograph of a circa. 20 micron diameter stainless steel fiber at1000X magnification.

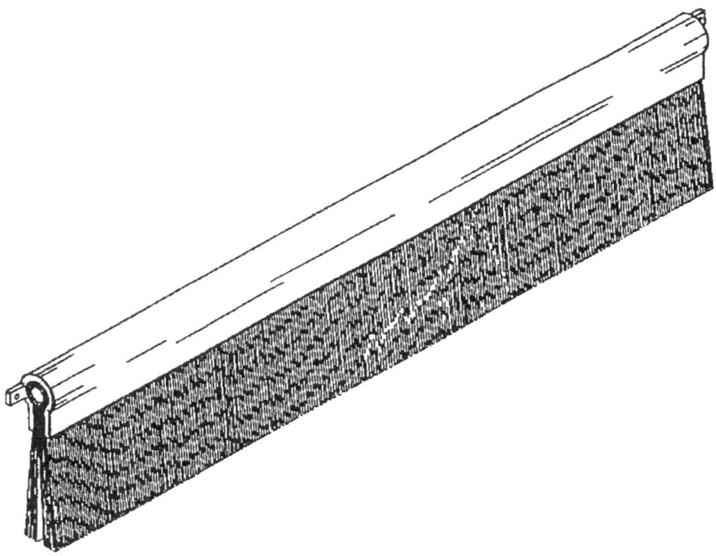

Figure 4. Static Eliminator Brush Device.

properties, independent of its conductivity, and rendering the composite fiber electrically conductive by the metal overcoating. Current transfer brushes that exhibit low contact force have been cited(*26*) as uses for the Type 1a fibers. Importantly, the layer-on-base fiber is a configuration common to many of the other fiber types.

The Type 1b fibers are a metal-on-polymer type where, in general, a metal layer is applied to a conventional textile fiber, such as, nylon, polyester, silk, cotton, or acrylic to impart electrical conductivity while preserving the other desirable properties of the base fiber. Nickel and silver electroplated polyester and nylon fibers have been converted into yarns suitable for fabric formation. The resultant cloth-like fabrics exhibit anti-static properties which are beneficial to medical operating room and cleanroom garments, for example(*33*). Very interesting, recent high technology applications of the Type 1b fibers are reported by Horch, et. al.(*34,35*). These involve use of Type 1b fibers as leadwires for neuroprosthetic devices and as biologic microelectrodes for monitoring neural activity. In this case, three metals; a Ti/W alloy, Au, and Pt comprise 0.3 micron thick metallization layers incorporated upon a monofilament polymer by a series of vacuum deposition processes to result in mechanical stability and an extremely high degree of flexibility not possible with the solid, single-metal type fiber.

Similar to the Types 1a and 1b fibers, the Type 1c fibers consist of a thin, conductive metal layer electrodeposited upon a carbon base fiber (see Figure 5). Their manufacture is described by Morin(*27*) and by Hall and Ando(*28*). The paper by Hall and Ando provides a good overview of their properties and characteristics. Nickel plated exPAN carbon fiber are typical of the Type 1c fibers that are readily commercially available. General uses for Type 1c fibers are in ESI shielding, conductive adhesives and paints, conductive fabrics, and high performance electric contacts(*29*). They are included in the Type 1 category because their conductivity is characteristically metallic. Thus, by this convention they appear in the Type 1 classification while the various other carbon fibers fall into the Type 2 category.

As is the case with Type 4 and some Type 3 fibers, the Type 1a, 1b, and 1c fibers are characterized as composite fibers because they consist of at least two components which can vary in proportions and shapes so as to combine and enhance the original properties of the constituents(*30*).

Type 2 Fibers (Carbon)

One of the first technical uses of carbon fiber dates back to the 1890s when Thompson(*38*) used a lightly metallized, carbonized cellulose fiber to make an electric contact having a large number of contact points requiring only a minimum of contact pressure to secure sufficient performance. Fanter, et. al.(*49*) describe the origins of modern, high strength and high modulus carbon fibers which have emerged to become a centerpiece of the reinforced composites industry during the period between 1960 and 1980. Due to the high cost of carbon fiber early in this period, applications were cost-constrained to those, such as aerospace and military, that were able to justify the demand driven costs. With the addition of new production capacity by new entrant suppliers into the market, costs have fallen steadily in a manner typical of a competitive market thereby permitting numerous commercial applications to surface. These include

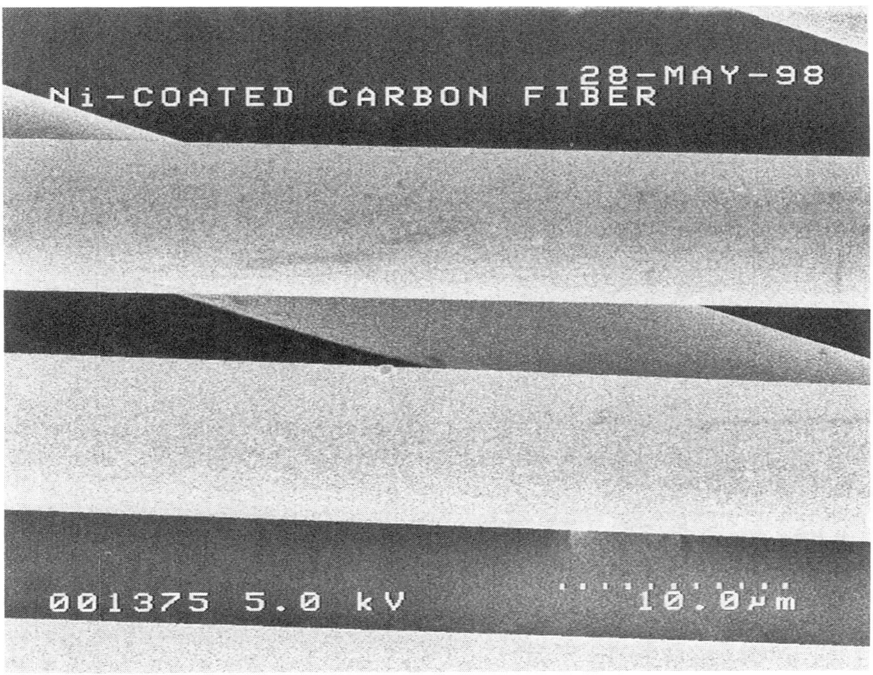

Figure 5. Type 1c, metallized (nickel coated) carbon fiber at 3000X magnification (INCO Europe Ltd).

a wide span of products, such as; components for sporting goods, recreational sailboats, medical apparatus, land transportation, etc.

As mentioned earlier, many of the passive type static eliminators used in the xerographic industry have now migrated to carbon fibers(*50,51*) because of the ability to selectively tailor their conductivity over the range of 10^3 to 10^{-5} S/cm that has been demonstrated by a number of researchers(*52, 54, 59, 65*). Still other applications have recently found use as unique, carbon fiber reinforced advanced composites that combine the properties of controlled electrical conductivity, high strength and low density. These include, for example, components for sensing, switching and biasing devices and low arcing electrical contacts which will be discussed somewhat later.

Thus, the 2nd fiber category encompasses carbon and carbonized polymers. The foundation materials of this group are the heat convertible, fiber forming polymers, the most common of which today is PAN (polyacrylonitrile). Others are rayon, PBI, and pitch tar. One of the earliest to report results of PAN pyrolysis was Goodhow, et. al.(*53*) in 1975. Later, in 1979, Fischbach and Komaki(*54*) and Brehmer, et. al.(*65*) in 1980 reported on the electrical properties of carbon fibers made from various polymers and described a dependency of resistivity upon the heat treat temperature (HTT) employed to carbonize the fiber which is now well known. The studies by Swift, et. al(*52*) in 1985 and more recently reported herein were undertaken to expanded upon this base of knowledge and to initiate studies of the stability of the electrical properties of fibers made on commercially viable platforms. These studies have led to a launch point for what is believed to have been the first, relatively large scale, commercial application for partially carbonized PAN fibers(*50*), as resistive carbon fiber based static eliminator brushes.

In the conventional fiber carbonization processes, the polymer precursor fiber is converted to a carbon fiber in a sequence of steps that, in general, involve low temperature stabilization in an oxygen-rich environment followed by higher temperature treatments in vacuum or nitrogen atmosphere. The works by Ehrburger and Donnet(*55*) and by Ishikawa, Nagaoki, and Lewis(*56*) provide comprehensive discussions of the heat conversion of PAN and other precursors to carbon fiber. Cernia(*57*) provides a basis for understanding the formation of polyacrylonitrile fiber.

One measure of the uniqueness of carbon fiber is the wide range of conductivity it can provide. Reported(*56,65*) as 10^5 to 10^{-14} S/m, it is obtained by full or partial carbonization over the heat temperature treatment (HTT) range of from about 350C to 3000C. Another measure of carbon fiber's uniqueness is the range of elastic modulus possible. The interrelation between modulus and structure has been well known for more than a decade(*66*) and is a covariant of conductivity that spans from about 100 to >350 Gpa(*55*). Importantly, the dependency of these properties as a function of HHT has been well outlined in 1981 by Lerner(*58*) who cites impurity formation as the basis for the semiconductive behavior and reports an activation energy of 88 kcal/mole for the conduction process in certain carbon fibers. Teoh, et. al.(*59*) discuss also the chemical structure of these fibers which underpin the observed range of conductivity. Their data illustrates that the conductivity of fibers from any given degree of pyrolysis is not only semiconductive, but also, fairly stable. Teoh reports that the conductivity increase with temperature is reversible. Yang and Butkus(*60*) in 1980 discussed structure, strength and conductivity as a function of HTT and indicated that tensile strength peaks at about 1800 C, while the modulus continues to increase up to the

3000 C mark cited earlier(*56*). An indicator of the research intensity in the field of modern carbon fibers and their applications can be illustrated by the fact that literally hundreds, if not thousands, of technical papers and publications have been written since 1960. The reader is advised that those cited herein are but a mere fraction of what exists on this subject.

Studies on carbon fibers began at Xerox Corporation's Wilson Research Center during the 1970s and led to a focus on the PAN-based fibers that were commercially available at that time. This resulted in commercialization of partially carbonized fibers within a range of conductivity between 10^{-8} and 10^{3} for the earlier mentioned static brush(*50*) as well as experimental probes into other application areas, such as; charge deposition brushes(*23,24*), sensors(*61*), and electric contacts(*62*).

Our most recent work has been conducted on partially carbonized, ex-PAN carbon fibers obtained from Amoco Performance Products of Alpheretta, GA. For some of our studies, a series of seven, experimentally produced samples of T-300 like, 12k fiber were prepared to span a range of initial conductivity from approximately 6×10^{-3} to 2 S/m. Environmental aging effects upon instantaneous conductivity as well as the long and short term temperature dependence were tracked for a period of more than a year. At regular intervals, each sample was analyzed by Leco Corporation of St. Joseph, Michigan for changes in elemental composition (viz. O, C, and N). The results, see Figure 6, were correlated to the instant conductance (measured and reported as resistance per unit length of fiber tow). No correlation with relative humidity could be made even though the R.H. in which the samples resided fluctuated in an uncontrolled fashion through the seasons over a range of 15% to 80%. Likewise, no significant correlation could be observed nor attributed to variations in the absolute nitrogen concentration as the nitrogen varied only from 19.4 to 20.8% for all these samples. This led to a focus upon the effect of oxygen uptake because this was found to correlate highly with decreased conduction. Figure 6 includes a highly conductive,T300, fiber as a reference to illustrate that the oxygen content model appears to apply over the entire range of partially and completely pyrolyzed PAN materials.

In earlier studies at Xerox, circa. 1983, electron spin resonance showed a decrease in spin density with time for exposure to oxygen in zero humidity indicating that oxygen diffusion occurs independent of the presence of moisture. These observations led to a working hypothesis that centered on the diffusion of oxygen into the bulk of the fiber and reoxidization as a primary determinant of the time-based, conductivity changes. Our hypothesis was that the included oxygen was acting as a scattering center, recombination center, or trap and thereby decreasing the mobility of charge carriers. This work parallels that of Lerner(*58*) who concludes that decreases in conductivity upon aging in air is caused by a decrease in the electron-phonon scattering time and not by a decrease in the number of charge carriers.

Venner and Ko, in a 1990 patent(*64*), claim that a secondary post processing heat treatment may be used to accelerate and perhaps stabilize the aging behavior and note that the untreated, partially carbonized fiber will continue to increase in resistivity, albeit slightly, for very long periods. This has become an area somewhat in dispute, as our experience with contemporary fibers indicates otherwise. Figure 7 is linear plot of room temperature resistance as a function of time for two samples of the above-described set. This data indicates that the room temperature resistance for these

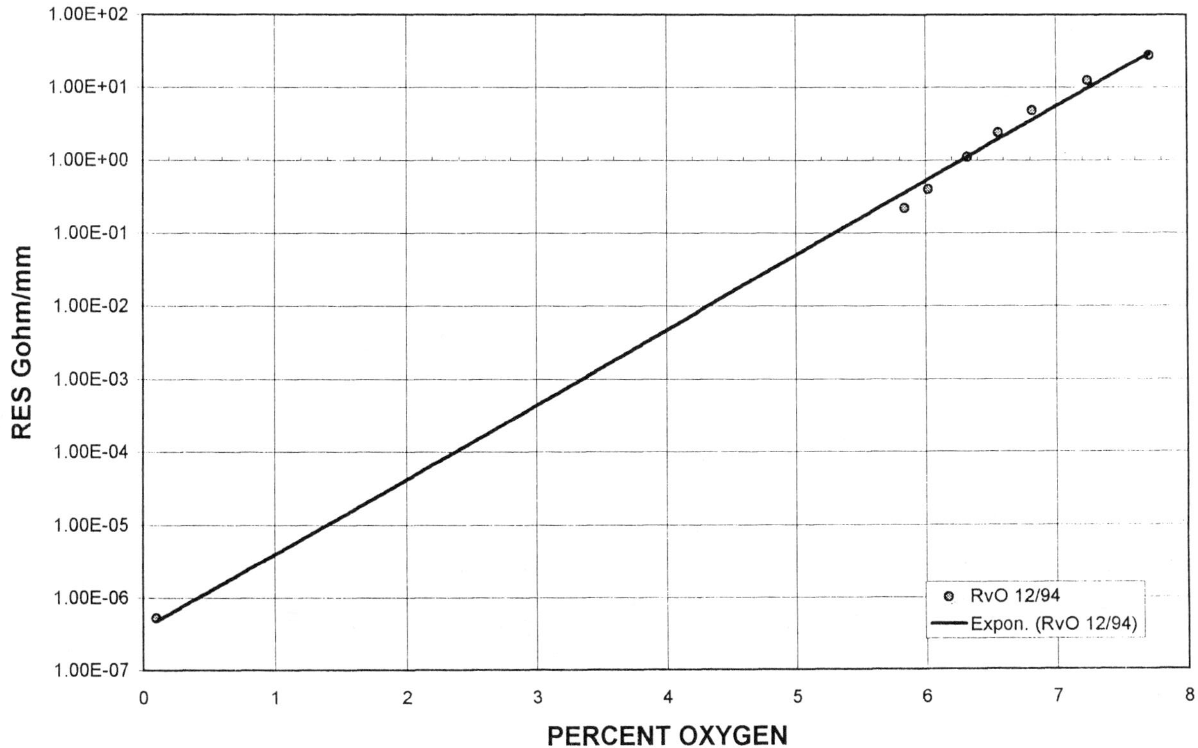

Figure 6. Graph of resistivity as a function of oxygen content for partially carbonized ex-PAN fibers.

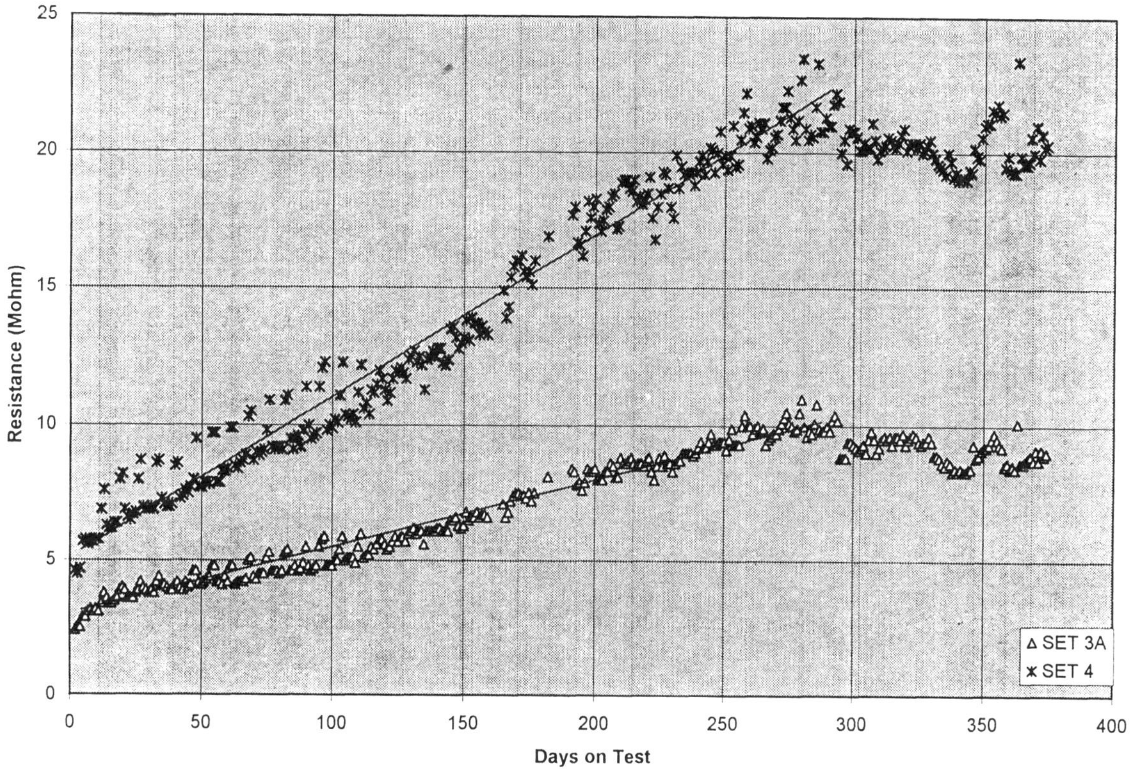

Figure 7 Graph of sample resistance as a function of aging time under ambient environmental conditions.

samples, aged under laboratory ambient, conditions increased monotonically for approximately 260 days and then leveled off to a mature stable value. The slight variation observed during the final 100 days or so is largely due to short term, ambient temperature fluctuations.

The behavior exhibited for the samples in Figure 7 is typical for all seven samples. Thus we believe that, even though the early resistance aging data can be fitted to a power function and will appear as a linear relationship on a log-log scale, the rate of change does indeed saturate, but, takes about nine months under ambient aging conditions. The mature resistance level stabilizes at about 3 to 5 times the initial value depending upon the degree of pyrolysis. At this matured level, the resistance values can be normalized for the observed, small variations in instantaneous ambient temperature. The values, adjusted for short term temperature fluctuations for the lowest resistance sample, viz. Set 1, appear in Figure 8 showing saturation occurs beyond about 260 days.

Owing to the fact that short term temperatures affected instant resistance measurements, a strong interdependency was suspected. The resistivity as a function of temperature was measured for all seven sample sets across the range of -10 to 100 °C in order to examine the semiconducting properties of the fibers. Using a diameter of 8 μ, the conductivity in SI units (Siemens/meter) was calculated and plotted as an Arrhenius function. The results, shown in Figure 9, reveal behavior that is typical of semiconductors. Energy gaps for these materials range from 0.35 to 0.55 eV, which agrees with those reported by Lerner(*58*). This work provides evidence that (based upon the facts that the target resistivities were readily and precisely achieved over a wide range and the aging behavior appears to be well characterized and predictable) supports the notion that the pyrolysis process is well understood and controlled relative to electrical properties.

The semiconductive properties are an important feature of electric contacts made from Type 2 carbon fibers. Over the past decade or so, devices made using Distributed Filament Contacts© (DFC©) have shown great promise in xerographic applications of interest here. These are described in detail elsewhere(*18*). In brief, DFCs are fabricated from advanced composite plastics where continuous carbon fiber tow is pultruded with a thermosetting host resin to form a suitably shaped, solid material having loadings of carbon fiber in the range of about 65 to approximately 90% by weight. A high performance, fiber rich electric contact is then manufactured by selectively removal of the polymer from the fibers by use of the waterjet or laser methodologies described elsewhere(*67-69*). DFCs that are configured into slip ring devices(*70*) have now been commercialized in a number of xerographic products. The tendency towards decreased contact resistance with increased temperature which, for example, can result from a transient current overload or overvoltage, is a potential mechanism for DFC-type contacts to self-regulate against potentially destructive overloads. Thus, as fiber contacts experience a hot spot from an overload, this semiconductive behavior could result in decreased local resistance, increased current flow, and nondestructive recovery from the event. The semiconductive nature of the Type 2 materials may serve as a factor in their adoption in temperature sensitive applications where metal may not be a suitable option. We, therefore, surmise that the

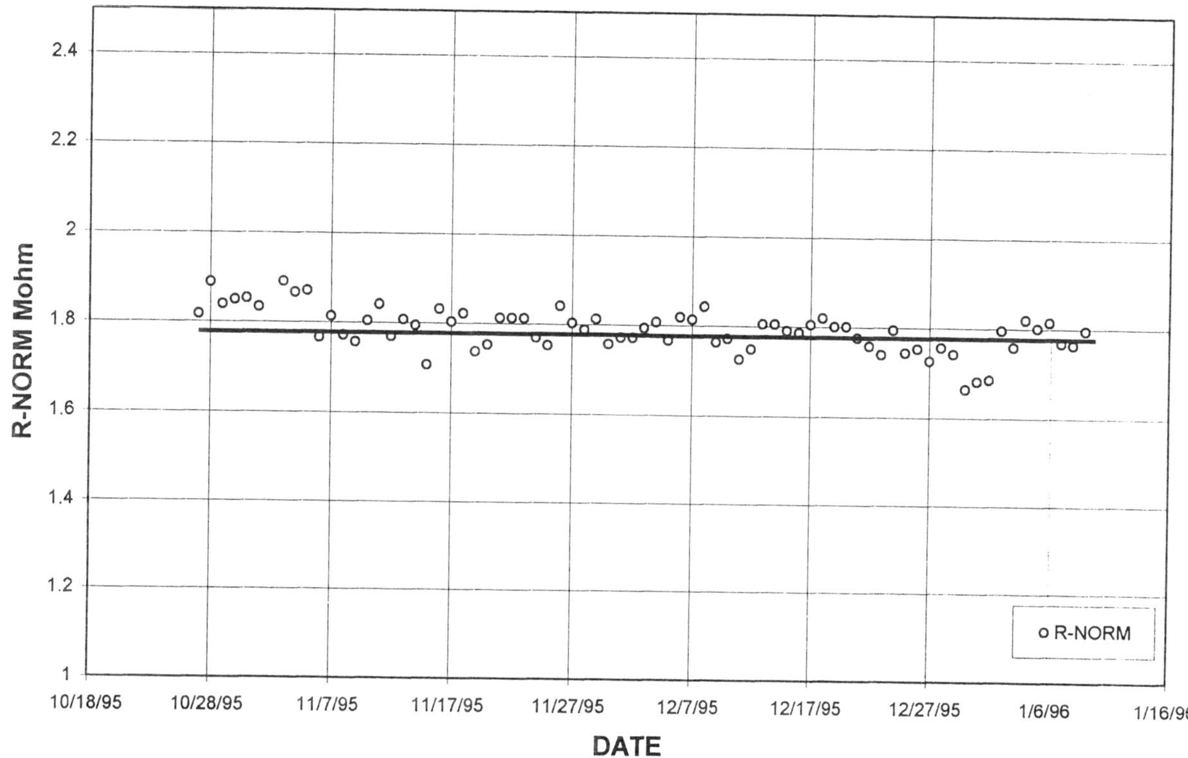

Figure 8. Linear plot of resistance as a function of time beyond 260 days adjusted for short term ambient temperature fluctuations.

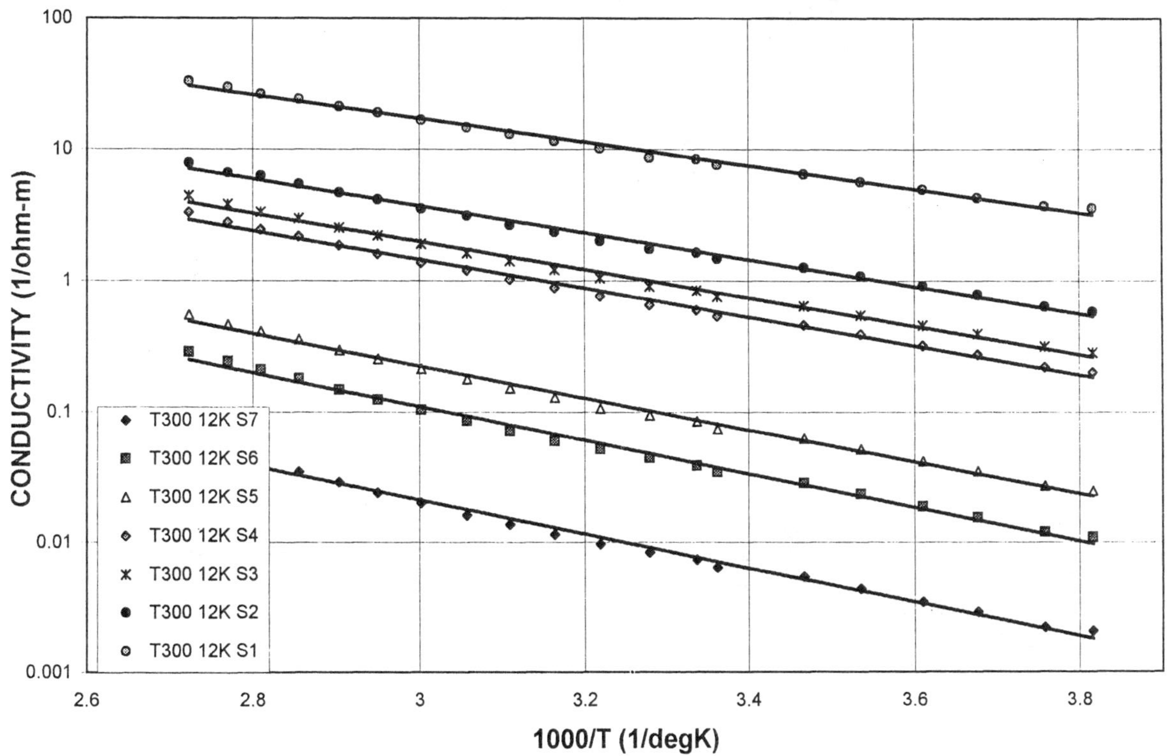

Figure 9. Ahrrenius plot of log conductivity as a function of 1/Temperature for partially carbonized PAN fibers.

use of carbon fibers at high temperature can benefit some applications by relying upon this thermal-conductivity enhancement.

Another intriguing application for carbon fiber is in the form of carbon fiber wire(*71*) used for carrying the low current, high voltages that are common in xerographic products. In this case, the controlled conductivity of 3k partially carbonized carbon fiber tow may be selected in the range of 10-3 to 10-6 S/cm. The fiber is then combined with an electrically insulating jacket via a conventional wire manufacturing process and formed into a stranded, sheathed wire. This carbon fiber wire has interesting and useful properties. Using the aforementioned semiconductive properties, it is an unusual configurational form that can respond to and thereby sense temperature. It can also provide a distributed, rather high, resistance along the entire length of the wire to yield a distributed resistance capability that is not possible with metal, which can provide desirable RFI/EME/EMI properties without additional components to a circuit. The carbon fiber wire can be combined into cable assemblies having DFC type connectors at the ends. In this case, the entire cable can be non-metal. Which, when coupled with the x-ray transmission or high radiation immunity properties of carbon conductor, may suggest other high technology or medical applications, for example, that will emerge to benefit mankind.

Two other examples of applications for Type 2 fiber include its use as a contact and a non-contact member for a surface sensing electrovoltmeter(*72,73*). In the first case, large numbers of carbon fibers create a sliding contact with a delicate photoreceptor surface and serve to detect localized electrostatic potentials that exist under various conditions. Again, the controlled conductivity property of the Type 2 fibers coupled with their low contact pressure and low abrasion characteristics provide a "soft touch" to the moving surface. In the non-contact sensor case, the extra-ordinary resiliency of carbon fiber enable its use as a unique fibrous electromechanical interconnect between a rapidly vibrating sensing element and a printed wiring board. These examples highlight the rapid advance of carbon fiber into a very wide spectrum of high technology components that is now in progress.

Type 3 Fibers (Conducting Polymer)

Fibers made from conducting polymer are the newest of the type classes. The accidental discovery of a polymer having conductive properties is reported to have occurred in Japan in the early 1970s, when a student at the Tokyo Institute of Technology added an excessive amount of catalyst to a sample batch of polyacetylene. The result, when later doped with iodine, an oxidizing agent, by MacDiarmid, et. al. at the University of Pennsylvania in 1977, was a new form of polyacetylene $(CH)_n$ that materialized as a silvery conductive film. The history is summarized nicely by Edelson(*39*). Today, the list of "electric plastics"(*40*) is substantial and growing. It includes: polyaniline, poly diarylanilines, polyfuran, polysulfur nitride, polypyrrole, poly-p-phenylenevinylene, polyparaphenylene, and polythiophene. One manner in which conductivity is derived in these conjugated polymers is by introduction of mobile charge carriers into the backbone Π-electron network via chemical doping with oxidation, reduction or protonation initiators(*75,76*). The paper by Miller(*46*) entitled "Conducting Polymers - Materials of Commerce" and the volume by Moulton and Smith(*47*) are excellent reviews of the underlying technology and the evolution and commercial applications of these materials which, without doubt, have been numerous. Since nearly all of the cited applications thusfar use the solid or film form of conducting polymers, we conclude that the use of conducting polymer fibers is believed to be rather limited as of this time.

The earliest conducting polymers were characteristically hard, intractable, insoluble, environmentally unstable, or difficult to process into fibrous forms. Although these limitations impeded their commercial adaptation at one time, they are now being overcome and all forms, including fibers, have now emerged. For example, recent developments of fibers based upon polymer blends(*77*), such as polyaniline + poly p phenyleneterephthalamide and poly N-methylpyrrole + poly bisphenol-A-carbonate(*48*), have led to improved mechanical properties and stable electricals. Epstein and Miller(*74*) in 1979 described conducting polymers as long-chain conductors. Seemingly, this long chain domain of conducting polymers should materialize, in the ideal case, in the form of fibers since, clearly, these have geometry dictated long-domain orientations. In spite of the aforementioned limitations to the emergence of conducting polymers, the tide now appears to be turning. Today, a variety of fibers made from conducting polymers have been reported(*41-43*), including demonstration of the viability of melt spinning(*41*) as a manufacturing process. At least one firm, Milliken of Spartanburg, SC, is making efforts to develop commercial markets for a Type 3a fiber based upon a surface reaction of a film forming aniline compound in aqueous solution with an oxidizing and a doping agent upon a base fiber or fabric(*44*). Figure 10 illustrates a Milliken fiber coated with a conducting polyaniline.

Fabric patterns having defined, high conductivity regions greater than 10^{-3} S/cm have been developed by DeAngelis et. al(*45*) at Milliken. The patterns are established by depositing a uniform conductive polymer film upon the textile then protecting the regions to remain on the fabric with a protective layer and then chemically etching and rinsing to remove unprotected conductive polymer regions.

The communication by Chiang, et. al.(*42*) describes a process of in situ polymerization within a carrier gel for producing continuous polyacetylene fibers having conductivity in the range of 1200 to 6000 S/cm. Alternatively, Bhattacharjee and Paley(*78*) describe anti-static cleanroom products, including wipers, swabs, and garments made from particulate pyrrole polymers impregnated in low concentration into the surfaces of the host fibers. Jousse, et. al(*79*) provide an excellent insight into processing conducting polymers into conductive fabrics which can act as microwave absorbing, antistatic, and EMI materials.

One of application areas considered for conducting polymers within the xerographic industry is its use as a conducting layer upon a seamless belt for ionographic imaging(*80*). A somewhat similar application is cited(*81*) for an antistatic layer of a photographic element. While the opportunity for Type 3 fibers to impact upon the xerographic industry may be substantial, as of the time of this writing, commercial applications of Type 3 fibers in xerographic products remain in the evolutionary stage.

Type 4 Fibers (Composite Fibers)

A well known and straightforward way of rendering a conventional, insulating polymer-fiber electrically conductive is to blend a suitable conductive filler, as randomly

Figure 10. Photomicrograph of a Type 3 Conducting Polymer fiber at 3000X magnification (Milliken).

distributed particles, with a fiber forming host polymer to form a composite. Particle size, shape, distribution, and conductivity are determinants of the minimum filler concentration within any specific polymer system that is needed to achieve particle to particle contact essential to electrical conduction. This critical loading is referred to as the percolation threshold and depending upon the specific gravity of the filler in relation to that of the polymer usually falls within the range of about 10-20% by weight of the polymer. This brief review of particulate conduction establishes the configurational framework for the Type 4 category of fibers which are characterized as filled polymer systems in the manner described in 1995 by Garboczi, et.al.(*36*). While any suitable conductive filler can be used with most fiber forming polymers, most commonplace fillers are those that are readily available, inexpensive, non-toxic, non-reactive, non-corrosive, and compatible with the host polymer and fiber manufacturing process. Conductive carbon blacks, short lengths of carbon fibers, fine metal particles, short lengths of metal fibers, as well as fine particles of conducting polymers have been used as conductivity enhancing fillers. The features of these systems are detailed in a 1996 work by Garboczi and Douglas(*37*).

Owing to the need during fiber formation to apply tension (for example, for spooling, unspooling, drawing and stretching the fibers), filler levels are typically held close to those minimally required for conductivity percolation. Most, if not all, particulate fillers can degrade the mechanical strength of the composite fiber thereby explaining why the manufacture of these tends to be complex and difficult to control. The difficulty of manufacture of a filled composite fiber is a reason why there are only a few Type 4 fibers commercially available. Of these, many are made by a solution or wet spinning methods which subject the fibers to only minimal tensions during its manufacture(*82,83*) and thereby minimizing in-process fiber breakage. Figure 11 illustrates the celery-like appearance of a 10-11 denier, wet spun, composite acrylic fiber.

The concept of a composite filled polymer layer on, or in, an insulating host fiber emerged as a variant that proved easier to mass manufacture. Referred to as "conjugate" and "bicomponent" fibers(*84*), many of the Type 4a fibers contain a co-axially configured, insulating core fiber that provides strength while a thin outer coating layer provides electrical conductivity. One of the earliest, commercially viable processes for depositing filled composite layers upon core fibers is attributed to Sanders(*85,86*). Sander's process uses a suitable solution to pre-swell an insulating core filament, then a solution containing a suitable polymer plus finely divided conductive particles is applied to a base fiber and dried. The result is a polymer substrate with finely dispersed conductive particles in an annular region upon the core fiber. Figure 12 illustrates this version provided by BASF of a Type 4a fiber that comprises a carbon black layer upon a nylon core. A similar, contemporary process for impregnating particulate materials into the surface of a core fiber is defined by Cintins(*87*) of DuPont.

Conductive bicomponent fibers can be made by a large number of melt spinning processes which are commonly employed throughout the fiber industry. The chapter by Hicks, et.al.(*84*) provides a thorough discussion on conjugate-fiber spinning. The list of variants of melt spinning process suitable for manufacture of conductive bicomponent fibers is enormous indicating that it has been a very active area for development. Some representative ones are attributed to; Dugan(*88*), Hodab and

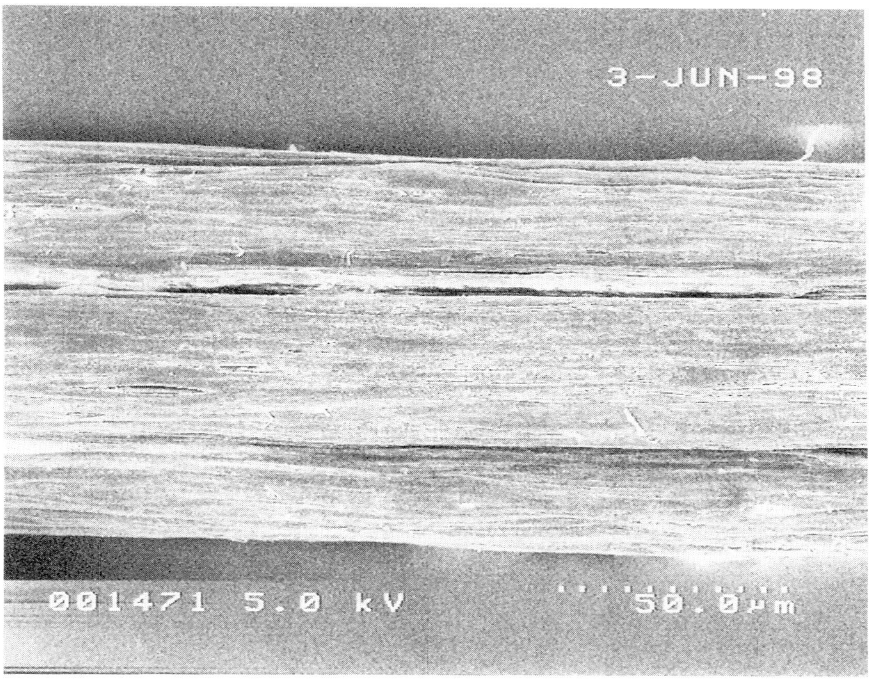

Figure 11. Photomicrograph of a Type 4 wet spun, conductive carbon filled acrylic fiber at 600X magnification (Toray).

Figure 12. Photomicrograph of a Type 4a solution coated, 17 denier, bi-layer composite fiber at 500X magnification (BASF).

Ilg(*89*), Okamoto(*90*), and Matsui(*91*). One very interesting method of spinning a conductive fiber is reported by Mitamure, et.al.(*92*) which involves co-spinning a low temperature melting, metal core and a thermoplastic sheath. Obviously, these inventors have successfully solved the problems associated with handling molten metals in a conjugate spinning process.

The final category of fibers centers upon those that achieve conductivity by inclusion of ionic salts within a polymer host. The fibers described by Iwao, et. al.(*93*) are typical of the Type 4b category.

The most successful application of the Type 4 fibers in the xerographic industry is their use as photoreceptor cleaner brushes(*94,95*). Conductive fibers are woven into high fiber fill density, plush pile fabrics that comprise the periphery of a cylindrical, brush that makes rotary contact with a moving photoreceptor surface. Residual toner is removed from this surface by bias voltages imparted upon the fibers which define an electrostatic field at the working interface sufficient to cause attraction between the toner particles and brush fibers. Upon rotation away from the surface, the toner particles move with the brush, held firmly by electrostatic forces until they are removed by another device. Fibers such as those shown in Figures 10 and 11 are used for these brushes and are selected to have an effective conductivity that is somewhat resistive, i.e. in the range of about10^1 to 10^{-5} S/cm. This range is sufficient to assure that charge relaxation fully occurs within the time of contact between the brush and receptor. We describe the electrical behavior of these materials as "pseudo-conductive" because they exhibit resistive characteristics and do not necessarily demonstrate the inverse temperature dependence as described earlier that is typical of true semiconductors.

Kandel(*96*) refers to another xerographic cleaner device, but, in this case, the fiber appears to be a Type 4b, ionic salt composite fiber.

Summary

In this chapter, we have provided details of the various types of conducting fibers which can span the entire conductivity spectrum. We have also developed a framework for their classification in a manner that is predicated upon their constituent materials and configurations. The topic of modern conducting fibers would, however, not be complete, without noting the existence of the superconducting fibers that have recently achieved notoriety. The research work reported by Hajime(*97*) plus the developments attributed to Costa, et.al.(*98*) are good starting points for those seeking further insights into this class of fiber.

Drawing upon the xerographic industry, which has been a leader in the use of electroconducting fibers, we presented details of applications that provide insights into how conductive fibers function in ways that are not possible with other material forms. The fundamental principles of electric contacts were used to set the stage for understanding fibers in this role. The contemporary work on Distributed Filament Contacts that exhibit semiconducting properties is particularly exciting because, it highlights a very successful case of where a structural material, carbon fiber, could be engineered to meet non-traditional, new requirements.

In sum, the future for electrically conductive fibers is indeed very bright. The applications that already exist serve to validate the uniqueness and value of fiber conductors. The cycle of new fiber materials feeding new applications will continue in a never ending fashion, limited only by the imagination and determination of the researcher.

Acknowledgments

The authors wish to thank the following individuals: A. Epstein (presently at Ohio State University), J. Drawe, J. Bell, and G.K. Thompson of Xerox Corporation as well as the following companies: Amoco, BASF, and Milliken for supplying some of the samples described in this chapter.

Literature Cited

1. Holm, R., *Electric Contacts: Theory and Application*, 4th edition, Springer-Verlag New York Inc., 1967.
2. Greenwood, J.A., *BRIT. J. APPL. PHYS.*, **1966**, Vol.17, p1621-1632.
3. Dorsey, G., Hayes, R., *Electric Contacts*, IEEE, 1997, p272-281.
4. Williamson, J.B.P., Hunt, R.T., *Proc. R. Soc. Lond.*, A., **1982**, 327, p147-157.
5. Wilson, R.W., *Proc. Phys. Soc.* **LXVIII**, 9, p 625-641.
6. Slade, P.G., *IEEE Transactions on C.H.&M.T.*, **1986**, vol. CHMT-9, No. 1, p 3-16.
7. Antler, M., *Plating and Surface Finishing*, **1991**, p 60-62.
8. Zhenbiao, L., Lichun, C., Guansheng, Z., *Electric Contacts*, IEEE, 1996, p 488-493.
9. Schneehans, O., Houze, F., Meyer, R., Boyer, L., *Electrical Contacts*, IEEE, 1996, p205-211.
10. Abbott, W.H., *Electric Contacts*, IEEE, 1996, p414-429.
11. Abbott, W.H., *Electric Contacts*, IEEE, 1995, p97-123.
12. Antler, M., *Electric Contacts*, IEEE, 1995, p83-97.
13. Harper, W.R., *Contact and Frictional Electrification*, 1967, p 25 ff.
14. Williamson, J.B.P., Greenwood, J.A., Harris, *J., Proc. Royal Soc.* A., **1965**, Vol 237, p560-573.
15. Tanii, T., Nakanishi, K., *Electric Contacts*, IEEE, 1997, p286-290.
16. Kuhlmann-Wilsdorf, D., Rijke, A.M., *Electric Contacts*, IEEE, 1997 p 291-302.
17. Kuhlmann-Wilsdorf, D., *Electric Contacts*, IEEE, 1995, p 295-314.
18. Swift, J.A., Wallace, S.J., *Electric Contacts*, IEEE, 1997, p 190-200.
19. Evanicsko, J., Deibel, C., Deibel, J., US Patent 3,254,189, 1966.
20. Wilsdorf, D., Wilsdorf, H.G.F., Adkins, C.M., US Patent 4,415,635 1993.
21. Wilsdorf, D., US Patent 4,358,699 1992.
22. Krause, K.A., Moradzadeh, Y., US Patent 3,691,993 1972.
23. Murray, H., Marks, L.M., US Patent 4,336,565 6/22/82
24. Ewing, J.R., Swift, J.A., US Patent 4,761,709 1988.
25. Crowley, J.M., *Fundamentals of Applied Electrostatics*, Wiley-Interscience, 1986 p 63 ff.
26. McNab, I.R., US Patent 3,668,451 1972.

27. Morin, L.G., US Patent 4,609,449 1986.
28. Hall, D.E., Ando, H., *31st Int. SAMPE Symposium*, **1986**, p 1340-1349.
29. Swift, J.A., Wallace, S.J., US Patent 5,599,615 1997.
30. Janszen, F., Vloemans, M., "Innovation and the materials revolution", *Technovations*, **1997**, Vol. 17. No. 10, p 549-555.
31. Reichner, P., *IEEE Transactions*, **1980**, Vol. CHMT-3, No. 1, p 21-25.
32. Reichner, P., *IEEE Transactions*, **1984**, Vol. CHMT-4, No. 1, p 2-4.
33. See for example, Ono, K., Okoshi, K., US Patent 4,546,497, "Antistatic clothing" 1985.
34. Horch, K.W., Mushahwar, V.K., *IEEE Trans Rehab. Eng.*, **1997**, V 5, p.237-243.
35. Horch, K.W., McNaughton, T.G., *J. Neurosci. Methods*, **1996**, V 70, p103-110.
36. Garboczi, E.J., Snyder, K.A., Douglas, J.F., *Phys. Review E*, July **1995**, Vol. 22, No 1., p 819 ff.
37. Garboczi, e.j., Douglas, J.F., *Phys. Review E*, July **1996**, Vol. 53, No 6., p 6061.
38. Thompson, E., US Patent 539,454 1895.
39. Edelson, E., *Popular Science*, June **1990**, p 90-93.
40. Ashley, S., *Mechanical Engineering*, April **1998**, p 62-64.
41. Yoshino, K., Nakajima, S., Fujii, M., Sugimoto, R., *Polymer Comm*unications, **1987**, Vol. 28, p 309-310.
42. Chiang, J.C., Smith, P., Heeger, A.J., Wudl, F., *Polymer Communications*, June **1988**, Vol 29, p161-3.
43. Kuhn, H.H., Gregory, R.V., Kimbrell, W.C., *3rd Int. SAMPE Electronics Conf.* **1989**, p570-577.
44. Huhn, H.H., Kimbrell, W.C., US Patent 4,981,718 1991.
45. DeAngelis, A.R., Child, A.D., Green, D.E., US Patent 5,624,736 1997.
46. Miller, J.S., *Advanced Materials*, **1993**, V 5, No. 9. P 671-67.
47. Moulton J., Smith, P., *High Technology Fibers, Handbook of Fiber Science and Technology: vol.III*, Part C, Marcel Dekker, NY, 1993 p275 - 315.
48. Geibler, U., Hallenseiben, Toppare, L., *Adv. Mater.*, **1991**, V3, NO. 2, p 104-6.
49. Fanter, D.L., Strandburg, D.B., Dry, A.C., *40th Int. SAMPE symp.* May **1995**, p 989-994.
50. Franks, W.S., Randall, J.M., Swift. J.A., US Patent 4,553,191 1985.
51. Swift, J.A., et. al US Patent 5,354,607 1994.
52. Swift, J.A., Epstein, A.J., Harbour, J., Walzak, J., Lin, W-P., "Physical Properties of Pyrolized PAN Fibers", *189th Nat'l. Mtg. ACS*, **1985**.
53. Goodhow, et. al., *Mat'l. Sci. and Engr.*, **1975**, Vol 17, No. 3.
54. Fischbach and Komaki, "Electrical Resistance of Carbon Fibers", *Biannual Conference on Carbon*, **1979**.
55. Ehrburger, P., Donner, J-B., *High Technology Fibers, Handbook of Fiber Science and Technology: vol.III*, Part C, Marcel Dekker, NY, 1993, p 169-220.
56. Ishikawa, T., Nagaoki, T., Lewis, I.C., *Recent Carbon Technology*, JEC Press, 1983, p 86ff.
57. Cernia, E., *Man-Made Fibers: Science and Technology*, Vol. 3, Interscience Publishers, p135-198.

58. Lerner, N.R., "Electrical Conductivity and Electron Spin Resonance in Oxidatively Stabilized Polyacrylonitrile Subjected to Elevated Temperature", *J. Appl. Phys*, Nov **1981**, Vol 52, No. 11.
59. Teoh, et. al., "Electrical Conductivity of Pyrolyzed Polyacrylonitrile", *Mol. Cryst. Liq. Cryst.* , **1982**, Vol 83, 1329.
60. Yang and Butkus, 'Carbon Fibers Conductivity Studies", NASA Contractor Report 166302, Aug 1980.
61. Swift, J.A., Werner, A.J., US Patent 5,220,481 1993.
62. Wallace, S.J., Jedlicka, J.E., Peck, W.M., US Patent 4,641,949 1987.
63. Swift, J.A., et.al., US Patent 5,410,386, "Hollow Pultruded Electrical Contact".
64. Venner, J.G. and Ko, Y.S., US Patent 4,938,941, "Partially Carbonized Polymeric Fibrous Material having an Electrical Resistivity of Enhanced Stability", 1990.
65. Brehmer, L., Pinnow, M., Ludwig, J., *Plaste und Kautschuk*, **1980**, Vol 27, No 6, p 309-313.
66. Bahl, O.P., Mathur, R.B., Kundra, K.D., Technical Note, *Fiber Science and Technology*, **1981**, 15, p 147-151.
67. Orlowski, T.E., Swift, J.A., Wallace, S.J., Peck, W.M., Courtney, J.E., Rollins, D.E., US Patent 5,396,044 1995.
68. Orlowski,T.E., et.al. US Patent 5,270,106 1993.
69. Rommelmann, H., Thompson, A.J., US Patent 5,282,310 1994.
70. Swift, J.A., (US Patents pending).
71. Swift, J.A., Wallace, S.J., Peck, W.M., US Patent 5,414,216 1995.
72. Swift, J.A., Werner, A.J., US Patent 5,220,481 1993.
73. Werner, A.J., Schlueter, E.L., Smith, J.F., US Patent 5,489,850 1996.
74. Epstein, A. J., Miller, J. S. *Scientific American.* **1979**, Vol 241 No. 4, p52-61.
75. Stejskal, J., Sapurina, I., Trochova, M., Prokes, J., Krivka, I.,Tobolkova,E., *Macromolecules*, **1998**, 31, p 2218-2222.
76. Ikkala, O., Passiniemi, P., US Patent 5,520,852 1996.
77. Kulkarni, V.G., Wessling, B., US Patent 5,290,483 1994.
78. Bhattacharjee, H.R., Paler, E., US Patent 5,736,469 1998.
79. Jousse, F., Delnaud, L., Olmedo, L., *40th Int SAMPE Symp.*, May **1995**, V. 40 p 360-365.
80. Facci, J.S., Yuh, H-J., Limberg, W.W., Badesha, S.S. US Patent 5,079,121 1992.
81. Gardner, S.A., Shaw-Klein, L.J., Brady, B.K., US Patent 5,716,550 1998.
82. Tanaka, H., Koseki, T., Fujii, S., US Patent 4,107,129 1978.
83. Tanaka, H., et.al., US Patent 3,963,803 1976.
84. Hicks, E.M., Tippetts, E.A., Hewett, J.V., Brand, R.H., *Man-Made Fibers: Science and Technology*, V 1, Interscience Pub. 1968, p375-408.
85. Sanders, J.H., US Patent 4,255,486 1981.
86. Sanders, J.H., US Patent 3,823,035 1974.
87. Cintins, M.M., US Patent 5,298,028 1994.
88. Dugan, J.S., US Patent 5,525,282 1996.
89. Hodab, J.A., Ilg, O.M., US Patent 5,575,063 1996.
90. Okamoto, M., US Patent 4,127,696 1978.

91. Matsui, M., Naito, H., Okamoto, K., US Patent 4,457,973 1984.
92. Mitamura, H., Yoshida, F., Shimura, T., US Patent 5,248,468 1993.
93. Iwao, A., Yasuyuki, H., Ishii, Y., US Patent 4,912,593 1990.
94. Swift, J.A., US Patent 4,706,320 1987.
95. Swift, J.A., US Patent 4,835,807 1989.
96. Kandel, T.G., US Patent 4,005,512 1977.
97. Hajime, K., J.*Ceramic Soc. Japan*, **1994**, Vol 102, p 195-201.
98. Costa, F.M., Goncalves, A.P., Abilio, C., *Physica C*, 1994, p 235ff.

Chapter 15

Organic and Polymeric Materials for the Fabrication of Thin Film Field-Effect Transistors

Zhenan Bao

Bell Laboratories, Lucent Technologies, 600 Mountain Avenue, Murray Hill, NJ 07974

Organic thin-film metal-insulator-semiconductor field-effect transistors (MISFETs) are potentially useful in low-cost large area flexible displays and low-end data storage such as smart cards. Much progress has been made recently in discovering new materials and using low-cost solution-based fabrication processes, such as screen printing techniques. In this paper different semiconducting matrials which have been studied for thin film transistors will be reviewed. Specifically, different aspects which affect the performance of these materials, such as molecular structures, film morphologies, and fabrication conditions, will be discussed.

Organic and polymeric thin-film metal-insulator-semiconductor field-effect transistors (MISFETs) have received increasing interest recently because of their potential applications in low-cost large area flexible displays and low-end data storage such as smart cards (*1-2*). Organic materials offer numerous advantages for easy processing (e.g. spin-coating, printing, evaporation), good compatibility with a variety of substrates including flexible plastics, and great opportunities in structural modifications. Extensive research has been carried out to identify new materials with promising properties, high charge carrier mobility and high current modulation (on/off ratio). Materials with extended π-conjugation, e.g. conjugated oligomers and polymers, have received the most attention. In order for organic MISFETs to be useful for liquid crystal displays, the field-effect mobility should be greater than 0.1 cm^2/Vs and the on/off ratio must be higher than 10^6. In this paper, we will review recent progress in material development and low-cost device fabrication techniques for thin film transistors.

Experimental

Vacuum deposited devices: The transistor device structure is shown in Figure 1a. An *n*-doped Si was used as substrate, with gold contact that functioned as the gate and an oxide layer of 3000 Å as the gate dielectric having a capacitance per unit area of 10 nF/cm^2. The channel lengths of the devices were 25, 12, 4, and 1.5 μm. Semiconducting thin films were prepared by vacuum deposition at a rate of 4 to 5 Å/s under a pressure of 2.0 x 10^{-6} Torr, and the thickness of the resulting films was between 500 to 600 Å. Different substrate temperatures for deposition were obtained by mounting the substrate to a heated copper block controlled by a temperature controller and measured by a thermocouple.

Printed plastic transistors: The transistor device structures is shown in Figure 1b. An ITO-coated poly(ethylene terephthalate) film (from Southwall Technologies) is chosen as the plastic substrate. A polyimide (OPTIMER AL 3046 from Japan Synthetic Rubber Co.) layer is then printed through a screen mask onto the ITO surface. The screen mask is made of a stainless steel fabric with 400 mesh count per inch; an emulsion thickness of about 7.5 μm is used. After being printed, the polyimide dielectric layer is baked at 120° C for an hour. An organic semiconductor layer consisting of regioregular poly(3-alkylthiophene)s (from Aldrich Chemical Co.) with different alkyl chain lengths is then put down by spin-coating, casting, or printing using chloroform as the solvent. Finally, the device is completed by printing the drain and source electrodes using a conductive ink (479SS from Acheson Co.) through a screen mask made of the same fabric and using the same thickness of emulsion. The drain and source electrodes are two strips 0.5 mm x 4 mm each, separated by a gap of 100 μm, and are about 10 μm thick.

Patterning electrodes using micromolding in capilliaries: Similar steps as described above were used to fabricate all layers except the drain and source electrodes. A elastomeric mold was made by casting and curing polydimethylsiloxane against patterned photoresist. The mold was then brought into conformal contact with a surface and generated a network of capillary channels; holes machined through the thickness of the elastomer or molded during the casting and curing step allow access to these channels. Solutions of polyaniline in m-cresol, or carbon particles in ethanol wick into the capillary channels when the access holes, which act as reservoirs for the solutions, are filled. Removal of the elastomer after the solvent evaporates yields an organic conductor patterned in the geometry of the mold.

The electric characteristics of these devices were measured under vacuum (10^{-3} Torr) unless otherwise specified. The current-voltage characteristics were obtained with a Hewlett-Packard (HP) 4145B analyzer. At the saturated region, I_{DS} (drain-source current) can be described using equation (1), where μ is the field-effect mobility, W is the channel width, L is the channel length, and C_i is the capacitance per

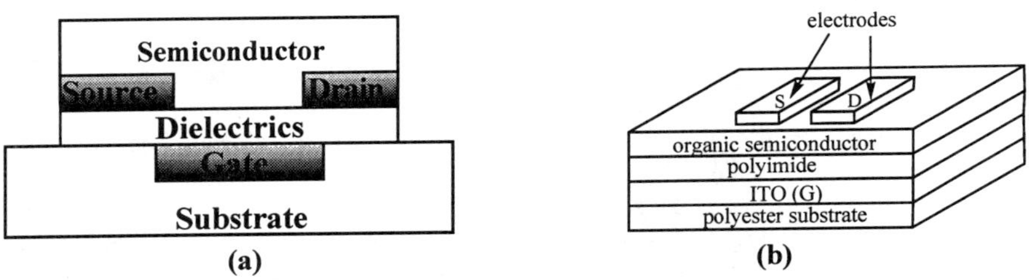

Figure 1. Device structures of organic transistors. (a) Bottom-contact device structure; (b) Top-contact device structure Copyright 1997 Wiley–VCH. (reproduced with permission from reference 33).

unit area of the insulating layer (SiO_2, 3000 Å, C_i = 10 nF/cm^2). A plot of $I_{DS}^{1/2}$ vs. V_G (gate voltage) can be used to obtain V_0, the extrapolated threshold voltage, after extrapolation to the V_G axis. The field-effect mobility can then be calculated from equation (1). X-ray diffractiograms were obtained in the reflection geometry using Ni-filtered CuK_α radiation. Electron microscopy and diffraction was conducted at 100 kV on C-coated films that had been shadowed with Pt at $\tan^{-1}1/2$ to increase contrast.

$$I_{DS} = \frac{WC_i}{2L}\mu(V_G - V_0)^2 \qquad (1)$$

Results and Discussion
***P*-channel materials**

Vacuum deposited p-channel materials. A number of conjugated oligomers and metallophthalocyanines have been studeied by different research groups as *p*-channel semiconducting materials (Table 1). These compunds have limited solubility in organic solvents and therefore vacuum evaporation has to be used to fabricate their thin films. The highest field-effect mobility has been reported with pentacene ca. 1.5 cm^2/Vs (*3*), which is in the same order of magnitute as amorphous Si (α-Si). Its high performance has been attributed by Laquindanum et al. to the ability of forming single-crystal-like filmsupon vacuum deposition onto gently heated (about 80 °C) substrates (*4*). α-Dihexyltetrathienyl (DH-α-4T) has also been found to form single-crystal-like films and high field-effect mobility ca. 0.2 cm^2/Vs has been reported by Katz et al. (*41*).

We have investigated the transistor behavior of different metallophthalocyanines (M = Cu,Zn,Pt,Ni,Sn,Fe,H_2) since they are commercially available in large quantity and high purity (*10-11*). They are also chemically and thermally stable and have been widly used in dye processing, spectral sensitization, chemical sensors, and optical data storage. They were found to function as *p*-channel accumulation-mode devices. The charge carrier mobilities of these devices are strongly dependent on the morphology of the semiconducting thin films. Highly ordered films are obtained by vacuum deposition at elevated substrate temperatures. Relatively high mobilities (ca. 0.02 cm^2/Vs for Cu-Pc and greater than 10^{-3} cm^2/Vs for Zn-Pc, Sn-Pc, and H2-Pc) and drain current on/off ratios greater than 10^4 can be achieved with optimized substrate temperature during deposition (T_D) (Table 2).

Soluble p-channel materials. To truly realize the advantages (i.e. processability and low-cost) of organic materials in device applications, liquid phase processing techniques by spin-coating, casting, or printing are strongly desired. Three methods have been used to fabricate polymer TFT devices form the liquid phase. In the first method, a semiconducting polymer layer is formed directly on the electrodes by electrochemical polymerization, and these electrodes are used subsequently as

Table 1. *p*-Channel organic materials fabricated by vacuum deposition.

compound	field-effect mobility (cm^2/Vs)	reference
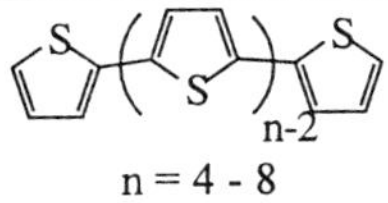 n = 4 - 8	0.002 - 0.02	(*5-7*)
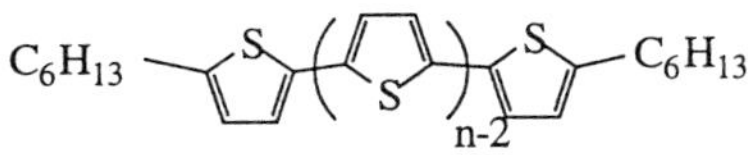 n = 4 - 8	0.01 - 0.2	(*5,7-9*)
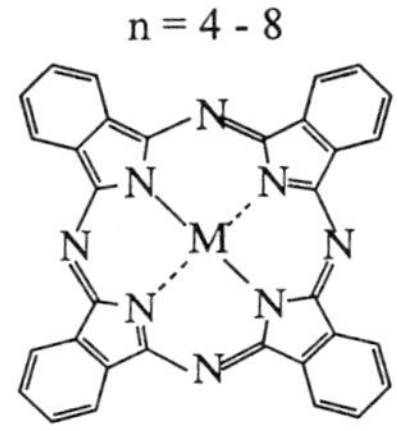 M = Cu, Sn, Zn, H2	0.003 - 0.02	(*10-12*)
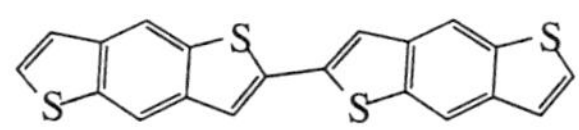	0.03 - 0.04	(*13*)
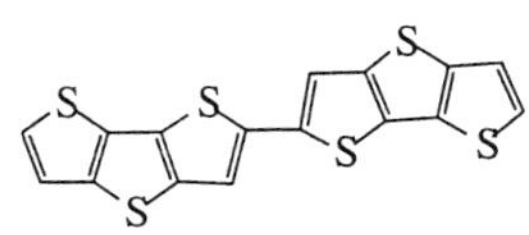	0.01 - 0.02	(*14-15*)
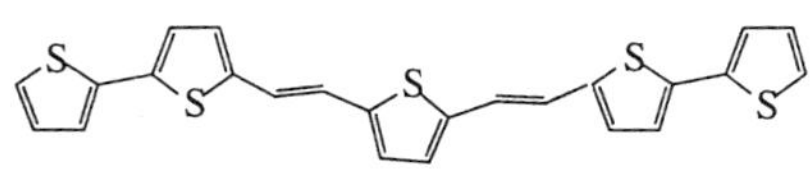	0.001 - 0.01	(*16*)
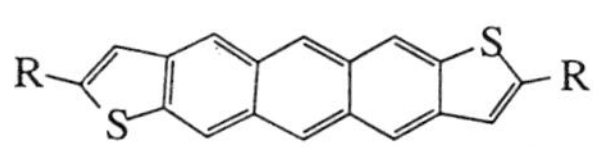 R = H, alkyl	0.015 - 0.17	(*17*)
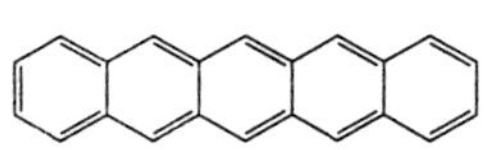	0.003 - 1.5	(*3-4,18-19*)

Table 2. Field-effect mobilities for samples deposited at different substrate temperatures (T_D) (reproduced with permission from reference 11).

M-Pc	T_D (30° C)	T_D (125° C)	T_D (200° C)
Cu-Pc	6.0×10^{-4}	2.0×10^{-2}	6.7×10^{-3}
Sn-Pc	7.3×10^{-5}	3.4×10^{-3}	no field-effect
H_2-Pc	1.3×10^{-3}	2.6×10^{-3}	5.6×10^{-7}
Zn-Pc	2.3×10^{-4}	2.4×10^{-3}	2.8×10^{-3}
Fe-Pc	3.6×10^{-5}	6.9×10^{-4}	1.1×10^{-5}
Pt-Pc	1.5×10^{-4}	1.5×10^{-4}	9.0×10^{-5}
Ni-Pc	7.0×10^{-6}	3.0×10^{-5}	5.4×10^{-5}

drain and source electrodes (*20-21*). In fact, the first organic TFT was fabricated in 1986 by this method using polythiophene as the semiconducting layer ($\mu = 10^{-5}$ cm^2/Vs) (*20*). Polypyrrole ($\mu = 1.2 \times 10^{-5}$ to 1.77 cm^2/Vs) and poly(N-alkylpyrrole)s ($\mu = 6.3 \times 10^{-4}$ to 1.74 cm^2/Vs) have also been prepared by electrochemical synthesis and their TFT properties were studied for these essentially doped films (*21*). The second technique involves the use of a soluble precursor polymer which can undergo subsequent chemical reactions to give the desired conjugate oligomer or polymer, such as pentacene ($\mu = 0.009$ cm^2/Vs) (*22*) and poly(thienylene vinylene) ($\mu = 0.22$ cm^2/Vs) (*23*). In these two methods, low field-effect mobilities have been reported except for poly(thienylene vinylene) and polypyrroles, which were doped to achieve high mobility. The low mobility in most of these materials is probably due to poor ordering and the amorphous nature of the thin films. The third technique utilizes soluble conjugated polymers and they are fabricated by spin-coating, casting, or printing techniques. We have studied different conjugated polymers. Examples including poly(2,5-dialkylpheneylene-*co*-phenylene)s, poly(2,5-dialkylphenylene-*co*-thiophene)s, poly(2,5-dialkylphenylene vinylene)s and the dialkoxyl derivatives of the above polymers (*24*). However, very low (less than 10^{-4} cm^2/Vs) or no field-effect mobilities have been found. More extensive effort has been directed towards soluble polythiophene derivatives since they are widely used as conducing and semiconducting materials (*20,25-29*). We have studied the electrical characteristics of field-effect transistors using solution cast regioregular poly(3-hexylthiophene) (P3HT, Figure 3) (*30*). It is demonstrated that both high field-effect mobilities (ca. 0.05 cm^2/Vs in the accumulation-mode and 0.01 cm^2/Vs in the depletion-mode), and relatively high on/off current ratios (greater than 10^3) can be achieved (Figure 2). It was also found that the film quality and field-effect mobility are strongly dependent on the choice of solvents (*30*). The field-effect mobility can range from 10^{-4} to 10^{-2} cm^2/Vs when different solvents are used for film preparation. In addition, treating a film with ammonia or heating to 100° C under N_2 can increase the on/off ratio without decreasing the mobility (*30*). Recently, Sirringhaus et al. have reported field-effect mobility in the reange of 0.05 to 0.1 cm^2/Vs for regioregular poly(3-hexylthiophene) using HMDS treated SiO_2 as dielectric layers (*42*).

Another class of liquid phase processible material is oligomer-based compounds which has low solubility in organic solvents but enough to form a well-ordered thin film with relatively high field-effect mobility. Such materials include DH-α-4T and DH-α-6T (*17,31*). The best mobility ca. 0.03 cm^2/Vs has been obtained with DH-α-6T from chlorobenzene solution.

Printed plastic transistors

The first printed transistor has been demonstrated by Garnier et al. (*32*). In these transistors, however, only the gate electrode and a pair of drain and source electrodes, were printed separately on each side of a sheet of polyester film (1.5 μm thick) which acts as the dielectric layer. This film with electrodes was then taped to a plastic substrate followed by vacuum deposition of an organic semiconductor layer of insoluble dihexyl-α-hexathienylene (DH-α-6T). For practical applications, it is desirable that all the necessary components may be printed in a continuous process.

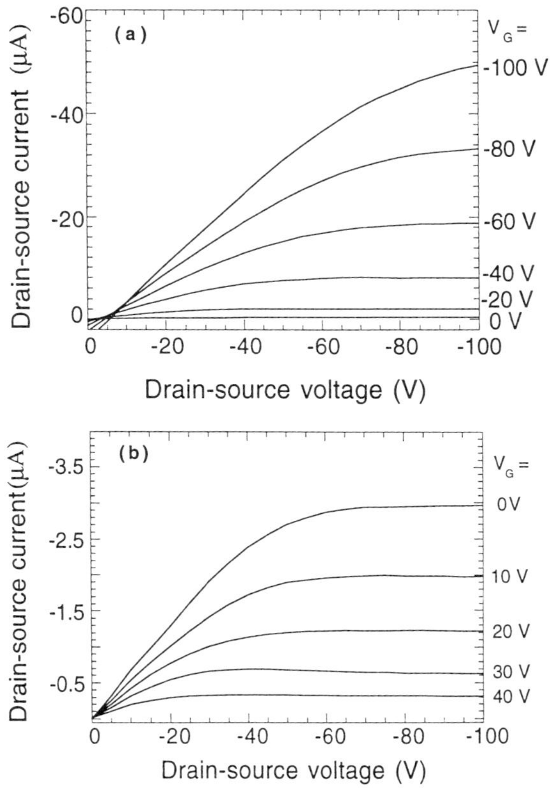

Figure 2. The current-voltage characteristics of a FET with regioregular poly(3-hexylthiophene) semiconducting layer operated in the accumulation mode (a) and depletion mode (b) at different gate voltages Copyright 1996 American Institute of Physics (reproduced with permission from reference 30).

Therefore, liquid-phase processible organic semiconductors need to be used so that low-cost large area electronics with flexible plastic substrates for display or data storage can be realized by using printing techniques.

With the above liquid phase processible materials, we have made the first transistor in which all the essential components (electrodes, dielectric, and semiconductor) are printed (*33*). It has been demonstrated that high performance transistors can be made by printing technique on a plastic substrate. These transistors (shown in Figure 1b) consist of a polymer dielectric, a semiconducting regioregular poly(3-alkylthiophene), and two silver electrodes, and all of which have been printed on an ITO-coated plastic substrate. The performance of these transistors are comparable to those from Si substrate and SiO_2 dielectric with lithographically defined Au electrodes (Figure 4). The field effect mobilities are in the order of 10^{-2} cm^2/Vs.

The smallest channel length can be acieved with the above screen printing technique is about 75 μm. However, much smaller channel length is desired for high current outputs. Recently, we have been able to demonstrate the fabrication of printed transisotrs with channel length as small as 1 μm using micromolding in capillaries (*34*). Different conducting materials, such as polyanilline and graphite ink, have been sucessfully applied as drain and source electrodes.

Air-stable *n*-channel matrials

Air-stable *n*-channel semiconducting materials are important components of *p-n* junction diodes, bipolar transistors, and complementary circuits. The existing *n*-channel materials are either air and moisture-sensitive or have relatively low field-effect mobilities (Table 3). Recently, we have modified metallophthalocyanines by adding strong electron-withdrawing groups such as -CN, -F, and -Cl to their outer rings (Figure 5) (*35*). By doing so, the LUMO levels of these molecules are significantly lowered compared to metallophthalocyanines and electron injection and transporting are made possible. Among them, the hexadecafluoro and hexadecachloro metallophthalocyanines were found to function as *n*-channel semiconductors (*35*). The best performance has been obtained with Copper hexadecafluoro-phthalocyanine with a field-effect mobility ca. 0.03 cm^2/Vs (Table 4). We have found that the high mobilities of these compounds are the results of highly ordered films upon vacuum deposition (*35*). The charge carrier mobilities of these devices are strongly dependent on the morphology of the semiconducting thin films. Highly ordered films with larger grain sizes, observed by X-ray diffraction and TEM have been obtained by vacuum deposition at elevated substrate temperatures (*35*). A complementary circuit has been fabricaed using $F_{16}CuPc$ as the *n*-channel transistor and pentacene as the *p*-channel transistor. In addition, all of the above materials pocess remarkable stability in air and their transistors can be operated both in vacuum and in air. These transistors without any packaging are still functional with high mobilities after stored in open air for several months. These metallophthalocyanine derivatives are by far the only materials which have been found to have longtime stability in air with mobilites greater than 10^{-2} cm^2/Vs.

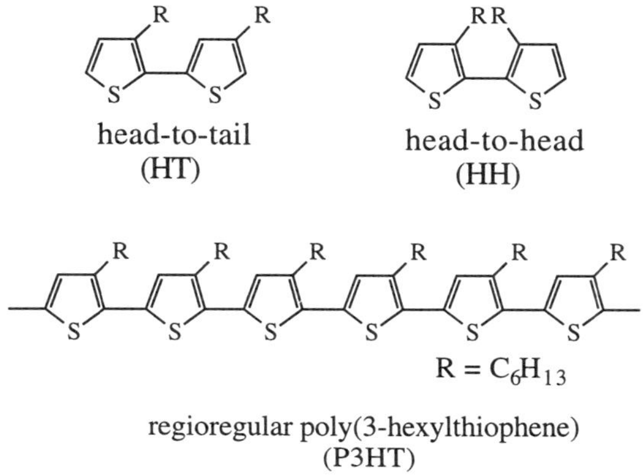

Figure 3. Chemical structure of regioregular poly(3-hexylthiophene) Copyright 1996 American Institute of Physics (reproduced with permission from reference 30).

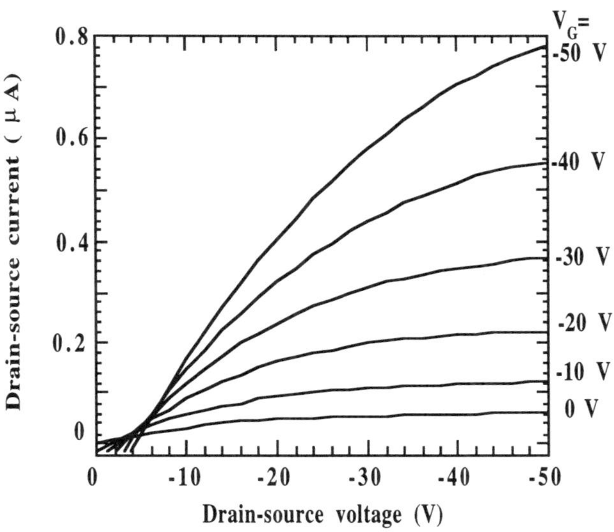

Figure 4. I-V characteristic of a printed plastic transistor with regioregular poly(3-hexylthiophene) semiconducting layer on polyimide dielectrics coated ITO plastic substrate with printed Ag drain and source electrodes.

Table 3. Organic n-channel semiconducting materials.

compound	field-effect mobility (cm^2/Vs)	dramatically decreased performance in air	reference
C_{60}/C_{70}	0.01-0.08	yes	(*36-37*)
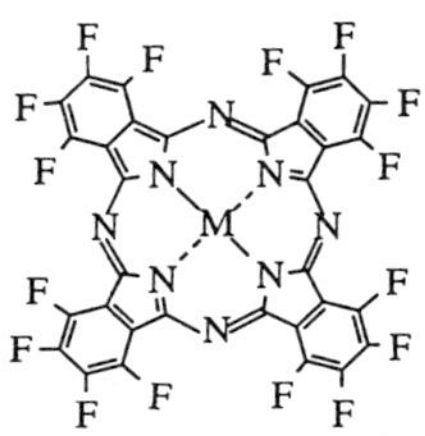	0.001 - 0.03	no	(*35*)
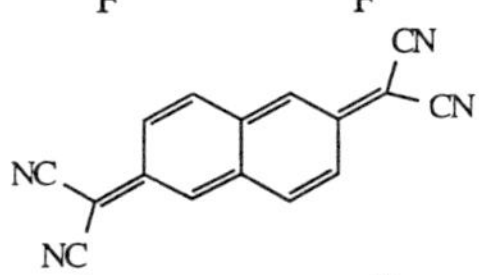	0.003	no	(*38*)
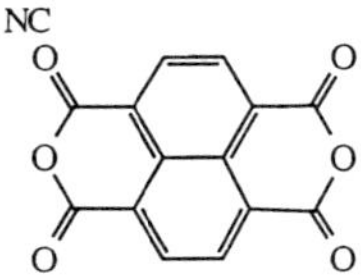	0.003	yes	(*38*)
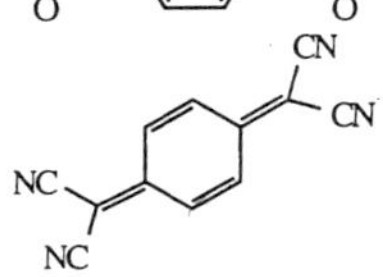	3×10^{-5}	no	(*39*)
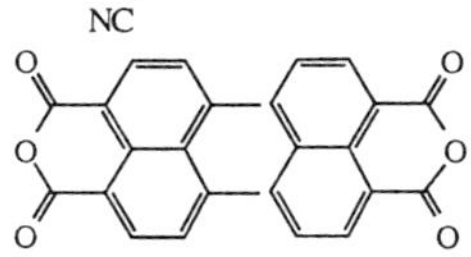	1.5×10^{-5}	yes	(*40*)

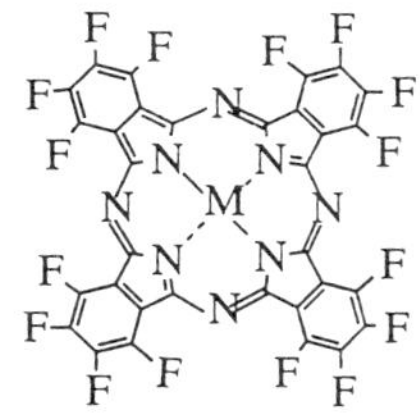

M = Cu,Zn,Fe,Co

F_{16}MPc

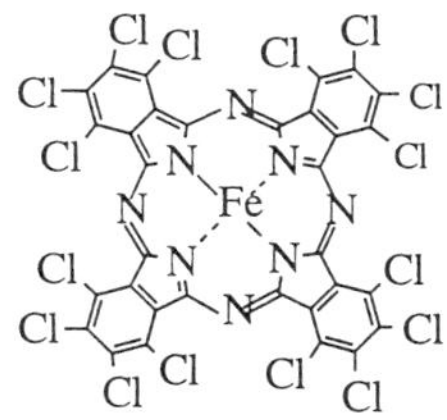

Cl_{16}FePc

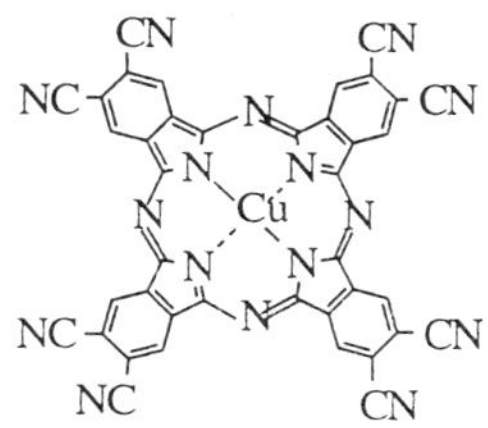

$(CN)_8$CuPc

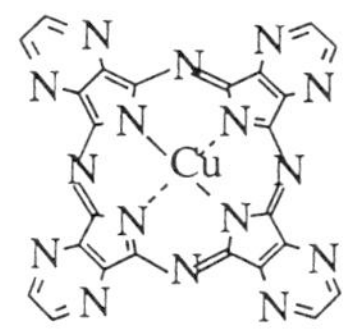

PyCuPc

Figure 5. Structure of electron-deficient metallophthalocyanines (*35*).

Table 4. Summary of field-effect mobilities (cm^2/Vs) for different substituted metallophthalocyanines (reproduced with permission from reference 35).

MPc	T_d (30 °C)	T_d (125 °C)	T_d (215 °C)
F_{16}CuPc	5×10^{-3}	0.03	0.02
F_{16}ZnPc	1.7×10^{-5}	4.6×10^{-4}	1.2×10^{-3}
F_{16}CoPc	1.8×10^{-6}	4.5×10^{-5}	4.3×10^{-5}
F_{16}FePc	5.5×10^{-4}	5.8×10^{-3}	2.1×10^{-3}
Cl_{16}FePc	no field-effect	2.7×10^{-5}	*a*
$(CN)_8$CuPc	*b*	*b*	*b*
PyCuPc	*b*	*b*	*b*

a. Compound desorbs at this temperature.

b. Compound can not be sublimed.

Conclusions

Promising transistor performance has been shown with organic and polymeric semiconducting materials with relatively high field-effect mobilities and on/off ratios. The demonstration of the first plastic transistor by continuous printing techniques is a new step towards printed plastic circuits. Nevertheless, there are still great needs for solution processible *p*-channel and *n*-channel organic materials with mobilites into the range of 10^{-1} cm^2/Vs. In addition, realiability and lifetime of organic transistor-based circuits need be investigated in the near future.

Literature Cited

(1) Lovinger, A. J.; Rothberg, L. J. *J. Mater. Res.* **1996**, *11*, 1581.

(2) Katz, H. E. *J. Mater. Chem.* **1997**, *7*, 369-376.

(3) Lin, Y., -Y.; Gundlach, D. J.; Nelson, S. F.; Jackson, T. N. *IEEE Trans. Elec. Dev.* **1997**, *44*, 1325.

(4) Laquindanum, J. G.; Katz, H. E.; Lovinger, A. J.; Dodabalapur, A. *Chem. Mater.* **1996**, *8*, 2542.

(5) Horowitz, G.; Deloffre, F.; Garnier, F.; Hajlaoui, R.; Hmyene, M.; Yassar, A. *Syn. Metal* **1993**, *54*, 435-445.

(6) Dodabalapur, A.; Torsi, L.; Katz, H. E. *Science* **1995**, *268*, 270-271.

(7) Hajlaoui, R.; Fichou, D.; Horowitz, G.; Nessakh, B.; Constant, M.; Garnier, F. *Adv. Mater.* **1997**, *9*, 557-561.

(8) Garnier, F.; Yassar, A.; Hajlaoui, R.; Horowitz, G.; Deloffre, F.; Servet, B.; Ries, S.; Alnot, P. *J.Am. Chem. Soc.* **1993**, *115*, 8716-8721.

(9) Katz, H. E.; Lovinger, A. J.; Laquindanum, J. G. *Chem. Mater.* **1998**, *10*, 457.

(10) Bao, Z.; Lovinger, A. J.; Dobabalapur, A. *Appl. Phys. Lett.* **1996**, *69,* 3066.

(11) Bao, Z.; Lovinger, A. J.; Dodabalapur, A. *Adv. Mater.* **1997**, *9*, 42-44.

(12) Guillaud, G.; Madru, R.; Al Sadoun, M.; Maitrot, M. *J. Appl. Phys.* **1989**, *166*, 4554-4556.

(13) Laquindanum, J.; Katz, H. E.; Dodabalapur, A.; Lovinger, A. J. *Adv. Mater.* **1997**, *9*, 36.

(14) Sirringhaus, H.; Friend, R. H.; Li, X. C.; Moratti, S. C.; Holmes, A. B.; Feeder, N. *Appl. Phys. Lett.* **1997**, *71*, 3871.

(15) Li, X. C.; Sirringhaus, H.; Garnier, F.; Holmes, A. B.; Moratti, S. C.; Feeder, N.; Clegg, W.; Teat, S. J.; Friend, R. H. *J. Amer. Chem. Soc.* **1998**, *20*, 2206-2207.

(16) Dimitrakopoulos, C. D.; Afzali-Aradakani, A.; Furman, B.; Kymissis, J.; Purushothaman, S. *Syn. Metal* **1997**, *89*, 193-197.

(17) Laquindanum, J.; Katz, H. E.; Lovinger, A. J. *J. Amer. Chem. Soc.* **1998**, *120*, 664-672.

(18) Horowitz, G.; Peng, X.-Z.; Fichou, D.; Garnier, F. *Syn. Metal* **1992**, *51*, 419-424.

(19) Dimitrakopoluos, C. D.; Brown, A. R.; Pomp, A. *J. Appl. Phys.* **1996**, *80*, 2501-2508.

(20) Tsumura, A.; Koezuka, H.; Ando, T. *Appl. Phys. Lett.* **1986**, *49*, 1210-1212.

(21) Kou, C.-T.; Liou, T.-R. *Syn. Met.* **1996**, *82*, 167-173.

(22) Brown, A. R.; Pomp, A.; Hart, C. M.; de Leeuw, D. M. *Science* **1995**, *270*, 972-974.

(23) Fuchigami, H.; Tsumura, A.; Koezuka, H. *Appl. Phys. Lett.* **1993**, *63*, 1372-1374.
(24) Bao, Z.; Lovinger, A. J. unpublished results.
(25) Koezuka, H.; Tsumura, A.; Ando, T. *Synth. Met.* **1987**, *18*, 699-704.
(26) Paloheimo, J.; Kuivalainen, P.; Stubb, H.; Vuorimaa, E.; Lahti, P. Y. *Appl. Phys.Lett.* **1990**, *56*, 157.
(27) Yoshino, K.; Takahashi, H.; Muro, K.; Ohmori, Y.; Sugimoto, R. *J. Appl. Phys.* **1991**, *70*, 5035.
(28) Ohmori, Y.; Muro, K.; Uchida, M.; Kawai, T.; Yoshino, K. *Jpn. J. Appl. Phys.* **1991**, *30*, L610.
(29) Assadi, A.; Svensson, C.; Willander, M.; Inganas, O. *Appl. Phys. Lett.* **1988**, *53*, 195.
(30) Bao, Z.; Dadabalapur, A.; Lovinger, A. J. *Appl. Phys. Lett.* **1996**, *69*, 4108-4110.
(31) Katz, H. E.; Laquindanum, J. G.; Lovinger, A. J. *Chem. Mater.* **1998**, *10*, 633-638.
(32) Garnier, F.; Hajlaoui, R.; Yassar, A.; Srivastava, P. *Science* **1994**, *265*, 1684.
(33) Bao, Z.; Feng, Y.; Dodabalapur, A.; Raju, V. R.; Lovinger, A. J. *Chem. Mater.* **1997**, *9*, 1299-1301.
(34) Rogers, J. A.; Bao, Z.; Raju, V. R. *Appl. Phys. Lett.* **1998**, in press.
(35) Bao, Z.; Lovinger, A. J.; Brown, J. *J. Amer. Chem. Soc.* **1998**, *120*, 207-208.
(36) Haddon, R. C.; Perel, A. S.; Morris, R. C.; Palstra, T. T. M.; Hebard, A. F.; Fleming, R. M. *Appl. Phys. Lett.* **1995**, *67*, 121.
(37) Paloheimo, J.; Isotalo, H.; Kastner, J.; Kuzmany, H. *Synth. Met.* **1993**, *55-57*, 3185.
(38) Laquindanum, J. G.; Katz, H. E.; Dodabalapur, A.; Lovinger, A. J. *J. Amer. Chem. Soc.* **1996**, *118*, 11331-11332.
(39) Brown, A. R.; De Leeuw, D. M.; Lous, E. J.; Havinga, E. E. *Synth. Met.* **1994**, *66*, 257-261.
(40) Horowitz, G.; Kouki, F.; Spearman, P.; Fichou, D.; Nogues, C.; Pan, X.; Garnier, F. *Adv. Mater.* **1996**, *8*, 242-244.
(41) Katz, H.E.; Lovinger, A.J.; Laquindanum, J.G. *Chem. Mater*. **1998**, *10*, 457-459.
(42) Sirringhaus, H.; Tessler, N.; Friend, R.H. *Science* **1998**, *280*, 1741-1744.

Chapter 16

New Photoluminescent Display Devices

Christoph Weder, Andrea Montali, Christian Sarwa, Cees Bastiaansen, and Paul Smith

Department of Materials, Institute of Polymers, ETH Zürich, UNO C14, CH–8092 Zürich, Switzerland

Liquid crystal displays represent the dominant flat panel display technology, despite their limitations in brightness and efficiency originating from the use of absorbing polarizers and color filters. This paper reviews novel concepts of photoluminescent liquid crystal displays that employ photoluminescent polarizers. These polarizers exhibit highly anisotropic absorption and / or emission and efficiently combine two separate features: the polarization of light and the generation of bright color. A new photophysical effect - polarizing energy transfer - can be used to drive the efficiency of these elements to the theoretical limit: it enables to produce photoluminescent polarizers which optimally harvest incident light by isotropic absorption but emit the absorbed energy in highly linearly polarized fashion. Experiments suggest that the novel concepts can simplify device design and substantially increase device brightness, contrast, and efficiency.

Despite the growing research interests in inorganic (*1*) and, more recently, organic (*2*) electroluminescent (EL) light-emitting devices, and other techniques, such as plasma display panels (*3*) or vacuum fluorescent displays (*4*), liquid crystal displays (LCDs) have maintained their dominant position in the field of flat-panel displays (*5,6*). The main advantages of LCDs are their versatility, low-voltage operation and semiconductor compatibility; but their limited brightness and energy efficiency, as well as the often unsatisfying viewing angle of LCDs, leave ample room for further improvement (*6*). The severe limitations in brightness and efficiency of LCDs arise chiefly from the use of dichroic sheet polarizers and - in case of color devices - absorbing color filters since these elements convert a large fraction (> 85 %) of the incident light into thermal energy. As an alternative to dichroic polarizers which typically transmit less than 40 % of unpolarized incident light, polarizers have recently been proposed that are based on selective reflection (*7*) or scattering (*8*) of one polarization and allow recycling of the reflected or scattered light. The ultimate efficiency of these polarizers is, in principle, unity, thus, doubled compared to absorbing polarizers. However, color applications based on these elements still rely on color filters which typically absorb at least 70 % of white light (*6*).

The use of photoluminescent (PL) materials, which act as "active" color filters and, therefore, might enhance the visual performance of LCDs, was previously suggested (*9-11*). Several principal possibilities exist to incorporate luminescent materials into LCDs, including the use of fluorescent LCs or the dissolution or dispersion of luminescent molecules in a conventional LC layer (*9*), or the application of PL plates (*10*) or front-face screens (*11*). However, the proposed devices suffer from a number of drawbacks, related to the limited stability of the fluorescent dyes or LCs or both, the difficulty to produce pixilated devices, depolarization effects, or the required thickness (> 1 mm) and (large) area of the luminescent layer (*10*). This paper reviews novel concepts of photoluminescent liquid crystal displays that comprise PL polarizers which efficiently combine two separate features: the polarization of light and the generation of bright color. Experiments suggest that this approach can simplify device design and substantially increase device brightness, contrast, and efficiency, and - in specific configurations - viewing angle.

Photoluminescent Polarizers

Uniaxially oriented, PL materials usually exhibit anisotropic, i.e., linearly polarized, absorption and emission. This phenomenon has been known for inorganic crystals for more than a century (*12*) and was reported for oriented blends of ductile polymers and low-molecular weight PL materials as early as the 1930's (*13*). Since, the effect has been shown in a variety of materials and using a diversity of orientation methods (*14*) including, for example, PL liquid crystal systems (*15*) or low molecular PL materials uniaxially grown onto orienting substrates, such as oriented poly(tetrafluoroethylene) (*16*). Direct deposition through friction or rubbing (*17*), or the Langmuir-Blodgett technique (*18*), as well as mechanical deformation of pure conjugated polymer films (*19*) have also been demonstrated to yield PL layers or films which exhibit anisotropic optical properties. However, the degree of orientation and, hence, the dichroic ratios obtained with these methods are usually only modest, typically well below 10. By contrast, the tensile deformation of guest-host systems, in which the guest molecules adopt the orientation of the host, was found to be a most promising technique for the production of PL polarizers with significant anisotropic optical properties. This concept has been used in the past by different research groups for the preparation of polarized PL films based on blends of various formanisotropic low-molecular and oligomeric compounds and a variety of matrix polymers (e.g. polystyrene or polyethylene) (*14,20,21*). Rather surprisingly, this method has only recently been adapted for blend films of conjugated polymers; a blend of poly(2-methoxy-5-(2'ethyl-hexyloxy)-*p*-phenylenevinylene (MEH-PPV) and ultra-high molecular weight polyethylene (UHMW-PE) was used in the initial experiment (*22*).

With the above described relevance of polarized photoluminescence for LCD applications in mind, we recently have systematically investigated the structure-property relations of oriented films based on blends of poly(2,5-dialkoxy-*p*-phenyleneethynylene)s (PPEs) and UHMW-PE (*23*). PPEs exhibit an ideal matrix of properties with respect to the preparation of such PL polarizers (*23-30*), including large PL quantum efficiencies in solution and solid state (*24*), and an extremely stiff, linear polymer backbone that enables maximum orientation. Reported here are yellow-green light-emitting PL polarizers based on EHO-OPPE ($\bar{M}_n$ ~ 10,000 and ~ 84,000 $gmol^{-1}$) (*24,27*), a highly soluble PPE derivative, substituted with linear and sterically hindered alkyloxy groups in an alternating pattern, as well as blue light-emitting PL polarizers based on PPE-copolymers (co-PPE blue) (*29*) of controlled conjugation length (Figure 1). Uniaxially oriented films with contents of typically 1

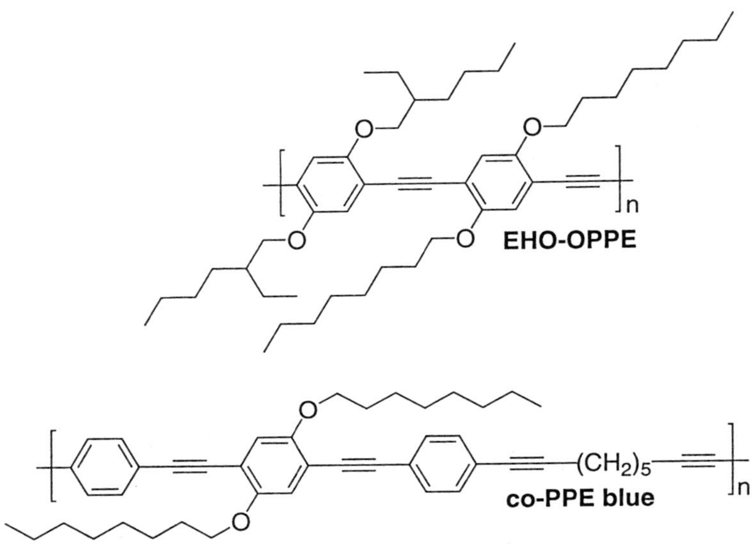

Figure 1. Chemical structures of the poly(2,5-dialkoxy-*p*-phenylene-ethynylene) derivatives EHO-OPPE and co-PPE blue.

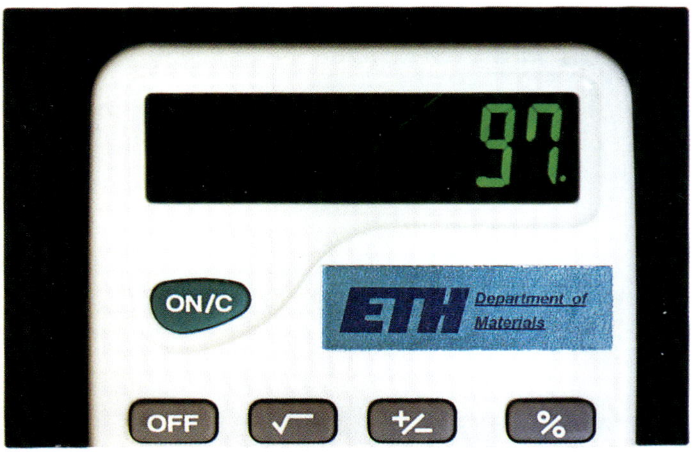

Figure 4. (a) Photograph of a PL display device according to Figure 3a. (b) Photograph of a bicolor device according to Figure 3b. (Figure 4a reproduced with permission from ref. *31*.)

Continued on next page.

Figure 4. *Continued*

to 2 % w/w of PPE in UHMW-PE were prepared by solution casting, drying and subsequent tensile drawing in the solid state (*23*). The *pristine*, solution-cast films were drawn at temperatures of 90 - 130 °C (i.e., in the narrow temperature window above the glass transition of the PPE guest, and below the melting point of the polyethylene) to a series of different draw-ratios (λ = final length / initial length) that ranges from 10 to 80. The thickness of drawn films of the maximum draw ratio of 80 was in the order of 2 μm.

The anisotropic photophysical behaviour of these drawn films was studied employing polarized UV/Vis absorption and steady-state PL spectroscopy, using unpolarized light for excitation (*23*). The anisotropic characteristics are expressed in terms of dichroic ratios, which are defined for absorption (DR_A) and emission (DR_E) as the ratio between the respective spectra measured with polarization parallel (*p*) and perpendicular (*s*) to the drawing direction. Note, that different phonon bands are observed for *p*- and *s*-polarized light, making the dichroic behavior wavelength-dependent, with maximum distortion at absorption and emission maxima; thus, to better reflect the 'average' dichroic behavior, all evaluations are based on integration of the respective spectra. Highly linearly polarized absorption and PL emission was observed for the oriented films, as visualized in Figure 2 for a film containing 2 % of EHO-PPE (M_n~ 84,000 gmol^{-1}) of a draw ratio of 80. The polarized emission spectra of this film reflect state-of-the-art optical anisotropy, characterized by dichroic ratios, DR_E, of 27 (*23*). Matching dichroic ratios were determined for absorption and emission experiments, suggesting that no molecular reorientation, and no energy transfer processes, which could eventually limit the emission dichroic ratio, are present. The significant influence of the draw ratio on the dichroic ratio in all systems under investigation is characterized by an initial linear increase of dichroic ratio, with a tendency to level off at $\lambda > 50$ (*23*). Comparing the properties of blends based on EHO-OPPE with M_n of 10,000 gmol^{-1} and 84,000 gmol^{-1} it was found, that the molecular weight of the conjugated polymer has a notable influence on the orientation; the high-molecular weight material exhibits maximum orientability, which is understood in terms of its more favourable molecular aspect ratio. It was also observed that the orientation of the PPE was not influenced by the blend composition in the concentration regime of 1 - 2 % (w/w) of the PPE. X-ray and electron diffraction experiments as well PL spectra indicate that the orientation process appears to induce a transformation of an initially phase-separated system into an apparent molecular dispersion of the conjugated polymer guest in the PE host (*23*). It is also noteworthy that the oriented films were found to be extremely stable: the materials could be stored under ambient conditions (exposure to air and light) for months without any noticeable change of their properties. The latter, most relevant, phenomenon is attributed to the encapsulation of the conjugated polymer in the highly crystalline PE matrix (*22,23*) and the outstanding intrinsic stability of the PPE backbone.

Photoluminescent Display Devices

PL polarizers can combine, and directly replace, the standard polarizer and color filter in conventional LCDs and, using an appropriate, for example ultraviolet (UV) backlight, result in efficient, colored PL LCDs (*31,32*). We have designed and fabricated devices that use the PL polarizers described above, based on blends of UHMW-PE and yellow-green light-emitting EHO-OPPE (2 % w/w, M_n of 84,000 gmol^{-1}, λ = 80), blue light-emitting co-PPE blue (10 % w/w, λ = 80) or orange-red light-emitting MEH-PPV (1 % w/w, λ = 80).

In backlit PL display devices either the light used to photoexcite the PL polarizer, or light emitted from the PL polarizer may be switched by a twisted nematic (TN) electrooptical (EO) light valve (Figure 3). Depending on the selected

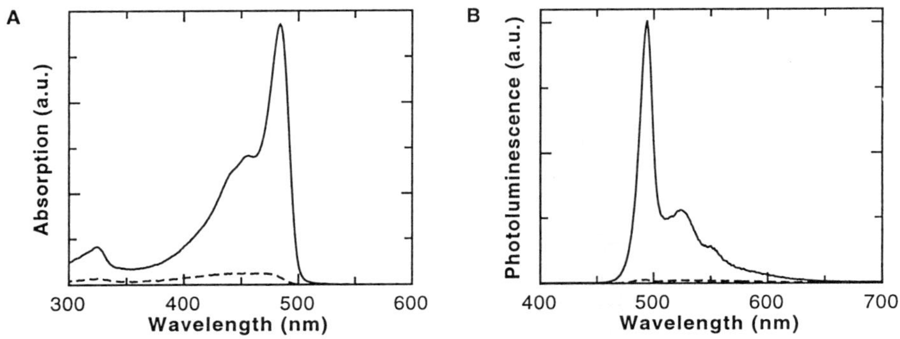

Figure 2. Polarized absorption (a) and photoluminescence (b) spectra of a PL polarizer (λ = 80) based on a 2 % w/w EHO-OPPE / UHMW-PE blend, recorded for absorption and emission parallel (solid line) and perpendicular (dashed line) to the drawing direction. (Adapted with permission from ref. *31*.) Copyright 1998 American Association for the Advancement of Science.

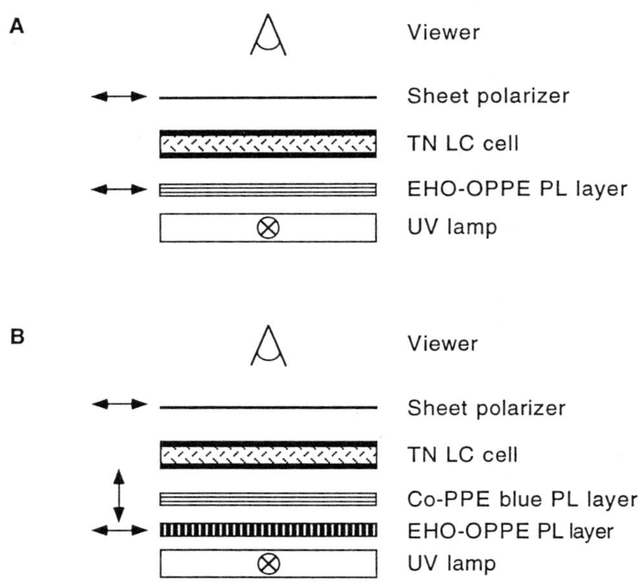

Figure 3. Schematic structures of PL display devices. (a) Device in which the light emitted by the polarized PL layer is switched. (b) Bicolor device in which the light emitted by the polarized PL layer is switched. (Reproduced with permission from ref. *31*.) Copyright 1998 American Association for the Advancement of Science.

configuration, the predominantly relevant polarization characteristic of the PL polarizer is either a high degree of anisotropy in absorption or PL emission. A monochromatic device structure (Device A, Figure 3a), in which light emitted from the polarized PL layer is switched, consists of a linear polarizer (here used as analyzer), an EO light valve, the PL polarizer, and a UV light source, emitting around 365 nm. The light radiated by the light source is at least partially absorbed in the PL polarizer, so that polarized light is emitted. This emitted light either passes the combination of EO light valve and polarizer (switching state "bright") or is blocked (switching state "dark"). Maximum contrast is obtained when (i) the PL polarizer is characterized by a high dichroic ratio for PL emission, and (ii) the portion of light visible to the human eye that is emitted by the light source, but not absorbed by the PL polarizer - and thus exits the device in the direction of the viewer - is minimal. This can be achieved by different means; for example, by using an additional cut-off-filter or, as in the experiment described here, using a UV light source. In the latter case the portion of UV light that is not absorbed by the PL polarizer, may be absorbed by the EO light valve and the polarizer, since both these elements are usually absorbing in the UV regime. Switching of the Device A was perceived by eye as a change from a very bright yellow-green to almost completely dark (Figure 4a) and was measured to be 30 and 4 cd/m^2 for the bright and dark states. The apparent limitations of Device A with respect to contrast and brightness are clearly related to the low optical density of the PL polarizer (isotropic absorption of the PL polarizer employed = 0.04) at the excitation wavelength and also to the experimental conditions, in particular the reflection of ambient light in the dark state. Thus, brighter devices can be obtained by increasing the optical density of the PL polarizer. This was also demonstrated by using an EHO-OPPE based PL polarizer of increased (but still by no means optimized) optical density (isotropic absorption = 0.18 at 365 nm) which yielded a device according to Figure 4a exhibiting a bright state of 65 cd/m^2.

In a bicolor device configuration (Device B, Figure 3b) the light emitted from the PL polarizers is switched. This device is similar to Device A but two PL polarizers, with their polarization directions perpendicular to each other, are included. We used PL polarizer based on EHO-OPPE, co-PPE blue, and MEH-PPV that emit yellow-green, blue, and orange-red light, respectively. The switching between the "on" and the "off" states changes from one bright color to another (Figure 4b).

Many other device configurations based on PL polarizers can be designed (*31,32*), including devices where the excitation light is switched, and the PL polarizer, positioned between the EO light valve and the viewer, functions as an analyzer. In this arrangement, a significantly improved viewing angle is obtained (*31*), since the emission of the PL polarizer is less angle-dependent than standard LC effects.

Polarizing Energy Transfer

The efficiency of PL polarizers is chiefly limited by the luminophore's quantum yield which, ultimately, can approach unity, but for PL polymers typically is up to 80 % (*24,33*). However, when used in a standard PL LCD configuration (see above), only ~ 50 % of light incident from the light source is used, since the absorption of these PL polarizers is also anisotropic (*14-23*). We recently have demonstrated a new concept for polymer-based PL polarizers which overcome this limitation and can be used in PL LCDs with, in principle, an ultimate efficiency of unity (*34,35*). These PL polarizers comprise a nearly randomly oriented sensitizer that maximally harvests light by isotropic absorption, efficiently transfers the energy to a uniaxially oriented PL polymer which, subsequently, emits highly linearly polarized light. Key step is a polarizing energy transfer which, to a certain extent, mimics the concept used by nature in photosynthesis to optimally use optical energy (*36*). While the reverse effect, i.e. PL depolarization, is well known (*36-38*), the polarizing energy transfer exhibited by the new materials is not only of technological relevance but also manifests a new photophysical phenomenon.

The PL films investigated were based on uniaxially oriented, ternary blends of UHMW-PE, EHO-OPPE (2 % w/w), and 7-diethylamino-4-methylcoumarin (DMC) (2 % w/w) as the sensitizer (Figure 5). The respective binary blends (UHMW-PE / EHO-OPPE and UHMW-PE / DMC) were used as reference systems. All blend films were prepared according to the methods outlined above. DMC was selected as the sensitizer because of its low form-anisotropy, suitable photophysical prerequisites and particularly beneficial phase-behavior. The melting temperature of 74 °C makes DMC compatible with the orientation process which requires mobility of the guest molecules during deformation; in addition, DMC and EHO-OPPE are miscible at elevated temperatures, which enables a most favourable morphology of the oriented blends (see below). The absorption of DMC around 364 nm optimally overlaps with the emission of common UV lamps that may be used as excitation source in PL LCDs (*31,32*). Importantly, DMC seems not to quench emission of EHO-OPPE and, mandatory for energy transfer (*39*), exhibits an own emission that favourably overlaps with the absorption of EHO-OPPE. Photophysical characteristics of PL films based on the ternary and the binary reference blends were investigated employing polarized UV-vis absorption and steady-state PL spectroscopy. Since the absorption at 440 nm is exclusively related to EHO-OPPE and at 365 nm principally due to DMC, experiments were performed at these two wavelengths to separately address the conjugated polymer and the sensitizer.

Polarized absorption spectra, acquired with *p*- and *s*-polarized incident light (Figures 2a and 6) show that the characteristics of the ternary blend are a combination of those of the two respective binary blends. The ternary blend exhibits high absorption dichroic ratios DR_A of up to 13 at 440 nm, resulting from a high degree of orientation of EHO-OPPE. By contrast, the absorption at 365 nm is essentially isotropic (DR_A = 1.5) and reflects the nearly random orientation of the sensitizer within the oriented UHMW-PE matrix.

Polarized emission spectra, obtained under isotropic excitation at 365 nm and polarized detection in either *p*- or *s*-mode are shown in Figures 2b and 7. In binary UHMW-PE / DMC films, the emission from DMC, centred around 400 nm, exhibits only minor polarization, expressed by an emission dichroic ratio, of 2.3, consistent with the low degree of orientation of the sensitizer. In the ternary blend, importantly, the DMC emission is almost fully suppressed, while the emission from EHO-OPPE is highly polarized (DR_E= 16). The fact that DR_E is somewhat lower in the ternary than in a comparable binary UHMW-PE / EHO-OPPE blend (DR_E= 27) is explained with a plastisizing effect of DMC on EHO-OPPE that reduces the efficiency of the orientation process.

Energy transfer from DMC to the conjugated polymer is evident when comparing the emission intensities (related to EHO-OPPE) of the ternary and the binary UHMW-PE / EHO-OPPE blend (Figures 8a, 8b) for isotropic excitation at 440 and 365 nm, respectively. The binary reference blend shows a significantly lower emission intensity when excited at 365 nm compared to excitation at 440 nm, due to the much lower absorption of EHO-OPPE at the shorter wavelength. The ternary blend, by contrast, shows similar emission intensities when excited at 365 and 440 nm, as a result of the sensitizing effect of DMC: the effective, isotropic absorption of the sensitizer, followed by energy transfer to the conjugated polymer, is the obvious rationalisation for the increased emission intensity. The polarizing characteristic of the energy transfer is demonstrated by the results presented in Figure 8c. The intensity of *p*-polarized emission from the ternary blend was found to be only

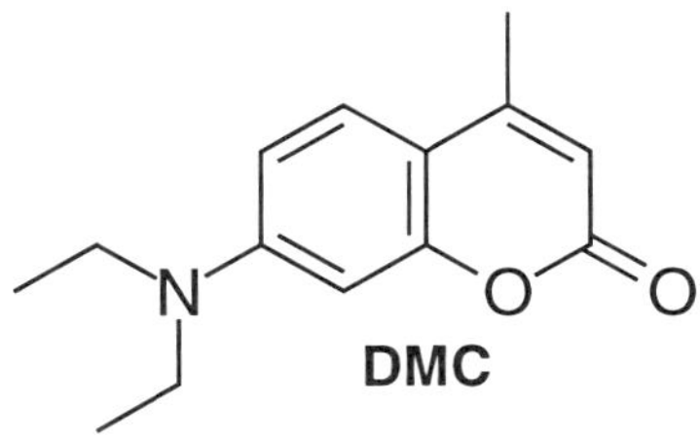

Figure 5. Molecular structure of 7-diethylamino-4-methylcoumarin (DMC).

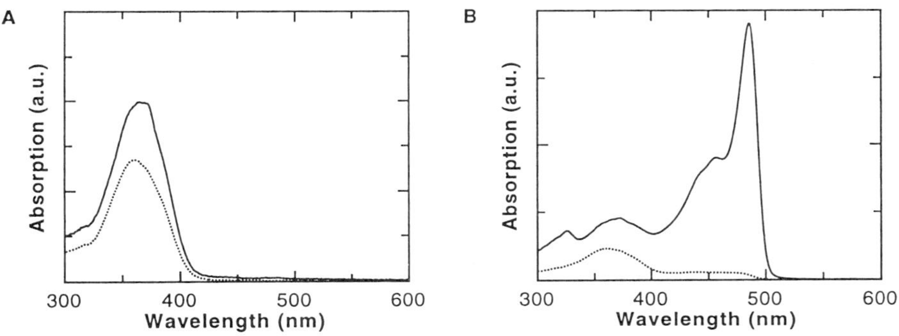

Figure 6. Polarized absorption spectra of oriented films obtained with *p*- (solid line) and *s*- (dashed line) polarized light; (a) Binary UHMW-PE / DMC blend; (b) Ternary UHMW-PE / EHO-OPPE / DMC blend. (Adapted with permission from ref. *34*.) Copyright 1998 Macmillan Magazines, Ltd.

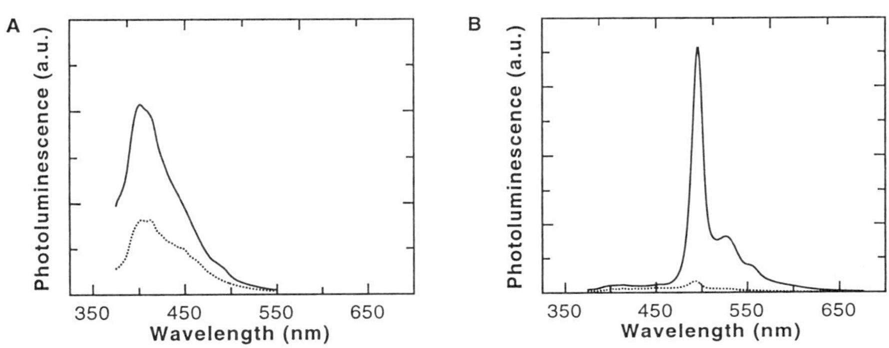

Figure 7. Emission spectra of oriented films obtained under isotropic excitation at 365 nm and polarized detection in *p*- (solid line) and *s*- (dashed line) mode; (a) Binary UHMW-PE / DMC blend; (b) Ternary UHMW-PE / EHO-OPPE / DMC blend. (Adapted with permission from ref. *34*.) Copyright 1998 Macmillan Magazines, Ltd.

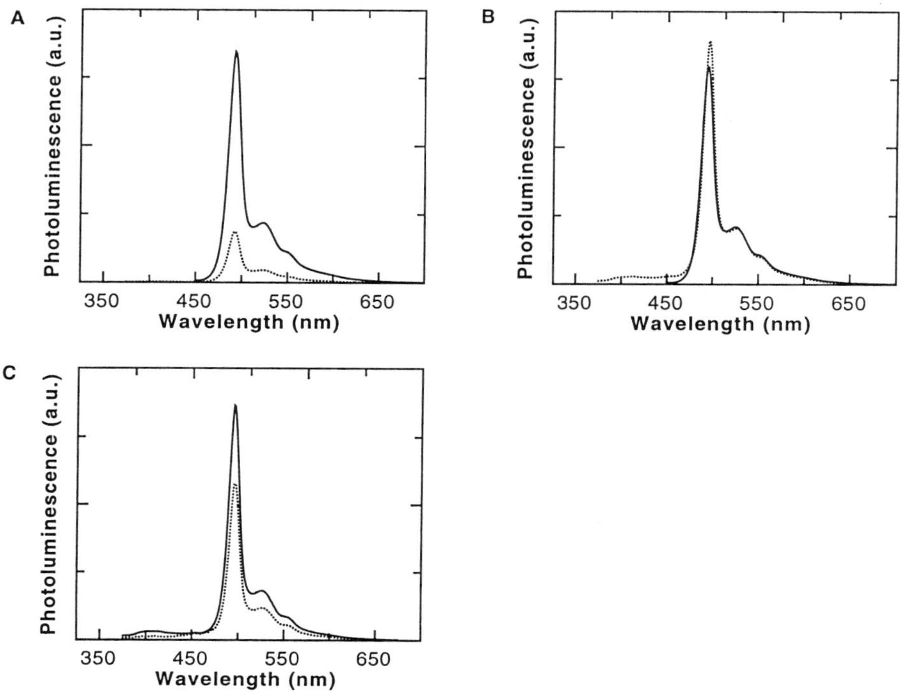

Figure 8. (a) and (b): Emission spectra of oriented films obtained under isotropic excitation at 440 (solid line) and 365 nm (dashed line), and polarized detection in *p*-mode; (a) Binary blend of UHMW-PE / EHO-OPPE; (b) Ternary blend of UHMW-PE / EHO-OPPE / DMC. (c): Emission spectra of an oriented UHMW-PE / EHO-OPPE / DMC ternary blend film obtained under polarized excitation at 365 nm in *p*- (solid line) and *s*- (dashed line) mode and polarized detection in *p*-mode. (Adapted with permission from ref. *34*.) Copyright 1998 Macmillan Magazines, Ltd.

weakly depending on the polarization of the incident light (when excited at 365 nm). In fact, the ratio of the emission intensities for excitation with *s*- and *p*-polarized light (1.5) is in gratifying agreement with the slightly dichroic absorption of the film at 365 nm ($DR_A = 1.5$). Thus, the ternary blend unambiguously exhibits the phenomenon of polarizing energy transfer: optical energy is isotropically absorbed by DMC, with similar efficiency for both absorption (excitation) polarizations transferred to EHO-OPPE, which subsequently emits polarized light. In the most unfavourably limit (Figure 8c) the new material converts fully *s*-polarized into highly *p*-polarized light.

The polarizing energy transfer process observed in the new PL materials may, principally, originate from either radiative (*39*) (trivial), long-range Coulombic (*40*) (Förster) or short-range electron-exchange (*41*) (Dexter) energy transfer between the DMC sensitizer as donor and the oriented EHO-OPPE as acceptor. The fact that energy is transferred between donor molecules that have been excited with *s*-polarized light and acceptor molecules which subsequently emit *p*-polarized light implicates a depolarization of the donor excited state, unless Dexter-type coupling is involved (*39*). This depolarization can derive from randomizing energy migration (*42*) or orientational relaxation of the donor (*36*) and is indeed observed when exciting the binary DMC reference blend with polarized light. The low optical density of the samples essentially excludes a radiative energy transfer (*36-38*). A nonradiative energy transfer might, on the other hand, point to a very particular phase-behavior of the oriented blends. As discussed above, it was found that in binary blends EHO-OPPE forms an apparent molecular dispersion in the UHMW-PE matrix (*23*). Thus, the incompatibility of DMC and UHMW-PE, and the demonstrated affinity of DMC and EHO-OPPE make the formation of DMC / EHO-OPPE aggregates very likely, in which a nonradiative energy transfer is enabled by the close proximity of donor and acceptor molecules.

As direct indication for the practical impact of DMC sensitization, the absolute brightness, measured under isotropic excitation with a 365 nm UV lamp, i.e., in a configuration of relevance to actual PL LCDs (*31,32*). The luminosity of a ternary blend film is dramatically increased (82 cd/m^2), compared to an unsensitized binary blend (22 cd/m^2) of similar optical density in the EHO-regime. Of course, the absolute brightness can be further enhanced by an increase in optical density.

Conclusions

In summary, we have introduced new concepts for the design of photoluminescent polarizers and application of the latter in photoluminescent liquid crystal displays. PL polarizers efficiently combine two separate features, i.e. the polarization of light, and the generation of bright color. In addition we have found and used a new photophysical effect - polarizing energy transfer - which can bring the efficiency of these elements to the theoretical limit. Hence, the new photoluminescent display devices can offer a substantial increase in device brightness and efficiency.

References

1. For a recent review see: Kobayashi, H. *Proc. SPIE* **1993**, *1910*, 15.
2. For a recent review see: Bradley, D. *Current Opinion in Solid State & Materials Science* **1996**, *1*, 789.
3. *Electronic Information Display Technologies*; Nelson, T.J.; Wullert II, J.R. Eds.; World Scientific: Singapore, **1997**, pp. 83-104.
4. Nakamura, T.; Kiyozumi, K.; Mito, S. In *Advances in Image Pickup and Display*; *Vol. 5*; Academic Press: New York, **1982**, pp. 200-280.

5. Chan, L.K.M. In *The Encyclopedia of Advanced Materials, Vol. 2*, Bloor, D.; Brook, R.J.; Flemings, M.C.; Mahajan, S. Eds.; Elsevier Science: Oxford, **1994**, pp. 1294-1304.
6. *Electronic Information Display Technologies*; Nelson, T.J.; Wullert II, J.R. Eds.; World Scientific: Singapore, **1997**, pp. 133-266.
7. Schadt, M.; Fünfschilling, *Jap. J. Appl. Phys.* **1990**, *29*, 1974. Broer, D.J.; Lub, J.; Mol, G.N. *Nature* **1995**, *378*, 467. Coates, D., Goulding, M.J., Greenfield, S., Hanmer, J.M.W., Jolliffe, E., Marden, S.A., Parri, O.L., Verrall, M. *SID 96 Applications Digest* **1996**, 67.
8. Dirix, Y.; Jagt, H.; Hikmet, R.; Bastiaansen, C. *J. Appl. Phys.* **1998**, *83*, 2927.
9. Funada, F.; Matsuura, M.; Wada, T. US Patent 4,336,980 (**1982**). Coles, H.J. *Liq. Cryst.* **1993**, *14*, 1039.
10. Baur, G.; Greubel, W. *Appl. Phys. Lett.* **1977,** *31*, 4. Bechtler, M.; Krüger, H. *Electronics* **1977**, *50*, 113.
11. Crossland, W.A.; Springle, I.D.; Davey, A.B. *Proc. SID Symp. Digest of Technical Papers* **1997**, *27*, 837.
12. Lommel, E. *Ann. d. Physik und Chemie* **1879**, *8*, 634.
13. Jablonski, A. *Acta Phys. Polon.* **1934**, *A14*, 421.
14. *Spectroscopy with polarized light*; Michl, J.; Thulstrup, E.W. Eds.; VCH Publishers: New York, **1986**.
15. Lüssem, G.; Festag, G.; Greiner, A.; Schmidt, C.; Unterlechner, C.; Heitz, W.; Wendorff, J.H.; Hopmeier, M.; Feldmann, J. *Adv. Mater.* **1995**, *7*, 923. Sariciftci, N.S.; Lemmer, U.; Vacar, D.; Heeger, A.J.; Janssen, R.A.J. *Adv. Mater.* **1996**, *8*, 651. Lüssem, G.; Geffarth, F.; Greiner, A.; Heitz, W.; Hopmeier, M.; Oberski, M.; Unterlechner, C.; Wendorff, J.H. *Liq. Cryst.* **1996**, *21*, 903. Conger, B.M.; Mastrangelo, J.C.; Chen, S.H. *Macromolecules* **1997**, *30*, 4049.
16. Wittmann, J.C.; Smith, P. *Nature* **1991**, *352*, 414. Pichler, K.; Friend, R.; Burn, P.L.; Holmes, A.B. *Synth. Met.* **1993**, *55-57*, 454. Fenwick, D.; Smith, P.; Wittmann, J.C. *J. Mat. Sci.* **1996**, *31*, 128. Fenwick, D.; Pakbaz, K.; Smith, P.; *J. Mat. Sci.* **1996**, *31*, 915. Gill, R.E.; Hadziioannou, G.; Lang, P.; Garnier, F.; Wittmann, J.C. *Adv. Mater.* **1997**, *9*, 331.
17. Hamaguchi, M.; Yoshino, K. *Appl. Phys. Lett.* **1995**, *67*, 3381. Tanigaki, N.; Yase, K.; Kaito, A. *Mol. Cryst. Liq. Cryst.* **1995**, *267*, 335.
18. Cimrova, V.; Remmers, M.; Neher, D.; Wegner, G. *Adv. Mat.* **1996**, *8*, 146. Bolognesi, A.; Bajo, G.; Paloheimo, J.; Östergard, T.; Stubb, H. *Adv. Mat.* **1997**, *9*, 121.
19. Dyreklev, P.; Berggren, M.; Inganäs, O.; Andersson, M.R.; Wennerström, O.; Hjertberg, T. *Adv. Mat.* **1995**, *7*, 43.
20. Hennecke, M.; Damerau, T.; Müllen, K. *Macromolecules* **1993**, *26*, 3411.
21. Damerau, T.; Hennecke, M. *J. Chem. Phys.* **1995**, *103*, 6232.
22. Hagler, T.W.; Pakbaz, K.; Moulton, J.; Wudl, F.; Smith, P.; Heeger, A.J. *Polymer Comm.* **1991**, *32*, 339. Hagler, T.W.; Pakbaz, K.; Heeger, A.J. *Phys. Rev.* **1991**, *44*, 8652.
23. Weder, Ch.; Sarwa, C.; Bastiaansen, C.; Smith, P. *Adv. Mat.* **1997**, 9, 1035.
24. Weder, Ch.; Wrighton, M.S. *Macromolecules* **1996**, *29*, 5157.
25. Weder, Ch.; Wrighton, M.S.; Spreiter, R.; Bosshard, C.; Günter, P. *J. Phys. Chem.* **1996**, *100*, 18931.
26. Weder, Ch.; Wagner, M.J.; Wrighton, M.S.; *Organic Solid State Materials III*, L.R. Dalton, Ed.; *Mat. Res. Soc. Symp. Proc.* **1996**, *413*, 77.
27. Steiger, D.; Smith, P.; Weder, Ch. *Macromol. Chem. Rapid. Commun.* **1997**, *18*, 643.
28. Montali, A.; Weder, Ch.; Smith, P. *Proc. SPIE* **1997**, *3148*, 298.
29. Dellsperger, S.; Weder, Ch. *in preparation.*

30. For a recent review on PPEs see: Giesa, R. *J. Macromol. Sci. Rev. Macromol. Chem. Phys.* **1996**, *C36*, 631.
31. Weder, Ch.; Sarwa, C.; Montali, A.; Bastiaansen, C.; Smith, P. *Science*, **1998**, *279*, 835.
32. Weder, Ch.; Sarwa, C.; Bastiaansen, C.; Smith, P. EP 97111229.7 (**1997**).
33. Tessler, N., Denton, G.J.; Friend, R *Nature* **1996**, *382*, 695.
34. Montali, A.; Bastiaansen, C.; Smith, P.; Weder, Ch. *Nature* **1998**, *392*, 261.
35. Weder, Ch.; Bastiaansen, C.; Montali, A.; Smith, P. EP 98101520.9 (**1998**).
36. Guillet, J. *Polymer Photophysics and Photochemistry*; Cambridge University Press, New York, **1985**.
37. Webber, S.E. *Chem. Rev.* **1990**, *90*, 1469.
38. Vekshin, N.L. *Energy Transfer in Macromolecules*; SPIE Optical Engineering Press, Washington: **1997**.
39. Gilbert, A.; Baggot, J. *Essentials of Molecular Photochemistry*; Blackwell Science, Cambridge: **1991**.
40. Förster, T. *Ann. Phys.* **1948**, *2*, 55.
41. Dexter, D.L. *J. Chem Phys.* **1953**, *21*, 836.
42. Knox, R.S. *Theory of Excitons*; Academic Press, New York: **1963**.

Chapter 17

Thermally Stable Intrinsically Conductive Polymer–Carbon Black Composites as New Additives for Plastics

Jamshid K. Avlyanov [1] and Sam Dahman [2]

[1] **Eeonyx Corporation, 750 Belmont Way, Pinole, CA 94564**
[2] **RTP Company, 580 East Front Street, Winona, MN 55987**

One of the main limitations of intrinsically conductive polymers (ICP's) towards their wide application as conductive additives for thermoplastics is their poor thermal-oxidative stability at typical melt processing temperatures (i.e., above 200 °C). On the other hand, the use of high surface area carbon blacks (CB) as conductive additives is limited due to the increased melt viscosity of their blends with thermoplastics. Eeonomers are a new class of thermally stable, chemically neutral, and electrically conductive composites made via *in-situ* deposition of conductive polyaniline (PANI) or polypyrrole (PPY) on CB substrates. Eeonomer composites are more stable (up to 300 °C) than pure ICP's and more easily processible with thermoplastics than CB. Use of Eeonomers as conductive additives for plastics lead to compounds with improved electrical, mechanical, and processing properties. By varying the conductive polymer to CB ratio, it is possible to fine tune the polarity of Eeonomer composites and achieve very low percolation thresholds. This control is possible because of preferred Eeonomer localization at the 2D phase boundary of the immiscible polymer blends.

The major difficulties involved in making electrically conductive thermoplastic blends using polyaniline, polypyrrole, or their composites, are two-fold. The first is the thermal instability of doped polyaniline and polypyrrole at melt processing temperatures (*1-3*). The second is the chemical incompatibility of acidic conductive polyaniline with acid sensitive polymers such as the nylons. Conductive polyaniline is quite acidic and the adjustment of its acidity to neutral pH values eliminates its high conductivity (*4,5*). The authors present here thermal aging studies of conductivity and thermal gravimetric analysis - mass spectroscopy (TGA-MS) of Eeonomers which show pH independence of conductivity in acidic to neutral environments. The tunable surface properties of Eeonomer composites allows one to optimize the processibility as well as the electrical and mechanical properties of their blends with various thermoplastics.

Experimental

Synthesis. All Eeonomers were prepared as described in U.S. Patent No 5,498,372 (*6*). All reagents were used as received from commercial suppliers.

pH Adjustment. The pH of the samples was adjusted by suspending one kilogram of Eeonomer in 12 L of de-ionized water and adjusting the pH of the suspension to a selected value with aqueous ammonium hydroxide. This adjusted suspension was stirred for twelve hours, and the pH was readjusted to the desired value. The suspensions were then filtered and rinsed twice with 4 L of de-ionized water. The solids were then dried in a convection oven at 110 °C for 18 hours. After drying, the "pH" of the material was determined as a 20% suspension in de-ionized water using an Orion 9207BN pH electrode.

Thermal Gravimetric Analysis. Thermal gravimetric analysis (TGA) was performed at the Analytical Sciences Center, Monsanto Corporation, St. Louis, MO. The thermal gravimetric analyses were run on a Mettler Thermoanalyzer-1 equipped with a Hewlett-Packard model 5970 quadrupole mass spectrometer. Weight loss was measured as a function of temperature. Evolved gasses were analyzed via mass spectroscopy (MS) and water and sulfur dioxide were monitored continuously. Approximately 11 mg was ramped between 25 °C to 500 °C at a heating rate of 10 °C/min in an open 8 x 20 mm alumina crucible with helium as the purge gas. A custom-built heated quartz capillary interface maintained at about 185 °C carried a portion of the purge gas stream to a Hewlett-Packard model 5970 quadrupole mass spectrometer. Teknivent Vector V software was used to collect data and produce plots.

Thermal Aging Studies. Thermal aging studies were run at 300 °C in air in a Ney 2-525 Series II muffle furnace. All samples of Eeonomer were pre-dried at 125 °C for one hour. The muffle furnace was allowed to thermally equilibrate at temperature for approximately one hour. Fifteen dried, pre-weighed samples were placed in the muffle furnace. The temperature of the muffle furnace returned to the set temperature within 5 minutes of closing the door. After one hour, three of these samples were removed and allowed to cool in a desiccator. The furnace was allowed to returne to the set temperature within 5 minutes. Another three samples were removed after an additional hour and so on until the last three samples were removed after a total elapsed time of 5 hours. After the samples had cooled, their final weight was recorded and their conductivity was measured as pressed pellets using the four-point-probe technique on a Loresta AP resistivity meter (Mitsubishi). The pellets were pressed using a Beckman KBr pellet die and a Carver hydraulic press. Pellets were pressed at 24000 psi for ten minutes, followed by five minutes at 5000 psi to alleviate cracking.

Surface Area. BET surface area analyses were performed at the Analytical Sciences Center, Monsanto Corporation, St. Louis, MO. The analyses were performed on a Micrometrics ASAP 2010 gas physisorption instrument using a standard 10 cm^3 sample tube with a 1.27 cm O.D. stem. The samples were degassed at 90 °C overnight using high vacuum. The analyses were completed at liquid nitrogen temperature using ultra-high-purity nitrogen as the adsorbing gas and alumina as a standard.

Compounding with Thermoplastics. Conductive thermoplastics were prepared by first physically mixing Eeonomer composites with thermoplastic resin pellets and then compounding using a twin-screw compounding extruder. Compound flow behaviors were analyzed with a capillary rheometer. All materials were dried according to established resin guidelines before compounding, capillary rheometry, and molding.

ASTM test specimens were prepared by injection molding. The corresponding ASTM test methods were used to generate mechanical properties for the injection molded specimens. For instance, ASTM D-638, D790 and D-256 were used respectively for tensile strength and elongation, flexual modulus and Izod impact. Electrical properties of the specimens were characterized by measuring volume resistivity. This was accomplished by contacting the leads of a multimeter to conductive silver paint electrodes at the ends of a test bar and then determining the DC resistance at different test voltages.

Results and Discussion

Properties of Eeonomer Composites. As can be seen in Figure 1, polyaniline/ carbon black Eeonomer shows excellent thermal stability at 300 °C in air (*7*).

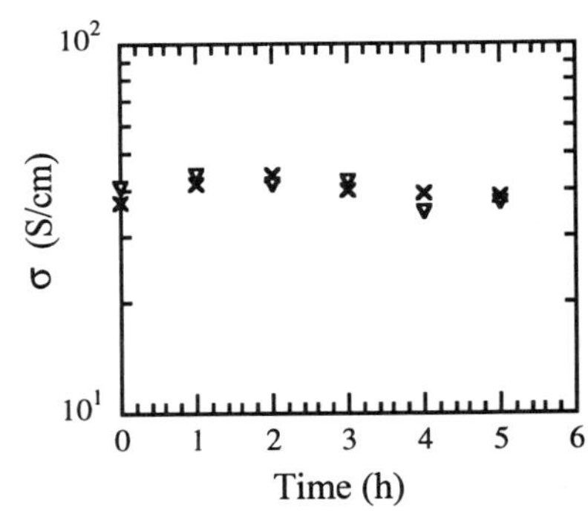

Figure 1. Conductivity (σ) stability in air at 300 °C of polyaniline/carbon black (x), and polypyrole/carbon black (∇) Eeonomers; ICP coating level 30 wt %. Reproduced with permission from ref. 7. Copyright 1998.

There is no statistically significant change in the conductivity of Eeonomer over the first three hours at 300 °C. Over the next two hours, there is a 13% drop in conductivity (from about 42 S/cm to about 37 S/cm). Versicon, a commercial conductive form of polyaniline, is reported to have a conductivity half-life of approximately 10 minutes at 225 °C in nitrogen (1). Therefore at 225 °C after 3 hours, the conductivity of 'doped' polyaniline should decrease by a factor of 262000 (18 half-lives of 10 minutes). Obviously, Eeonomer is much more stable at 300 °C in air than pure polyaniline at 225 °C in nitrogen.

Highly stable polypyrrole/carbon black Eeonomers can also be synthesized by depositing conductive polypyrrole *in-situ* on carbon black. There is no loss in the conductivity of polypyrrole/carbon black Eeonomer after five hours at 300 °C in air (Figure 1). An initial increase in conductivity is observed in the first hour, and the conductivity drops to its original value in the next hour. Similar behavior has been observed for polypyrrole coated fabric at 125 °C (2).

From the TGA profile (Figure 2), it can be seen that there is no detectable weight loss for PANI Eeonomer below 300 °C, which means there is no appreciable degradation of the doped polyaniline within this temperature range. The mass loss which begins around 300 °C corresponds to concurrent degradation of the polyaniline backbone and the dopant ion. The mass spectrometry data indicates the loss of aniline starting at 270 °C with a maximum at 330 °C. For a sample of polypyrrole Eeonomer, the onset of decomposition is observed at approximately 320 °C, as can be seen in the total ion current in the MS profile (Figure 3).

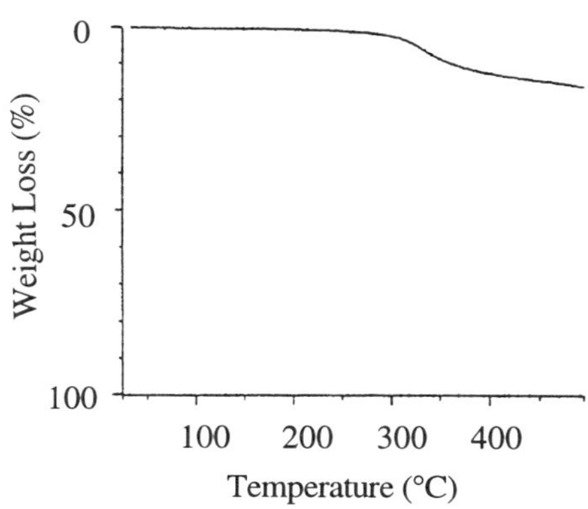

Figure 2. TGA profile of a doped polyaniline/carbon black Eeonomer under helium at a scan rate of 10 °C/min. Eeonomer contains 25 wt % of doped polyaniline. Reproduced with permission from ref. 7. Copyright 1998.

There are several possible reasons for the increased thermal stability of the conductive polymers in Eeonomers. The improved thermal stability of doped, conjugated polymers on the carbon black can be explained by multiple π-π interactions between the unsaturated backbone of the conductive polymer, the aromatic rings of the carbon substrate, and the planar aromatic dopant ions. These interactions should limit the rotational freedom of the conjugated polymer segments and therefore enhance the thermal stability of the polymer. Knowing the surface area of the original carbon black (1400 m^2/g), the amount of the conductive polymer in the composite, and assuming uniformity of coating, the thickness of the deposited layer of conductive polymer was calculated to be approximately 20 Å for a 22 % conductive polymer / 78% carbon black composite (*8*). This value corresponds to only a few molecular layers of polymer. These calculations are supported by transmission electron microscopy (TEM) studies. No evidence of free conductive polymer was detected by TEM. There was no observed change in the dimensions and morphology of the individual carbon black particles as a result of conductive polymer deposition. Therefore, the thickness of the conductive polymer layer has to be less than the inherent resolution of the TEM. Since the layer is only a few molecules, thick, each individual polymer chain most likely interacts strongly with the carbon black substrate, resulting in the enhanced stability of Eeonomer.

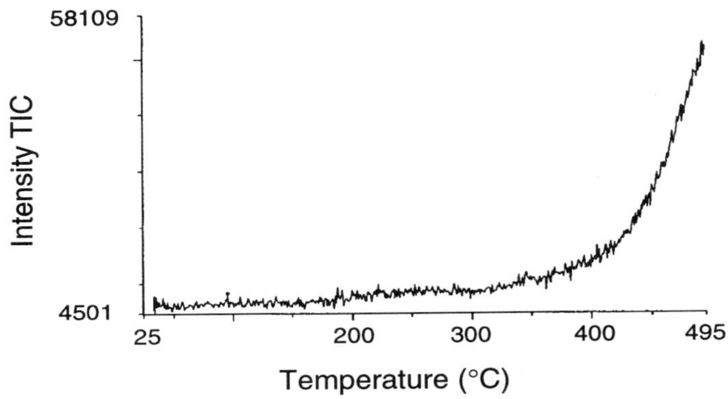

Figure 3. TGA-MS profile of the polypyrrole/carbon black Eeonomer under helium at scan rate of 10 °C/min. Eeonomer contains 30 wt % of doped polypyrrole.

Conductive Thermoplastic Compounds. Conductive thermoplastic compounds were prepared using Eeonomers. Volume resistivity as a function of Eeonomer content is shown in Figure 4 for nylon-6 and ABS compounds. The volume resistivity is observed to fall several orders of magnitude between 5 and 10 wt % of composite concentration. This indicates that the resistivity is very responsive to Eeonomer loading level and that percolation is occurring within this range.

The properties of some thermoplastic compounds that were prepared with Eeonomer additives are given in Table I. Also, listed as a control for comparison is a nylon-6 compound with uncoated carbon black. The mechanical properties of the Eeonomer - nylon compounds are observed to be similar or better than the control. Note that in the nylon-6 case, the polypyrrole - Eeonomer compound offers improved mechanical properties over the polyaniline composite while still retaining the same conductivity level.

In addition to high thermal stability, the conductivity of Eeonomer composites was found to be independent of pH. In Figure 5, the conductivity remains relatively unchanged over a pH range of 0 to 8 for a doped polyaniline - carbon black composite. Acid doped polyaniline by itself, on the other hand, will typically begin to exhibit a conductivity loss at a pH greater than 3 (*4*). Acid sensitive thermoplastics incorporating neutral Eeonomer composites should exhibit little or no degradation. This is shown in Figure 6 in which the viscosities of nylon-6 and PBT compounds remain relatively unchanged for at least 20 minutes at elevated temperature.

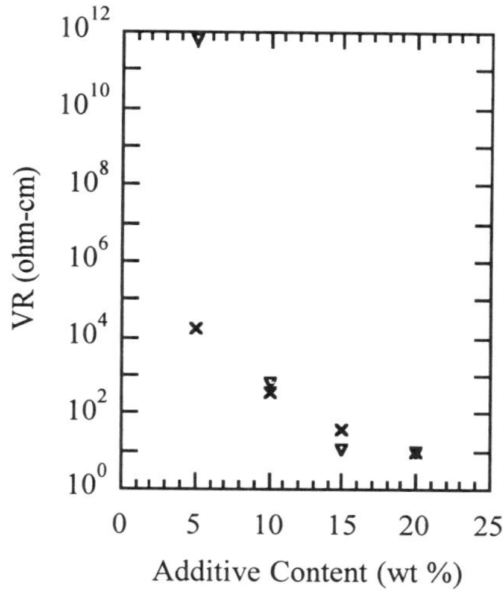

Figure 4. Volume resistivity (VR) as a function of Eeonomer content in nylon-6 (∇) and ABS (x). Reproduced with permission from ref. 7. Copyright 1998.

Another key advantage of the deposition of ICPs onto carbon black substrates is the reduction of surface area and pore volume as shown in Figure 7. Due to its high surface area, it is difficult to compound highly structured carbon black into thermoplastics. At a coating level of 10 wt % of polyaniline onto highly structured carbon black, the surface area is observed to decrease by half. The resulting composites were found to process much easier than the uncoated carbon blacks.

Table I. Properties of Some Eeonomer/Thermoplastic Compounds

Thermoplastic	Nylon-6	Nylon-6	Nylon-6	ABS	PBT
ICP Coating Type on CB	PPY	PANI	None, neat CB	PANI	PPY
Additive Content (wt %)	15	15	15	15	15
Tensile Strength (MPa)	54.7	45.5	31.2	44.3	43.3
Flexural Modulus (GPa)	2.41	2.82	2.41	3.31	3.45
Elongation (%)	2.9	1.6	1.3	1.7	1.5
Notched Izod Impact (J/m)	43.2	22.9	28.8	16	22
Volume Resistivity (ohm-cm)	$1.0x10^{2}$	$1.1x10^{1}$	$4.1x10^{1}$	$3.9x10^{1}$	$3.0x10^{1}$

Reprinted with permission from ref. 7. Copyright 1998.

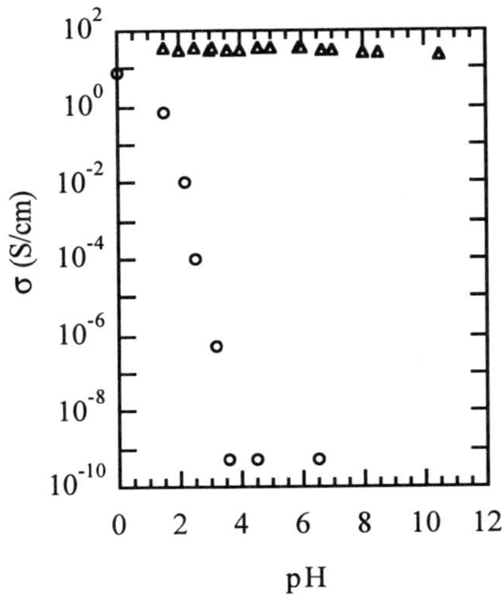

Figure 5. Conductivity as a function of the pH of the equilibrium solution for polyaniline (o) and polyaniline/carbon black Eeonomer (Δ). Reproduced with permission from ref. 7. Copyright 1998.

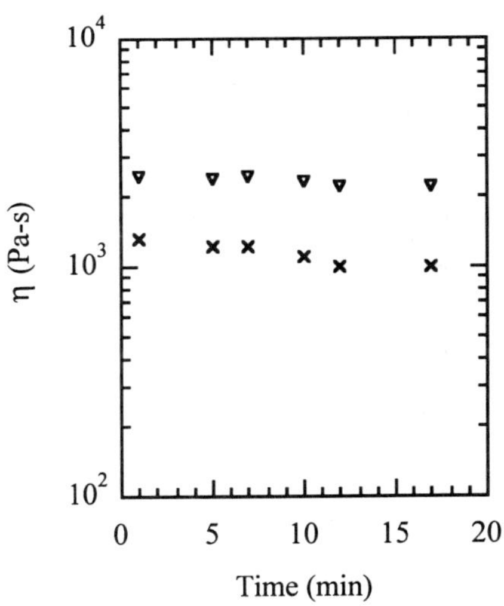

Figure 6. Melt flow stability of acid sensitive compounds loaded with neutral Eeonomer: nylon-6 at 235 °C (∇) and PBT at 250 °C (x). Reproduced with permission from ref. 7. Copyright 1998.

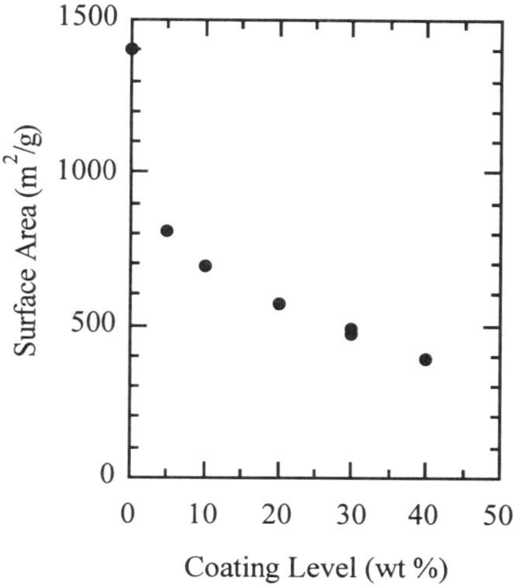

Figure 7. Surface area of Eeonomer as a function of polyaniline coating level on high structure carbon black. Reproduced with permission from ref. 7. Copyright 1998.

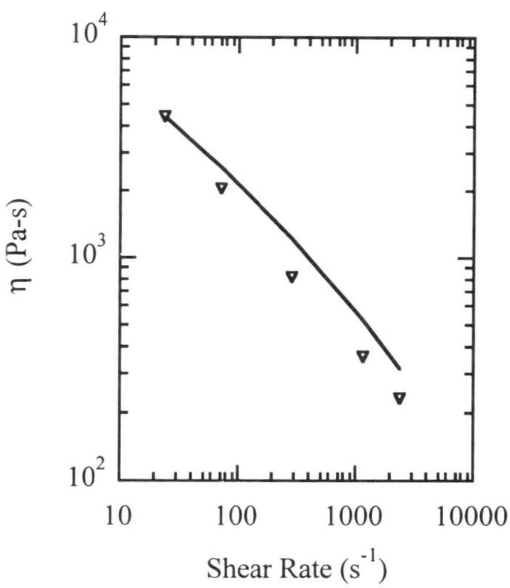

Figure 8. Melt flow behavior of nylon-6 with 20 wt % of additive: 20 wt % polypyrrole coating level on high structure carbon black (∇) and uncoated low structure carbon black (solid line). Reproduced with permission from ref. 7. Copyright 1998.

Additionally, the surface area reduction improves the melt flow behavior of conductive compounds when compared to uncoated carbon black loaded compounds. Figure 8 shows that the viscosity of an ICP – high structure carbon black composite is even lower than that of an uncoated low structure carbon black at the same additive content.

Eeonomer in Two Phase Plastic Blends. The flexibility of Eeonomer technology allows one to fine-tune its properties to each thermoplastic, thermoset, or polymer blend. Such tuning might be achieved by varying: (a) the type of conductive polymer; (b) the amount of conductive polymer in Eeonomer; (c) the nature of doping ions; and (d) the grade of carbon black. For instance, in this section a method for adjusting the polarity of the Eeonomer additive is presented which involves simple variation of doped polypyrrole to carbon black ratio.

Double percolation phenomenon has been reported previously for two phase polymer blends loaded with carbon black (*9-11*). Very low percolation thresholds were reported when conductive carbon black was preferentially localized at the phase boundary (*9*). Figure 9 shows the volume resistivity of phase-separated nylon/poly-

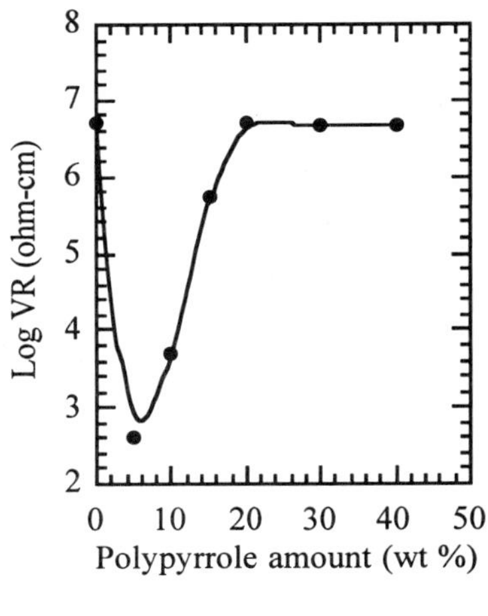

Figure 9. Volume resistivity (VR) of polypropylene/nylon-6 blend as a function of polypyrrole content in Eeonomer. Conductive additive loading is 4 wt % for all compounds.

propylene (PA/PP) blends loaded with Eeonomer composites containing different amounts of doped polypyrrole. It is important to note that all the compounds in Figure 9 contain the same loading level of conductive additive. As can be seen, there is a clearly defined minimum in the resistivity of blends at approximately 5wt % of polypyrrole on CB. Such a resistivity change could be explained as due to controlled localization of conductive additive within the co-continuous PA/PP blend. Solvent extraction experiments indicated that less polar CB is preferentially located in the PP phase. On the other hand, Eeonomer with a high coating level (more than 20 %) of the more polar polypyrrole, is located in the nylon phase. In this case, when conductive compounds were ultrasonically treated in formic acid (is a solvent for nylon-6 and non-solvent for polypropylene), the entire black additive was washed away with the soluble nylon phase. The PP phase which remained after the extraction was white in color, which indicates that the entire additive was located within the nylon phase. DSC tests proved efficient nylon phase extraction, as confirmed by the absence of the peak corresponding to melting of nylon.

Eeonomer with intermediate polarity would tend to be localized at the phase boundary of a plastic blend. As mentioned earlier, such a segregation of the conductive additive leads to increased conductivity of plastic compounds (*9-11*). The possibility for adjusting Eeonomer polarity should allow one to select the optimal conductive additive for a variety of immiscible plastic blends.

Conclusions

Eeonomers are a new class of conductive additives for thermoplastics made via *in-situ* deposition of intrinsically conductive polyaniline or polypyrrole on carbon black. Eeonomers are highly thermally stable, pH neutral conductive materials that are compatible with the chemistry and melt processing conditions of acid sensitive polymers. Compounding studies with thermoplastics indicate better electrical, mechanical, and melt flow properties of Eeonomer blends as compared to blends with traditional carbon blacks. In co-continuous plastic blends it was possible to fine tune the polarity of Eeonomer by varying the conductive polymer to CB ratio. The same variation affords very low percolation thresholds due to preferred Eeonomer localization at the 2D phase boundary.

Acknowledgments

The authors grateful to Bruce Frushour from Monsanto Company for TGA-MS tests. We also thank to Martin Cooper and Keith Schwartz for technical assistance.

References

1. Shacklette, L.W.; Han, C.C. *Synth. Met.* **1993,** *55-57*, 3532.
2. Child, A.D.; Kuhn, H.H. *Synth. Met.* **1997,** *84*, 141.
3. Thieblemont J.C., Planche M.F., Petrescu C., Bouvier J.M.; Bidan G. *Synth. Met.* **1993,** *59*, 81.
4. Chiang, J.C.; MacDiarmid, A.G. *Synth. Met.* **1993**, *13*, 193,.
5. MacDiarmid, A.G.; Epstein A.J. *Faraday Discuss. Chem. Soc.* 1989, 88, 317.
6. Hedges, W.L., United States Patent 5,498,372, **1996.**
7. Dahman, S.; Avlyanov J.K. ANTEC'98 **1998,** 1313.
8. Du, G.; Avlyanov J.K.; Wu C.Y.; Reimer K.G.; Benatar A.; MacDiarmid A.G.; Epstein A.J. *Synth. Met.* **1997,** *85*, 1339.
9. Gubbels, F.; Jerome, R.; Teyssie, Ph.; Vanlathem, E., Deltour, R.; Calderone, A.; Parente, V.; Bredas, J.L. *Macromolecules* **1994,** *27*, 1972.
10. Tchoudakov, R.; Breuer, O.; Narkis, M; Siegmann, *A. Polym. Eng. and Sci.* **1996,** *36*, 1336.
11. Gueskens, G.; Gielens, G. L.; Geshef, D.; and Deltour, R. *Eur. Polym. J.* **1987,** *23*, 993.

Chapter 18

Review of Poly(bis(dialkylamino)phenylene vinylene)s as Corrosion-Inhibiting Materials

Peter Zarras, John D. Stenger-Smith [1], Gregory S. Ostrom, and Melvin H. Miles

Research and Technology Group, Naval Air Warfare Center Weapons Division, China Lake, CA 93555–6100

A brief history of research on conducting polymers, with specific emphasis on results in the area of conducting polymers as corrosion-protective coatings is presented. The synthesis and corrosion inhibiting properties of poly(bis(dialkylamino)phenylene vinylenes at isotonic sea water conditions are presented and discussed.

On a historical note, polyaniline was first made as far back as 1862 by Letheby[1]. Known as "aniline black", this material was formed by oxidation of aniline under mild conditions[2,3]. Aniline black was an important material for dyeing and printing[4]. Conducting polymer research has roots back to the 1960s when Pohl, Katon, and others first synthesized and characterized semiconducting polymers[5-7] and conjugated polymers[8]. The discovery of the high conductivity of poly(sulfurnitride) $(SN)_x$, a polymeric inorganic explosive[9], and its interesting electrical properties[10-16] was a step towards conducting polymers as they are known today.

The beginning of conducting polymer research began nearly a quarter of a century ago, when films of polyacetylene were found to show tremendous increases in electrical conductivity when exposed to iodine vapor[17,18]. This was the first report of polymers with high-electrical conductivity. The procedure for synthesizing polyacetylene was based upon a route discovered in 1974 by Shirikawa through serendipitous addition of a thousand times the normal amount of catalyst during the polymerization of acetylene[19]. Over the past two decades, there have been several excellent reviews on conducting polymers[20-34]. Conducting polymer research is evolving rapidly enough that yearly reviews are almost a necessity. Today, there are hundreds of articles on conducting polymer research published every year.

The conducting forms of conducting polymers are usually classified as the cation salts of highly conjugated polymers. The cation salts are obtained by electrochemical oxidation and electrochemical polymerization[35] or chemical oxidation (removal of an electron). It is also possible to obtain the anion salts of the same highly conjugated polymers, which are also conducting but much less stable than the cation counterparts, by either electrochemical reduction or by treatment with reagents such as solutions of sodium naphthalide [36,37]. In general, a conjugated backbone and/or a backbone that has a low enough oxidation potential is necessary but not sufficient for the electrically conducting form of a polymer to remain stable in the

[1] Corresponding author.

presence of air and water or end-use conditions such as inside an automobile or home. Figure 1 lists some examples of conducting polymers.

Polyaniline, one of the most studied conducting polymers (in general as well as for corrosion protection), is usually obtained by protonation of what is called the emeraldine base form, shown in Figure 2. The protonation reaction does not change the number of electrons in the polymer backbone. However, starting at the leucoemeraldine form of polyaniline, one would obtain the emeraldine salt (conducting) form of polyaniline by an oxidation reaction. Protonic doping has also been observed in the case of alkoxy substituted poly(para-phenylene vinylene) (PPV)[38,39].

Corrosion Protection

In the area of potential applications, conducting polymers are used as replacements for metals because the conducting polymers have potentially unique and or superior properties, or because the metals are toxic or damage the environment. Current methods of corrosion protection (particularly marine coatings)[40] do not last very long and are coming under increased scrutiny by the Environmental Protection Agency (EPA). For example, the use of chromium (especially hexavalent chromium) and cadmium for anti-corrosion coatings will soon be banned. A mechanism for corrosion protection involves the use of a sacrificial electrode, such as a zinc coating, which will corrode (oxidize) in the place of the substrate. However, the coatings do not last very long. The oxidized zinc metal is dissolved by water or moisture. For this reason there are extreme environmental concerns since toxic metals are being released. Barrier coatings such as epoxy are employed extensively but are not very durable/robust once a pit or hole in the coating has been formed. The corrosive species then attacks the underlying metal and, thereby, increases the exposed surface, accelerating the corrosion process.

The corrosion inhibiting properties of conducting polymers were suggested by MacDiarmid in 1985. Initial studies on the protection of metal surfaces against corrosion by conducting polymers was reported in the literature that same year[41]. Much of the work on corrosion protection has focused on polyaniline (PANI)[42-46], but also has been extended to other conjugated polymers[47-50]. A major type of corrosion occurs by oxidation of a metallic surface by a saltwater medium to produce oxides and hydroxides. As these form, soluble species are produced, the surface pits increase the surface area, and the rate of decomposition accelerates. One way to provide corrosion protection is to coat the metal with a barrier to prevent the reactive species from reaching the surface. Galvanization with zinc (or other metal with low enough oxidation potential) prevents corrosion via the creation of an interfacial potential at the metal:zinc interface. The zinc will corrode preferentially. While the reactive species may come in contact with the metal, the increased oxidation potential causes the metal to be unreactive.

Prior work utilizing PANI as a corrosion-protection coating shows that it works quite well. In fact, exposed metal surfaces adjacent to conducting polymer coatings (scratches or edges) are unreactive to corrosion as reported by Thompson and co-workers[51,52]. The corrosion protection properties of PANI on aluminum in acidic media have also been studied[53]. Therefore, corrosion protection is an area where conducting polymers have great potential. If superior performance of conducting polymers can be demonstrated under maritime conditions (which vary greatly depending upon use and environment), there is a potential for multi-billion dollar savings. For example, estimates from the United States Navy indicate that corrosion costs of aircraft (presented at the Third International Conference on Aircraft

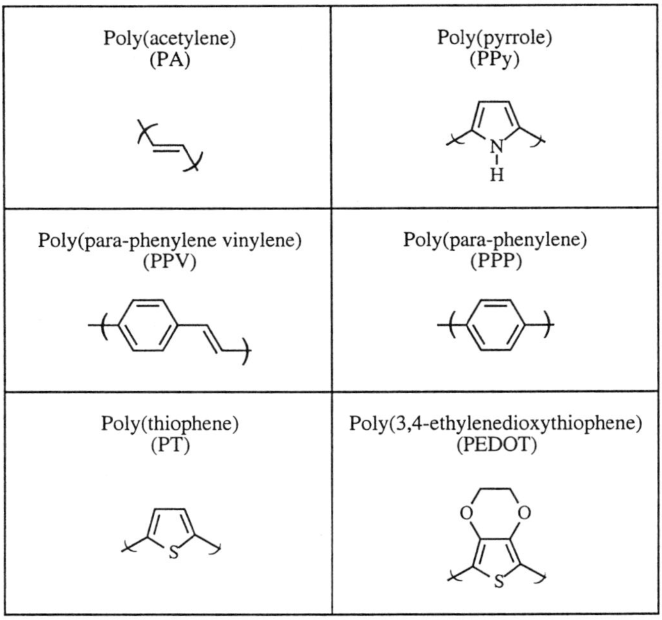

Figure 1. Structures of Conjugated Polymers in Their Respective Neutral Forms.

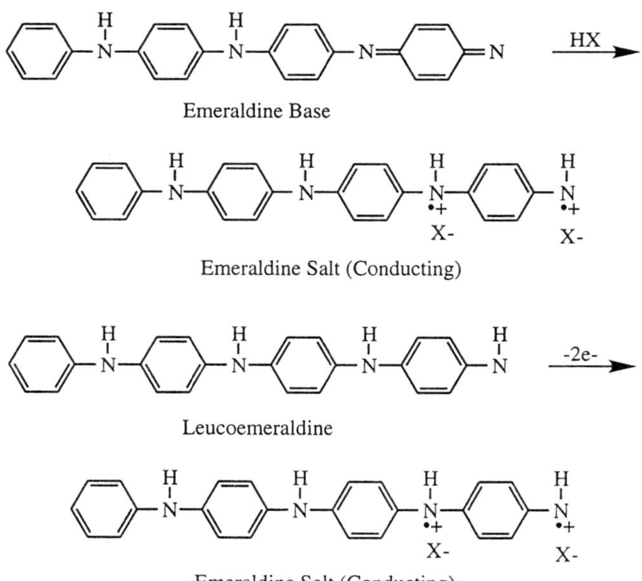

Figure 2. Structures of Polyaniline.

Corrosion, August 25-27, Solomons Island MD) are several billions of dollars per year and corrosion costs of ships are tens of billions of dollars per year[54].

There are several proposed mechanisms for corrosion protection, one or more of which could be occurring at any time. The first is a simple galvanic process by which the polymer has a lower oxidation potential than the metal it is protecting; the polymer is preferentially oxidized. Because oxidized polymers are usually insoluble and, therefore, do not dissolve away as zinc does, corrosion protection with conducting polymers should last longer. Another proposed mechanism is that the polymer reacts with the surface of the metal, requiring that the polymer have an oxidation potential higher than that of the metal. The surface of the metal reacts with the polymer and forms a passivating layer[55] which inhibits further corrosion by either setting up a barrier or by changing the surface potential or both.

One possible disadvantage to using polyaniline is that the corrosion-protection ability is pH dependent. In acidic media, polyaniline-coated mild steel coupons corrode a hundred times slower than counterparts, while in pH 7 media, the PANI-coated material corrodes twice as slowly[56,57]. Because the pH of seawater is around 8.0 to 9.4 depending upon season and location, it is unclear or unproven that PANI will provide any additional corrosion protection for ocean-going vessels. This could be explained by the pH dependence of the structure of PANI. At low pH, the conducting emeraldine salt is the predominant form; at high pH, the non-conducting emeraldine base is the predominant form. It appears that the conducting form is required for the formation of the passivation layer. In summary, the amount of corrosion protection is controlled by the type of polyaniline (emeraldine base versus emeraldine salt) and the characteristics of the corrosion environment (acidic medium, aqueous sodium chloride, or seawater) and also by adhesion to the substrate. Studies on the marine application of corrosion protection capability of conducting polymers will need to be performed in solutions isotonic with seawater and/or a salt fog according to American Society for Testing of Materials (ASTM) methods[58]. For corrosion protection, it may be necessary to develop conducting polymers that do not have the pH dependence of conductivity that PANI has or to formulate PANI in such a way that it provides corrosion protection at basic pH.

There may be some concern that all conducting polymers behave as polyaniline, that is, that all conducting polymers exhibit the same pH dependence of corrosion protection that polyaniline does. However, since polyaniline is made conducting by treatment with acid whereas most other conducting polymers are not, it is extremely unlikely that all conducting polymers will exhibit similar pH dependence of corrosion inhibition.

Synthesis of Poly(2,5-bis(n-methyl-n-propyl)amino phenylene vinylene) (BAMPPV)

The addition of bis(dialkyl amino) substituents onto the PPV backbone is of interest for several reasons. First, the amino groups are generally stronger electron donors than alkoxy groups (provided that the resulting amino substituted polymer structure is planar) and should bring the oxidation potential of the polymer down around 0 volts vs. SCE, making the conducting polymer even more stable. Also, amine functionalized polymer should adhere fairly well to aluminum. Finally, amino groups can also be quarternized, which could be exploited to make the polymer water soluble.

The synthesis of amino functional PPV presented some serious synthetic challenges[59]. Usually, a radical halogenation step is used in making the precursors to PPV. For example, 1,4-dimethyl benzene would be chlorinated or brominated to make a precursor to PPV. Although this method can be adapted well for the alkoxy derivatives, it cannot be used for amino derivatives because the reaction is dangerously exothermic. The danger of this type of reaction was established back in 1957, when a fatal accident occurred[60]. Another method, used generally for dialkoxy derivatives, is to chloromethylate the dialkoxy substituted compounds. This reaction will not work with amines because the acidic conditions used will protonate the amine making it unreactive. Therefore, another synthetic strategy was developed. This method, shown in Scheme 1, does not involve any of the problem steps mentioned above.

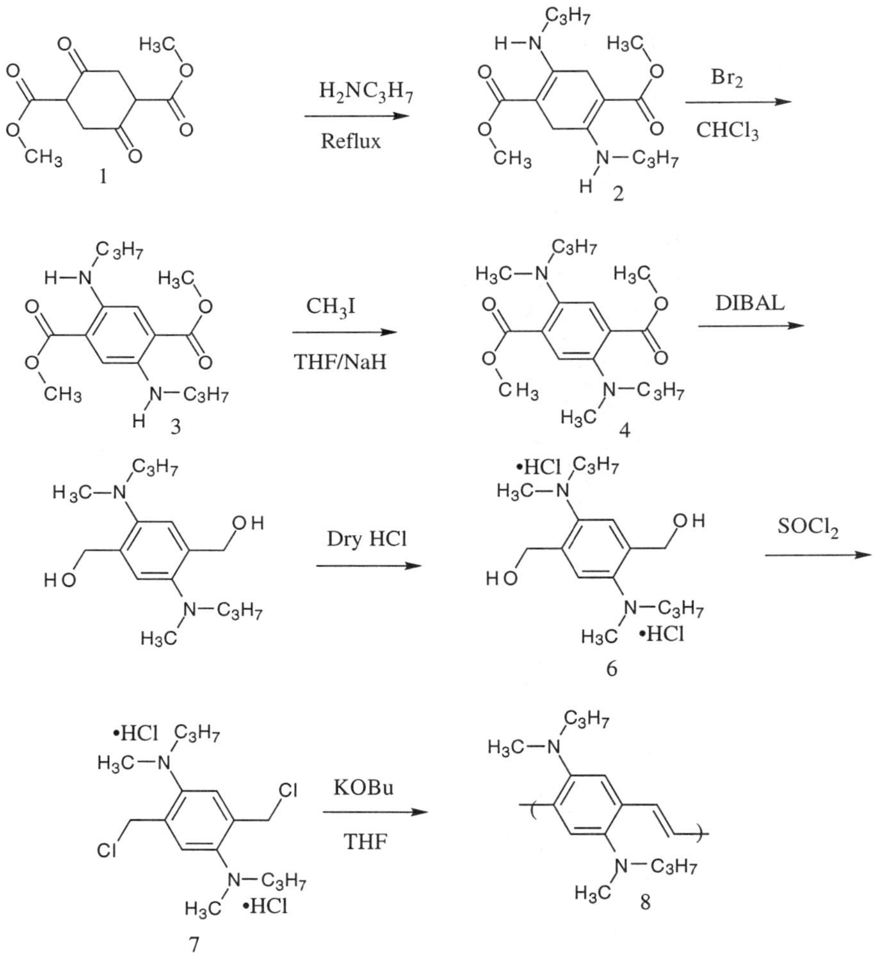

Scheme 1.

This method allows the synthesis of fairly pure polymer[61] and is being improved to allow for scaleup[62]. The electrical and electrochemical properties of this polymer are currently being studied. The neutral form of the polymer is orange-red in color and the absorption maximum is 460 nanometers. This absorption maximum is much higher in energy than expected, so it is possible that the polymer backbone is non-planar.

Corrosion Studies of BAMPPV

BAMPPV was coated onto Type II anodized aluminum T3 plates (5.1 x 5.1 x 0.15 cm, A = 55 cm^2) in saltwater for a 12 month immersion study. Constant current (galvanostatic and constant potential (potentiostatic) methods were used to investigate the corrosion of these aluminum plates. The electrochemical studies were conducted in concentrated saltwater solutions by dissolving 70.13 grams of sea salt (Bio-Sea Marine Mix) in 1200 milliliters of deionized water. This produced a solution of approximately 1 molar in NaCl with a specific gravity of 1.04257, with a pH of approximately 8.1, which contained the trace elements of seawater. Potentiostatic and galvanostatic studies of an (unanodized) aluminum plate (immersed for 1 month) in this salt solution are shown in Figure 3. The two methods give approximately the same results, showing the pitting potential (the potential at which the current increases rapidly due to corrosion) near -0.6 volt vs. Ag/AgCl.

The electrochemical behavior is very different for anodized-aluminum plates (immersed for 1 month), as shown in Figure 4. There is no measurable current for this potentiostatic study to within 0.001 milliampere from -0.600 to -0.300 volts. The pitting potential is shifted markedly from that shown in Figure 3 to about -0.28 volts vs. Ag/AgCl. Despite the anodic potentials applied in Figure 4, currents larger than 15 milliamperes are not observed. Further increases in potentials out to 3.00 volts were investigated; the largest current obtained was only 25 milliamperes. Detailed examination of the aluminum plate indicated that a few isolated regions on the edges of the plate were the main contributors to the anodic current. During the time period of this potentiostatic study, the anodic current gave a yield of 1.380 coulombs. Also shown in Figure 4 are exactly the same potentiostatic measurements for an anodized aluminum plate coated with BAMPPV (immersed for 1 month). The striking feature is that very little corrosion current is observed. There is no measurable current to within 0.001 milliampere from -0.600 to +0.45 volts. The corrosion potential, if there is one, is near 0.525 volts vs. Ag/AgCl. This increase in overvoltage corresponds to an activation energy barrier increase of 57.9 kilojoules/mole. The current increases only slightly to 0.071 milliampere at 0.80 volts and then decreases. There is no further increase in the current even out to 3.00 volts. The coulombic measurements during this potentiostatic study yielded only 0.00358 coulombs. Therefore, based upon the coulombic measurements, the polymer-coated anodized-aluminum plate yielded only 0.26% of the corrosion measured for the uncoated anodized-aluminum plate. This is quantitative evidence that polymer coatings can substantially reduce pitting corrosion of aluminum. Figure 4 also shows a potentiostatic study after the plates were immersed in isotonic sea water for an additional 5 months (6 months total). The pitting potential is still much higher for the coated film (0.225 volts vs. Ag/AgCl), which is about 75% of the initial value, giving an activation energy barrier for corrosion of 43.4 kilojoules/mole. In addition to the studies shown in Figures 4 and 5, several long-term (days) constant-current electrolysis experiments were conducted with polymer-coated and uncoated-aluminum plates. In each long-term study, the corrosion pits that developed were always significantly less for the polymer-coated plates.

Further studies of the BAMPPV plates extending to a 12 month study were accomplished using the same methods for the 6 month study. Figure 5 shows similar behavior for the anodized-aluminum plates as in Figure 4. There was no measurable current for this 12 month potentiostatic study to within 0.0001 milliampere from -

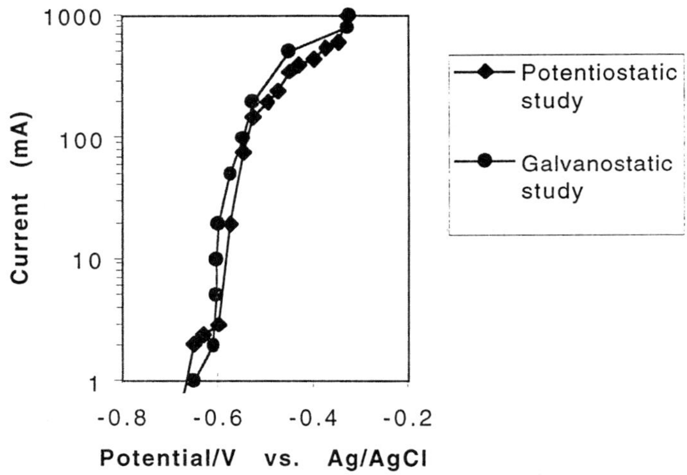

Figure 3. Potentiostatic and Galvanostatic Electrochemical Studies of an Unanodized Aluminum Plate in the Sea Salt Solution.

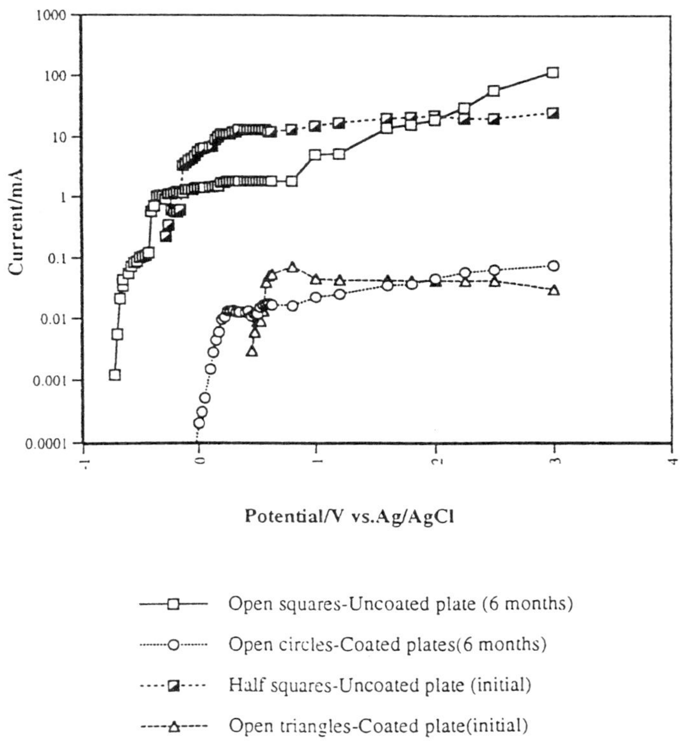

Figure 4. Potentiostatic Electrochemical Studies of Anodized-Aluminum Plates (1 and 6 month).

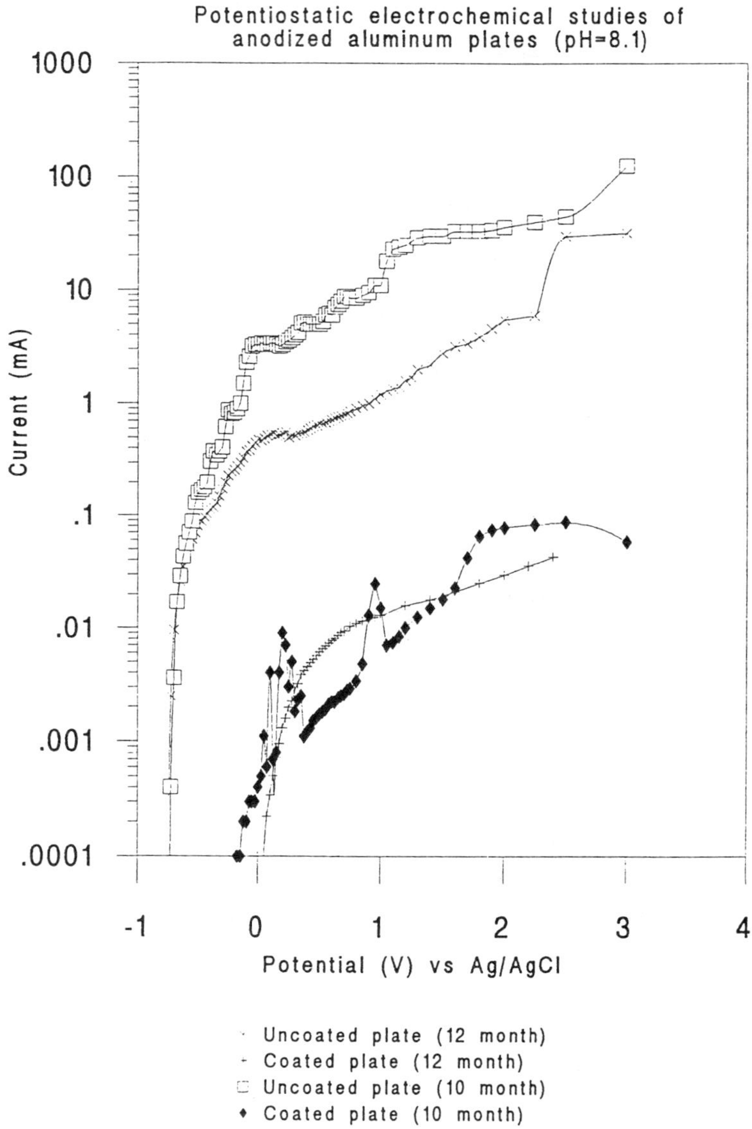

Figure 5. Potentiostatic Electrochemical Studies of Anodized Aluminum Plates (10 and 12 months).

0.600 to -0.300 volts and detailed examination of the aluminum plate still indicated that only a few isolated regions of the plate due to edge effects were the main contributors to the anodic current. Furthermore, the pitting potential is still much higher for the coated film (around 0.2 volts vs. Ag/AgCl), giving an activation energy of around 38 kilojoules/mole, which is 65% of the initial value. The results from the 10 month study show some very small current spikes, which could be due to edge effects or instrument artifacts. These spikes make quantitative interpretation of the 10 month data difficult, but generally, the 10 month results are consistent with the overall study. Scanning electron microscopy (SEM) was also performed on anodized aluminum plates and BAMPPV-coated aluminum plates (both immersed for 6 months). The results from the SEM, shown in Figure 6, show that the polymer coating provides a significant degree of corrosion protection.

(a)

(b)

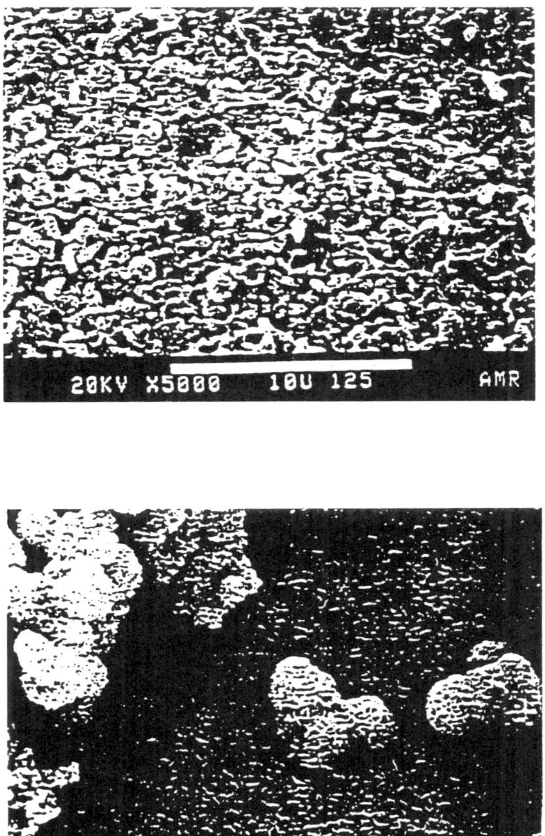

Figure 6. (a) BAMPPV-Coated Aluminum with Minimal Corrosion and (b) Uncoated Aluminum with Severe Corrosion.

Summary and Dicussion

A new conducting polymer, poly(2,5-bis(N-methyl-N-propyl)amino phenylene vinylene), was synthesized and characterized. This conducting polymer is a very promising candidate for corrosion protection of aluminum in isotonic sea water. The time is ripe to expand research in this area, with specific focus on shipboard applications and neutral to slightly basic pH, and incorporation of polyaniline for use in shipboard acidic environments. This conducting polymer program should have some focus on corrosion protection at neutral to basic pH, include new materials synthesis to meet these needs, and formulations of polyaniline that might inhibit corrosion at higher pH.

Acknowledgements

The authors wish to thank the Office of Naval Research and NAWC for financial support, and to Mr. J. Nelson and Mrs. C. Webber for technical assistance.

Liturature Cited

1. H. Letheby. *J. Chem. Soc.*, Vol. 15 (1862), p. 161.
2. O. Piequet. *Bull. Soc. ind Amiens*, Vol. 47, No. 10 (1910).
3. A. G. Green. *Chem. Zentr.*, Vol. I (1914), p. 535.
4. E. Noelting. *Scientific and Industrial History of Aniline Black.* New York, Wm. J. Matheson, 1889.
5. H. A. Pohl, J. A. Bornmann, and W. Itoh. *Am. Chem. Soc. Div. Polym Chem. Preprints*, Vol., 2, No. 1 (1961), p. 211.
6. H. A. Pohl. *Chem. Eng.*, Vol. 68, No. 22 (1961), p. 105.
7. J. E. Katon and B. S. Wildi. *J. Chem. Phys.*, Vol. 40, No. 10 (1964), p. 2977.
8. H. A. Pohl and E. H. Engelhardt. *J. Phys. Chem.*, Vol. 66 (1962), p. 2085.
9. F. B. Burt. *J. Chem. Soc.*, 1910, p. 1171.
10. V. V. Walatka, Jr., M. M. Labes, and J. H. Perlstein. *Phys. Rev. Lett.*, Vol. 31 (1973), p. 1139.
11. M. J. Cohen, A. F. Garito, A. J. Heeger, A. G. MacDiarmid, C. M. Mikulski, M. S. Saran, and J. Kleppinger. *J. Am. Chem. Soc.*, Vol. 98, No. 13 (1976), p. 3844.
12. R. H. Baughman, P. A. Apgar, R. R. Chance, A. G. MacDiarmid, and A. F. Garito. *J. Chem. Soc., Chem. Comm.*, Vol. 2 (1977), p. 49.
13. R. J. Nowak, H. B. Mark, Jr., A. G. MacDiarmid, and D. Weber. *J. Chem. Soc., Chem. Comm.,* Vol. 1 (1977), p. 9.
14. M. M. Labes, P. Love, and L. F. Nichols. *Chem. Rev.*, Vol. 79, No. 1 (1979), p. 1.
15. M. Whango, R. Hoffman, R. B. Woodward. *Proc. Roy. Soc. Ser. A.*, Vol. 366 (1979), p. 23.
16. M. Akhtar, C. K. Chiang, A. J. Heeger, and A. G. MacDiarmid. *J. Chem. Soc., Chem. Comm.,* Vol. 23 (1977), p. 846.
17. H. Shirikawa, E. J. Louis, A. G. MacDiarmid, C. K. Chiang, and A. J. Heeger, *J. Chem. Soc., Chem. Comm.*, Vol. 16 (1977), p. 578.
18. C. K. Chiang, C. R. Fincher, Y. W. Park, A. H. Heeger, H. Shirikawa, E. J. Louis, S. C. Gau, and A. G. MacDiarmid. *Phys. Rev. Lett.*, Vol. 39, No. 17 (1977), p. 1098.
19. T. Ito, H. Shirakawa, and S. Ikeda. *J. Poly. Sci. Polym. Chem. Ed.*, Vol. 12 (1974), p. 11.
20. A. G. MacDiarmid and A. J. Heeger. *Synthetic Metals*, Vol. 1 (1978), p. 1013.

21. G. Wegner. *Angew. Chem. Int. Ed. Engl.*, Vol. 20 (1981), p. 361.
22. K. J. Wynne and G. B. Street. *I&EC Prod. Res. Dev.*, Vol. 21 (1982), p. 23.
23. R. H. Baughman. *Contemp. Topics Poly. Sci.*, Vol. 5 (1984), p. 321.
24. R. L. Greene and G. B. Street. *Science*, Vol. 226 (1984), p. 651.
25. J. L. Bredas and G. B. Street. *Acc. Chem. Res.*, Vol. 18 (1985), p. 309.
26. J. R. Reynolds. *J. Molec. Elec.*, Vol. 2 (1986), p. 1.
27. A. J. Epstein and T. A. Skotheim, eds. *Handbook of Conducting Polymers. Vol. 2.*, New York, Marcel Dekker, 1986. P. 1041.
28. R. S. Potember, R. C. Hoffman, H. S. Hu, J. E. Cocchiaro, C. A. Viands, R. A. Murphy, and T. O. Poehler. *Polymer*, Vol. 28 (1987), p. 574.
29. A. O. Patil, A. J. Heeger, and F. Wudl. *Chem. Rev.*, Vol. 88 (1988), p. 183.
30. J. R. Reynolds. *Chemtech*, Vol. 18 (1988), p. 440.
31. M. Kanatzidis, *C&E News*, Vol. 68, No. 49, December 3, 1990, p. 36.
32. J. R. Reynolds and M. Pomerantz. *Electroresponsive Molecular and Polymeric Materials*, ed. by T. A. Skotheim. New York, Marcel Dekker, 1990.
33. J. R. Reynolds, A. D. Child, and M. B. Gieselman. *Kirk-Othmer Encyclopedia of Chemical Technology*. 4th ed. New York, John Wiley, Vol. 9, 1994. P. 61.
34 J. D. Stenger-Smith, *Progress in Polymer Science*, 23(1), 57, (1997).
35. P. C. Searson and T. P. Moffat. *Crit. Rev. Surf. Chem.*, Vol. 3, No. 3-4 (1994), p. 171.
36. D. Chen, M. J. Winokur, Y. Cauo, A. J. Heeger, and F. E. Karasz. *Phys. Rev. B*, Vol. 45, No. 5 (1992), p. 2035.
37. J. H. Simpson, D. M. Rice, F. C. Rossitto, P. Lahti, and F. E. Karasz. *Polymer*, Vol. 34 (1993), p. 4595.
38. C. C. Han and R. L. Elsenbaumer. *Synthetic Metals*, Vol. 30 (1989), p. 123.
39. Y. Xia, A. G. MacDiarmid, and A. J. Epstein. *Adv. Mater.* Vol. 6, No. 4 (1994), p. 293.
40. S. M. Cohen. *Corrosion Eng.*, Vol. 51, No. 1 (1995), p. 71.
41. D. W. DeBerry. *J. Electrochem. Soc.*, Vol. 132 (1985), p. 1022.
42. A. Akelah. *J. Material Science*, Vol. 21, No. 9 (1986), p. 2977.
43. P. May. *Physics World*, Vol. 8, No. 3 (1995), p. 52.
44. A. G. MacDiarmid. *Short Course on Electrically Conductive Polymers*. New York, New Platz, 1985.
45. Bernhard Wessling. *Conducting Polymer Film Coatings*. 1990. (German Patent DE 3834526 A1.)
46. S. Sathiyanarayanan and K. Balakrishnan. *Br. Corros. J.*, Vol. 29, No. 2 (1994), p.152.
47. S. D. Dhawan and D. C. Trivedi. *Synthetic Metals*, Vol. 60, No. 1 (1993), p. 67.
48. S. Hettaiarachichi, Y. W. Chan, R. B. Wilson, and V. S. Agawala. *Corrosion*, Vol. 45, No. 1 (1989), p. 30.
49. V. B. Miskovic-Stankovic, D. M. Drazic, M. J. Teodorovic, *Corrosion Science*, Vol. 37, No. 2 (1995), p. 241.
50. F. Beck. *Metalloberflaeche*, Vol. 46, No. 4 (1992), p. 177.
51. D. A. Wrobleski, B. C. Benicewicz, K. G. Thompson, and C. J. Bryan. *ACS Polymer Preprints*, Vol. 35 (1994), p. 265.
52. Los Alamos National Laboratory. *Corrosion-Protective Coatings From Electrically Conducting Polymers*, by D. A. Wrobleski, B. C. Benicewicz, K. G. Thompson, and C. J. Bryan. Los Alamos, New Mexico, LANL. (LANL Report LA-UR-92-360, publication UNCLASSIFIED.)
53. R. Racicot, R. L. Clark, H-B. Liu, S. C. Yang, N. M. Alias, and R. Brown. "Thin Film Conductive Polymers on Aluminum Surfaces: Interfacial Charge-Transfer and Anti-Corrosion Aspects" in *Optical and Photonic Applications of Electroactive and Conducting Polymers, Vol. 2528* . San Diego, Calif., International Society for Optical Engineering, 1995. P. 251.

54. K. G. Thompson, C. J. Bryan, B. C. Benicewicz, and D. A. Wrobleski. "Corrosion-Protective Coatings From Electrically Conducting Polymers," in *Symposia Proceedings*, NASA Conference Publication 3136, Vol. 1, December 1991.
55. B. Wessling. *Advanced Materials*, Vol. 6, No. 3 (1994), p. 226.
56. W-K. Lu, R. L. Elsenbaumer, and B. Wessling. *Synthetic Metals*, Vol. 71 (1995), p. 2163.
57. Y. Wei, J. Wang, X. Jia, J-M. Yeh, and P. Spellane. *Polymer*, Vol. 36, No. 23, (1995), p. 4535.
58. ASTM Standards G 85, G 4, B 117, G 44, G 16, G 61, G52, G31, D 2776, D 2688, G2, G50, and G60.
59. J. D. Stenger-Smith, A. P. Chafin, and W. P. Norris. *J. Org. Chem. Commun.*, Vol. 59, No. 20 (1994), p. 6107.
60. R. H. Martin. *Nature*, Vol. 168 (1957), p. 32.
61. P. Zarras, J. D. Stenger-Smith, et al. *ACS Polymeric Materials Science and Engineering Preprints*, Vol. 75 (1996), p. 310.
62 J. D. Stenger-Smith, P. Zarras, L. H. Merwin, S. Shaheen, B Kippelen and N. Peyghambarian, *Macromolecules Communication*, in press.

Chapter 19

Water-Containing Ionically Conductive Polymers: Applications and Chemistry

James K. Young, Rosa Uy, Timothy M. Dietz, Michael R. Engel, and Steven S. Kantner

The 3M Company, 3M Center, 201-2W-17, St. Paul, MN 55144–1000

This chapter reviews the chemistries, properties, and commercial uses water-containing ionically conductive polymer systems. In medical applications, these polymers serve as the conductive interface between the patient's skin and the medical equipment. These electrolyte systems are commercially produced in gel, paste, or sheet form using either natural or synthetic polymers. Regardless of the physical form, these systems are typically formulated to a conductivity range of 10^{-3} to 10^{-5} S cm^{-1} to provide acceptable performance. A new application of this type of polymer is reported recently in the prevention of steel rebar corrosion in concrete structures.

Water-containing, ionically conductive polymers are uniquely versatile and hold high economic value in industry. The patent literature contains a plethora of examples of novel chemistries and uses. Historically ionic conductivity was viewed as an undesirable property, which interfered with the intended insulating properties of polymers. As a consequence, the literature suffers from a relative paucity of informative articles or reviews (*1*). In this paper we review the chemistries, properties, and applications of several patented and commercially available conductive polymer systems.

A glance at the chemistry of water-containing, ionically conductive polymer systems reveals a variety of approaches that can be used to tailor the range of properties to a specific application. Since the commercial importance of these materials lies with the ability to modify the chemistry to meet specific performance characteristics, an understanding of the applications is requisite to appreciating the chemistry. For this reason an emphasis is placed on describing the areas where water-containing, ionically conductive polymer systems play a key role, before the details of the chemistries are discussed.

Applications

In medical applications, ionically conductive polymers serve as the critical part of the conductive interface between the patient's skin and an electronic medical device as shown in Figure 1. Suitable ionic conductivity is achieved by swelling a polymer with

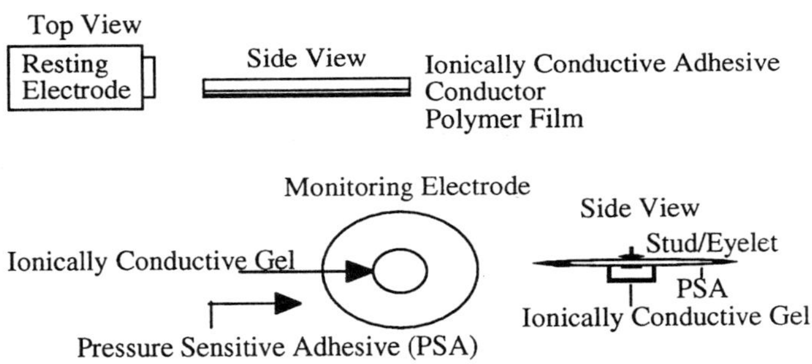

Figure 1. Components of Resting and Monitoring Biomedical Electrodes.

an aqueous electrolyte solution or by employing a polymeric salt or a solution of salt in an ion-solvating polymer. The introduction of ions into a high mobility polymeric environment must be accomplished in a way that minimizes the likelihood for irritation or sensitization of the patient's skin.

Transcutaneous Electrical Nerve Stimulation. Water-containing, ionically conductive polymers are used in transcutaneous electrical nerve stimulator (TENS) treatment to deliver electrical pulses to the underlying nerve and muscle tissue to either relieve acute and chronic pain or to repair and regenerate damaged tissue (*2*). TENS electrodes typically consist of a nonmetallic conductor, an adhesive tape patch, and a conductive gel medium between the skin and the conductor.

Iontophoresis. Iontophoresis uses similar types of conductive polymers to deliver drugs through the skin. Transdermal iontophoresis usually employs continuously applied DC to achieve enhanced transport of ionic species through the interaction of charged ions with the imposed field (*3*). In "reverse" iontophoresis, the imposition of an electric current across the skin results in the extraction of a substance (e.g., glucose) via the convective movement of solvent through a charged channel in response to the preferential passage of counter ions (*4*).

Electrosurgery. In electrosurgery, where high radio frequency electrical current is used to cut tissue and/or coagulate blood vessels, conductive polymers are used in grounding plates to disperse the current back to the generator. The high voltage, high frequency current is concentrated in the active electrode producing arcing and high localized temperatures. Several modes of use of High Frequency (HF) current in electrosurgery are shown in Figure 2. During surgery the electrical circuit to the power generating unit is completed by using a return electrode in electrical contact with the patient's body. This electrode provides a relatively high area, low impedance interface which prevents burn damage to the skin by producing a minimum of localized heating.

The technology of grounding plates (as these return electrodes are often referred) progressed from large metal plates under the patient, to metal foil "electrodes" with adhesive borders and a wet conductive gel which was applied to the foil surface just prior to use. The "wet gel" electrodes were improved with the development of "pre-gelled" electrodes which had the gel impregnated in a pre-attached sponge. This provided more convenience but did not solve the problem of "tenting" whereby the center of the electrode could lose contact with the skin -- effectively reducing the area of electrical contact. Tenting is most likely when the electrode is applied to an irregular or curved area of the body.

The use of conductive adhesives was a significant advance in the construction of electrosurgical grounding plates. An ionically conductive adhesive has the characteristics of a conducting hydrogel as well as those of a pressure sensitive adhesive. The adhesive properties allow the plate to be made thinner and more flexible and effectively prevent tenting. The cohesive integrity of the conductive adhesive prevents migration of conductive material (gel, paste) which can cause loss of electrical isolation and arcing to operating tables.

Biomedical Electrodes. Biomedical electrodes are also used in the diagnosis and monitoring of cardiovascular activity. These electrodes employ a conductive polymer system that acts as the medium to convert ionic to electronic current in order to obtain electrical activities of the heart during the depolarization and repolarization events. A complete cardiac cycle is represented by the atrial contraction (P wave), ventricular contraction (QRS complex) and the ventricular repolarization (T wave) as shown in Figure 3. The amplitude of these signals ranges from 0.1 to 2.5 millivolts. These traces are analyzed by the shape, size, and interval between the different wave signals

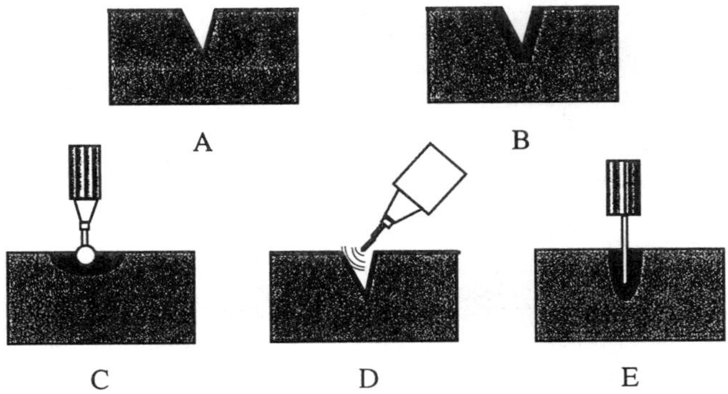

Figure 2. High Frequency Surgery: A) Smooth Cut--undamped, unmodulated HF-current produces a smooth cut with very little tissue necrosis, B) Sloughed Cut--impulse modulated HF-current simultaneously cuts and coagulates, C) Coagulation--deep coagulation occurs when using a spherical or plane electrode, D) Fulguration--a superficial carbonization of tissue is achieved through spark discharge at very high impulse modulated HF-current, and E) Desiccation--deep tissue desiccation is achieved through the use of a needle electrode with impulse modulated HF-current.

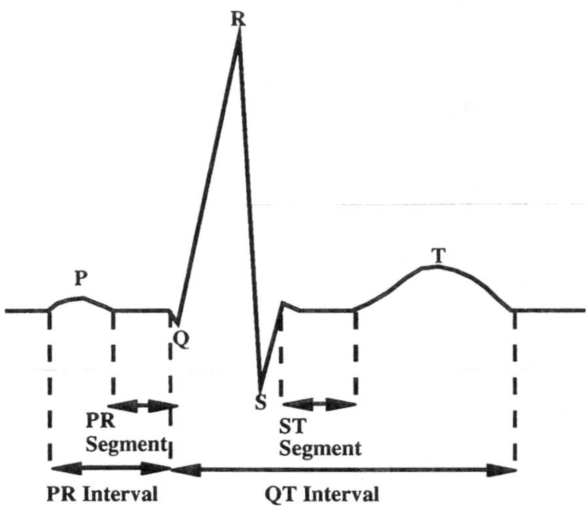

Figure 3. Depiction of a Complete Cardiac Cycle.

to interpret the patient's heart condition. It is very important that the conductive polymer be in intimate contact with the skin to provide stable and artifact free signals.

ECG electrodes vary in size, design, and nature of conductive polymer system depending on the intended use. Diagnosis using "resting" electrodes requires that the patient remain motionless and relaxed for a relatively short time (< 0.5 h). In contrast, patient monitoring may require extended wear capability (ca. 5 days) combined with good shear holding characteristics for ambulatory patients.

Defibrillator pads are related to electrodes and grounding plates and are used to deliver high voltage pulses to a patient's heart during cardiac arrest. External defibrillators deliver a voltage pulse through a pair of ionically conductive polymer-based electrodes placed on a patient's chest or back by attending medical personnel (*5*). External pacing electrodes use relatively high DC current (10-100 mA) to stimulate cardiac muscle located deep within the chest. New multipurpose electrodes combine several functions such as pacing and monitoring and are also suitable for TENS applications (*6*).

Anodic Tapes. Recently, galvanically active metal/adhesive anodic tape systems have been developed for cathodic protection of reinforcing bars in concrete. Environments high in moisture and salt can corrode the rebars in reinforced concrete resulting in deterioration of the concrete structure. Anodic tape systems have been developed which utilize water containing ionically conductive polymeric systems in combination with a metal such as zinc and are designed to sacrificially protect the rebars in concrete. The anodic tape system is affixed to the concrete and a galvanic circuit is created by electrical connection of the reinforcing bars in concrete to the zinc (*7*). The chemistries needed for all of these applications are discussed below.

Chemistry

The wide range of performance demands for ionically conductive polymers drives the continued development of new chemistries. In general, water-containing ionically conductive polymer systems consist of polar polymer(s), water, salt, and humectant. The main function of the polar polymer is to serve as a thickening agent to contain the electrolytes and to impart a degree of physical integrity to the device. The humectant is added to slow the water loss thereby maintaining high conductivity. In most cases, the humectant also acts as a plasticizer to increase tack for adhesion purposes.

The use of water is common because it is biocompatible and has the additional benefit of being a suitable plasticizer. Water has high permittivity (e=81.0) and high Gutmann donor and acceptor numbers (DN=18.0, AN=54.8), which is critical for the solvation of small cations and anions, respectively (*8*). The non-negligible vapor pressure of water presents the disadvantages of often requiring the use of additional humectants and/or special packaging.

These electrolyte systems are commercially produced in gel, paste, or sheet form using either natural or synthetic polymers. Regardless of the physical form, these systems are typically formulated to a conductivity range of 10^{-3} to 10^{-5} S cm^{-1} to provide acceptable performance. In medical applications, it is critical that the polymer electrolyte system does not cause any irritation, sensitization, or other trauma to the patient's skin. The most commonly used components in the preparation conductive polymer systems are listed in Table I.

Naturally Occurring Polymers. The use of ionically conductive polymers derived from natural polymers is claimed in several patents (*5,9,10*). There are many advantages of using natural gums such as guar, karaya, arabic, and agar as the polymer network for the electrolytes. Due to the high molecular weights, the corresponding high water content (ca., 80%) electrolyte systems are attainable.

Furthermore, these natural polymer are relatively inexpensive. One example is the use of guar gum. This natural carbohydrate is crosslinked with potassium tetraborate in the presence of KCl and propylene glycol to produce a high water content, crosslinked conductive gel. The natural polymers tend to have limitations based on their physical and chemical inconsistencies, undesirable support for microbial growth, potential for creating adverse skin sensitivities, and the amount of impurities present in the raw material.

Sheet Form Gels. The most common and useful fabrication of ionically conductive polymers for medical applications is the sheet form. The old fashioned gel or paste type formulations are deemed to be too messy for most modern applications. The polymers are synthesized from water soluble monomers in the presence of a chemical crosslinker, water, salt, and humectant by free radical initiated processes. One significant advantage of this approach is that these materials can be formed by a solventless process in substantially final form. For example, acrylic acid, or mixture of acrylic acid and N-vinyl-pyrrolidone are polymerized to form hydrophilic conductive adhesives that have good adhesive properties on human skin. These polymers also exhibit excellent conductivity properties (*11,12*).

Free radical polymerization of 2-acrylamido-2-methyl propane sulfonic acid (Lubrizol's AMPs" monomer) or its soluble salt also yields sheet form, ionically conductive adhesive for use in biomedical electrodes (*13*). This material exhibits uniform conductivity preventing the "hot spots" associated with localized current. Additionally, the AMPs"-based materials are inherently electrically conductive, not requiring the use of additional additives (e.g., salts).

The use of methacrylate ester monomers such as 2-sulfoethylmethacrylate, potassium-3-sulfopropylmethacrylate, or methacryloxyethyl trimethylammonium chloride is recently described (*14*). The adhesive properties of these systems are significantly improved when a second water soluble monomer such as acrylic acid or N-vinyl pyrrolidone is copolymerized with the methacrylate ester monomer.

Free radical generation with photochemical, thermal, or redox initiators yields polymeric networks from telechelic poly(ethylene glycol) oligomers end-capped with ethylenic unsaturation. The functionalized oligomer is prepared in a two step process as shown in Figure 4. Poly(ethyleneglycol) diamine reacts with isophorone disocycante (IPDI) to give a higher molecular weight diamine through chain extension. The chain-extended diamine is then reacted with vinyl dimethyl azlactone (VDM) to give the terminal free-radically polymerizable functionality. These reactive oligomers yield hydrophilic pressure sensitive and ionically conductive adhesives when formulated with water, salt, and humectant and cured (*15*). Adhesion, modulus, and tack can be tailored by adjusting the molecular weight and amount of oligomer and the nature of the other additives present.

Radiation Crosslinking. Crosslinking of conductive polymer precursors can also be achieved without the use of additional reactive additives. Irradiation crosslinking of the polymer network to increase cohesive strength is reported in several patents. In general the irradiation energy required is at least 100,000 electron volts. The exact dosage depends largely on the concentration of the uncrosslinked polymer in the solubilizing plasticizer and on the molecular weight of the uncrosslinked polymer. Ionizing radiation crosslinking of a polyvinylpyrrolidone mixture yields a hydrophilic conductive adhesive (*16*), while novel tacky conductive gels can be produced via the irradiation of polyethylene oxide solutions (*17,18*). Gamma irradiated polyvinylpyrrolidone (PVP) particles can be mixed with a humectant and water to produce a lightly crosslinked conductive adhesive (*19*). In this example the solvating characteristics of PVP are exploited. Poly(N-vinyl pyrrolidone) particles in the size range of 10 to 75 mm were exposed to gamma radiation of 155 kGys using a cobalt-60 high energy source to produce the crosslinked solid. The conductive adhesive formulations based on the crosslinked solid have suitable ionic conductivity for biomedical electrode applications over a broad range of water contents.

Blends. Jevne has patented the blending of polymers to increase cohesive properties through non-covalent bonding interactions (*20*). In this work polyvinyl alcohol (PVA) is extruded with polyvinylpyrrolidone (PVP) to produce a firm sheet. The polyvinyl alcohol is at least 75% hydrolyzed and acts, in relatively low concentration, to crosslink the PVP through hydrogen bonding interactions. Lorenz reports on the blending of partially hydrolyzed PVP with polyethylenimine (*21*). Effective crosslinking occurs between partially hydrolyzed polyvinylpyrrolidone and amine-containing polymers such as polyethylenimine or amine terminated polyethylene oxide. At least 1.5×10^{-2} milliequivalents of free carboxylic groups per gram of polymer are generated through the ring opening of the lactam moiety of the polyvinylpyrrolidone as shown in Figure 5. Mixing the partially hydrolyzed polyvinylpyrrolidone with a multifuntional amine-containing polymer produces an ampholyte salt that is water-swellable and water-insoluble.

The adhesion of conductive polymers can be improved by mixing heterogeneous phases. The mixing of a non-polar polymeric adhesive phase with a polar conductive aqueous phase comprised of a water receptive polymer humectant and an electrolyte has been disclosed (*22,23*). The two phase composite consists of a continuous phase of ionically conductive, hydrophilic pressure-sensitive adhesive (PSA) composition and a discontinuous phase of domains of hydrophobic PSA and provides for enhanced adhesion to mammalian skin, while retaining the critical alternating current impedance.

Crosslinking via Covalent or Ionic Bonding. Hydrophilic polymeric gels can also be generated from aqueous solutions of linear polymers or oligomers through chemical crosslinking reactions. Approaches which have been widely used to cause such crosslinking include covalent or ionic bond formation with an added chemical species and condensation or addition reactions of functionality present on the linear polymer. For instance, the hydroxyl moieties of polyvinyl alcohol (*24*) or polysaccharides such as guar gum (*25*) react with borate anion to crosslink through multiple borate ester linkages yielding high modulus hydrogels. Polysilicic acid can also be used to crosslink polyhydroxylated polymers (*26*).

Mixing carboxylated polymer solutions with di-, tri-, or tetravalent cations gives ionically crosslinked hydrogels. A hydrogel-forming skin coating material which has good water resistance consists of a blend of a carboxy-containing polymer and a second hydrophilic polymer also containing an aluminum or iron salt to crosslink the carboxy-containing polymer (*27*). This material, initially flowable, passes through a tacky stage but ultimately becomes tack free. Using metal chelates such as aluminum, zirconium, or titanium acetylacetonates as crosslinking agents for carboxy-group containing polymers provides materials which are useful for wound secretion absorbing hydrogels (*28,29*) and ionically conductive pressure sensitive adhesives for biomedical electrodes (*30*).

Partially neutralized carboxylated polymer solutions can also be reacted with polyepoxides to crosslink via chemical bond formation yielding tacky electrically conductive hydrogels for biomedical electrodes (*31*). Low molecular weight di-, tri-, or tetraaldehydes react with N-(3-aminopropyl)methacrylamide/N-vinyl pyrrolidone copolymers to form chemically bonded crosslinks giving hydrogels, adhesives, and coatings (*32*).

The high reactivity of the isocyanate moiety has been exploited to yield polyurethane/urea hydrogels. A polyisocyanate prepolymer obtained by functionalizing a moderate molecular weight polyether diol or triol with a diisocyanate will crosslink on exposure to an aqueous environment due to partial hydrolysis and urea formation (*33*). A similar hydrogel results by reacting a polyisocyanate

Table I. Commonly Used Components in Water-Containing Ionically Conductive Polymer Systems

Monomers	acrylic acid, AMPS, vinyl pyrrolidone, 2-carboxyethylacrylate, methacrylic acid, acrylamide, methacrylamide
Polymers	PVP, PVA, PEO, polyacrylamide, natural carbohydrate
Humectants	glycerin, propylene glycol, polyethylene glycol, sorbitol
Salts	KCl, NaCl, LiCl, Mg acetate
Crosslinkers	tetra(ethylene glycol)-bis-methacrylate, methylene bis(acrylamide)

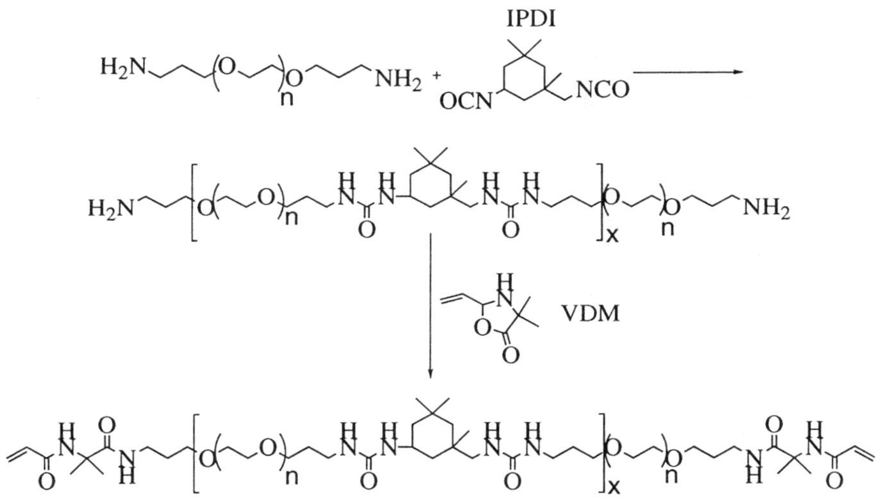

Figure 4. Telechelic Poly(ethylene oxide) Oligomers.

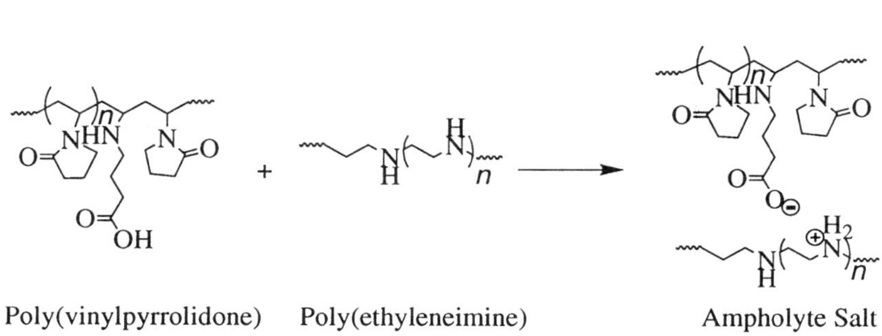

Figure 5. Crosslinking via Ionic Bonds.

prepolymer with polyether diamine (34). Reaction of polyurethane polyol prepolymers with polyisocyanates in the absence of water also can give useful hydrophilic, ionically conductive polymer networks (*35*).

Thermal or photochemically initiated free-radical cure of oligomers or polymers containing terminal or pendant polymerizable double bonds represents another approach to network formation. Several methods of synthesizing such functionality have been described. Reaction of a polyoxyalkylene diol monoalkyl ether and a polyfunctional isocyanate yields an isocyanate prepolymer which is further reacted with hydroxyethyl (meth)acrylate. When formulated with photoinitiator and irradiated an inherently tacky, water absorbing gel results (*36*). Telechelic functionalization of polyethylene glycol with ethylenically unsaturated moieties yields an oligomeric precursor curable to a poly(ethylene oxide) derived network which, when plasticized, provides a hydrophilic pressure sensitive adhesive (*37*). Reaction of alkyl vinyl ether/maleic anhydride copolymers with aminoalkenes such as allyl amine gives pendant vinyl groups which can be cured with redox initiation in the presence of water to yield a hydrogel (*38*). Reaction of a hydrophilic carboxy-functional polymer with an epoxy functional (meth)acrylate, such as glycidyl methacrylate, yields a free-radically crosslinkable polymer which when plasticized gives a pressure sensitive adhesive on cure (*39*). Copolymerization of a vinyl functional photoinitiator with a polar hydrophilic comonomer gives a photocrosslinkable polymer that provides a hydrophilic pressure sensitive adhesive when cured in the presence of plasticizer (*40*).

Conclusion

This brief sampling of chemistries illustrates many of the approaches used to encompass the diverse range of performance requirements for this important class of materials. Upon review, the chemistry of water-containing, ionically conductive polymer systems reveals a variety of approaches that can be used to tailor the range of properties to a specific application. The commercial importance of water-containing ionically conductive polymer systems requires the continued development of new chemistries and processes to meet the ever increasing performance demands.

References

1. Owen, J. Ionic Conductivity in Comprehensive Polymer Science: The Synthesis, Characterization, Reactions, & Applications of Polymers; Booth, C.; Price, C., Eds.; Pergamon Press: New York, NY, 1989, Vol. 2; pp 669-686.
2. Mannheimer, J. S.; Lampe, G. N. Clinical Transcutaneous Electrical Nerve Stimulation; Davis: Philadelphia, PA, 1984.
3. Green, P.G.; Hinz, R.S.; Cullander, C.; Yamane, G.; Guy, R.H. Int. J. Pharm. **1991**, 44, p. 1113.
4. Delgado-Charro, M.B.; Guy, R.H. Pharm. Res. **1994**, 11, p. 929.
5. Morgan, C.B. Defibrillator Electrode System. U.S. Patent 5,466,244, Nov. 14, 1995.
6. Montecalvo, D.A.; Rolf, D. Multipurpose Medical Electrode. U.S. Patent 5,330,527, July 19, 1994.
7. Bartholomew, J.J.; Bennett, J.E.; Martin, B.L.; Mitchell, T.A. Method and Apparatus for Cathodically Protecting Reinforced Concrete Structures. U.S. Patent 5,650,060, March 8, 1994.
8. Burger, K. Solvation, Ionic and Complex Formation Reactions in Non-Aqueous Solvents; Elsevier: Amsterdam, 1983.

9. Hymes, A.C. Monitoring and Stimulation Electrode. U.S. Patent 4,125,110, Nov. 14, 1978.
10. Rolf, D. Medical Electrode for Monitoring and Diagnostic Use. U.S. Patent 4,674,512, June 23, 1987.
11. Engel, M.R. Conductive Adhesive and Biomedical Electrode. U.S. Patent 4,539,996, Sept. 10, 1985.
12. Engel, M.R. Conductive Adhesive and Biomedical Electrode. U.S. Patent 4,554,924, Nov. 26, 1985.
13. Cahalan, P.T.; Coury, A.J. Tape Electrode. U.S. Patent 4,391,278, July 5, 1983.
14. Tang, J.; Mruk, N.J. Polyacrylate and Polymethacrylate Ester Based Hydrogel Adhesives. U.S. Patent 5,674,275, Oct. 7, 1997.
15. Kantner, S.S.; Rustad, N.J.; Stefely, J.S. Hydrophilic Pressure Sensitive Adhesives. U.S. Patent 5,536,768, July 16, 1996.
16. Sieverding, D.L. Hydrophilic, Elastomeric, Pressure-Sensitive Adhesive. U.S. Patent 4,699,146, Oct. 13, 1987.
17. Keusch, P.; Essmyer, J.L. Adhesive Polyethylene Oxide Hydrogel Sheet and Its Production. U.S. Patent 4,684,558, Aug. 4, 1987.
18. Keusch, P.; Essmyer, J.L. Conductive Adhesive Medical Electrode Assemblies. U.S. Patent 4,777,954, Oct. 18, 1988.
19. Uy, R.; Dietz, T.M. Solid State Conductive Polymer Compositions, Biomedical Electrodes Containing Such Compositions, and Method of Preparing Same. U.S. Patent 5,385,679, Jan. 31, 1995.
20. Jevne, A.H.; Vegoe, B.R.; Holmblad, C.M.; Cahalan, P.T. Hydrophilic Pressure Sensitive Biomedical Adhesive Composition. U.S. Patent 4,593,053, June 3, 1986.
21. Lorenz, D.H. Skin Adhesive Hydrogel, Its Preparation and Uses. U.S. Patent 5,306,504, April 26, 1994.
22. Dietz, T.M.; Asmus, R.A.; Uy, R. Two-Phase Composites of Ionically-Conductive Pressure-Sensitive Adhesive, Biomedical Electrodes Using the Composites, and Methods of Preparing the Composite and the Biomedical Electrodes. U.S. Patent 5,338,490, Aug. 16, 1994.
23. Mruk, N.J.; Vaughan, R.C.; Eddy, Jr., A.R. Pressure Sensitive Adhesive Compositions for Medical Electrodes. U.S. Patent 4,588,762, May 13, 1986.
24. Anderson, C.J.; Gumbusky, J.T. Bio-Medical Electrode Conductive Gel Pads. U.S. Patent 3,998,215, Dec. 21, 1976.
25. Carim, H.M. Cohesive Nonsticky Electrically Conductive Gel Composition. U.S. Patent 4,406,827, Sept. 27, 1983.
26. Marks, A.M.; Marks, M.M. Electrically Conductive Transparent Materials. U.S. Patent 3,357,930, Dec 12, 1967.
27. Murray, D.G.; Smith, D.C.; Guillet, J.E. Hydrogel-Forming Wound Dressing or Skin Coating Material. U.S. Patent 4,920,158, Apr. 24, 1990.
28. Czech, Z.; Seeger, K. Cross-Linked Hydrogels and Their Use as Wound Dressings. U.S. Patent 5,135,755, Aug. 4, 1992.
29. Czech, Z.; Seeger, K. Cross-Linked Hydrogels and Their Use as Wound Dressings. U.S. Patent 5,336,501, Aug. 9, 1994.
30. Czech, Z. Electrically Conductive Transparent Pressure-Sensitive Adhesive Films; Process for their Production; and Their Use in the Production of Biomedical Electrodes. U.S. Patent 5,433,892, July 18, 1995.
31. Yamamoto, K.; Matsumoto, K.; Okada, Y. Electrode Pad. U.S. Patent 4,515,162, May 7, 1985.
32. Login, R.B.; Shih, J.S.; Chuang, J.-C. Crosslinkable, Functional Polymers of Vinylpyrrolidone and Aminoalkyl Acrylamides. U.S. Patent 5,206,322, Apr. 27, 1993.

33. Cordts, H.P.; Karloske, J.E. Aqueous Liquid Filled Polyurethane Gels and Method of Making the Same. U.S. Patent 4,517,326, May 14, 1985.
34. Riazzi, T.J.; Wolf, M.L.; Allaire, M.J. Biomedical Electrode Having a Secured One-Piece Conductive Terminal. U.S. Patent 5,406,945, Apr. 18, 1995.
35. Shikinami, Y.; Hata, K.; Ohmura, Y. Sticking Agent of Ionic-Conductive Polymer. U.S. Patent 4,855,077, Aug. 8, 1989.
36. Ansell, C.W.G. Adhesives, Their Preparation and Use. European Patent 271,292, June 15, 1988.
37. Kantner, S.S.; Rustad, N.J.; Stefely, J.S. Hydrophilic Pressure Sensitive Adhesives. U.S. Patent 5,660,178, Aug.26,1997.
38. Weber, F.J. Hydrogel Material and Method of Preparation. U.S. Patent 5,393,798, Feb 28, 1995.
39. Kantner, S.S.; Kuester, W. Use of Pendant Free-Radically Polymerizable Moieties with Polar Polymers to Prepare Hydrophilic Pressure Sensitive Adhesive Compositions. WO Patent 97/24378, July 10, 1997.
40. Kuester, W.; D'Haese, F.C.; Kantner, S.S. Use of Pendant Photoreactive Moieties on Polymer Precursors to Prepare Hydrophilic Pressure Sensitive Adhesives. WO Patent 97/24376, July 10, 1997.

SYNTHESIS AND FABRICATION

Chapter 20

Construction of Oriented and Patterned Conjugated Polymers

Liming Dai, Berthold Winkler, Shaoming Huang, and Albert W. H. Mau

CSIRO Molecular Science, Bag 10, Clayton South, Victoria 3169, Australia

Conjugated polymers have recently attracted a great deal of interest because of their potential applications in electronic and photonic devices. The construction of oriented and/or patterned conjugated polymers is a key prerequisite for most of these applications (e.g., in optoelectronic displays, integrated circuits, field-effect transistors, and optical memory storages). Although unfunctionalized conjugated polymers are generally intractable, efforts to synthesize processable conjugated polymers have allowed attempts to fabricate them into devices with desirable characteristics. The progress towards advanced synthesis and construction of oriented and patterned conjugated polymers is reviewed. Recent developments in the field have clearly indicated the possibility of producing a wide range of new materials with exotic physicochemical properties and devices of novel features.

DURING THE PAST 20 YEARS or so, various conjugated polymers have been synthesized with unusual electrical (*1*), magnetic (*2, 3*), and optical properties (*4-6*) due to the substantial π-electron delocalization along their backbones. These properties allow conjugated polymers to be used in electronic and photonic devices (e.g., in optoelectronic displays, integrated circuits, field-effect transistors, and optical memory storages) (*7-9*), in which they play an active role in regulating the performance of the device. However, in order to fabricate high performance electronic and/or photonic devices based on conjugated polymers, one often needs to align and/or pattern the functional polymers.

As electronic conductors, conjugated polymers can provide extended π-orbitals for charge carriers to move through. However, since most organic polymers do not have intrinsic charge carriers, the charge carriers must be provided by partial oxidation (*p*-doping) of the polymer chain with electron-acceptors (e.g., I_2, AsF_5, etc.) or by partial reduction (*n*-doping) with electron-donors (e.g., Na, K, etc.) (*1, 10*). Through such a doping process, charged defects (e.g., polaron, bipolaron, and soliton) are introduced (*1, 11*), which may then be available as the charge carriers. The report in 1977 from Shirakawa et al. (*10, 12*) that the conductivity of polyacetylene can be increased by up to ca. 13 orders of magnitude through doping provoked considerable excitement, from which a new research field of conducting polymers emerged. However, some conjugated polymers, such as polyaniline, can also acquire high conductivities through the protonation of imine nitrogen atoms without any electron transfer between the polymer and "dopants" to occur (*13*).

The bulk conductivity of conducting polymers should consist of contributions, at least, from the intrachain and interchain electron transportations (*14*). While the intrachain diffusion of the charge carriers along the conjugated backbone plays a dominant role in the charge transporting process, the interchain transport (via hopping, tunnelling, etc.) also has a significant influence on the bulk conductivity since a single chain does not extend throughout the entire length of a sample in most practical applications. The competition between the intra- and inter-chain transports has been discussed based on various theoretical models. Using scaling arguments, for example, de Gennes (*15*) and Heeger and Smith (*16*) have independently demonstrated that if the mean lifetime of the charge carrier on the polymer chain (τ_c) is much greater than the time required for a charge carrier to completely explore the polymer chain (τ_i,) the conductivity (σ) is limited by interchain hopping and is given by:

$$\sigma \propto (ne^2/kT)a(L/\tau_c) \qquad (1)$$

where, a is the persistence length of the polymer chain (*17*), L is the total length of the macromolecule, n is the charge density, e is the charge per carrier, k is the Boltzmann constant, and T is the absolute temperature.

On the other hand, if $\tau_i > \tau_c$, the conductivity is independent of molecular weight and σ is given by:

$$\sigma \propto (ne^2/kT)a(D_i/\tau_c)^{1/2} \qquad (2)$$

with $D_i \propto L^2/\tau_i$ being the diffusion coefficient for the charge carriers along the polymer chain.

Pearson et al. (*18*) derived the same equations through a quantitative analysis and obtained a reasonable agreement with some existing experimental data. From equations 1 and 2, it can be seen that σ increases with increasing persistence length. This can be caused, for example, by alignment of the polymer chains (*vide infra*).

The necessity for aligning and/or patterning conjugated polymers can also been seen from some recent developments in electroluminescent displays (*19-21*) and nonlinear optical (NLO) devices (*22, 23*), although in these areas conjugated polymers are used in the non-conducting state. For instance, polymeric light-emitting diodes (LEDs) with polarized light emissions have been made using stretch-oriented (*24, 25*), rubbing-aligned (*26*), Langmuir-Blodgett (LB) deposited (*27*), and specifically synthesized liquid crystalline (*28, 29*) films of appropriate electroluminescent (EL) polymers, while patterned (multiple-color) emissions have been reported for LEDs based on certain photolithographically patterned (*30-32*), and phase separated conjugated polymers (*33*). Due to their delocalized π-electron orbitals, some conjugated polymers have also been demonstrated to possess large, ultrafast nonlinear optical responses (*34-38*). Nonlinear optical processes occur when a medium is polarized by an intense electric field (E) (e.g., the one associated with a strong pulse of light), which creates an induced dipole moment (μ_{ind}). μ_{ind} can be expressed as a power series at the molecular level (*35*):

$$\mu_{\text{ind}} = \alpha E + \beta EE + \gamma EEE + \cdots \tag{3}$$

At the bulk level, an analogous equation is used for the induced polarization (P):

$$P = \chi^{(1)}E + \chi^{(2)}EE + \chi^{(3)}EEE + \cdots \tag{4}$$

where, $\chi^{(1)}$ is the linear susceptibility which describes the linear response associated with ordinary refraction and absorption; $\chi^{(2)}$ and $\chi^{(3)}$ are the second and third-order nonlinear optical susceptibilities.

As $\chi^{(2)}$ is a third-rank tensor, it is effective only for noncentrosymmetric media. Therefore, the medium used for the second-order nonlinear optical effect must have a noncentrosymmetric ordering of the NLO dipoles, possibly by spontaneous ordering or by electric field poling (*39*). In contrast, order is not required for third-order nonlinear materials (e.g., conjugated polymers), but the $\chi^{(3)}$ of a given material is given by (*40*):

$$<\chi^{(3)}> = \chi_{\text{xxxx}}{}^{(3)}<\cos^4\theta> \tag{5}$$

where $\chi_{\text{xxxx}}{}^{(3)}$ is the third order susceptibility of an individual macromolecule along the polymer chain direction (x), θ is the angle between the polymer chain and the exciting optical electric field, and $< >$ denotes an average over all polymer chain orientations. For certain special cases, the $<\cos^4\theta>$ is given by (*40*):

$$<\cos^4\theta> = \begin{cases} 1 & \text{for a uniaxially aligned polymer system} \\ 3/8 & \text{for a bidimensionally disordered system} \\ & \text{(all polymer chains are parallel to a plane but} \\ & \text{randomly disoriented within the plane)} \\ 1/5 & \text{for a three-dimensional disorder} \end{cases} \tag{6}$$

Equations 5 and 6 imply that an increase of a factor of 5 in cubic susceptibility (and a factor of at least 25 in efficiency) can be obtained by aligning a completely disordered system into a uniaxially oriented one (e.g., single crystal) (*40*).

As can be seen from the above discussion, microconstruction of conjugated polymers is both technologically important and scientifically interesting. As such, a variety of molecularly aligned conjugated polymer systems have been developed, for example, through ordered-matrix-assisted syntheses (*41, 42*), template syntheses (*43-45*), mechanical stretching (*46-48*), and Langmuir-Blodgett manipulation of the polymer chains (*49*). On the other hand, various techniques including photolithography (*50*), self-assembly (*51-53*), polymeric phase separation (*33*), and plasma treatment/polymerization (*54*) have now been used for microlithographic patterning of conjugated polymers.

The aim of this paper is to spotlight some of the important issues on construction of oriented and/or patterned conjugated polymers. As no attempt has been made for a comprehensive literature survey, the examples presented are not exhaustive.

Oriented Conjugated Polymers

Synthesis-Induced Orientation

Ordered-Matrix or Template Assisted Syntheses. Due to its simple conjugated structure, polyacetylene has served as the prototype for other conducting polymers. The first synthesis of polyacetylene in a film form was reported as early as 1974 (*55*). Initially, the Ziegler-Natta polymerization of acetylene could only produce polyacetylene films with a randomly-oriented fibrillar morphology (Figure 1) (*10, 55*). In 1987, however, Naarmann et al. (*56, 57*) reported a method to prepare highly oriented polyacetylenes by aging the catalyst mixture of $Ti(OC_4H_9)_4/Al(C_2H_5)_3$ in silicone oil, which was used as a reaction medium. The resulting polyacetylene film showed a copperlike conductivity (up to 10^5 S/cm) after having been stretched and doped with iodine (*58*). By replacing the silicone oil with a nematic liquid crystal phase, Shirakawa et al. (*41, 59*) have also produced partially aligned, highly conducting polyacetylene films. These authors subsequently obtained polyacetylene films with a highly oriented fibrillar morphology (Figure 2) by carrying out the polymerization of acetylene on a vertical glass wall of a flask over which the catalyst-containing nematic liquid crystal solution flowed down under the influence of the gravity (*41, 60*). The high infrared (IR) anisotropies ranging from 2.0 to 3.8, together with the high I_2-doping-induced DC conductivities in directions both parallel ($\sigma_{||}$ = 4600 S/cm) and perpendicular ($\sigma_{\perp}$ = 3900 S/cm) to the orientation axis, indicate significant orientation of the polyacetylene chains along the flow direction. Aiming for superior optoelectronic properties, several groups have used a magnetic field (typically, 2-14 kG) to further align the active catalyst sites within the nematic liquid crystal phase (*41, 42, 61-64*). Although polyacetylene films thus prepared have a quite similar morphology as that shown in Figure 2, higher average IR dichroic ratios in the range of 3.3 - 3.5 and the I_2-doped conductivities as high as $\sigma_{||}$ = 12000 S/cm and $\sigma_{\perp}$ = 4800 S/cm were recorded (*41*).

In addition to the oriented polyacetylenes discussed above, highly oriented poly(*p*-phenylenevinylene) (PPV; - PPV and its derivatives represent the class of conjugated polymers most studied for electroluminescent devices (*20, 21, 65, 66*) - and its composites with nanometer-scale architectures have been polymerized within self-organizing liquid-crystal matrixes through the Wessling route (Scheme 1) (*67, 68*). Furthermore, scanning tunnelling microscopy (STM) / scanning electrochemical microscopy (SECM) have also been used to prepare conjugated polymers with nanoscale structures. While the region specific polymerization of pyrrole onto

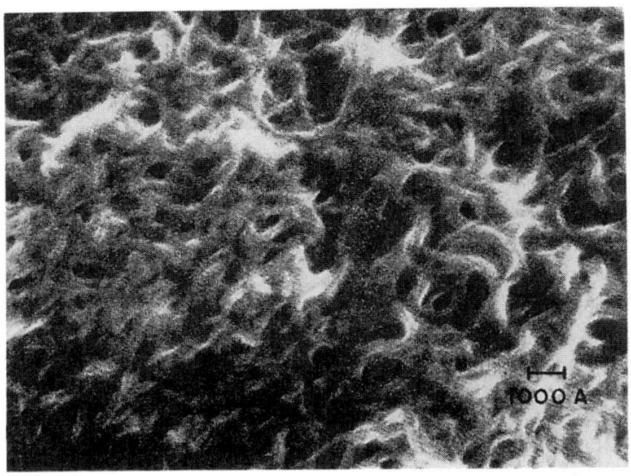

Figure 1. A typical SEM micrograph of thin polyacetylene films prepared by the Ziegler-Natta catalyst (Reproduced with permission from reference 55. Copyright 1974 John Wiley & Sons, Inc.).

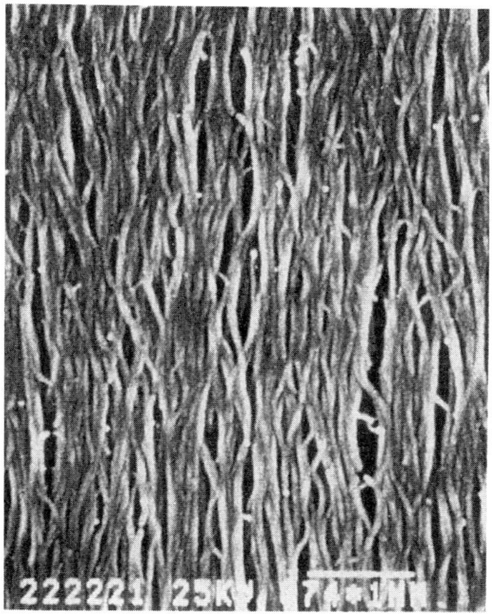

Figure 2. SEM micrograph of a polyacetylene film prepared by the flow method. The flow direction is vertical (Reproduced with permission from reference 41. Copyright 1988 Marcel Dekker, Inc.).

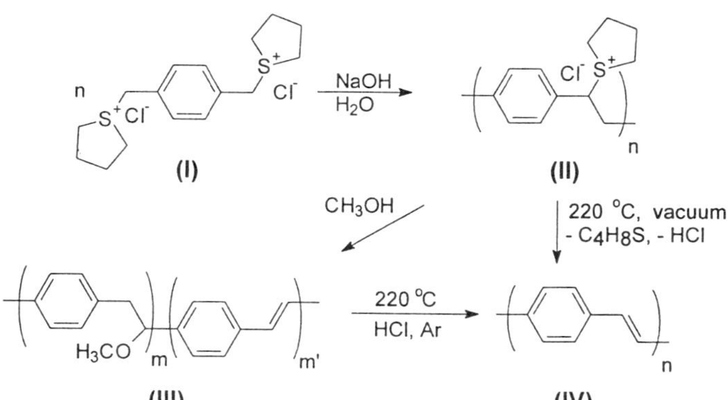

Scheme 1. Sulfonium precursor route to poly(p-phenylenevinylene).

graphite substrates at a submicron resolution was performed by using the STM tip as an electrode (*69, 70*), polypyrrole strips with a linewidth of 50 μm and length of 1 mm have been produced on a SECM (*71*), as have micro-sized polypyrrole towers (*72*). By spin-coating of anilinium sulfate-containing Nafion onto a Pt electrode in a SECM unit, Wuu et al. (*73*) performed the polymerization of aniline in a micro-structured fashion. More recently, Nyffenegger and Penner (*74*) have produced electrochemically active polyaniline particles with a size ranging from 100 to 600 Å in diameter and 10 to 200 Å in height by using a Pt STM tip as an electrode for the electropolymerization of aniline. Using the tip of a SECM as an electrode for region-specific oxidation of bromide into bromine, followed by the diffusion of the bromine into a conductive substrate covered with a thiophene derivative for a localized oxidative polymerization of thiophene monomers, Borgwarth et al. (*75*) have successfully prepared a polythiophene line of width 20 μm.

More generally, conjugated polymer fibers and tubules with improved order and fewer defects can be synthesized within the pores of a nanoporous membrane (*43, 44, 76, 77*) or the nanochannels of a mesoporous zeolite (*77, 78*), a process which has been widely known as template synthesis (*43-45*). The template synthesis often allows the production of monodispersed polymeric rods or tubules with controllable diameters and lengths. Template syntheses of conjugated polymers including polyacetylene, polypyrrole, polythiophene, polyaniline, and PPV may be achieved by electrochemical or chemical oxidative polymerization of the corresponding monomers. While the electrochemical template synthesis can be carried out within the pores of a membrane that is pre-coated with metal on one side as an anode (*79*), the chemical template synthesis is normally performed by immersing the membrane into a solution of the desired monomer and oxidizing agent with each of the pores acting as if a micro reaction vessel. It is noted with interest that if polycarbonate membranes are used as the template, highly ordered polymeric tubules are produced by preferential polymerization along the pore walls (*80-82*) due to the specific solvophobic and/or electrostatic interactions between the polymer and the pore wall (*43, 44, 76*). These conducting polymer tubules show a wide range of conductivities, increasing with decreasing pore diameter (*80, 83-85*) because the alignment of the polymer chains onto the pore wall can enhance conductivity and the smaller tubules contain proportionately more of the ordered material.

The preparation of carbon nanotubes by high-temperature graphitizing the polyacrylonitrile (PAN) nanotubules synthesized within the pores of an alumina template membrane or zeolite nanochannels is of particular interest (*86, 87*). This is because the discoveries of carbon nanotubes by Ijima (*88*) in 1991 and a method for the large-scale synthesis of nanotubes by Ebbesen and Ajayan (*89*) in 1992 have opened up a new era in material science and nanotechnology (*90*). Since then, various nanotubes, with or without encapsulated metals, having a straight, curved, planar-spiral, single- and double-helical shape have been experimentally produced either by the metal (e.g., Ni, Co, Fe) catalyzed pyrolysis of organic vapors (e.g., acetylene, benzene, methane) or by co-vaporising cobalt with carbon in an arc fullerene generator (*91, 92*). These nanotubes can behave as a semiconductor or metal depending on their network topology, and dissimilar carbon nanotubes may be joined together allowing the formation of molecular wires with interesting electrical, magnetic, nonlinear optical, and mechanical properties (*93-105*). In order to test properties of single nanotubes and to demonstrate their applications, however, carbon nanotubes need to be prepared with certain controllable macrostructures (e.g., in aligned form) so that individual nanotubes can be probed with existing techniques and be effectively incorporated into devices. In this context, the template synthesis technique has recently been used to prepare aligned carbon nanotubes (*106, 107*), apart from being used for producing nanotubes filled with guest materials (*108*). Aligned carbon nanotubes have also been prepared either by slicing a nanotube-dispersed polymer composite (*103, 109*) or by rubbing a nanotube-deposited plastic surface with a thin Teflon sheet or aluminium foil (*110*). More recently, Kroto and Walton (*96*) reported a method to align carbon nanotubes by pyrolysis of 2-amino-4,6-dichloro-*s*-triazine on a silica substrate pre-patterned with cobalt catalyst (Figure 3). Unlike most of the aforementioned cases where nanotubes were aligned in the plane parallel to the substrate, we have recently prepared large-scale aligned carbon nanotubes perpendicular to the substrate surface by pyrolysis of appropriate Fe-containing organic-metal complexes (Figure 4). The feasibility for a region-specific deposition of the aligned nanotubes has also been demonstrated. These newly-synthesized semiconducting nanotubes of controllable diameters and lengths could open avenues for testing properties of single nanotubes and to rationalize various potential applications ranging from reaction vessels for spatially confined chemical reactions to nanoelements in advanced optoelectronic devices.

Liquid Crystalline Conjugated Polymers. Another important technique for preparing ordered conjugated polymers is to chemically link liquid crystalline (LC) mesogens to conjugated backbones. Interest in liquid crystalline materials has a long history (*111*). However, the syntheses of liquid crystalline conjugated polymers is a very recent development. Since there are many monographs and review articles already available for conventional polymer liquid crystals (*112-119*), we will focus our attention on the syntheses and properties of liquid crystalline conjugated polymers.

Substitution of conjugated polymers, such as poly(thiophenes), poly(*p*-phenylene) (PPP), PPV, and poly(*p*-phenyleneethynylene), by certain long or bulky side groups combines the order of conjugated backbones with the chain mobility, lead to liquid crystalline behavior (*120-123*). For instance, a typical Schlieren texture has been observed for poly(3-hexylthiophene) (P3HT) after annealing at 145 °C for 6 h (*124*) and a picture of the nematic morphology has been reported for poly(3-decylthiophenes) (P3DT) with a low molecular weight (M_w = 5700) at 70 °C (*125*). Lyotropic liquid crystalline order was also observed in camphorsulfonic acid (HCSA) doped polyaniline in *m*-cresol (*126*). Furthermore, well-organized layered structures with an average interlayer distance of about 30 Å have been produced by thermal doping of polyaniline with dodecylbenzenesulfonic acid (*127*). The resulting electroactive complexes exhibited a higher maximum conductivity than HCl doped polyaniline by two orders of magnitude, attributable to the well-defined anisotropic order. As mentioned above, the general approach to liquid crystalline conjugated polymers, however, is to graft LC mesogen-containing side chains onto the conjugated backbone. Table I summarises most of the side chain liquid crystalline conjugated polymers reported so far.

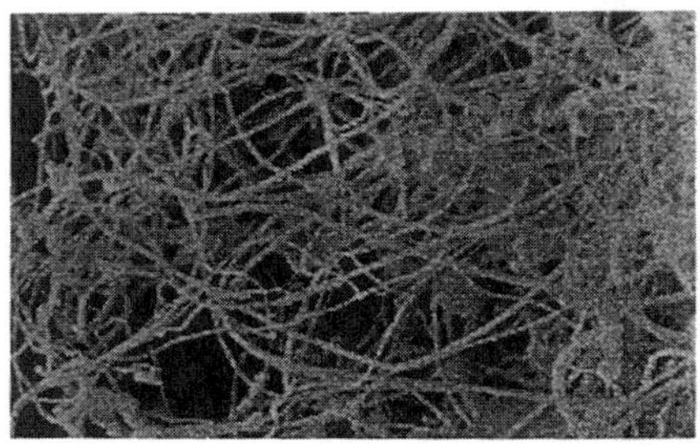

(a)

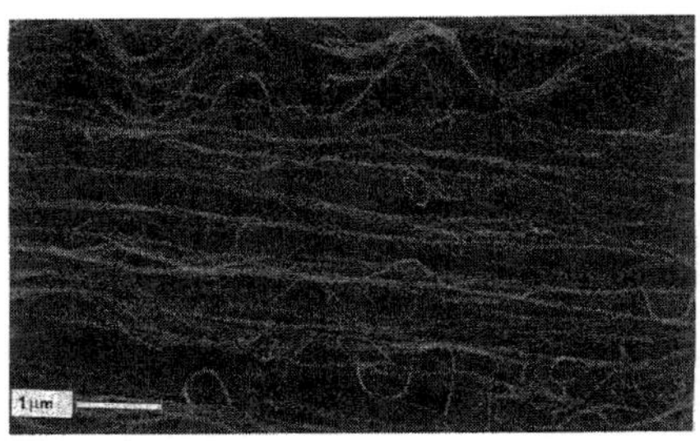

(b)

Figure 3. (a) SEM micrograph of nanotubes prepared by pyrolysis of 2-amino-4,6-dichloro-s-triazine over non-patterned cobalt surfaces (Kroto, H. W.; Walton, D. R. M. Univ. of Sussex, personal communication, 1997); (b) SEM image of aligned nanotubes prepared by pyrolysis of 2-amino-4,6-dichloro-s-triazine over patterned cobalt surfaces (Reproduced with permission from reference 96. Copyright 1997 Macmillan Magazines Ltd.).

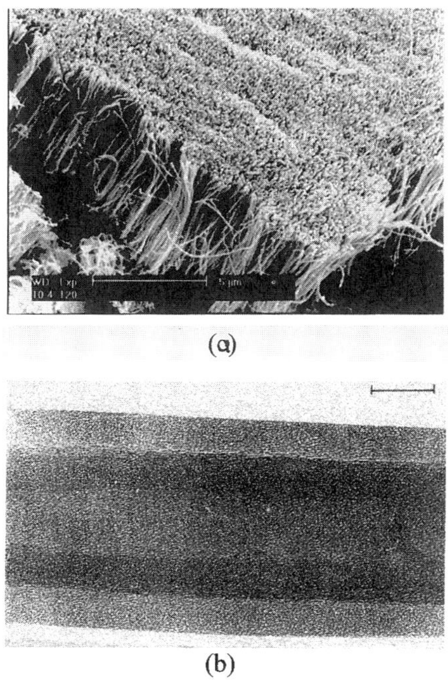

Figure 4. (a) SEM image of the highly aligned nanotubes prepared by pyrolysis of Fe-containing organic-metal complex (Huang,, S.; Dai, L.; Mau, W. H. M. to be submitted for publication); (b) The high-resolution TEM image of a single nanotube. The scale bar represents 25 nm.

Table I. Survey of some π-conjugated polymers with liquid crystalline side chains.

	Polymer	Substituent R		Ref.
A	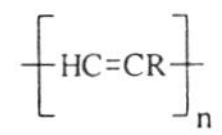	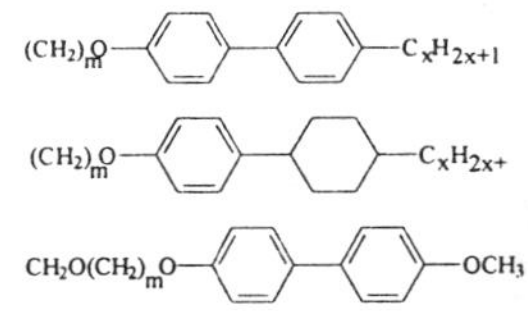 (CH₂)₈COO-cholesteryl	m = 3, 4, 6 x = 0, 2, 3, 5-8, C_9H_{20}* m = 4-8	(129-131) (132-136) (137) (138)
B	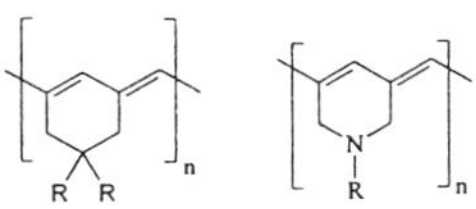	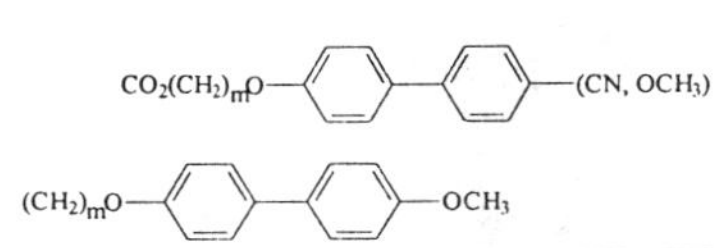		(139, 140) (141)
C	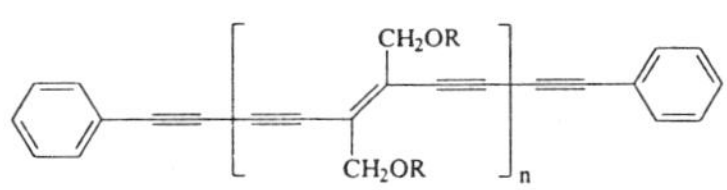	$C_{17}H_{35}CO$, $(C_{12}H_{25}O)_3C_6H_2CO$		(142)
D	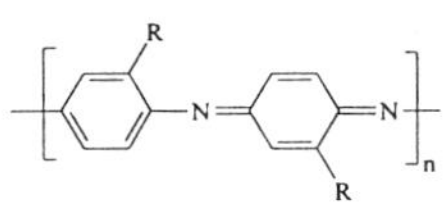	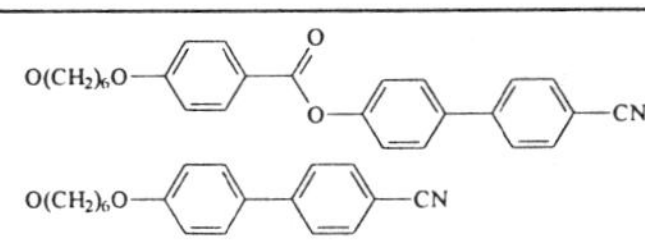		(143)
E	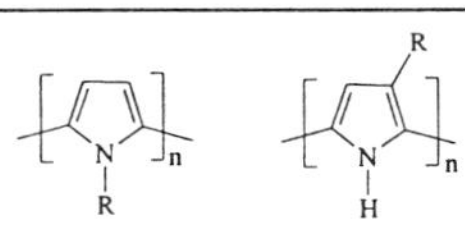	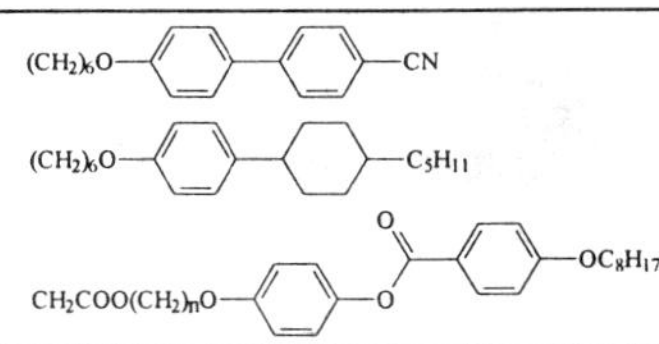	n = 3, 8	(144) (145) (146)
F	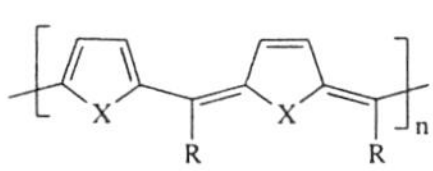 X=NH,S	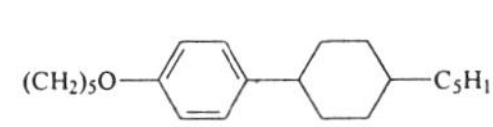		(147)
G	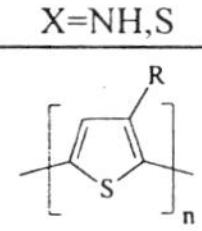	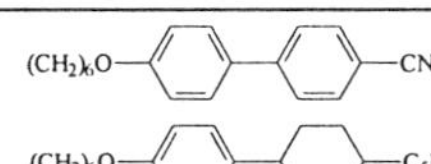		(148)
H	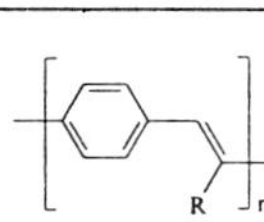	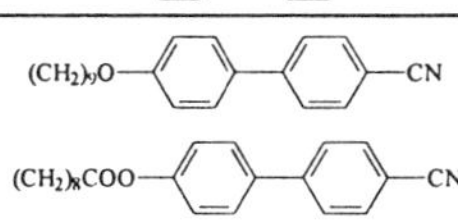	**H1** **H2**	(149, 150)

* optical active

Like most other functional polymers (*128*), liquid crystalline conjugated polymers can be prepared either by direct polymerization of appropriate functional monomers or by chemical modification of preformed conjugated polymers. Both approaches have been used to synthesize those liquid crystalline conjugated polymers listed in Table I, and the detailed information for each of them can be found in the corresponding reference (*129-150*). In particular, Winkler et al. have recently extended their work on non-conjugated liquid crystalline polynorbornenes and polyoctenylenes (*151-153*) to synthesize PPV-type liquid crystalline polymers by chemically grafting them with side chains containing biphenyl-4-carbonitrile mesogens (i.e., Polymer **H** in Table I). The commercial availability coupled with a high NLO hyperpolarizability (*154*) makes biphenyl-4-carbonitrile the mesogen of choice. These yellow-coloured LC-PPVs show following transition temperatures for phase transitions from the isotropic (i) melt to glass (g) state through the nematic (n) and smectic (s) phases **H1:** i 109 °C n 105 °C s 29 °C g; **H2:** i 165 °C n (s) 28 °C g. The transition from the isotropic melt to the nematic phase for both polymers is very clear, as reflected by corresponding structural evolution from the droplet to Schlieren texture under crossed polarisers. Further evidence for the smectic phase of a quenched **H1** sample is obtained from X-ray diffraction measurements, which show, in good agreement with its molecular structure, a value of ca. 4.5 Å for the separation between adjacent mesogenic units and ca. 19±2 Å for the distance from the terminal cyano group to the conjugated backbone. The LC-PPVs have similar backbone optical absorption and fluorescence emission as the unsubstituted PPV. However, the liquid crystalline behavior associated with the LC-PPV derivatives facilitates the alignment of the polymer chains by spontaneous ordering or external poling. For instance, Winkler et al. (*149, 150*) have obtained highly ordered LC-PPV films by simply casting the polymer solution onto a substrate, which was unidirectionally pre-rubbed with a polytetrafluoroethylene (Teflon) bar (*155, 156*). In contrast, in order to align even a five-ring oligo(*p*-phenylenevinylene) with two (non-LC) *n*-octyl side chains in the *para-* position of the central phenyl ring onto a highly preoriented Teflon substrate, vacuum evaporation has to be applied (*157*). Polarized UV/VIS and IR measurements on the aligned LC-PPV films show optical-absorption anisotropies ($A_{\parallel}/A_{\perp}$) ranging from 2 to 3 for both the mesogens and conjugated backbone (Figure 5), confirming significant orientation of the mesogenic groups and the main chain parallel to the rubbing direction (*149, 150*). Thus, the biphenylene groups in the bulky side chains are energetically preferred to align parallel to the aromatic PPV main chain through strong π-π-stacking, leading to a dense packing for the polymer film. Polarized adsorption and emission have also been reported for some conjugated main-chain LC polymers, as exemplified by rubbing-aligned polyfluorene macromolecules grafted with octyl side chains (*158*).

Post-Synthesis Orientation

Mechanical Stretching. The most straightforward way to stretch-orient conjugated polymers is to mechanically stretch a free-standing conjugated polymer film. In this regard, polyacetylene has been stretch-oriented and subsequently doped into a highly conducting form (*46, 159-161*), and a stretching-induced crossover from a semiconducting to metallic state in iodine-doped polyacetylene films was observed (*47*). Highly aligned electrochemically-synthesized polypyrrole films with conductivities along the alignment direction up to 1005 S/cm have also been prepared by stretching them to more than twice their original length (*162, 163*). Elongations up to 5 times the initial length were demonstrated for P3HT (*164*) and poly(3-octylthiophene) (P3OT) upon stretching (*165*). The stretched P3OT gels showed highly anisotropic and reversible size changes with temperature, solvent composition, and doping level. Simultaneous heat treatment and application of stress on films of polyaniline emeraldine base and salt led to the formation of stretch-oriented, partially crystalline polyaniline films with $\sigma_{||}/\sigma_{\perp}$ = 3.0 and 3.5 for the undoped emeraldine base and emeraldine hydrochloride salt, respectively, along with an up to 10-fold increase in $\sigma_{||}$ for the salt sample as compared with the DC conductivity of an unoriented counterpart (*166*).

Instead of stretching preformed conjugated polymers, orientation can also be achieved by stretching certain non-conjugated precursor polymers, in either a film or fiber form, followed by thermal conversion of the stretched materials into their corresponding conjugated structures (*167-172*). Table II lists the stretch ratios and the conductivities along the stretching direction for some aligned conjugated conducting polymers thus prepared.

The ability of certain conjugated polymers to form polymer blends with common non-conjugated polymers or elastomers provides additional possibilities for stretch-alignment of conjugated polymers. For example, Heeger et al. (*48*) have prepared highly oriented polymer blends of poly(2-methoxy-5-(2'-ethyl-hexoxy)-*p*-phenylenevinylene) (MEH-PPV) and ultrahigh-molecular-weight polyethylene (UHMW-PE) by processing the composite material into a gel intermediate state and followed by tensile drawing. The highly anistropic absorption and photoluminescence emission from the resulting materials indicate a strong orientation of the conjugated polymer guest along the draw axis. Similarly, highly polarized photoluminescent layers based on highly uniaxially stretched blends of a poly(2,5-dialkoxy-*p*-phenylene ethynylene) derivative and the UHMW-PE have been used as both the polarize light and bright color generator in liquid crystal displays, leading to substantially enhanced device brightness, contrast, efficiency, and viewing angle (*173*). In a separate study, poly(3-alkythiophene)/ethylene-propylene elastomer composite has been stretched up to a stretch ratio of 12, causing a blue shift for the absorption peak with a high anisotropy of ca. 3 (*174*). However, a better oriented poly(3-alkythiophene) was obtained by spinning the polymer solution onto a stretchable substrate (e.g. polyethylene) and followed by stretching the substrate film (*175*). Using polyethylene films as the stretchable substrate, Theophilou et al. (*176*) have obtained highly aligned polyaniline films with $\sigma_{||}$ as high as 2500 S/cm, while Naarmann et al. (*177*) produced a polyacetylene film with $\sigma_{||}$ = 150000 S/cm and $\sigma_{||}/\sigma_{\perp}$ = 1000. Furthermore, Inganäs et al. (*178, 179*) constructed the first organic LED that could emit polarized light under a voltage as low as 2 V by using a stretch-oriented polythiophene derivative as the electroluminescent material.

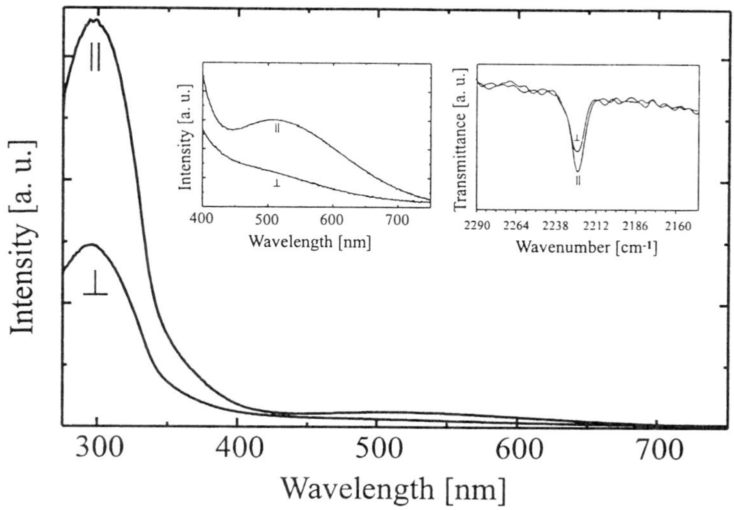

*Figure 5. Polarized UV/VIS and IR spectra of the oriented liquid crystalline PPV-derivative **H1** in the parallel (//) and perpendicular (⊥) directions to the rubbing axis (Reproduced with permission from reference 149. Copyright 1996 Elsevier Science Ltd.).*

Table II. Typical electrical conductivities of aligned π-conjugated polymers prepared from precursor polymers.

Precursor Polymer	Conjugated Polymer	Stretch Ratio	Dopant	$\sigma_{max, 25\,°C}$ (S/cm)	Ref.
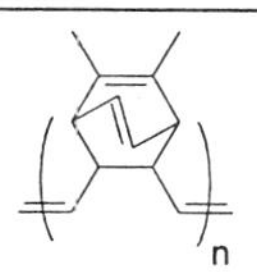		7.5	AsF_5	1250	(168)
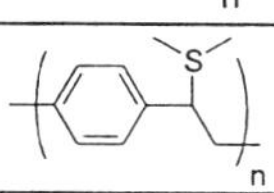	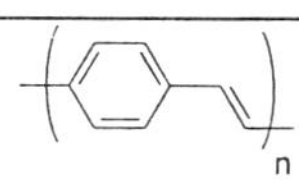	7	AsF_5	500	(169)
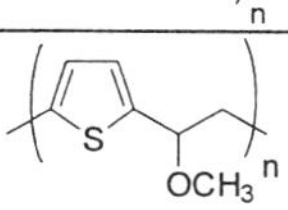	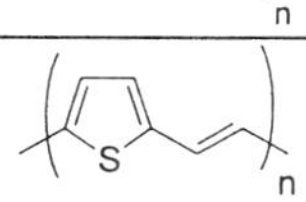	20	I_2	2000	(170)
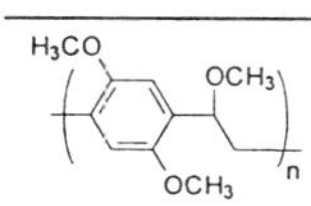	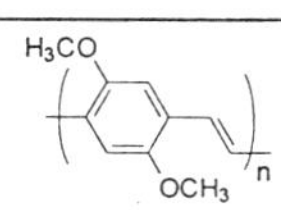	8	I_2	1200	(170, 171)

Extended conformations can also be obtained for conjugated polymer chains through certain chemical and/or physical interactions (e.g., with dopants). While the chain stretching in end-adsorbed polymer brushes has been widely investigated both theoretically and experimentally (*180-184*), Bumm et al. (185, 186) demonstrated the use of a STM tip to probe electrical properties of individual conjugated conducting molecules ("molecular wires") dispersed into a self-assembled monolayer film of nonconducting alkanethiolate molecules. Dai and White (*187*) have also reported a conformational transition from a random coil to a rod-like structure caused by the I_2-induced conjugation of polyisoprene chains, while MacDiarmid and Epstein (*188, 189*) demonstrated that the interaction of a HCSA-doped polyaniline sample with *m*-cresol could cause a conformational transition of the polymer chain from the so-called "compact coil" to "expanded coil" with a concomitant increase in conductivity by up to several orders of magnitude. The latter process has been widely known as "secondary doping" (*188, 189*). Further to our work on C_{60}-containing polymers (*190-192*), we have recently used hydrogensulfated fullerene derivatives with multiple -(O)SO_3H groups as protonic acid dopants for polyaniline emeraldine base (*193, 194*). It was found that the hydrogensulfated fullerene derivatives can act simultaneously as both a primary and secondary dopant, and the resulting material showed metallic characteristics with room-temperature conductivities up to 100 S/cm, about 6 orders of magnitude higher than the typical value for fullerene-doped conducting polymers (*194, 195*).

The Use of Langmuir-Blodgett (LB) and Self-Assembling Techniques. Although the stretch-orientation and rubbing alignment techniques discussed above provide simplified approaches for ordering conjugated polymers in certain systems, the severe difficulty associated with the stretching method for preparing highly oriented thin films (e.g., $< 1\mu m$) that retain structural integrity, together with the general lack of a uniform film thickness and adequate adhesion to the substrate for rubbing aligned thin polymer films, may preclude the general application of these techniques. In contrast, the number of molecular layers (and hence the thickness of films) can be controlled at the molecular level by the LB deposition. Besides, the LB technique has another major advantage for controlling the film structure in three-dimension (*196*) as the orientation of polymer chains in each of the layers can be independently controlled by anisotropic compression of the corresponding Langmuir film during the dipping process (*27*). The oriented and/or patterned LB films (*197*) can be further used as substrates, for example, for controlling liquid crystal alignment (*198*). As a consequence, the LB technique has been widely used for making mono- and multi-layer films with well defined structures and ordered molecular organizations from amphiphilic conjugated polymers, non-conjugated precursor polymers, and even certain non-surface-active conjugated polymers.

LB films of highly anisotropic conducting polymers can be prepared through: (a) the manipulation of amphiphilic macromolecules comprising an alkyl chain as the hydrophobic group and with a conjugated polymer being attached to its hydrophilic end (*199*); (b) the manipulation of mixed monolayers consisting of a surface active agent (e.g., dopant) and non-surface active conjugated polymer (*200*); or (c) the direct synthesis of conducting polymers at the air-water interface or within a preformed multilayer film (*49, 200*). To mention but a few examples: Shimidzu et al. (*49*) have prepared conducting LB films containing polypyrrole (*201, 202*), polyaniline (*49*) or poly(3-alkylthiophene) (*199*). Among them, a multilayer LB films (200 layers) of a substituted polypyrrole showed an anisotropic conductivity as high as 10^{10} (i.e., $\sigma_{||} = 10^{-1}$ S/cm and $\sigma_{\perp} = 10^{-11}$ S/cm) (*49*). Well characterized LB films of various conjugated polymers including poly(3-alkylthiophene) have also been reported by Rubner et al. (*200, 203-205*), for which the readers are referred to the excellent review by Rubner and Skotheim (*51*).

Using LB films of the sulphonium precursor of PPV as the starting material, Era et al. (*206*), and Nishikata et al. (*207*) obtained LB films of PPV by thermal conversion (Scheme 1) although PPV-type LB films have also recently been prepared from some of its soluble derivatives (*208-210*). Values of 2.9×10^{-10} and 0.75×10^{-10} esu for $\chi^{(3)}$ in the directions parallel and perpendicular to the dipping direction have been obtained from a highly oriented LB film of poly(2,5-dimethoxy-*p*-phenylenevinylene). Also, LEDs with polarized EL emissions have been fabricated using certain LB films as the EL material (*27, 211*). On this basis, Grüner et al. (*212*) have precisely determined the location of the exciton recombination zone (*20*) in LEDs by making multilayer LB films of soluble PPP derivatives with two regions of perpendicular macroscopic orientation, followed by quantifying the emission profile through the analysis of the EL anisotropy.

Another interesting related development is the construction of multilayer conducting films from various bilayer combinations, such as conjugated/nonconjugated polyion bilayers or precursor polymer/conjugated polyion bilayers, through a layer-by-layer self-assembly process developed by Rubner et al. (*213*). The principle of the layer-by-layer process lies in the alternate spontaneous adsorption of oppositely charged polymers from dilute solutions. This technique has been successfully used to generate multilayer thin films comprised of sequentially adsorbed layers of partially doped polyaniline and a polyanion (e.g., sulphonated polystyrene), leading to conductivities comparable to those obtained with spin cast films (0.5-1.0 S/cm) after having been doped with strong acids such as HCl and methanesulfonic acid (*214*). Based on the layer-by-layer adsorption process, Rubner et al. (*215-217*) have also constructed multilayer LEDs from PPV and poly(styrene sulfonic acid) (SPS) or poly(methacrylic acid) (PMA), and found that the type of polyanion used has a significant effect on the performance of the LED devices fabricated with Al and ITO as electrodes. In particular, LEDs fabricated from PMA/PPV multilayers were found to exhibit luminance levels in the range of 20-60 cd/m^2 with a thickness dependent turn-on voltage and rectification ratios greater than 10^5, whereas the SPS/PPV counterparts showed nearly symmetric I-V curves with a thickness independent turn-on voltage and much lower luminance levels. The observed difference in the device performance can be attributed to a doping effect associated with the sulfonic acid groups in SPS. Furthermore, these authors have recently extended the layer-by-layer adsorption process to include the hydrogen-bonding interactions between the polyaniline and poly(vinylpyrrolidone), poly(vinyl alcohol), poly(acrylamide), or poly(ethylene oxide) (*218*). By using pre-ordered / pre-patterned substrates, the layer-by-layer absorption process should, in principle, lead to the construction of oriented / patterned conjugated polymers, as we shall see later.

Patterned Conjugated Polymers

Photolithographic Patterning of Conjugated Polymers. A few approaches for photolithographic patterning of conjugated polymers have been reported in the literature (*50*). For instance, Clarke et al. (*219*) reported a photochemical doping method using onium salts (e.g., triarylsulfonium and diaryliodonium salts) as the photochemical dopants, which allows doping to occur only in the UV illuminated areas of a polyacetylene film. Holdcroft et al. (*220-223*), Dao et al. (*224*), and Cai et al. (*225*) obtained conducting patterns from the soluble form of conjugated polymers (e.g., poly(3-alkylthiophenes)) through patterned photo-crosslinking (or crosslinking by electron beam), followed by a solution-based development process. Angelopoulos et al. (*226*) and Venugopal et al. (*227*) demonstrated that conducting patterns could also be made from a mixture of the soluble base form of polyaniline and a photoacid generator, which photochemically generated the acidic dopants required for generating conducting patterns of the insoluble polyaniline salts. The solubility of the conducting polymers used in these approaches was acquired by covalently bonding conjugated polymers with soluble side groups or polymeric block chains, which often reduced the conductivity (*1, 128, 228*). Besides, the poor solubility of most unfunctionalized conjugated polymers in common organic solvents may limit the general application of the above techniques. Hence, Bargon et al. (*229*) and Baumann et al. (*230*) produced conducting patterns through the polymerization of monomers, such as pyrrole, aniline and thiophene, initiated by HCl which was generated photochemically in a patterned fashion from chlorine-containing polymer matrices, including poly(chloro-acrylonitrile), poly(vinylchloride), poly(chloroprene), and poly(chlorostyrene). The conducting regions generated in these cases, however, are composite materials containing newly formed conducting polymer chains within the non-conducting polymer matrices. A particularly attractive option is the formation of conducting patterns from a processable (soluble and/or fusible) insulating polymeric matrix through, for example, a photochemical transition which can directly convert the microlithographically exposed regions into unsubstituted conjugated sequences.

In 1988, Thakur reported that the conductivity of *cis*-1,4-polyisoprene can be increased by about 10 orders of magnitude upon "doping" with iodine (*231*). On this basis, he concluded that a conjugated structure is not always necessary for a polymer to be electrically conducting. In view of the ease with which conducting rubbers can be made and the potential challenge of Thakur's claim to modern theory of conducting polymers (*1*), Thakur's work received complimentary comments from several scientific journals (*232*). We were the first to demonstrate that "I_2-doping" of 1,4-polydienes produces conjugated sequences of unsaturated double bonds through reactions shown in Scheme 2 (*233-237*). Further investigation on the I_2-induced conjugation of polydienes indicated that for *cis*-1,4-polybutadiene the reaction sequence given in Scheme 2 terminated, at room temperature, at product (II), whereas for the *trans*-isomer the reaction sequence proceeded towards conjugated products (III) and/or (IV) with electrical conductivity (*236*). This discovery provides means for

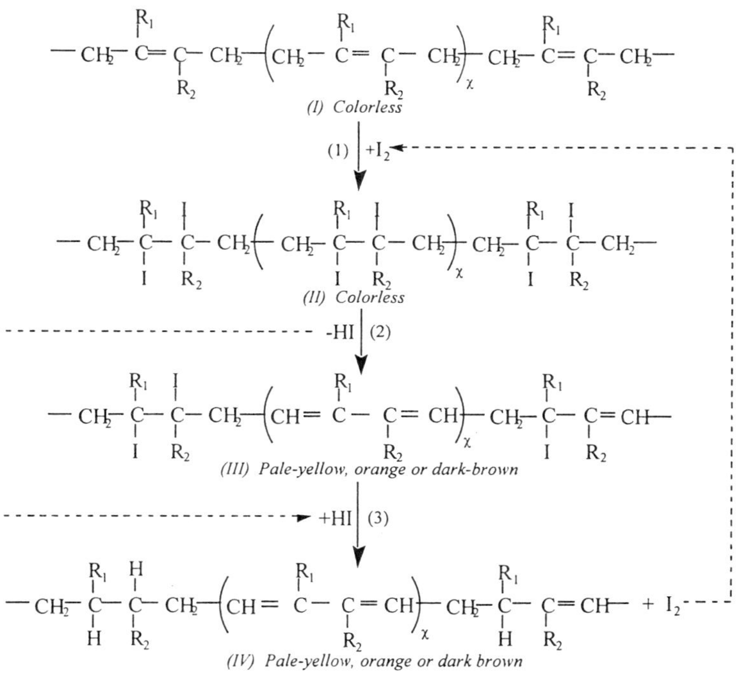

R_1, R_2 refer to either H or CH_3 group.

Scheme 2. Reactions of polydiene with iodine leading to the formation of conjugated sequences. (Reproduced with permission from reference 239. Copyright 1996 The American Chemical Society) photolithographic generation of conducting patterns from *cis*-1,4-polybutadiene as the *cis*-isomer can be photoisomerized into its *trans*-counterpart. Using a microlithographic mask, we have recently found that, *cis*-1,4-polybutadiene films can be photoisomerized into the *trans*-isomer in a patterned fashion without significant lateral diffusion (*238, 239*). As a result, only the photoisomerized regions are capable of generation of conjugated double bonds upon exposure of the entire polybutadiene film to iodine at room temperature, resulting in the formation of conducting patterns in an insulating matrix of iodinated *cis*-1,4-polybutadiene. The conducting patterns thus formed are coloured and show strong fluorescence emission, which enables visualization of the conducting polymer regions. An example of the conducting patterns thus generated is shown in Figure 6a. It is a close replication of the photomask structure, and conducting wires on a micrometer scale are clearly evident. A corresponding fluorescence microscopic image of the conducting pattern is given in Figure 6b. It shows the same features as the optical micrograph (Figure 6a), but with inverse intensities in the image. The dark regions characteristic of the "I_2-doped" *trans*-1,4-polybutadiene in Figure 6a gave rise to bright fluorescence emission in Figure 6b, consistent with the fluorescence emission originating from the conjugated structures (*1, 239*). The dark regions in Figure 6b represent non-fluorescent components associated with the *cis*-isomer. While the conducting patterns thus prepared may find applications in certain electronic and photonic devices, the photolithographic method has also been applied successfully to micropattern electroluminescent conjugated polymers (*240*) including PPV (*241*).

For instance, Wei et al. (*241*), and DeAro et al. (*242*) have recently demonstrated the use of near-field optical microscopy for writing / reading images on conjugated polymer films including PPV with a resolution of ca. 0.1 μm. On the other hand, Cho et al. (*243*) have discovered that the methoxy-substituted PPV precursor polymer (i.e., Product (III) in Scheme 1) decomposed at a lower temperature (80-200 °C) after an acid-catalyzed UV irradiation than the non-irradiated precursor polymer (> 220 °C). On this basis, these authors prepared PPV conducting micropatterns (10^{-2} - 10^{-3} S/cm) by using triphenylsulfonium salts as both an acid catalyst and photochemical dopant for a selectively photoinitiated thermolysis of the methoxy-substituted PPV precursor polymer at a relatively low temperature. Similarly, micropatterns of PPV with patterned emissions have also been prepared from several copolymers containing PPV precursors through a selectively thermolysis, with or without involving a photoinitiator (*31, 32*). Besides, PPV has also been microstructured in a non-doped form by UV interferometry (*30*), as the transformation from the unsubstituted PPV precursor polymer (i.e., Product (I) in Scheme 1) to the conjugated state (i.e., Product (IV) in Scheme 1) can occur not only through the conventional thermal conversion shown in Scheme 1 but also upon UV irradiation (*30, 244*). Compared with the conducting PPV micropatterns, the non-doped PPV microstructures thus formed seem to be more suitable for EL applications because of the possible doping-induced fluorescence quenching (*20*). More recently, using an ink-jet printer, Yang et al. (*245, 246*) created polymer-based light-emitting logos by patternwise printing a water-soluble polythiophene, which provides a patterned electrical connection from ITO electrode to an overlying layer of MEH-PPV. On the

(a)

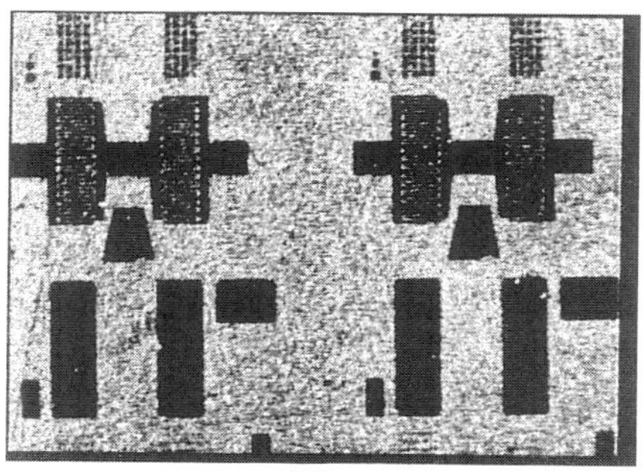

(b)

Figure 6. (a) Optical microscopy image of a pattern obtained by reacting the patternwise photoisomerized cis-1,4-polybutadiene film on chromium-coated glass with an I_2-saturated methanol solution. The dark areas are regions of the conjugated polybutadiene, and the width of the white rectangles at the bottom part of the picture is 18 μm; (b) Fluorescence micrograph of the conducting pattern (λ_{exc} = 410 nm) (Reproduced with permission from reference 239. Copyright 1996 The American Chemical Society).

other hand, Noach et al. (*247*) have demonstrated that a PPV copolymer (i.e., poly(1,4-phenylenevinylene-*co*-2,6-pyridylenevinylene)) based light emitting diode array consisting of micron sized emitting pixels can be produced through a selective laser ablation process. In addition to the achievement of patterned emissions, these authors also found that emission intensities increased significantly due to the strongly enhanced electric fields at the edges of the conducting (Al/ITO) pixels.

Pattern Formation by Self-Assembling. Self-assembled monolayer films (SAMs) form, notably, when materials such as alkanethiols and organosilanes are physically or chemically adsorbed onto gold, silver, aluminium, copper, or silicon dioxide surfaces (*248, 249*). Pattern formation / recognition by the SAMs has recently become a very promising technique for microstructuring organic materials (*250-254*), and various strategies including photolithographic (or beam-induced) transformation (*255-257*), mechanical scraping (*249, 258*), microcontact printing (μCP) (*250, 259-261*), scanned-probe based micro-machining (*262*), and locally-confined plasma surface modification (*263*) have been developed for microlithographic fabrication and subsequent transformation of SAM patterns. Recently, the possibility for multi-dimensional micro-construction by the SAM technique has also been explored (*264-266*).

In particular, the SAM technique has been employed by Rozsnyai and Wrighton (*52, 53*) to pattern conducting polymers of polypyrrole, poly(3-methylthiophene), and polyaniline through selective electropolymerization of the corresponding monomers onto photochemically patterned disulfide SAMs on gold. Meanwhile, Ferreira et al. (*213, 267*) have used a positively-charged SAM layer (formed from aminopropyl dimethylethoxy silane or poly(ethyleneimine)) for surface immobilization of poly(thiophene-3-acetic acid). While Garnier et al. (*268*) have produced a highly ordered, densely packed, electroactive SAM of thiol-functionalised oligothiophene on platinum surfaces, Kim et al. (*269*) prepared polydiacetylene monolayers by first self-assembling diacetylene group-containing *n*-alkanethiols onto gold surfaces, then photopolymerizing the SAM with UV light. The photopolymerization involved in the latter work suggests that the micropattern formation of polydiacetylene monolayer should be straightforward by use of a photomask, though this has not been reported. Indeed, photofabrication of azobenzene-functionalised polydiacetylenes has recently led to the formation of regular surface relief gratings with interesting NLO properties (*270*).

Microcontact printing (μCP) is a very convenient patterning technique, in which an elastomeric stamp (typically, a polydimethylsiloxane stamp) is used to region-specifically transfer molecular "inks" (e.g., alkanethiol, alkylsiloxane, etc.) (*259*). It has been widely used for generating SAM patterns onto certain metal (e.g., Au, Ag, Cu, etc.) and Si surfaces with a minimum feature size of ca.100 nm (*271*). The use of μCP has recently been extended to patterning of conducting polymers. For instance, MacDiarmid et al. (*272, 273*) produced hydrophobic patterns of $C_{18}H_{37}SiCl_3$ on hydrophilic glass substrates through a μCP stamp. Then, polypyrrole and polyaniline were adsorbed preferentially onto the hydrophobic regions from dilute aqueous solutions of the polymerizing monomer to form thin film patterns. The patterned conducting polymers on the hydrophobic surface were found to take an extended conformation with an enhanced conductivity (*274*), which have been demonstrated to be useful as electrodes in display devices based on polymer dispersed liquid crystals (*274, 275*). Furthermore, recent reports on the use of certain patterned SAMs for region-specific alignment of liquid crystal molecules (*276, 277*) should have an important implication for patterned alignment of the LC conjugated polymers discussed above.

Pattern Formation by Polymer Phase Separation. Like the SAM technique, polymer phase separation plays an important role in fabrication of microstructures, as it provides the opportunity for nanoscale patterning which otherwise is difficult by lithographic techniques (*278, 279*). The conditions necessary for microphase separation in immiscible polymer mixtures depend on their molecular architectures, nature of monomers, compositions, and molecular weights (*280*). Briefly, for linear homopolymer mixtures of A and B, the free energy of mixing per unit volume is given by (*280, 281*):

$$\Delta G/kT = (\phi_A/N_A)\ln\phi_A + (\phi_B/N_B)\ln\phi_B + \chi\phi_A\phi_B \qquad (7)$$

where k is the Boltzmann constant, T is the absolute temperature, ϕ_i and N_i are the volume fraction and degree of polymerization for species i, respectively. χ is the Flory-Huggins segmental interaction parameter. $\phi_A + \phi_B = 1$ is a result of incompressible mixing. From equation 7, the conditions for equilibrium, stability, and criticality can be derived as follows (*280-282*):

$$\text{Equilibrium:} \quad \partial\Delta G\,(\phi_A')/\partial\phi_A = \partial\Delta G\,(\phi_A'')/\partial\phi_A \qquad (8)$$
$$\text{Stability:} \quad \partial^2\Delta G/\partial\phi_A^{\,2} = 0 \qquad (9)$$
$$\text{Criticality:} \quad \partial^3\Delta G/\partial\phi_A^{\,3} = 0 \qquad (10)$$

where the superscripts ' and " refer to separated phases. Taking the symmetric case of $N_A = N_B = N$ as a simplified example (*280-282*), it can be derived from above equations that phase separation occurs only when χ is larger than a threshold value $\chi_c = 2/N$. The large value of N, and hence very small value of χ_c, for macromolecules rationalizes the strong immiscibility often observed in polymer mixtures. For detailed discussion on the theory of polymer phase separation, the interested readers are referred to specialized books and reviews (*280-286*).

Although the above description is ideal and equation 7 may fail to provide a quantitative description of real polymer mixtures (*286*), the general trends governed by equations 7-10 are even applicable to block copolymer systems, where the phase separation may be reduced to some extent by the covalent linkages between the constituent blocks. Generally speaking, the phase behavior of block copolymers can be classified into two categories according to the χN value (*281, 282*): $\chi N >> 10$, strong segregation limit; $\chi N \leq 10$, weak segregation region. In the weak segregation region, entropic factors dominate and diblock copolymers exist in a spatially homogeneous state (*284*). When $\chi N >> 10$, however, well-segregated microdomain structures are formed with relatively sharp interfaces as the number of contacts between the two constituent segments decreases at the cost of chain stretching (*180, 285*). In fact, seven well-defined microdomain structures with sizes typically in the range of a few tens of nanometers have been observed in polystyrene-polyisoprene copolymers in the bulk (*282, 286*), which represents a strongly segregated system. Similar phase separation has been demonstrated to also occur in thin films of adsorbed block copolymer chains, forming lateral periodical micropatterns (*287, 288*). Furthermore, the chain segregation has also been either experimentally observed or theoretically predicted for block copolymers confined on patterned surfaces (*289-291*), between two constituent homopolymer layers (*292*), two foreign walls (*293-298*), or end-attached at the liquid/solid interface (*181-184, 293, 294*). This suggests a considerable scope for patterning of polymer chains through polymer phase separation. However, the fabrication of conjugated polymer patterns by polymer phase separation is still a research field in infancy, though a few reported examples are reviewed below.

Interestingly, Inganäs et al. (*33*) demonstrated that self-organization (i.e., phase separation) in polymer blends consisting substituted polythiophenes with different band gaps allows the formation of submicrometer-sized domains having a range of compositions and emission characteristics. As a result, the emission color can be varied by controlling the operating voltage. The ease with which a voltage-controlled multiple-color emission can be achieved suggests an attractive way for making future polymeric electroluminescent color screens.

It now appears that polyacetylene may be solubilized by making copolymers with polyisoprene, polystyrene or polybutadiene via a so-called anionic to Ziegler-Natta route (*299-303*). The polyisoprene-polyacetylene (PI-PA) diblock copolymers thus prepared present an unusual situation where the high flexibility and solubility associated with polyisoprene chains and the stiffness and conductivity characteristic of insoluble conjugated polyacetylene chains are combined into one macromolecule. Due to the large difference in physical properties between these two blocks, the copolymers of PI-PA may be expected to show phase separation in solution and domain structure in the solid state (*299-307*). Indeed, Dai et al. (*306, 307*) have recently found that I_2-doping of the PI-PA copolymer films leads to the formation of a pseudo interpenetrating polymer network (PIPN). As shown in Figure 7a, an almost featureless surface was observed for a freshly synthesized and freshly cast PI-PA diblock copolymer film. Upon doping, however, dispersed worm-like micelles developed (Figure 7b) due to the doping-induced aggregation of polyacetylene chains. As the doping proceeded further, the static electrical interactions between the doped polyacetylene chains and their counterions brought the neighbouring micelle cores together to form a micro-sized thermoplastic PIPN network structure, as shown in Figure 7c. Closer inspection of Figure 7c under a higher magnification (Figure 7d) clearly shows the interpenetrating network structure. Small angle X-ray scattering measurements indicate the formation of polyacetylene "rods" at the molecular level with an average length of ca. 40 Å, which is equivalent to the polyacetylene chain length, suggesting that polyacetylene blocks in the PI-PA films are fairly stretched, facilitating the formation of the PIPN network (*307*). The aggregation structures (especially the PIPN network) play an important role in regulating the electrical

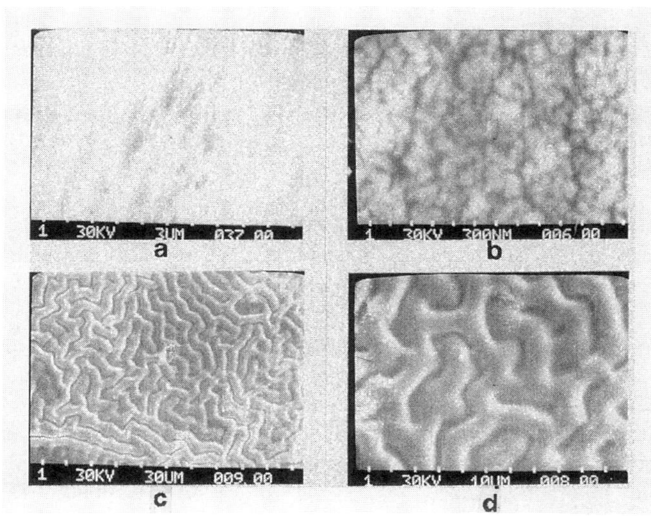

Figure 7. SEM micrographs of a polyisoprene-polyacetylene film before and after I_2-doping. I_2-doping time: (a) 0 h; (b) 0.2 h; (c) 2 h; (d) as for (c) under a higher magnification (Reproduced with permission from reference 307. Copyright 1977 Elsevier Science Ltd.).

properties of the PI-PA copolymers (*307, 308*), and the doping-induced phase separation process may have potential implications for making various multiphase conducting polymers with different ordered structures ranging from the nanometer to micrometer scale.

Plasma Patterning of Conjugated Polymers. Vapor deposition techniques, including plasma polymerization and chemical vapor deposition, are solvent-free, low temperature thin film forming processes with a clean working environment, which can be used to circumvent problems possibly associated with various solution-processing methods (e.g., deterioration of mechanical properties of conjugated polymer films by trapped solvent, and difficulty in choosing suitable solvent(s) for preparing pinhole-free single layer or multilayer polymer thin films). In fact, the chemical vapor deposition technique has been widely used for sublimation of low molecular weight dyes (*309, 310*), and recently for forming polymer films in EL devices (*311-313*). The plasma polymerization technique, however, seems to be much less discussed for the same purpose. For this reason, the remainder of this paper will be mainly concerned with the plasma technique.

Plasma treatment and plasma polymerization can be carried out with the gas molecules being at ambient temperature, and thus without thermal degradation of the sample surface and/or the plasma-polymerized layer (*314*). Furthermore, plasma polymerization has been widely demonstrated to generate thin, cohesive, adhering, pinhole-free polymer films, which are often used as protective coatings or adhesion promoting layers and have potential applications in optical wave guides, sensor technology, biomedical and optoelectronic devices (*315-321*). Although plasma polymerization, in most cases, produces an electrically insulating organic film, a few approaches to semiconducting plasma polymer films have been reported. For example, Bhuiyan and Bhoraskar (*322*) prepared semiconducting organic thin films through plasma polymerization of acrylonitrile, followed by pyrolysis. The electrical conduction in these plasma-polymerized polymer films was shown to be dominated by the variable-range hopping mechanism (*307*). Without involving a plasma precursor polymer, Kawakami (*323*) prepared partially conjugated conducting polymers by polymerization of halogenated benzene and thiophene, respectively. Semiconductive thin organic polymer films were also prepared by Tanaka et al. (*324*) through plasma polymerization of 1-benzothiophene. In this case, it was demonstrated that the aromatic skeletons of the monomer molecules are largely retained in the plasma films, which, after doping with iodine, show an increase in the electrical conductivity by eight to nine orders of magnitude (*325*). Later, these authors extended the plasma monomers to include 3,4,9,10-perylenetetracarboxylic dianhydride (PTCDA), 3,4,9,10-perylene-tetracarboxylic diimide (PTCDI), and perylene (*326*). The highest conductivity (0.1 S/cm) reported so far for plasma-polymerized organic films has been obtained with the as-synthesized PTCDA plasma film. In a separate study, Xie et al. (*327*) prepared conducting organic films by plasma polymerization of a mixture of 7,7,8,8-tetracyanoquinodimethane (TCNQ) and quinoline, and found that the composite plasma polymers have a higher value of conductivity (ca. 10^{-5} S/cm) than those of the constituent organic analogues ($<10^{-12}$ S/cm for quinoline plasma polymer and 10^{-9} S/cm for TCNQ plasma polymer) due to a higher conjugation of the π-electrons in the composite film.

Although plasma polymerization of aniline has previously been reported (*328-331*), the conducting characteristics and detailed characterization of the plasma-polymerized polyaniline films have not been reported until Gong et al. (*332*) have addressed this point recently. These authors found that the atomic ratios of C/N both higher and lower than that of aniline (C/N = 6) can be obtained by choosing appropriate discharge conditions, indicating the possibility of tailoring the structure and properties of the plasma polyaniline films by optimising the discharge conditions, and that the plasma polyaniline films they prepared are smooth and free of oxidant and solvent. The presence of conjugated sequences in the plasma-polyaniline films was confirmed by UV/VIS, FTIR, and ESR measurements. Preliminary results indicate that the conductivity of the plasma-polyaniline films can be enhanced by 3 orders of magnitude through HCl treatment.

In a closely related study, Kang et al. (*333*) have recently performed a plasma treatment on polyaniline films in order to produce surface properties demanded for specific applications (e.g., as bio-sensors). It was found that Ar plasma treatment caused some oxidation of the carbon atoms. Although no direct oxidation of the nitrogen heteroatoms was observed, the plasma treatment led to a steady decrease in the intrinsic oxidation state (i.e., the [=N-]/[-NH-] ratio) of the polyaniline films. Similarly, an oxygen or nitrogen plasma treatment of a PPV layer at the PPV/Al interface in a LED device with the Cr/PPV/Al structure has been demonstrated to cause a disappearance of the rectifying behavior, along with an increase in the current by many orders of magnitude (*334*). Thus, the plasma treatment is believed to be useful for creating a charge carrier confinement layer to facilitate the charge injection, as is the case with the electron-transporting layer. More than that, several papers on the use of the plasma polymerization technique for the fabrication of LED devices have appeared (*335, 336*). There is no doubt that the plasma techniques for semiconducting organic film formation and surface modification demonstrated above will have potential implications for making fascinating electronic and photonic devices, especially when coupled with various plasma patterning techniques to be discussed below.

Recently, Dai et al. (*54*) have demonstrated that H_2O-plasma can cause plasma etching of substrate surfaces under certain discharge conditions. On the basis of this finding, these authors have successfully created high resolution surface patterns of hydrophilic regions on, for example, certain hydrophobic polymer films through a patterned plasma treatment using a mask. Furthermore, Dai et al. (*54, 337*) have also demonstrated that surface patterns of various specific functionalities, including both hydrophilic and hydrophobic groups, on a micrometer scale could be produced by plasma polymerization of appropriate monomers in a patterned fashion. Figure 8a shows the surface pattern generated by the patterned plasma polymerization of *n*-hexane on a gold-coated mica surface, where the dark areas represent *n*-hexane plasma polymer and the bright regions are associated with the plasma-polymer-free gold surface. By extension, these authors developed a versatile method for obtaining patterned conducting polymers by first depositing a thin patterned nonconducting (e.g., *n*-hexane) plasma polymer layer onto a metal-sputtered electrode, and then performing electropolymerization of monomers, such as pyrrole and aniline, within the regions not covered by the patterned plasma polymer layer. Figure 8b represents a typical reflection light microscopic image of a polypyrrole pattern electrochemically polymerized onto platinum-coated mica sheets pre-patterned with the freshly prepared *n*-hexane plasma polymer. It shows the same features as the plasma pattern of Figure 8a, but with inverse intensities. The bright regions characteristic of the uncovered metal surface in Figure 8a become dark in Figure 8b due to the presence of a dark layer of the newly electropolymerized polypyrrole film. The bright regions in Figure 8b represent a more reflective surface associated with the *n*-hexane plasma polymer. The cyclic voltammogram measurements clearly indicates that the polypyrrole patterns thus prepared are electrochemically active.

Furthermore, some plasma-patterned surfaces can be directly used to deposit conducting polymers under certain conditions. For example, Vargo et al. (*338*) have region-specifically deposited polycations including polyaniline and polypyrrole onto a H_2/CH_3OH plasma patterned perfluorinated ethylene-propylene copolymer (FEP) surface. These authors have also demonstrated the possibility for fabrication of both homogeneous and patterned conducting polyaniline multilayer structures with reasonable electrical conductivities. In view of the preparation of semiconducting organic films by plasma polymerization of aromatic monomers discussed above, a patterned plasma polymerization of these monomers should allow the formation of semiconducting micro-patterns. This approach can be viewed as potentially promising for the fabrication of future electronic and/or photonic devices with region-specific characteristics.

Concluding Remarks

Conjugated polymers show interesting electronic and photonic properties, which make them attractive for a wide range of potential applications. In order to fabricate optoelectronic devices from conjugated polymers, however, they often need to be aligned in certain manners or patterned in the same way as for silicon or metals. Therefore, construction of oriented and patterned conjugated polymers is a key prerequisite for many applications involving conjugated polymers. This field is both technologically important and scientifically interesting. We have focused our attention here on the orientation and microlithographic fabrication of conjugated polymers in relation to electronic and photonic device applications. Even this brief survey has revealed the usefulness of chain orientation methods for producing materials with exotic physicochemical properties and advanced microfabrication techniques for making devices with good characteristics. Recent developments in the field have clearly indicated the possibility of producing a wide range of new materials and devices with novel features from existing conjugated polymers.

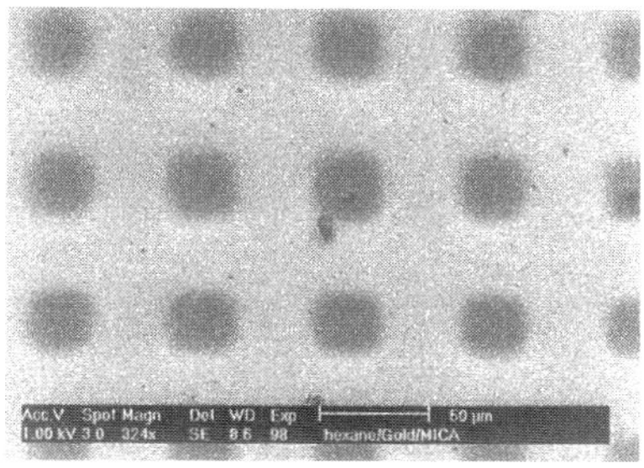

(a)

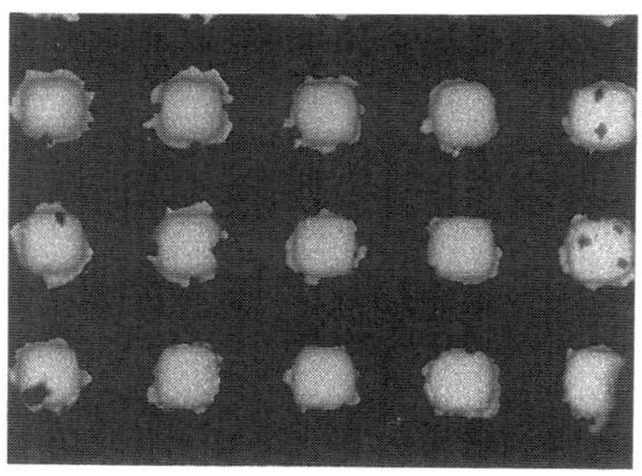

(b)

Figure 8. (a) A typical SEM micrograph of gold-coated mica sheets patterned by the n-hexane plasma polymer with a TEM grid consisting of square windows as the mask; (b) Optical microscopy image of a polypyrrole pattern electrochemically polymerized onto a platinum-coated mica surface pre-patterned by the n-hexane plasma polymer (Reproduced with permission from reference 54. Copyright 1997 The American Chemical Society).

Abbreviations

DC	direct current
EL	electroluminescent
ESR	electron spin resonance
FTIR	Fourier transform infrared
HCSA	camphorsulfonic acid
ITO	indium-tin oxide
LB	Langmuir-Blodgett
LC	liquid crystalline
LEDs	light-emitting diodes
NLO	nonlinear optical
P3DT	poly(3-decylthiophene)
P3HT	poly(3-hexylthiophene)
P3OT	poly(3-octylthiophene)
PI-PA	polyisoprene-polyacetylene
PIPN	pseudo interpenetrating polymer network
PMA	poly(methacrylic acid)
PPP	poly(*p*-phenylene)
PPV	poly(*p*-phenylenevinylene)
PTCDA	3,4,9,10-perylenetetracarboxylic dianhydride
SAM	self-assembled monolayer
SECM	scanning electrochemical microscopy
SPS	polystyrene sulfonic acid
SEM	scanning electron microscopy
STM	scanning tunnelling microscopy
TCNQ	tetracyanoquinodimethane
TEM	transmission electron microscopy
μCP	microcontact printing
UV/VIS/NIR	ultraviolet/visible/near infrared

Acknowledgments

The authors would like to thank their former and present colleagues. The financial supports from the *Austrian Science Fund*, Vienna, for B. W. (FWF Project J1426CHE) and the *Department of Industry, Science and Technology* (DIST), Australia, for S. H. are gratefully acknowledged.

References

1. *Handbook of Conducting Polymers*; Skotheim, T. A., Ed.; Marcel Dekker: NewYork, 1986.
2. *Molecule-Based Magnetic Materials: Theory, Techniques, and Applications*; Turnbull, M. M.; Sugimoto, T.; Thompson, L. K., Eds.; ACS Symp. Ser. 644; ACS: Washington, 1996.
3. Miller, J. S.; Epstein, A. J. *Angew. Chem. Int. Ed. Engl.* **1994**, *33*, 385.
4. Pattil, A. O.; Heeger, A. J.; Wudl, F. *Chem. Rev.* **1988**, *88*, 183.
5. *Nonlinear Optical Properties of Organic Materials and Crystals*; Zyss, J., Ed.; Academic Press: Orlando, 1987.
6. *Materials for Nonlinear Optics: Chemical Perspectives*; Stucky, D. G., Ed.; ACS Symp. Ser. 455, ACS: Washington, 1991.
7. *Electronic and Photonic Applications of Polymers*; Bowden, M. J.; Turner, S. R., Eds.; ACS: Washington, 1988.
8. Sariciftci, N. S. *Prog. Quant. Electr.* **1995**, *19*, 131.
9. Bao, Z.; Feng, Y.; Dodabalapur, A.; Raju, V. R.; Lovinger, A. J. *Chem. Mater.* **1997**, *9*, 1299.
10. Chien, J. C. W. *Polyacetylene: Chemistry, Physics and Materials Science*; Academic Press: New York, 1984.
11. Brédas, J. L.; Street, G. B. *Acc. Chem. Res.* **1985**, *18*, 309.
12. Shirakawa, H.; Louis, E. J.; MacDiarmid, A. G.; Chiang, C. K.; Heeger, A. J. *J. Chem. Soc., Chem. Commun.* **1977**, 578.
13. Epstein, A. J.; MacDiarmid, A. G. *Synth. Met.* **1995**, *69*, 179.
14. Avlyanov, J. K.; Min, Y.; MacDiarmid, A. G.; Epstein, A. J. *Phys. Rev. B* **1995**, *72*, 65.
15. de Gennes, P.-G. C. R. *Acad. Sci. Paris Ser. 2* **1985**, *302*, 1.
16. Heeger, A. J.; Smith, P. In *Conjugated Polymers*; Bredas, J. L., Silbey, R., Eds.; Kluwer Academic Publishers: Dordrecht, 1991, pp 141-210.
17. Dai, L. *Eur. Polym. J.* **1993**, *29*, 645.
18. Pearson, D. S.; Pincus, P. A.; Heffner, G. W.; Dahman, S. J. *Macromolecules* **1993**, *26*, 1570.
19. Burroughes, J. H.; Bradley, D. C. C.; Brown, A. R.; Marks, R. N.; Mackay, M. K.; Friend, R. H.; Burn, P. L.; Holmes, A. B. *Nature* **1990**, *347*, 539.
20. Greenham, N. C.; Friend, R. H. *Solid State Phys.* **1995**, *49*, 1.
21. Bradley, D. D. C.; Brown, A. R.; Burn, P. L.; Friend, R. H.; Holmes, A. B.; Kraft, A. In *Electronic Properties of Polymers*; Kuzmany, H.; Mehring, M., Eds.; Springer Ser. Solid-State Sci., 107; Springer: Heidelberg, 1992, pp 304-309.
22. *Nonlinear Optical Properties of Organic and Polymeric Materials*; Williams, D. J., Ed.; ACS: Washington, DC, 1983.
23. *Nonlinear Optical Effects in Organic Polymers*; Messier, J.; Kajzar, F.; Prasad, P.; Ulrich, D., Eds.; Kluwer Academic: Boston, 1989.
24. Dyreklev, P.; Berggren, M.; Inganäs, O.; Andersson, M. R.; Wennerström, O.; Hjertberg, T. *Adv. Mater.* **1995**, *7*, 43.

25. Lemmer, U.; Vacar, D.; Moses, D.; Heeger, A. J.; Ohnishi, T.; Noguchi, T. *Appl. Phys. Lett.* **1996**, *68*, 21.
26. Hamaguchi, M.; Yoshino, K. *Appl. Phys. Lett.* **1995**, *67*, 23.
27. Cimrová, V.; Remmers, M.; Neher, D.; Wegner, G. *Adv. Mater.* **1996**, *8*, 146.
28. Lüssem, G.; Festag, R.; Greiner, A.; Schmidt, C.; Unterlechner, C.; Heitz, W.; Wendorff, J. H.; Hopmeier, M.; Feldmann, J. *Adv. Mater.* **1995**, *7*, 923.
29. Lüssem, G.; Geffarth, F.; Greiner, A.; Heitz, W.; Hopmeier, M.; Oberski, M.; Unterlechner, C.; Wendorff, J.H. *Liq. Cryst.* **1996**, *21*, 903.
30. Schmid, W.; Dankesreiter, R.; Gmeiner, J.; Vogtmann, T.; Schwoerer, M. *Acta Polym.* **1993**, *44*, 208.
31. Renak, M. L.; Bazan, G. C.; Roitman, D. *Adv. Mater.* **1997**, *9*, 392.
32. Halliday, D. A.; Burn, P. L.; Bradley, D. D. C.; Friend, R. H.; Gelsen, O. M.; Holmes, A. B.; Kraft, A.; Martens, J. H. F.; Pichler, K. *Adv. Mater.* **1993**, *5*, 40.
33. Berggren, M.; Inganäs, O.; Gustafsson, G.; Rasmusson, J.; Andersson, M. R.; Hjertberg, T.; Wennerström, O. *Nature* **1994**, *372*, 444.
34. Nalwa, H. S. *Adv. Mater.* **1993**, *5*, 341.
35. *Organic Materials for Nonlinear Optics*; Hann, R.A.; Bloor, D., Eds.; Royal Society of Chemistry: London, 1989.
36. Arbogast, J. W.; Darmanyan, A. P.; Foote, C. S.; Rubin, Y.; Diederich, F. N.; Alvarez, M. M.; Anz, S. J.; Whetten, R. L. *J. Phys. Chem.* **1991**, *95*, 11.
37. Mukamel, S.; Takahashi, A.; Wang, H. X.; Chen, G. *Science* **1994**, *266*, 250.
38. Dagani, R. *C&E News* **1996**, *March 4th*, 22.
39. Prasad, P.N. In *Frontiers of Polymer Research*; Prasad, P.N.; Nigam, J.K., Eds.; Plenum Press: New York, 1991; pp 45-62.
40. Kajzar, F.; Le Moigne, J.; Thierry, A. In *Electronic Properties of Polymers*; Kuzmany, H.; Mehring, M., Eds.; Springer Ser. Solid-State Sci.; Springer: Heidelberg, 1992, pp 202-208.
41. Shirakawa, H.; Akagi, K.; Katayama, S.; Araya, K.; Mukoh, A.; Narahara, T. J. *Macromol. Sci., Chem.* **1988**, *A25*, 643.
42. Aldissi, M. *J. Polym. Sci., Part C: Polym. Lett.* **1989**, *27*, 105.
43. Martin,C. R. *Acc. Chem. Res.* **1995**, *259*, 957.
44. Cahalane, W.; Labes, M. M. *Chem. Mater.* **1989**, *1*, 519.
45. Martin, C. R. *Chem. Mater.* **1996**, *8*, 1739.
46. Shirakawa, H.; Ito, T.; Ikeda, S. *Polym. J.* **1973**, *4*, 460.
47. Joo, J.; Du, G.; Tsukamoto, J.; Epstein, A. J. *Synth. Met.* **1997**, *88*, 1.
48. Hagler, T. W.; Pakbaz, K.; Voss, K. F.; Heeger, A. J. *Phys. Rev. B* **1991**, *44*, 8652.
49. Ando, M.; Watanabe, Y.; Iyoda, T.; Honda, K.; Shimidzu, T. *Thin Solid Films* **1989**, *179*, 225.
50. Schanze, K. S.; Bergstedt, T. S.; Hauser, B. T. *Adv. Mater.* **1996**, *8*, 531.
51. Rubner, M. F.; Skotheim, T. A. In *Conjugated Polymers*; Brédas, J. L.; Silbey, R., Eds.; Kluwer Academic Publishers: Dordrecht, 1991, pp 363-403.
52. Rozsnyai, L. F.; Wrighton, M. S. *J. Am. Chem. Soc.* **1994**, *116*, 5993.
53. Rozsnyai, L. F.; Wrighton, M. S. *Langmuir* **1995**, *11*, 3913.
54. Dai, L.; Griesser, H. J.; Mau, A. W. H. *J. Phys. Chem. B* **1997**, *101*, 9548.

55. Ito, T.; Shirakawa, H.; Ikeda, S. *J. Polym. Sci. Polym. Chem.* **1974**, *12*, 11.
56. Naarmann, H.; Theophilou, N. *Synth. Met.* **1987**, *22*, 1.
57. Theophilou, N.; Aznar, R.; Munardi, A.; Sledz, J.; Schue, R.; Naarmann, H. *Synth. Met.* **1987**, *17*, 223.
58. Schimmel, T.; Denninger, G.; Riess, W.; Voit, J.; Schwoerer, M.; Schoepe, W.; Naarmann, H. *Synth. Met.* **1989**, *28*, D11.
59. Akagi, K.; Katayama, S.; Ito, M.; Shirakawa, H.; Araya, K. *Synth. Met.* **1989**, *28*, D51.
60. Araya, K.; Shirakawa, H. *Synth. Met.* **1986**, *14*, 199.
61. Shirakawa, H.; Akagi, K.; Suezaki, M. *Synth. Met.* **1989**, *28*, 1.
62. Aldissi, M. *J. Polym. Sci. B, Polym. Lett.* **1985**, *23*, 167.
63. Montaner, A.; Rolland, M.; Sauvajol, J. L.; Galtier, M.; Almairac, R.; Ribet, J. L. *Polymer* **1988**, *29*, 1101.
64. Montaner, A.; Rolland, M.; Sauvajol, J. L.; Meynadier, L.; Almairac, R.; Ribet, J. L.; Galtier, M.; Gril, C. *Synth. Met.* **1989**, *28*, D19.
65. Hsieh, B. R.; Yu, Y.; Forsythe, E. W.; Schaaf, G. M.; Feld, W. A. *J. Am. Chem. Soc.* **1998**, *120*, 231.
66. Dagani, R. *C&E News* **1998**, *January 19th*, 9.
67. Smith, R. C.; Fischer, W. M.; Gin, D. L. *J. Am. Chem. Soc.* **1997**, *119*, 4092.
68. Jacoby, M. *C&E News* **1997**, *May 19th*, 44.
69. Sasano, K.; Nakamura, K.; Kaneto, K. *Jpn. J. Appl. Phys. B* **1993**, *32*, L863.
70. Yang, R.; Evans, D. F.; Hendrickson, W. A. *Langmuir* **1995**, *11*, 211.
71. Kranz, C.; Gaub, H. E.; Schuhmann, W. *Adv. Mater.* **1996**, *8*, 634.
72. Kranz, C.; Ludwig, M.; Gaub, H. E.; Schuhmann, W. *Adv. Mater.* **1995**, *7*, 38; 568.
73. Wuu, Y.-M.; Fan, F. R. F.; Bard, A. J. *J. Electrochem. Soc.* **1989**, *136*, 885.
74. Nyffenegger, R. M.; Penner, R. M. *J. Phys. Chem.* **1996**, *100*, 17041.
75. Borgwarth, K.; Ricken, C.; Ebling, D. G.; Heinze, J. *Ber. dt. Bunsen. Ges. Phys. Chem.* **1995**, *99*, 1421.
76. Martin, C. R. *Science* **1994**, *266*, 1961.
77. Wu, C. G.; Bein, T. *Science* **1994**, *264*, 1757.
78. Wu, C. G.; Bein, T. *Science* **1994**, *266*, 1013.
79. Van Dyke, L. S.; Martin, C. R. *Langmuir* **1990**, *6*, 1123.
80. Cai, Z.; Martin, C. R. *J. Am. Chem. Soc.* **1989**, *111*, 4138.
81. Martin, C. R. *Adv. Mater.* **1991**, *3*, 457.
82. Martin, C. R.; Van Dyke, L. S.; Cai, Z.; Liang, W. *J. Am. Chem. Soc.* **1990**, *112*, 8976.
83. Cai, Z.; Lei, J.; Liang, W.; Menon, V.; Martin, C. R. *Chem. Mater.* **1991**, *3*, 960.
84. Martin, C. R.; Parthasarathy, R.; Menon, V. *Synth. Met.* **1993**, *55-57*, 1165.
85. Parthasarathy, R. V.; Martin, C. R. *Chem. Mater.* **1994**, *6*, 1627.
86. Parthasarathy, R. V.; Phani, K. L. N.; Martin, C. R. *Adv. Mater.* **1995**, *7*, 896.
87. Kyotani, T.; Nagai, T.; Inoue, S.; Tomita, A. *Chem. Mater.* **1997**, *9*, 609.
88. Iijima, S. *Nature* **1991**, *354*, 56.
89. Ebbesen, T. W.; Ajayan, P. M. *Nature* **1992**, *358*, 220.
90. *C&E News* **1992**, *July 20th*, 4.

91. Motojima, S.; Kawaguchi, M.; Nozaki, K.; Iwanaga, H. *Appl. Phys. Lett.* **1990**, *56*, 321.
92. Motojima, S.; Itoh, Y.; Asakura, S. *J. Mater. Sci.* **1995**, *30*, 5049.
93. Ihara, S.; Itoh, S. *Carbon* **1995**, *33*, 931.
94. Chico, L.; Crespi, V. H.; Louie, S. G.; Cohen, M. L. *Phys. Rev. Lett.* **1996**, *76*, 971.
95. Romero, D. B.; Carrard, M.; Heer, W. D.; Zuppiroli, L. *Adv. Mater.* **1996**, *8*, 899.
96. Terrones, M.; Grobert, N.; Olivares, J.; Zhang, J. P.; Terrones, H.; Kordatos, K.; Hsu, W. K.; Hare, J. P.; Townsend, P. D.; Prassides, K.; Cheetham, A. K.; Kroto, H. W.; Walton, D. R. M. *Nature* **1997**, *388*, 52.
97. Collins, P. G.; Zettl, A.; Bando, H.; Thess, A.; Smalley, R. E. *Science* **1997**, *278*, 100.
98. Gadd, G. E.; Blackford, M.; Moricco, S.; Webb, V.; Evans, P. J.; Smith, A. M.; Jacobsen, G.; Leung, S.; Day, A.; Hu, Q. *Science* **1997**, *277*, 933.
99. de Heer, W. A.; Bonard, J.-M.; Fauth, K.; Châtelain, A.; Forró, L.; Ugarte, D. *Adv. Mater.* **1997**, *9*, 87.
100. Dai, H.; Wong, E. W.; Lieber, C. M. *Science* **1996**, *272*, 523.
101. Ugarte, D.; Châtelain, A.; de Heer, W. A. *Science* **1996**, *274*, 1897.
102. Jirage, K. B.; Hulteen, J. C.; Martin, C. R. *Science* **1997**, *278*, 655.
103. Ajayan, P. M. *Adv. Mater.* **1995**, *7*, 489.
104. Tenne, R. *Adv. Mater.* **1995**, *7*, 965.
105. Terrones, M.; Hsu, W. K. *Chem. Soc. Rev.* **1995**, 341.
106. Li, W. Z.; Xie, S. S.; Qian, L. X.; Chang, B. H.; Zou, B. S.; Zhou, W. Y.; Zhao, R. A.; Wang, G. *Science* **1996**, *274*, 1701.
107. Kyotani, T.; Tsai, L.-F.; Tomita, A. *Chem. Mater.* **1996**, *8*, 2109.
108. Hoyer, P. *Adv. Mater.* **1996**, *8*, 857.
109. Ajayan, P. M.; Stephan, O.; Colliex, C.; Trauth, D. *Science* **1994**, *265*, 1211.
110. de Heer, W. A.; Bacsa, W. S.; Châtelain, A.; Gerfin, T.; Humphrey-Baker, R.; Forró, L.; Ugarte, D. *Science* **1995**, *268*, 845.
111. Collings, P. J. In *Kirk-Othmer Encyclopedia of Chemical Technology*; Vol. 15; 4th ed.; Wiley: New York, 1995; pp 372-409.
112. McArdle, C. B. *Side Chain Liquid Crystalline Polymers*; Blackie: Glasgow-London, 1989.
113. Cifferi, A; Kriegbaum, W. R.; Meyer, R. B. *Polymer Liquid Crystals*; Academic Press: New York, 1982.
114. Donald, M.; Windle, A. H. *Liquid Crystalline Polymers*; Cambridge University Press: Cambridge 1992.
115. *Liquid Crystals and Plastic Crystals*; Gray, G. W.; Winsor, P. A., Eds.; Ellis Horwood: Chichester, 1974.
116. *Advances in Liquid Crystals*; Brown, G. H., Ed.; Academic Press: New York, 1975.
117. Freemantle, M. *C&E News* **1996**, *December 16th*, 33.
118. Wiederrecht, G. P.; Yoon, B. A.; Wasielewski, M. R. *Adv. Mater.* **1996**, *8*, 535.
119. Zentel, R.; Galli, G.; Ober, C. K. (Eds.) *Macromol. Symp.* **1997**, *117*.
120. Gill, R. E.; Meetsma, A.; Hadziioannou, G. *Adv. Mater.* **1996**, *8*, 212.

121. Steiger, D.; Smith, P.; Weder, C. *Macromol. Rapid Commun.* **1997**, *18*, 643.
122. Babudri, F.; Cicco, S. R.; Farinola, G. M.; Naso, F.; Bolognesi, A.; Porzio, W. *Macromol. Rapid. Commun.* **1996**, *17*, 905.
123. Ringsdorf, H.; Schlarb, B.; Venzmer, J. *Angew. Chem. Int. Ed. Engl.* **1988**, *27*, 113.
124. Qian, R.; Chen, S.; Song, W.; Bi, X. *Macromol. Rapid Commun.* **1994**, *15*, 1.
125. Bolognesi, A.; Porzio, W.; Provasoli, F.; Ezquerra, T. *Makromol. Chem.* **1993**, *194*, 817.
126. Cao, Y.; Smith, P. *Polymer* **1993**, *34*, 3139.
127. Zheng, W.-Y.; Wang, R.-H.; Levon, K.; Rong, Z. Y.; Taka, T.; Pan, W. *Macromol. Chem. Phys.* **1995**, *196*, 2443.
128. Dai, L. *Chin. J. Mater. Res. (Eng.)* **1995**, *9*, 397.
129. Akagi, K.; Goto, H.; Kadokura, Y.; Shirakawa, H.; Oh, S.-Y.; Araya, K. *Synth. Met.* **1995**, *69*, 13.
130. Goto, H.; Akagi, K.; Shirakawa, H.; Oh, S.-Y.; Araya, K. *Synth. Met.* **1995**, *71*, 1899.
131. Akagi, K.; Goto, H.; Shirakawa, H. *Synth. Met.* **1997**, *84*, 313.
132. Oh, S.-Y.; Ezaki, R.; Akagi, K.; Shirakawa, H. *J. Polym. Sci., Polym. Chem.* **1993**, *31*, 2977.
133. Oh, S.-Y.; Akagi, K.; Shirakawa, H.; Araya, K. *Macromolecules* **1993**, *26*, 6203.
134. Akagi, K.; Shirakawa, H. *Macromol. Symp.* **1996**, *104*, 137.
135. Yoshino, K.; Kobayashi, K.; Myojin, K.; Ozaki, M.; Akagi, K.; Goto, H.; Shirakawa, H. *Jpn. J. Appl. Phys.* **1996**, *35*, 3964.
136. Lino, K.; Goto, H.; Akagi, K.; Shirakawa, H.; Kawaguchi, A. *Synth. Met.* **1997**, *84*, 967.
137. Koltzenburg, S.; Winkler, B.; Stelzer, F.; Nuyken, O. *Designed Monomers and Polymers* **1998**, *1*, 207.
138. Moigne, J. L.; Hilberer, A.; Kajzar, F. *Makromol. Chem.* **1991**, *192*, 515.
139. Jin, S.-H.; Kim, S.-H.; Cho, H.-N.; Choi, S.-K. *Macromolecules* **1991**, *24*, 6050.
140. Jin, S.-H.; Choi, S.-J.; Ahn, W.; Cho, H.-N.; Choi, S.-K. *Macromolecules* **1993**, *26*, 1487.
141. Choi, S.-J.; Cho, H.-N.; Choi, S.-K. *Polym. Bull. (Berlin)* **1994**, *32*, 179.
142. Nierengarten, J.-F.; Guillon, D.; Heinrich, B.; Nicoud, J.-F. *J. Chem. Soc., Chem. Commun.* **1997**, 1233.
143. Gabaston, L. I.; Foot, P. J. S.; Brown, J.W. *J. Chem. Soc., Chem. Commun.* **1996**, 429.
144. Ibison, P.; Foot, P. J. S.; Brown, J. W. *Synth. Met.* **1996**, *76*, 297.
145. Hasegawa, H.; Kijima, M.; Shirakawa, H. *Synth. Met.* **1997**, *84*, 177.
146. Vicentini, F.; Barrouillet, J.; Laversanne, R.; Mauzac, M.; Bibonne, F.; Parneix, J. P. *Liq. Cryst.* **1995**, *19*, 235.
147. Goto, H.; Akagi, K.; Shirakawa, H. *Synth. Met.* **1997**, *84*, 385.
148. Kijima, M.; Akagi, K.; Shirakawa, H. *Synth. Met.* **1997**, *84*, 237.
149. Winkler, B.; Tasch, S.; Zojer, E.; Ungerank, M.; Leising, G.; Stelzer, F. *Synth. Met.* **1996**, *83*, 177.

150. Winkler, B. *Ph.D.-Thesis*, Technical University of Graz (Austria), 1997.
151. Ungerank, M.; Winkler, B.; Eder, E.; Stelzer, F. *Macromol. Chem. Phys.* **1995**, *196*, 3623. *Idem. ibid.* **1997**, *198*, 1391.
152. Winkler, B.; Ungerank, M.; Stelzer, F. *Macromol. Chem. Phys.* **1996**, *197*, 2343.
153. Winkler, B.; Rehab, A.; Ungerank, M.; Stelzer, F. *Macromol. Chem. Phys.* **1997**, *198*, 1417.
154. Katz, H. E.; Dirk, C. W.; Singer, K. D.; Sohn, J. E. *Mol. Cryst. Liq. Cryst.* **1988**, *157*, 525.
155. Wittmann, J. C.; Smith, P. *Nature* **1991**, *352*, 414.
156. Pooley, C.M.; Tabor, D. *Proc. R. Soc. Lond.* **1972**, *A329*, 251.
157. Gill, R. E.; Hadziioannou, G.; Lang, P.; Garnier, F.; Wittmann, J. C. *Adv. Mater.* **1997**, *9*, 331.
158. Grell, M.; Bradley, D. D. C.; Inbasekaran, M.; Woo, Ed P. *Adv. Mater.* **1997**, *9*, 798.
159. Leising, G. *Polym. Commun.* **1984**, *25*, 201.
160. Leising, G. *Polym. Bull. (Berlin)* **1984**, *11*, 401.
161. Naarmann, H. *Springer Ser. Solid State Sci.* **1986**, *76*, 12.
162. Satoh, M.; Imanishi, K.; Yasuda, Y.; Tsushima, R.; Aoki, S.; Yoshino, K. *Jpn. J. Appl. Phys.* **1985**, *24*, 1423.
163. Ogasawara, M.; Funahashi, K.; Demura, T.; Hagiwara, T.; Iwata, K. *Synth. Met.* **1986**, *14*, 61.
164. Hotta, S.; Soga, M.; Sonoda, N. *J. Phys. Chem.* **1989**, *93*. 4994.
165. Yoshino, K.; Nakao, K.; Onoda, M. *Jpn. J. Appl. Phys.* **1989**, *28*, L1032.
166. Cromack, K. R.; Józefowicz, M. E.; Ginder, J. M.; Epstein, A. J.; McCall, R. P.; Du, G.; Leng, J. M.; Kim, K.; Li, C.; Wang, Z. H. *Macromoleculs* **1991**, *24*, 4157.
167. Gagnon, D.; Karasz, F. E.; Thomas, E. L.; Lenz, R. W. *Synth. Met.* **1987**, *20*, 85.
168. Townsend, P. D.; Pereira, C. M.; Bradley, D. D. C.; Horton, M. E.; Friend, R. H. *J. Phys. C: Solid State Phys.* **1985**, *18*, L283.
169. Karasz, F. E.; Capistran, J. D.; Gagnon, D. R. *Mol. Cryst. Liq. Cryst.* **1985**, *118*, 327.
170. Andreatta, A.; Tokito, S.; Smith. P.; Heeger, A. J *Mol. Cryst. Liq. Cryst.* **1990**, *189*, 169.
171. Tokito, S.; Smith, P.; Heeger, A. J. *Polymer* **1991**, *32*, 464.
172. Swager, T. M.; Grubbs, R. *Synth. Met.* **1989**, *28*, D57.
173. Weder, C.; Sarwa, C.; Montali, A.; Bastiaansen, C.; Smith, P. *Science* **1998**, *279*, 835.
174. Yoshino, K.; Onoda, M.; Sugimoto, R. *Jpn. J. Appl. Phys.* **1988**, *27*, L2034.
175. Dyreklev, P.; Gustafsson, G.; Inganäs, O.; Stubb, H. *Solid State Commun.* **1992**, *82*, 317.
176. Theophilou, N.; Manohar, S. *US Patent No. 4935181*, 1993.
177. Basescu, N.; Liu, Z.-X.; Moses, D.; Heeger, A. J.; Naarmann, H.; Theophilou, N. *Nature* **1987**, *327*, 403.

178. Dyreklev, P.; Berggren, M.; Inganäs, O.; Andersson, M. R.; Wennerström, O.; Hjertberg, T. *Adv. Mater.* **1995**, *7*, 43.
179. Dagani, R. *C&E News* **1995**, *January 23th*, 28.
180. Halperin, A.; Tirrell, M.; Lodge, T.P. *Adv. Polym. Sci.* **1992**, *100*, 31.
181. Dai, L.; Toprakcioglu, C. *Macromolecules* **1992**, *25*, 6000.
182. Dai, L.; Toprakcioglu, C. *Europhys. Lett.* **1991**, *16*, 331.
183. Dai, L.; Toprakcioglu, C.; Hadziioannou, G. *Macromolecules* **1995**, *28*, 5512.
184. Field, J. B.; Toprakcioglu, C.; Stanley, H. B.; Dai, L.; Barford, W.; Penfold, J.; Smith, G.; Hamilton, W. *Macromolecules* **1992**, *25*, 434.
185. Bumm, L. A.; Arnold, J. J.; Cygan, M. T.; Dunbar, T. D.; Burgin, T. P.; Jones II, L.; Allara, D. L.; Tour, J. M.; Weiss, P. S. *Science* **1996**, *271*, 1705.
186. Dagani, R. *C&E News* **1996**, *March 25th*, 7.
187. Dai, L.; White, J. W. *Eur. Polym. J.* **1994**, *30*, 1443.
188. MacDiarmid, A. G.; Epstein, A. J. *Synth. Met.* **1994**, *65*, 103.
189. MacDiarmid, A. G.; Epstein, A. J. *Synth. Met.* **1995**, *69*, 85.
190. Dai, L.; Mau, A.; Griesser, H.; Spurling, T.; White, J. W. *J. Phys. Chem.* **1995**, 99, 17302.
191. Dai, L.; Mau, A. W. H. *Synth. Met.* **1997**, *86*, 2277.
192. Dai, L.; Mau, A. W. H.; Zhang, X. *J. Mater. Chem.* **1998**, *8*, 325.
193. Lu, J.; Dai, L.; Mau, W. H. A. *Acta Polym.* (in press).
194. Dai, L.; Lu, J.; Matthews, B.; Mau, A. W. H. *J. Phys. Chem.* **1998**, *102*, 4049; and references cited therein.
195. Dai, L.; Huang, S.; Lu, J.; Mau, A. W. H.; Zhang, F. *Polym. Prep.* **1998**, *39*, 171.
196. Schwiegk, S.; Vahlenkamp, T.; Xu, Y.; Wegner, G. *Macromolecules* **1992**, *25*, 2513.
197. Aoki, A.; Miyashita, T. *Adv. Mater.* **1997**, *9*, 361.
198. Suzuki, M.; Ferencz, A.; Iida, S.; Enkelmann, V.; Wegner, G. *Adv. Mater.* **1993**, *5*, 359.
199. Sagisaka, S.; Ando, M., Iyoda, T.; Shimidzu, T. *Thin Solid Films* **1993**, *230*, 65.
200. Cheung, J. H.; Punkka, E.; Rikukawa, M.; Rosner, R. B.; Royappa, A. T.; Rubner, M. F. *Thin Solid Films* **1992**, *210/211*, 246.
201. Shimidzu, T.; Iyoda, T.; Ando, M.; Ohtani, A.; Kaneko, T.; Honda, K. *Thin Solid Films* **1988**, *160*, 67.
202. Iyoda, T.; Ando, M.; Kaneko, T.; Ohtani, A.; Shimidzu, T.; Honda, K. *Langmuir* **1987**, *3*, 1169.
203. Hong, K.; Rosner, R. B.; Rubner, M. F. *Chem. Mater.* **1990**, *2*, 82.
204. Watanabe, I.; Hong, K.; Rubner, M. F. *J. Chem. Soc., Chem. Commun.* **1989**, 123.
205. Punkka, E.; Rubner, M. F. *Thin Solid Films* **1992**, *213*, 117.
206. Era, M.; Shinozaki, H.; Tokito, S.; Tsutsui, T.; Saito, S. *Chem. Lett.* **1988**, 1097.
207. Nishikata, Y.; Kakimoto, M.-A.; Imai, Y. *Thin Solid Films* **1989**, *179*, 191.
208. Kamiyama, K.; Era, M.; Tsutsui, T.; Saito, S. *Jpn. J. Appl. Phys.* **1990**, *29*, L840.

209. Vahlenkamp, T.; Wegner, G. *Macromol. Chem. Phys.* **1994**, *195*, 1933.
210. Remmers, M.; Schulze, M.; Wegner, G. *Macromol. Rapid Commun.* **1996**, *17*, 239.
211. Bolognesi, A.; Bajo, G.; Paloheimo, J.; Östergard, T.; Stubb, H. *Adv. Mater.* **1997**, *9*, 121.
212. Grüner, J.; Remmers, M.; Neher, D. *Adv. Mater.* **1997**, *9*, 964.
213. Ferreira, M.; Rubner, M. F. *Macromolecules* **1995**, *28*, 7107.
214. Cheung, J. H.; Stockton, W. B.; Rubner, M. F. *Macromolecules* **1997**, *30*, 2712.
215. Fou, A. C.; Onitsuka, O.; Ferreira, M.; Rubner, M. F.; Hsieh, B. R. *J. Appl. Phys.* **1995**, *79*, 7501.
216. Onitsuka, O.; Fou, A. C.; Ferreira, M.; Hsieh, B. R.; Rubner, M. F. *J. Appl. Phys.* **1996**, *80*, 4067.
217. Ferreira, M.; Onitsuka, O.; Fou, A. F.; Hsieh, B. R.; Rubner, M. F. *Mat. Res. Soc. Symp. Proc.* **1996**, *413*, 49.
218. Stockton, W. B.; Rubner, M. F. *Macromolecules* **1997**, *30*, 2717.
219. Clarke, T. C.; Krounbi, M. T.; Lee, V. Y.; Street, G. B. *J. Chem. Soc., Chem. Commun.* **1981**, 384.
220. Abdou, M. S. A.; Diaz-Guijada, G. A. Arroya, M. I.; Holdcroft, S. *Chem. Mater.* **1991**, *3*, 1003.
221. Arroyo-Villan, M. I.; Diaz-Quijada, G. A.; Abdou, M. S. A.; Holdcroft, S. *Macromolecules* **1995**, *28*, 975.
222. Abdou, M. S. A.; Xie, Z.; Leung, A.; Holdcroft, S. *Synth. Met.* **1992**, *52*, 159.
223. Lowe, J.; Holdcroft, S. *Macromolecules* **1995**, *28*, 4608.
224. Dao, L. H.; Nguyen, M. T.; Do, T. N. *Polym. Prepr.* **1992**, *33*, 408.
225. Cai, S. X.; Kanskar, M.; Nabity, J.; Keana, J. F. W.; Wybourne, M. N. *J. Vac. Sci. Technol. B* **1992**, *10*, 2589.
226. Angelopoulos, M.; Shaw, J. M.; Lee, K.-L.; Huang, W.-S.; Lecorre, M.-A.; Tissier, M. *Polym. Eng. Sci.* **1992**, *32*, 1535.
227. Venugopal, G.; Quan, X.; Johnson, G. E.; Houlihan, F. M.; Chin, E.; Nalamasu, O. *Chem. Mater.* **1995**, *7*, 271.
228. Dai, L.; Mau, A. W. H.; Qian, R. *Chin. Polym. Bull.* **1995**, *1*, 1.
229. Bargon, J.; Behnck, W.; Weidenbruck, T.; Ueno, T. *Synth. Met.* **1991**, *41-43*, 1111.
230. Baumann, R.; Lennarz, K.; Bargon, J. *Synth. Met.* **1993**, *54*, 243.
231. Thakur, M. *Macromolecules* **1988**, *21*, 661.
232. See comments in: *Nature* **1988**, *333*, 296. *New Sci.* **1988**, *July 28th*, 39. *C&E News* **1990**, *May 7th*, 53. *Sci. Am.* **1988**, *Aug.*, 12.
233. Dai, L.; White, J. W. *Polymer* **1991**, *32*, 2120.
234. Dai, L. *J. Phys. Chem.* **1992**, *96*, 6469.
235. Dai, L.; Mau, A. W. H. *Synth. Met.* **1997**, *86*, 1893.
236. Dai, L.; Mau, A. W. H.; Griesser, H. J.; Winkler, D. A. *Macromolecules* **1994**, *27*, 6728.
237. Dai, L.; White, J. W. In *Polymers and Organic Solids*; Shi, L.; Zhu, D., Eds.; Science Press: Beijing, 1997, pp 287-321.

238. Dai, L.; Mau, A. W. H.; Griesser, H. J.; Winkler, D.; Spurling, T. H.; Hong, X.; Yang, Y.; White, J. W. *Synth. Met.* **1995**, *69*, 563.
239. Dai, L.; Griesser, H. J.; Hong, X.; Mau, A. W. H.; Spurling, T. H.; Yang, Y.; White, J. W. *Macromolecules* **1996**, *29*, 282.
240. Lidzey, D. G.; Pate, M. A.; Weaver, M. S.; Fisher, T. A.; Bradley, D. D. C. *Synth. Met*. **1996**, *82*, 141.
241. Wei, P. K.; Hsu, J. H.; Hsieh, B. R.; Fann, W. S. *Adv. Mater*. **1996**, *8*, 573.
242. DeAro, J. A.; Weston, K. D.; Buratto, S. K.; Lemmer, U. *Chem. Lett*. **1997**, *277*, 532.
243. Cho, I.; Kim, J.; Lee, S. *Macromol. Rapid Commun*. **1995**, *16*, 851.
244. *Lambda Highlights* **1996**, *42*, 4.
245. Yang, Y.; Bharatham, J. *Polym. Prep.* **1998**, *39(1)*, 98.
246. See comment in: *Science* **1998**, *279*, 1135.
247. Noach, S.; Faraggi, E. Z.; Cohen, G.; Avny, Y.; Neumann, R.; Davidov, D.; Lewis, A. *Appl. Phys. Lett*. **1996**, *69*, 3650.
248. Ulman, A. *Chem. Rev*. **1996**, *96*, 1533.
249. Terfort, A.; Bowden, N.; Whitesides, G. M. *Science* **1992**, *257*, 1380.
250. Kumar, A.; Whitesides, G. M. *Science* **1994**, *263*, 60.
251. Lopez, G. P.; Biebuyck, H. A.; Whitesides, G. M. *Langmuir* **1993**, *9*, 1513
252. Akari, S.; Horn, D.; Keller, H.; Schrepp, W. *Adv. Mater*. **1995**, *7*, 549.
253. Frisbie, C. D.; Rozsnyai, L. F.; Noy, A.; Wrighton, M. S.; Lieber, C. M. *Science* **1994**, *265*, 2071.
254. Schierbaum, K. D.; Weiss, T.; Thoden van Velzen, E. U.; Engbersen, J. F. J.; Reinhoudt, D. N.; Gopel, W. *Science* **1994**, *265*, 1413.
255. Dulcey, C. S.; Georger, J. H.; Kleinfeld, V.; Stenger, D. A.; Fare, T. L.; Calvert, J. M. *Science* **1991**, *252*, 551.
256. Wrighton, M. S. *Langmuir* **1993**, *9*, 1517
257. Dulcey, C. S.; Georger, J. H.; Chen, M.-S.; McElvany, S. W.; O'Ferrall, C. E.; Benezra, V. I.; Calvert, J. M. *Langmuir* **1996**, *12*, 1638.
258. Ross, C. B.; Sun, L.; Crooks, R. M. *Langmuir* **1993**, *9*, 632.
259. Kumar, A.; Whitesides, G. M. *Appl. Phys. Lett*. **1993**, *63*, 2002.
260. Jacoby, M. *C&E News* **1997**, *October 6th*, 34.
261. Kumar, A.; Biebuyck, H. A.; Whitesides, G. M. *Langmuir* **1994**, *10*, 1498.
262. Xu, S.; Liu, G.-Y. *Langmuir* **1997**, *13*, 127.
263. Vargo, T. G.; Thompson, P. M.; Gerenser, L. J.; Valentini, R. F.; Aebischer, P.; Hook, D. J.; Gardella, J. A. *Langmuir* **1992**, *8*, 130.
264. Terfort, A.; Bowden, N.; Whitesides, G. M. *Nature* **1997**, *386*, 162.
265. Zhao, X. M.; Xia, Y.; Whitesides, G. M. *Adv. Mater*. **1996**, *8*, 837.
266. Jeon, N. L.; Clem, P.; Jung, D. Y.; Lin, W.; Girolami, G. S.; Payne, D. A.; Nuzzo, R. G. *Adv. Mater*. **1997**, *9*, 891.
267. Ferreira, M.; Cheung, J, H.; Rubner, M. F. *Thin Solid Films* **1994**, *244*, 806.
268. Michalitsch, R.; Lang, P.; Yassar, A.; Nauer, G.; Garnier, F. *Adv. Mater.* **1997**, *9*, 321.
269. Kim, T.; Ye, Q.; Sun, L.; Chan, K. C.; Crooks, R. M. *Langmuir* **1996**, *12*, 6065.

270. Sukwattanasinitt, M.; Wang, X.; Li, L.; Jiang, X.; Kumar, J.; Tripathy, S. K.; Sandman, D. J. *Chem. Mater.* **1998**, *10*, 27.
271. Xia, Y.; Qin, D.; Whitesides, G. M. *Adv. Mater.* **1996**, *8*, 1015.
272. MacDiarmid, A. G. *Synth. Met.* **1997**, *84*, 27.
273. MacDiarmid, A. G.; Epstein, A. J. In *Photonic and Optoelectronic Polymers*; Jenekhe, S. A.; Wynne, K. J., Eds.; ACS Symp. Ser. 672, ACS: Washington 1997. pp 395-407.
274. Huang, Z.; Wang, P-C.; MacDiarmid, A. G. *Langmuir* **1997**, *13*, 6480.
275. Harlev, E.; Gulakhmedova, T.; Rubinovich, I.; Aizenshtein, G. *Adv. Mater.* **1996**, *8*, 994.
276. Gupta, V. K.; Abbott, N. L. *Science* **1997**, *276*, 1533.
277. Evans, S. D.; Allinson, H.; Boden, N.; Flynn, T. M.; Henderson, J. R. *J. Phys. Chem. B* **1997**, *101*, 2143.
278. Fredrickson, G. H.; Bates, F. S. *Annu. Rev. Mater Sci.* **1996**, *26*, 503
279. Park, M.; Harrison, C.; Chaikin, P. M.; Register, R. A.; Adamson, D. H. *Science* **1997**, *276*, 1401.
280. de Gennes, P. G. *Scaling Concepts in Polymer Physics*; Cornell University Press: Ithaca, 1979.
281. Bates, F. S. *Science* **1991**, *251*, 898.
282. Lodge, T. P.; Muthukumar, M. *J. Phys. Chem.* **1996**, *100*, 13275.
283. Flory, P. J. *Principles of Polymer Chemistry*; Cornell University Press: Ithaca, NY, 1953.
284. Leibler, L. *Macromolecules* **1980**, *13*, 1602.
285. Helfand, E.; Wasserman, Z. R. In *Developments in Block and Graft Copolymers - 1*; Goodman, L., Ed.; Applied Science: New York, 1982.
286. Bates, F. S.; Fredrickson, G. H. *Annu. Rev. Phys. Chem.* **1990**, *41*, 525.
287. Spatz, J. P.; Sheiko, S.; Möller, M. *Adv. Mater.* **1996**, *8*, 513.
288. Li, Z.; Zhao, W.; Liu, Y.; Rafailovich, M. H.; Sokolov, J. *J. Am. Chem. Soc.* **1996**, *118*, 10892.
289. Li, Z.; Qu, S.; Rafailovich, M. H.; Sokolov, J.; Tolan, M.; Turner, M. S.; Wang, J.; Schwarz, S. A.; Lorenz, H.; Kotthaus, J. P. *Macromolecules* **1997**, *30*, 8410.
290. Fasolka, M. J.; Harris, D. J.; Mayes, A. M. *Phys. Rev. Lett.* **1997**, *79*, 3018.
291. Böltau, M.; Walheim, S.; Mlynek, J.; Krausch, G.; Steiner, U. *Nature* **1998**, *391*, 877.
292. Jandt, K. D.; Dai, C.-A.; Kramer, E. J. *Adv. Mater.* **1996**, *8*, 660.
293. Zhulina, E. B.; Singh, C.; Balazs, A. C. *Macromolecules* **1996**, *29*, 6338.
294. Pickett, G. T.; Balazs, A. C. *Macromolecules* **1997**, *30*, 3097.
295. Birshtein, T. M.; Amoskov, V. M.; Mercurieva, A. A.; Pryamitsyn, V. A. *Macromol. Symp.* **1997**, *113*, 151.
296. Böhmer, M. *Langmuir* **1996**, *12*, 5747.
297. Koneripalli, N.; Levicky, R.; Bates, F. S. *Langmuir* **1996**, *12*, 6681.
298. Jo, W. H.; Nam, H. K.; Cho, J. C. *J. Polym. Sci.: Part B, Polym. Phys.* **1996**, *34*, 2169.
299. Galvin, M. E.; Wnek, G. E. *Polym. Bull. (Berlin)* **1985**, *13*, 109.
300. Bates, F.S.; Baker, G.L. *Macromolecules* **1983**, *16*, 704.

301. Aldissi, M. *J. Chem. Soc., Chem. Commun.* **1984**, 1347.
302. Armes, S.; Vincent, B.; White, J. W. *J. Chem. Soc., Chem. Commun.* **1986**, 1525.
303. Dai, L. *Macromol. Chem. Phys.* **1997**, *189*, 1723.
304. Dai, L.; White, J. W. *J. Polym. Sci., Part B, Polym. Phys.* **1993**, *31*, 3.
305. Dai, L.; White, J. W.; Kerr, J.; Thomas, R. K.; Penfold, J.; Aldissi, M. *Synth. Met.* **1989**, *28*, 69.
306. Dai, L. *Synth. Met.* **1997**, *84*, 957.
307. Dai, L.; White, J. W. *Polymer* **1997**, *38*, 775.
308. Hamnett, A.; Kerr, J. C. H.; White, J. W.; Dai, L. *J. Chem. Soc., Faraday Trans.* **1993**, *89*, 277.
309. Tang, C. W.; VanSlyke, S. A. *Appl. Phys. Lett.* **1987**, *51*, 913;
310. Väterlein, C.; Ziegler, B.; Gebauer, W.; Neureiter, H.; Sokolowski, M.; Bäuerle, P.; Weaver, M. S.; Bradley, D. D. C.; Umbach, E. *Synth. Met.* **1996**, *76*, 133.
311. Weaver, M. S.; Bradley, D. D. C. *Synth. Met.* **1996**, *83*, 61.
312. Meghdadi, F.; Tasch, S.; Winkler, B.; Fischer, W.; Stelzer, F.; Leising, G. *Synth. Met.* **1997**, *85*, 1441.
313. Winkler, B.; Meghdadi, F.; Tasch, S.; Müllner, R.; Resel, R.; Saf, R.; Leising, G.; Stelzer, F. *Opt. Mater.* **1998**, *9*, 159.
314. d'Agostino, R. *Plasma Deposition, Treatment, and Etching of Polymers*; Academic Press: San Diego, 1990.
315. *Techniques & Applications of Plasma Chemistry*; Hollan, J. R.; Bell, A.T., Eds.; Wiley: New York, 1974.
316. Yasuda, H. *Plasma Polymerization*; Academic Press: Orlando, 1978.
317. Dai, L.; Zientek, P.; StJohn, H.; Pasic, P.; Chatelier, R.; Griesser, H. J. In *Surface Modification of Polymeric Biomaterials*; Ratner, B. D.; Castner, D., Eds.; Plenum Press: New York, 1996; pp 147-156.
318. Dai, L.; StJohn, H.; Bi, J.; Zientek, P.; Chatelier, R.; Griesser, H. *Surf. Interf. Anal.* (in press).
319. Griesser, H.; Gengenbach, T.; Dai, L.; Li, S.; Chatelier, R. *J. Adhesion Sci. Tech.* (in press).
320. Chatelier, R.; Dai, L.; Griesser, H. J.; Li, S.; Zientek, P.; Lohmann, D.; Chabrecek, P. *PCT Int. Appl. WO 94 06, 485*, 1994.
321. Liang, W. B.; Masse, M. A.; Karasz, F. E. *Polymer* **1992**, *35*, 3101.
322. Bhuiyan, A. H.; Bhoraskar, S. V. *J. Mater. Sci.* **1989**, *24*, 3091.
323. Kawakami, S. *JP 01165603*, 1987.
324. Tanaka, K.; Yamabe, T.; Takeuchi, T.; Yoshizawa, K.; Nishio, S. *J. Appl. Phys.* **1991**, *70*, 5653.
325. Nishio, S.; Takeuchi, T.; Matsuura, Y.; Yoshizawa, K.; Tanaka, K.; Yamabe, T. *Synth. Met.* **1992**, *46*, 243.
326. Tanaka, K.; Nishio, S.; Matsuura, Y.; Yamabe, T. *J. Appl. Phys.* **1993**, *73*, 5017.
327. Xie, X.; Thiele, J.-U.; Steiner, R.; Oelhafen, R. *Synth. Met.* **1994**, *63*, 221.
328. Hernandez, H.; Diaz, A. F.; Waltman, R.; Bargon, J. *J. Phys. Chem.* **1984**, *88*, 3333.

329. Fally, F.; Riga, J.; Verbist, J. J. In *Polymer-Solid Interface*; Pireaux, J. J.; Bertrand P.; Bredas, J. L., Eds.; IOP Publ. Ltd: London, 1992, pp 455-462.
330. Augestine, M.; Sebastian, M.; Samkumar, M. K. A.; Joseph, M. J.; Joseph, T.; Sakthikumar, D.; Jayalakshmi, S.; Rasheed, T. M. A.; Anantharaman, M. R. *Indian J. Pure Appl. Phys.* **1996**, *34*, 966.
331. Samal, S.; Mohanty, B. C.; Nayak, B. B. *Polym. Sci. (India)* **1994**, *1*, 222.
332. Gong, X.; Dai, L.; Mau, A. W. H.; Griesser, H. J. *J. Polym. Sci., Part A, Polym. Chem.* **1998**, *36*, 633.
333. Kang, E. T.; Kato, K.; Uyama, Y.; Ikada, Y. *J. Mater. Res.* **1996**, *11*, 1570.
334. Nguyen, T. P.; Le Rendu, P.; Amgaad, K.; Cailler, M.; Tran, V. H. *Synth. Met.* *1995*, **72**, 35.
335. Xu, X.; Luo, X.; Xie, Y.; Zhou, H. *Proc. Int. Workshop Electroluminescence (EL'94)*; Science Press: Beijing, 1995.
336. Sun, R.; Peng, J.; Kobayashi, T.; Ma, Y.; Zhang, H.; Liu, S. *Jpn. J. Appl. Phys.* **1996**, *35*, L1506.
337. Dai, L.; Mau, A. W. H.; Gong, X.; Griesser, H. J. *Synth. Met.* **1997**, *85*. 1379.
338. Vargo, T. G.; Calvert, J. M.; Wynne, K. J.; Avlyanov, J. K.; MacDiarmid, A. G.; Rubner, M. F. *Supramol. Sci.* **1995**, *2*, 169.

Chapter 21

Tuning the Extent of Conjugation in Processable Polythiophenes Through Control of Side Chain Density and Regioregularity

Seth C. Rasmussen, Bennett D. Straw, and James E. Hutchison [1]

Department of Chemistry and Materials Science Institute, 1253 University of Oregon, Eugene, OR 97403–1253

Introduction of alkane side chains is a powerful approach to promoting the solubility of conjugated organic polymers. However, the unintended effects of side chains upon the properties of the polymers can not be ignored. In the case of polythiophenes, steric interactions involving the side chains can introduce tortional strain between thiophene rings that disrupt π-orbital overlap along the polymer backbone, thus decreasing the degree of conjugation in the chain. Eliminating or minimizing the steric interactions that distort the planarity of the polymer backbone can increase the extent of conjugation, making these materials more desirable for a number of applications. Here, a short review of the general principles employed to increase conjugation lengths in n-alkyl side chain substituted polythiophenes is followed by a discussion of specific examples that illustrate these principles. Because the reduction of side chain density is a powerful approach to increasing the conjugation length, particular attention is paid to substituted polythiophenes resulting from polymerization of partially substituted monomers (e.g. monoalkylbithiophenes, mono- and di-alkylterthiophenes and dialkyl-quaterthiophenes). Regioregular polymers with low side chain density exhibit an extent of conjugation that approaches that predicted from model studies for an ideal infinite polythiophene chain.

Conjugated organic polymers have been studied extensively over the last twenty years, in part, because of the fundamental and technological interest in their optical and electronic properties. Applications that have been demonstrated for these materials include their use as batteries, sensors, electrochromic devices, light-emitting diodes, and field effect transistors.(*1-9*) A unique advantage of conjugated polymer materials for these applications is that the properties of the polymer can be

[1] Corresponding author. E-mail: hutch@oregon.uoregon.edu.

tuned at the molecular level. Polythiophenes, in particular, are versatile materials because of their environmental stability(*3,10*) and the ease with which they can be synthetically modified.(*4-8*)

A key step in the development of polythiophenes as a class of useful conjugated materials was the preparation of soluble derivatives. The unsubstituted parent polymer is insoluble, however in 1986 it was found that substitution of long ($n > 4$) alkyl chains in the β-position of the thiophene rings resulted in polymers that were both soluble and fusible.(*5*) Soluble samples can be easily processed and more thoroughly characterized. Through the investigation of a large number of soluble polythiophene derivatives, the relationships between polymer structure and properties are starting to emerge.(*6,8,11*)

In order to rationally design conjugated polymers with tunable, well-defined electrical and optical properties, the influence of the side chains (number, position, and type) on the properties must be understood.(*5-7*) For polythiophene, tuning is typically accomplished through variation of side chain number and structure.(*4-7,11,12*) Judicious selection of alkane side chains can result in enhanced polymer characteristics including: enhanced solubility,(*5-7,12-14*) solid-state organization through chain self alignment,(*14-16*) and improved optical and electronic properties.(*8,17*)

In addition to inert hydrocarbon chains, functionalized chains can be used to promote specific interactions with ions or molecules.(*5-7,12,16,18,19*) Ionizable side chains (e.g. sulfonates or carboxylates) provide internal charge compensation(*20*) (or self-doping)(*4,5,21-23*) as well as impart water solubility, a desirable property for technical applications due to increasing restrictions on the use of organic solvents.(*4*) Alcohol- or ether-containing side chains could potentially control cation transport properties in solid state batteries, ion-selective electrodes, or membranes.(*5-7,12,24-26*) Chiral functionalities such as amino acids are of interest for use in materials capable of molecular recognition and asymmetric electrosynthesis.(*5-7,12,19*) However, the structural and electronic properties of many desirable substituents are often incompatible with conventional synthetic approaches and can adversely affect polymer properties.(*4-7,11-14*)

Side Chain Influence on the Extent of Conjugation in Soluble Polythiophenes

A number of studies have addressed the general goal of rationally tuning the properties of conjugated polymers through synthetic modification. Driven by both fundamental interest and technological applications, numerous studies have focused on extending conjugation lengths. The results of these studies have greatly improved our understanding of the fundamental structural parameters that strongly influence conjugation. In addition to the primary interest in understanding the relationship between conjugation and structure, a high degree of conjugation is essential for a number of the applications mentioned above. For example, optimizing the conjugation length is the key to designing low band gap materials(*17*) (including transparent conductors), enhancing electrical conductivity,(*8*) and maximizing polymer non-linear optical properties.(*5*) The ability to tune the degree of conjugation is also important in applications where it is desirable to control the optical absorption and emission such as light emitting diodes(*27*) and thermochromic devices.(*8*)

Although side chains can be used to enhance solubility and tune polymer properties, there are a number of unintended consequences that result from the substitution of side chains along the backbone. Thermal instability of the doped (conducting) state(*8,28*) and steric effects that twist the polymer backbone out of planarity are among the most profound of these effects. Since the conjugation length is dependent on the overlap of neighboring thiophene rings, it can easily be influenced by the introduction of tortional strain along the polymer backbone. If the backbone is twisted significantly from planarity (i.e. > 40° between rings)(*29*) the conjugation length will be decreased.(*5-7,12-14,28-34*) The number, type, and placement of side chains along the backbone can all contribute to increased steric effects and thus tortional strain.

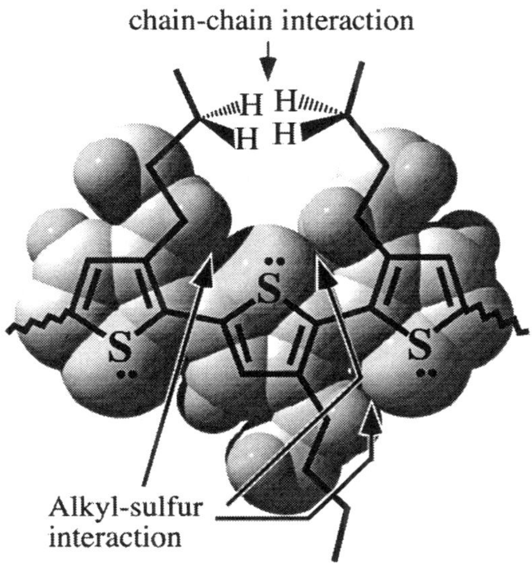

Figure 1. Steric interactions in a head-to head, head-to-tail triad of an alkyl substituted polythiophene. Steric clashes between alkyl side chains and between side chains and the sulfur lone pairs are indicated.

As depicted in Figure 1, there are two types of steric effects that can influence the planarity of the polymer backbone. Interactions between alkyl chains (chain-chain interactions) are most pronounced for longer alkane chains in instances where ring-ring couplings are regiorandom. The second type of steric interactions involves sulfur lone pair-alkyl side chain steric clashes. These interactions are found in all substituted polythiophenes, but are most pronounced for polymers containing high side chain densities and/or head-to-head couplings (see Figure 1). Regiorandom poly(3-alkylthiophene)s suffer a loss of conjugation caused by both of these types of steric effects.

Optical Absorption Spectroscopy of Substituted Polythiophenes. Many investigations have been conducted that probe the effect of the type and number of various substituents on the effective conjugation length.(*18*) The extent of conjugation in substituted polythiophenes is typically inferred from UV-vis spectroscopic data. The wavelength of the π-π* transition (usually recorded in $CHCl_3$) is taken as an indication of the conjugation length in solution. As the conjugation length increases, the energy of the transition decreases (giving a bathochromic (red) shift). It is important to use the same solvents when comparing solution optical absorption spectra of different samples because most polythiophenes exhibit solvatochromism.(*35*) Comparisons between spectra recorded in good solvents (e.g. $CHCl_3$) and poor solvents (e.g. hexane) must be avoided if one is interested in the relative conjugation lengths in the samples. Little attention has been paid to controlling/determining the extent of aggregation in solutions of substituted polythiophenes. Aggregation will likely affect the solution state spectra. Investigations aimed at understanding the effects of aggregation upon these spectra would be very useful.

The optical spectra of the polymeric samples are also measured as thin films, usually on glass substrates. In comparison to the solution spectra, the transitions in the solid state are often red-shifted. This shift is due to an intrachain coil-to-rod conformational change that takes place upon removal of the solvent and results in extended chains of coplanar thiophene rings.(*36-39*) This conformational change coupled with interchain interactions in the solid state results in increased electronic delocalization and a shift of the UV-vis transition to lower energy. Care must be taken in comparing the extent of conjugation in thin film samples because the absorption spectra depend strongly on sample preparation,(*40*) especially film thickness and casting solvent.

Approaches to More Highly Conjugated Polythiophenes

A number of studies have been undertaken to determine the extent of conjugation that should be possible in an ideal polythiophene chain. Numerous theoretical studies on polythiophene predict that conjugation length will increase with increasing number of coplanar rings in the chain.(*41*) In support of this work, alkyl substituted and unsubstituted oligothiophenes have been used as soluble model compounds in attempts to correlate spectroscopic properties with known conjugation lengths.(*42-46*) For each of the studies reported, plots of absorption energy as a function of the reciprocal of the chain length are linear. Extrapolation of the data to infinite chain length predicts absorption energies for the ideal extended chains. In the case of end-capped oligothiophenes(*43-44*) and alkyl substituted oligothiophenes(*42,46*) these extrapolations yield values in a narrow range of approximately 2.2-2.3 eV (540-560 nm in solution).(*43-45*) Based upon these studies, an ideal, fully conjugated alkyl-substituted polythiophene sample would be expected to have an absorption maximum in this range. Analogous studies recently reported on fully substituted, regioregular 3-alkylthiophene oligomers yield a value of 2.54 eV (488 nm) upon extrapolation to infinite chain length.(*47*) The higher energy of this absorption maximum suggests that high side chain densities can decrease conjugation lengths (*vide infra*).

A number of approaches have been taken to try to increase the conjugation length in soluble polythiophenes through the decrease of side chain induced steric interactions. Generally, the goal is to eliminate clashes between side chains as well as decrease steric repulsion between side chains and the polymer backbone. To date, three specific approaches have been investigated that eliminate or minimize the deleterious steric interactions illustrated in Figure 1: (i) eliminate steric clashes between side chains, (ii) reduce the magnitude of the steric interactions between the side chains and the polymer backbone, and (iii) reduce the total number of side chains. First we will review important examples of approaches (i) and (ii). The balance of this chapter will focus on soluble polythiophenes that have reduced side chain density including recent work on polymers derived from monosubstituted bithiophenes. Throughout the remaining discussion, the influence of the side chains on the optical absorption spectra in solution will be emphasized.

It is important to keep in mind that manipulation of alkane side chains will affect polymer solubility. In a fully substituted polymer, there is good solubility, but steric interactions due to the side chains decrease conjugation. On the other hand, the unsubstituted polymer avoids these steric interactions but is insoluble. The removal of side chains along the polymer has also been shown to decrease solubility. Thus, there will likely be a balance between enhanced conjugation and good solubility.

Elimination of Steric Clashes Between Alkane Side Chains. There are three main approaches that potentially eliminate the side chain interactions that decrease the degree of conjugation in processable polythiophenes: use shorter alkane chains, control the regiochemistry of the polymer, and increase the distance between side chains along the polymer backbone.

Copolymers of Mixed Alkylthiophenes. The use of shorter side chains potentially eliminates side chain clashes, however, poly(3-alkylthiophene)s with side chains shorter than butyl are not soluble.(*15*) An alternative approach is to use a copolymer where long, solubilizing side chains are diluted with short chains such as methyl.(*48-54*) For a series of random copolymers, the ratio of side chain alkyl substituents (R=octyl/R'=methyl) was varied and the optical spectra recorded(*48*) (Table I). The regioregularity of these samples was not determined. The effects observed are quite small (< 10 nm red shifts in the λ_{max}). Copolymers involving two non-methyl side chains are less conjugated than either of the analogous homopolymers.(*55*)

A series of regioregular (head-to-tail), random copolymers (R=dodecyl/R'=methyl) has been prepared by McCullough and Jayaraman.(*54*) Small red shifts in the optical spectra (450 $\rightarrow$ 458 $\rightarrow$ 470 $\approx$ 466 nm) occur upon increasing the fraction of methyl side chains in the polymer. In this case, the copolymer formed from two non-methyl monomers has an absorption maximum comparable to the homopolymers.(*54*) Unfortunately, the spectra for the regioregular copolymers cannot be directly compared to the those of the regiorandom forms because the spectra were not measured in the same solvents (*vide supra*).

Table I. Visible Spectroscopic Data for Random Copolymers of Various 3-Alkylthiophenes[a]

R	R′	Ratio of R/R′	Regio-regular	λ_{max} (Solution)[b]	λ_{max} (Solid)[c]	Ref
C_8	Me	100/0	No	439	---	48
		90/10		442	---	48
		80/20		447	---	48
		70/30		445	---	48
		60/40		443	---	48
C_{12}	Me	1/0	Yes	450 (xylene)	526, 561*, 609	54
		2/1		458 (xylene)	526, 565*	54
		1/1		470 (xylene)	520, 550*	54
		1/2		466 (xylene)	510, 545*, 600	54
C_{12}	C_6	1/0	Yes	450 (xylene)	526, 561*, 609	54
		1/1		450 (xylene)	525, 556*, 600	54

[a] All side chains are straight chain alkyl where C_n is the number of carbons in the chain. [b] Spectra recorded in $CHCl_3$ solution unless otherwise indicated. [c] Where multiple peaks are observed, wavelengths are given for all well-defined peaks. The absorption maximum is indicated by an *.

Regioregular Polyalkylthiophenes. Traditional synthetic methods for polymerization of 3-alkylthiophenes lead to regioisomers with undesirable steric interactions.(*13,14,40,56*) Three types of coupling (head-to-head (HH), tail-to-tail (TT), or head-to-tail (HT)) lead to the four regioisomer triads shown in Figure 2. Regiorandom linkages in the polymer backbone lead to steric interactions between alkyl side chains as seen in the trans configuration of the HT-HH triad.

McCullough(*14,38*) and Rieke(*13,56*) have developed chemical coupling methods in order to eliminate HH and TT regioisomers. These synthetic methods employ transition metal catalysts to chemically control the coupling of asymmetric monomers thereby producing polymers with only HT couplings.(*13,14,40,56*)

The improvement in conjugation lengths versus the percent HT coupling can be seen in representative poly(3-alkylthiophene)s shown in Table II. As the amount of HT coupling in the polymer increases from the completely random polymer (50% HT couplings) to those made by chemical oxidation with $FeCl_3$ (~70% HT couplings),(*28,33,60,61*) the λ_{max} for the polymers show a 7-13 nm red shift in $CHCl_3$ solution and a 61-78 nm red shift in the solid state. The regioregular polymers of McCullough and Rieke (>98% HT couplings) show a 9-21 nm red shift in solution and a 50-66 nm red shift in the solid state compared to the $FeCl_3$ polymerized polymers. From this data it can be seen that the HT content makes a more drastic impact on the red shift of the visible λ_{max} for the films than for the polymers in solution. This has been attributed to nonplanarity of the polymer chains caused by greater steric strain introduced by HH couplings.

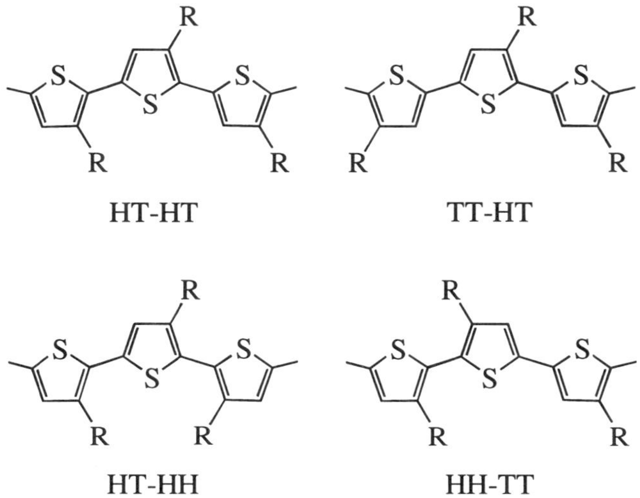

Figure 2. Regioisomer triads in a regiorandom poly(3-alkylthiophene)

An alternative approach to eliminating steric clashes between side chains involves the use of disubstituted 4,4'-dialkyl-2,2'-bithiophenes or 3,3'-dialkyl-2,2'-bithiophenes as monomers.(*30,33,57-59*) These monomers can only couple in a head-to-head or tail-to-tail fashion respectively and are thus considered regioregular. Although this approach is effective in eliminating side chain clashes, the optical absorbance spectra suggest that the large extent of head-to-head couplings greatly distorts the planarity of the π backbone.(*30,33*) Solution and solid state values for poly(dialkylbithiophene)s range from 390 to 400 nm(*30,33,57-59*) compared to ~435 nm for the regiorandom poly(3-alkylthiophene)s in solution.(*33,35,60,61*)

A final approach to eliminating steric clashes between side chains involves increasing the distance between side chains along the polythiophene backbone. This approach – decreasing the side chain density – will be explored fully, later in this chapter.

Reduction of Steric Clashes Between Side Chains and the Polythiophene Backbone. Each side chain has steric contacts with the lone pairs on the sulfur atom in the backbone (Figure 1). These steric interactions induce rotation about the thiophene-thiophene bonds and decrease π orbital overlap along the backbone. Head-to-head couplings involve two side chain-sulfur lone pair interactions and result in greater steric strain.(*30-32,58*) There are three main methods for decreasing the steric interactions between the side chains and the polymer backbone. The first involves eliminating head-to-head couplings within the polymer. The second approach is to use side chains that are sterically less bulky than alkyls (e.g. alkoxy).

These two approaches will be described in this section. A final method, decreasing the density of side chains in the polymer chain, will be discussed in the next section.

Table II. Comparison of Visible λ_{max} (nm) for Selected Polyalkylthiophenes in Solution and Films with Respect to Percent HT Coupling.[a]

Polymer	Random[b] 50% HT		FeCl$_3$ ~70% HT		McCullough[c] >98% HT		Rieke[b] > 98% HT	
	Solution	Film	Solution	Film	Solution	Film[h]	Solution	Film[h]
pC$_4$T	428	433	440[d]	494[d]	450	525 560* 608	449	522 556* 605
pC$_6$T	428	438	435[e]	505[f]	442	525 555* 610	456	526 556* 610
pC$_8$T	428	438	441[g]	506[g]	460	525 559* 610	451	522 556* 608

[a] The percentage HT coupling is the number of HT couplings divided by the sum of HT and HH couplings as determined by NMR. All solution spectra recorded in CHCl$_3$. pC$_4$T = poly(3-butylthiophene), pC$_6$T = poly(3-hexylthiophene), pC$_8$T = poly(3-octylthiophene). [b] Ref. 13. [c] Ref. 14. [d] Ref. 33. [e] Ref. 60. [f] Ref. 61. [g] Ref. 28. [h] Where multiple peaks are observed, wavelengths are given for all well-defined peaks. The absorption maximum is indicated by an *.

Approaches that Eliminate Head-to-Head Coupling. As described in the last section, regioregular poly(3-alkylthiophene)s have more extended conjugation lengths than their regiorandom counterparts. The regular placement of alkane side chains eliminates both alkane side chain clashes and head-to-head couplings. For a series of poly(3-hexylthiophene)s containing varying amounts of head-to-tail couplings (Table III), the absorption maxima shift to the red with increasing head-to-tail content.(*13*) In these polymers, the increased conjugation found in the regioregular forms derives from elimination of both types of steric effects. In order to isolate the effects of head-to-head couplings, one can compare poly(3,3'-dihexyl-2,2'-bithiophene)(*57,59*) with regioregular poly(3-hexylthiophene).(*13*) These two polymers have the same side chain length and density and are free of side chain clashes. The λ_{max} for the head-to-tail polymer is 60 nm higher than for the head-to-head isomer (both measured as CHCl$_3$ solutions) demonstrating the deleterious effect of the head-to-head couplings upon conjugation length.

An approach that eliminates head-to-head couplings in regiorandom polymers involves the use of certain "internally" substituted bithiophenes and terthiophenes as monomers (e.g. 3-alkyl-2,2'-bithiophenes and 3'-alkyl-2,2';5',2"-terthiophenes). Chemical oxidation or coupling results in regiorandom polymers possessing no head-to-head couplings. Because these polymers have decreased side chain density, they are considered in the next section.

Table III. Visible Spectroscopic Data for Poly(3-hexylthiophene)s with Varied Percentages of Head–to–Tail Coupling.

Polymer[a]	% HT	Solution λ_{max}(nm)[b]	Film λ_{max}(nm)[c]	Ref
$p(C_6)_2BT$	0	396	396	55, 57
pC_6T	50	428	438	13
	65	428	446	13
	70	429	451	13
	80	440	505	60
	>98	456	526, 560*, 610	13

[a] $p(C_6)_2BT$= poly(3,3'-dihexyl-2,2'-bithiophene), pC_6T = poly(3-hexylthiophene). [b] Spectra recorded in $CHCl_3$ solution unless otherwise indicated. [c] Where multiple peaks are observed, wavelengths are given for all well-defined peaks. The absorption maximum is indicated by an *.

Alkoxy Substituted Thiophenes and Alkoxy Substituted Bithiophenes. The incorporation of long chain alkoxy substituents has been another approach to increasing the conjugation of functionalized polythiophenes. Polyalkoxythiophenes often retain the solubility characteristics of alkyl functionalized polythiophenes, but exhibit increased conjugation due to a reduction of side chain induced steric interactions and the electron donating character of the alkoxy group.(*40,64-66*) The smaller van der Waals radius of the oxygen atom (1.4 Å) in comparison to the methylene unit (2.0 Å) results in less steric repulsion between the side chain oxygen and the thiophene sulfur.(*65*) The increased conjugation of these polymers was first demonstrated by poly(3-butoxy-4-methylthiophene) which exhibits a low energy solid state absorbance maxima of 545 nm.(*63*)

The ability to reduce steric interactions through the use of alkoxy side chains has been clearly demonstrated by Cloutier and Leclerc who polymerized 3,3'- and 4,4'-disubsituted-2,2'-bithiophenes containing both mixed alkyl/alkoxy, and dialkoxy substituents.(*65*) Poly(dialkylbithiophene)s contain all head-to-head couples that result in low conjugation lengths.(*30-32*) Stepwise substitution of alkoxy side chains for alkyl side chains extends the conjugation length as demonstrated by dramatic shifts of the visible transitions to lower energies as shown in Table IV. Unfortunately, the polythiophenes containing only alkoxy side chains resulted in only partially soluble materials. In particular, the alkoxy substitutents do not promote the solubilization of the higher molecular weight fractions. The similar solution and solid state absorption maxima suggest that rigid-rod conformational structures prevail in both states.

The red-shifted absorption maxima found for the alkoxy substituted polymers are the result of both the steric and electronic attributes of the alkoxy side chain. The relative contributions of sterics and electronics have not been unraveled.

Table IV. Visible Spectroscopic Data for Alkoxy Functionalized Polythiophenes.[a]

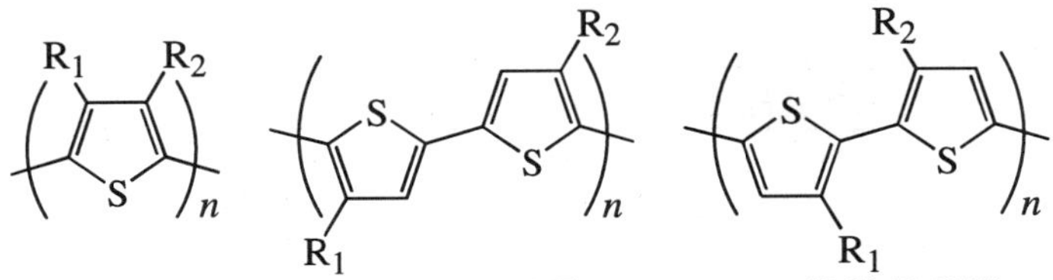

Polymer	R_1	R_2	λ_{max}(solution)[b]	λ_{max}(film)	Ref
p(3,4-R$_2$T)	H	OC_4	530	520	64
	H	OC_7	590	---	66
	OC_4	OC_4	480	460	63, 64
	C_1	OC_4	420	545	63, 64
	C_1	OC_7	408	420	66
	C_1	OC_8	440	545	64
p(4,4'-R$_2$BT)	C_{10}	C_{10}	390	390	31
	C_{10}	OC_4	480	500	65
	OC_4	OC_4	574	600	65
p(3,3'-R$_2$BT)	C_{10}	OC_4	460	506	65
	OC_4	OC_4	545	582	65

[a] All side chains are straight chain alkyl where C_n is the number of carbons in the chain. [b] Spectra recorded in $CHCl_3$ solution.

Reduction of the Number of Side Chains on the Polythiophene Backbone. In polymers of symmetrical dialkylbithiophenes interactions between alkyl chains are eliminated, but sulfur-alkyl steric repulsions are still present and are pronounced due to the high number of head-to-head couplings in these polymers.(*5,30-33*) Regioregular head-to-tail poly3-alkylthiophenes(*13,14,40,56*) lack the undesirable head-to-head couplings, but sulfur-alkyl steric interactions are still present. In an effort to increase the conjugation within polyalkylthiophenes, a number of polymers have been prepared with a reduced number of alkyl side chains. Decreasing the density of side chains reduces the number of sulfur-alkyl interactions, resulting in more extended conjugation lengths. The most common approaches to the reduction of side chain density in polyalkylthiophenes is the polymerization of partially substituted oligothiophenes and the copolymerization of thiophene/alkylthiophene mixtures. Oligothiophenes investigated include monosubstituted bithiophenes, (*28,34,67-69*) mono- and di-substituted terthiophenes, (*28,70-75*) and disubstituted quaterthiophenes.(*42,76*) Copolymers of thiophene and 3-alkylthiophene also result in polymers with reduced side chain density.(*15,77*)

Partially Substituted Terthiophenes and Higher Oligomers. The initial studies on partially substituted oligothiophenes were reported in 1991 by Zerbi and co-workers who chemically ($FeCl_3$) polymerized 3,3"-dihexyl-2,2':5',2"-terthiophene.(*70,71*) This polymer contains one third fewer side chains in comparison to poly(3-hexylthiophene). The reduction in side chains results in low energy solution and solid state UV-vis absorption maxima (Table V) indicated by a 20 nm red shift compared to the fully substituted analog.(*70,71*) The dioctylterthiophene analog, later polymerized by Collard and co-workers,(*42*) yields a polymer with slightly higher energy maxima possibly due to increased interaction between the longer alkyl side chains.

Table V. Visible Spectroscopic Data for Polyalkylterthiophenes.

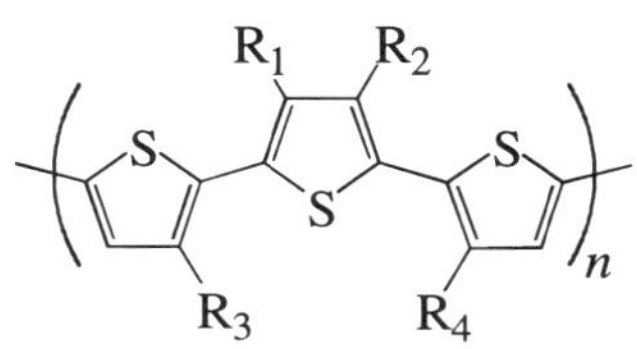

Polyalkylterthiophene[a]

R_1	R_2	R_3	R_4	$\lambda_{max}(CHCl_3)$	λ_{max}(film)	Ref
C_7	H	H	H		540	72
C_8	H	H	H	468	510	28
C_{18}	H	H	H		540	72
H	H	C_6	C_6	456	535	70, 71
H	H	C_8	C_8	437	517	42
C_4	C_4	H	H	499	522	73

[a] All side chains are straight chain alkyl where C_n is the number of carbons in the chain.

The success of this early work led others to prepare and polymerize 3'-alkyl-2,2':5',2"-terthiophenes in an effort to further decrease the side chain density of the resulting polymers.(*28,72*) These polyalkylterthiophenes contain side chains on every third thiophene ring and exhibit increased conjugation as evidenced by the reported 468 nm absorption maxima for poly(3'-octyl-2,2':5',2"-terthiophene) in solution (Table V).(*28*) Compared to the solution data, the solid state maxima exhibited little to no shifts in energy. Although the polymerization of 3'-alkyl-2,2':5',2"-terthiophenes cannot result in head-to-head couples, the monomer lacks two-fold symmetry and polymerization results in the regiorandom placement of the alkyl side chains as shown in Figure 3. The regiorandom nature can inhibit ordered packing in the solid state, diminishing polymer-polymer interactions and resulting in higher energy transitions in the visible spectrum.

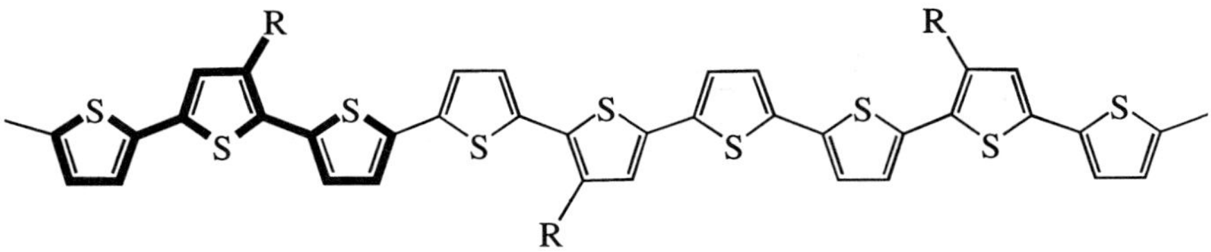

Figure 3. Regiorandom segment of poly(3'-alkyl-2,2':5',2"-terthiophene) (monomer unit in bold).

Kanatzidis and co-workers investigated a similiar approach utilizing the symmetrical monomer 3',4'-dibutyl-2,2':5',2"-terthiophene.(*73*) This allowed the regioregular placement of two alkyl chains on every third ring of the resulting polymer. This polymer exhibits absorption maxima at 499 nm in solution and 522 nm in the solid state.(*73*) This relatively small difference (23nm) suggests a greater predominance of the rod-like conformation in solution. This conformation is favored both by reduced side chain density and increased chain-chain interactions promoted by the regioregular structure of the polymer.

More recently, partially substituted quarterthiophenes have been investigated by Collard and co-workers who polymerized 3,3'''-dioctyl-2,2':5',2":5",2'''-quaterthiophene.(*42*) Polymerization of the quaterthiophene by $FeCl_3$ was inefficient: even at elevated temperatures (50°C), the majority of the material recovered (73%) was the dimerized monomer (the octathiophene). However, approximately 10% of the oxidized material was polymeric and exhibited absorption maxima of 463 nm in $CHCl_3$ and 514 nm as thin films. Although the $FeCl_3$ polymerization of the dimethyl analog has also been reported without the difficulties described above, the polymerized product was completely insoluble.(*76*)

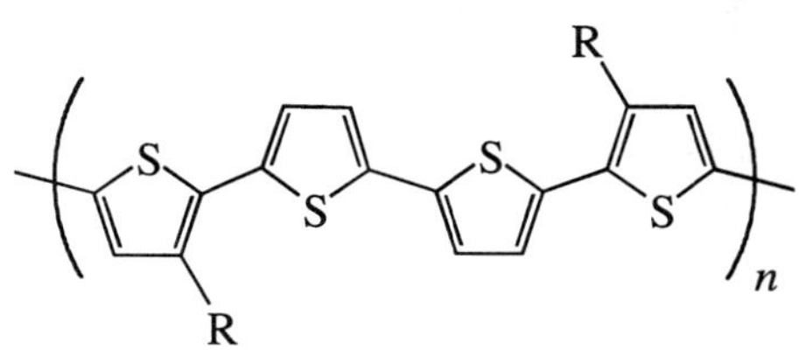

Figure 4. Poly(3,3'''-dialkyl-2,2':5',2":5",2'''-quaterthiophene)

In polymers of dialkylquaterthiophenes (Figure 4), one half of the thiophenes are alkyl functionalized, providing a side chain density falling between that of the doubly and singly substituted terthiophene polymers. The trend in solution absorption maxima as a function of side chain is as follows: 437 to 456 nm for two side chains per three rings; 463 nm for one side chain per two rings; and 468 nm for one side chain per three rings (Table V and ref 42).

An elegant study which illustrates the ability to tune conjugation through side chain density has been reported by Hadziioannou and co-workers.(*74,75*) Using the series of symmetrical dioctyloligothiophenes shown in Table VI, the authors were able to prepare regiospecific polyoctylthiophenes in which all octyl substituents exist in head-to-head couples. As the series progresses, each oligomer increases by one thiophene ring while maintaining a constant number of octyl side chains, thus reducing both the side chain density and the fraction of head-to-head content of the resulting polymer. The side chain density and head-to-head content of the polymer correlate directly to the oliogmer length by the fractions 2/n and 1/n respectively, where n is the number of thiophenes in the monomer. The effect of both side chain density and head-to-head content on conjugation and the visible spectra could then be measured in a systematic manner. Comparison of the solution and solid state UV-visible data in Table VI and Figure 5 shows a linear decrease in the energy of the absorption maxima in relation to both the side chain density and head-to-head content.

Table VI. Visible Spectroscopic Data for a Series of Regiospecific Head-to-Head Polyoctylthiophenes.

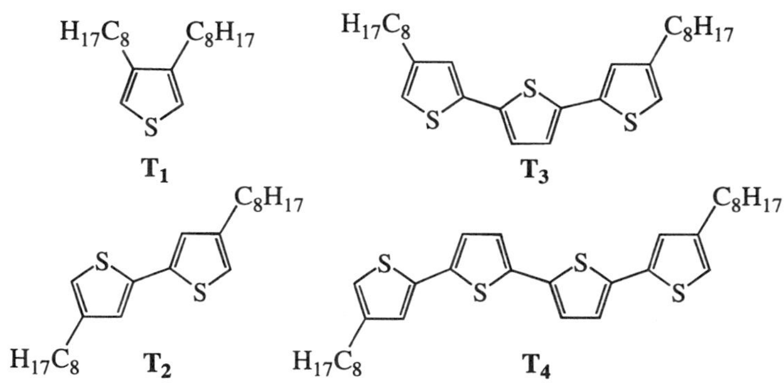

Polymer	$\lambda_{max}(CHCl_3)$[a]	λ_{max}(film)[b]
$p(T_1)$	326	465
$p(T_2)$	390	509
$p(T_3)$	422	523
$p(T_4)$	446	532

[a] Ref. 74. [b] Ref. 75.

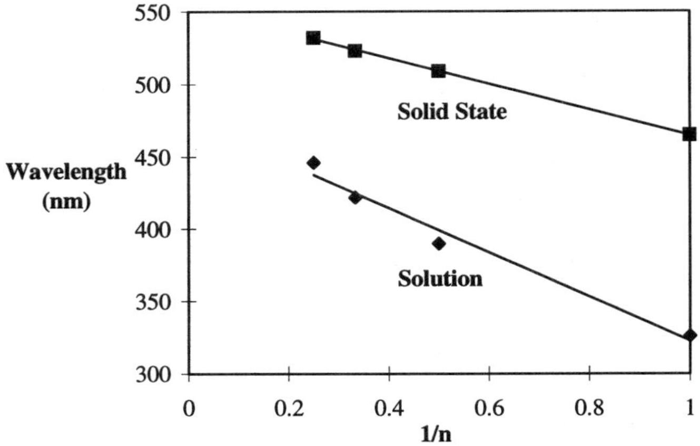

Figure 5. Optical absorption maxima for regiospecific head-to-head polyoctylthiophenes versus the fraction of head-to-head couples (1/n) where n is the number of rings in the repeat (see text).

Monosubstituted Bithiophenes. Recently, a small number of polymers derived from monosubstituted bithiophenes have been prepared.(*28,34,67-69*) Their reduced side chain density leads to a higher degree of conjugation than in their fully-substituted counterparts as evidenced by the longer wavelength absorption maxima found in the optical spectra (Table VII) of the alkylbithiophene polymers.

In the case of poly(3-alkyl-2,2'-bithiophene)s there are no head-to-head couplings or side chain clashes even for the regiorandom polymer (Figure 6). Oxidative polymerization of 4-alkyl-2,2'-bithiophenes with $FeCl_3$ results in a completely regiorandom polymer which contains head-to-head couplings (17%),(*67*) but no side chain clashes (Figure 6). The absorption maximum for the poly(3-octyl-2,2'-bithiophene) is red shifted relative to poly(4-octyl-2,2'-bithiophene) as expected (Table VII) due to its lack of head-to-head linkages. In the solid state spectra, there are few differences between polymers derived from the 3-alkyl and 4-alkyl isomers.(*28,34,67*)

Regiospecific polymerization of a 4-alkyl-2,2'-bithiophene is expected to produce a polymer with a higher degree of conjugation due to the elimination of head-to-head couplings (Figure 6). The regioregular polymer of the 4-octyl-2,2'-bithiophene, RR-p(4-C_8BT), has recently been prepared and its optical absorption spectrum recorded. It has the highest reported λ_{max} value (510 nm) for an alkyl substituted polythiophene in solution.(*68*) A regioregular poly(3-undecyl-2,2'-bithiophene) has also been prepared(*69*) although solution spectra were not reported. Spectra of thin film samples of RR-p(4-C_8BT) display absorption maxima that are modestly red shifted (30 nm) from the solution values. In the *regiorandom* poly(monoalkylbithiophene)s a red shift of ca. 50-60 nm is observed. The smaller red shift observed in the regioregular case possibly indicates a more rod-like character for these polymers in solution. The poorer solubility of the regioregular polymer(*68*) is also consistent with a rod-like conformation in solution.

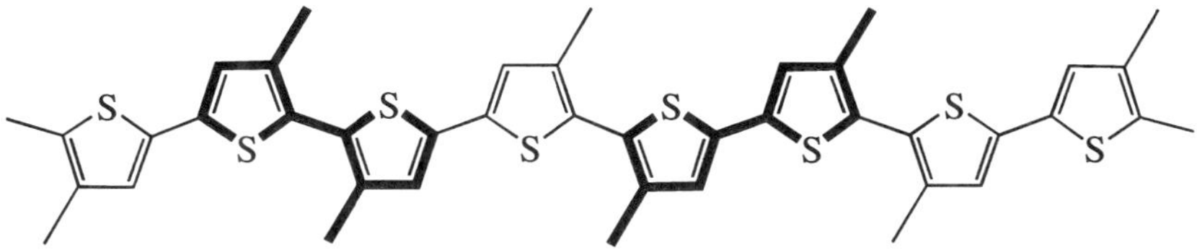

Regioregular p(R_2BT) (50% HH content)

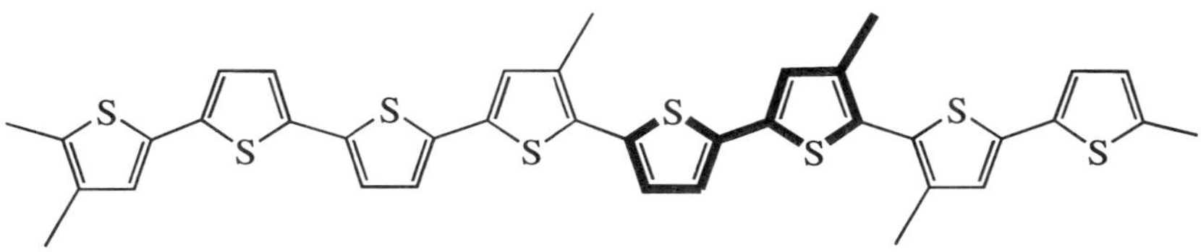

Regiorandom p(4-RBT) (~17% HH content)

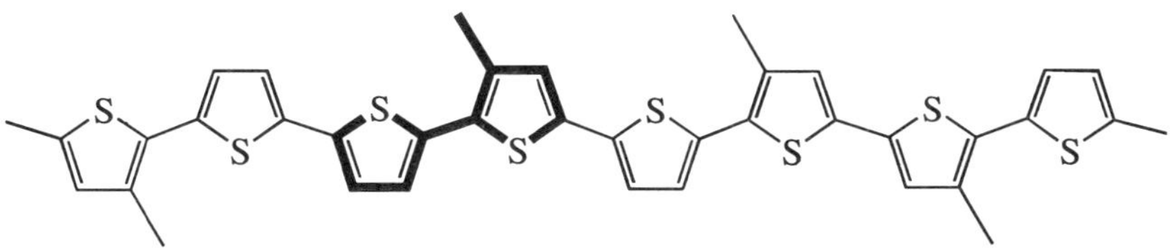

Regiorandom p(3-RBT) (0% HH content)

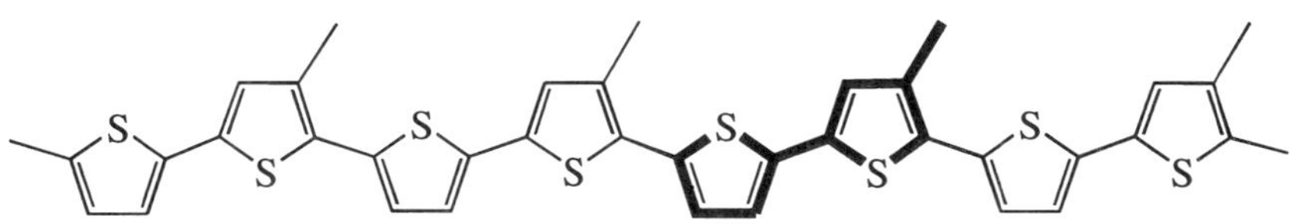

Regioregular p(4-RBT) (~0% HH content)

Figure 6. Regioisomers resulting from polymerization of monosubstituted bithiophenes. Regioregular poly(dialkylbithiophene) (p(R_2BT)) is included for comparison.

Thiophene/Alkylthiophene Copolymers. Another class of polymers that have been prepared in an attempt to minimize the steric effects of alkane side chains are random copolymers of thiophene and an alkylthiophene.(*15,77*) Only scattered reports of the optical properties of these polymers are available. Hotta found that the random copolymer of 3-hexylthiophene and thiophene gave a λ_{max} of 440 nm in solution which compares to 434 nm for poly(3-hexylthiophene) prepared under the same conditions.(*77*) This 3:1 (3-hexylthiophene/thiophene) copolymer exhibited red shifted absorbances in thin solid films relative to the fully substituted polymer. Although the side chain density is decreased in these copolymers, there is still the possibility of head-to-head couplings. McCullough and Jayaraman prepared regioregular copolymers of thiophene and 3-dodecylthiophene.(*15*) In xylene solution, the 1:1 copolymer gave an absorbance maximum of 510 nm. Although these data can not be compared to the other data found for polymers in $CHCl_3$ solution due to the solvatochromism of polythiophenes, they can be compared to the fully substituted regioregular poly(3-dodecylthiophene) in xylene which has an absorbance maximum of 450 nm.(*15*)

Table VII. Visible Spectroscopic Data for Regiorandom and Regioregular (RR) Monosubstituted Bithiophenes

Polymer[a]	λ_{max} (solution)	λ_{max} (film)[b]	Ref
p(3-C_{11}BT)	457	512, 530	34
p(3-C_8BT)	468	525	28
p(4-C_8BT)	454	520	67
RR-p(4-C_8BT)	510	540*, 583, 630	68
RR-p(3-C_{11}BT)	---	571*, 624	69

[a] All side chains are straight chain alkyl where C_n is the number of carbons in the chain. [b] Where multiple peaks are observed, wavelengths are given for all well-defined peaks. The absorption maximum is indicated by an *.

Due to the small amount of work that has been reported on the preparation and optical properties for copolymers of thiophene and alkyl thiophenes, it is difficult to draw many conclusions. It appears that small increases in the extent of conjugation result from eliminating some of the side chains. If more side chains are eliminated and head-to-head couplings are avoided, further increases in conjugation may result.

Conclusions and Outlook

A number of approaches have been investigated for increasing the effective conjugation length in processable, alkyl-substituted polythiophenes. Through these studies, much has been learned about the relationship between polythiophene structure and the extent of conjugation. Three approaches seem most effective, using alkoxy side chains, employing regioregular head-to-tail couplings, and reducing alkyl side chain density. Use of alkoxy side chains results in the lowest energy transitions (as low as 2.1 eV (590 nm) for poly(3-heptyloxythiophene)) due to reduced steric interactions with between the side chain and the sulfur lone pair and because of the electron donating character of the alkoxy group. For the alkyl substituted polythiophenes, the use of regioregular, head-to-tail poly(3-alkylthiophene)s results in modest gains in conjugation length (in the case that R = octyl, 460 nm is found for 98% HT vs. 428 nm for the completely random case). Through the use of reduced side chain density (e.g. poly(3',4'-dibutyl-2,2':5',2"-terthiophene), λ_{max} = 499 nm) the extent of conjugation can be significantly increased. The combination of reduced side chain density and regioregularity yields the most conjugated alkyl substituted polythiophenes (510 nm for regioregular poly(4-octyl-2,2'-bithiophene)).

Based upon extrapolation of data obtained from oligomer model systems, further tuning of polymer properties in alkyl substituted polythiophenes to enhance the extent of conjugation may be limited. The model systems suggest a limit of about 540-560 nm for a defect free chain. Possible hurdles to attaining polymers that have these optical absorption characteristics are the solubility problems that are likely to occur as the polymer chains become increasingly rod-like in solution. In fact, both the alkoxy-substituted polythiophenes and the regioregular poly(monoalkylbithiophene)s already suffer from poor solubility. To enhance solubility of polythiophenes with reduced side chain densities, it may be advantageous to employ polar or charged side chains.(*7,12,18,20-26*)

Although the structure/property relationships have become more evident for alkyl substituted polythiophenes, the synthesis and characterization of new heterofunctionalized polythiophenes continues to be an active research area because there are a number of potential applications for new materials. There is still much to be learned about the structure/property relationships in these functionalized polymers. The effects of these side chains on the stability, solubility and self-assembly of these polymers as well as their effects on the electronic, optical and electrooptical properties are largely unknown. An improved understanding of the structure/property relationships will, in turn, drive the rational synthesis of new, functionalized polymers with optimized properties.

Acknowledgements

We gratefully acknowledge support from the National Science Foundation (NSF-CAREER; CHE-9702726), the Camille and Henry Dreyfus Foundation, and the University of Oregon.

Literature Cited

1. Kaner, R. B.; MacDiarmid, A. G. *Sci. Am.* **1988**, *258* (2), 106.
2. Yam, P. *Sci. Am.* **1995**, *273* (1), 82.
3. Tourillon, G.; "Polythiophene and Its Derivatives," In *Handbook of Conducting Polymers;* Skothem, T. A., Ed.; Marcel Dekker: New York, NY 1986, pp 350.
4. Nguyen, M. T.; Leclerc, M.; Diaz, A. F. *Trends Polym. Sci.* **1995**, *3*, 186.
5. Roncali, J. *Chem. Rev.* **1992**, *92*, 711.
6. Schopf, G.; Kossmehl, G. *Adv. Polym. Sci.* **1997**, *129*, 1.
7. Garnier, F. *Angew. Chem. Int. Ed. Engl.* **1989**, *28*, 513.
8. Leclerc, M.; Faid, K. *Adv. Mater.* **1997**, *9,* 1087.
9. Horowitz, G. *Adv. Mater.* **1998**, *10*, 365.

10. Baker, G. L., "Progress Toward Processable, Environmentally Stable Conducting Polymers," In *Electronic and Photonic Applications of Polymers;* Bowden, M. J., Turner, S. R., Eds.; ACS Advances in Chemistry Series 210; American Chemical Society: Washington, DC, 1985.
11. (a) Waltman, R. J.; Diaz, A. F.; Bargon, J. *J. Electrochem. Soc.* **1984**, *131,* 1452. (b) Waltman, R. J.; Bargon, J. *Can. J. Chem.* **1986**, *64*, 76.
12. Roncali, J.; Garreau, R.; Delabouglise, D.; Garnier, F.; Lemaire, M. *Synth. Met.* **1989**, *28*, C341.
13. Chen, T. A.; Wu, X.; Rieke, R. D. *J. Am. Chem. Soc.* **1995**, *117*, 233.
14. McCullough, R. D.; Lowe, R. D.; Jayaraman, M.; Anderson, D. L. *J. Org. Chem.* **1993**, *58*, 904.
15. McCullough, R. D.; Jayaraman, M. *J. Chem. Soc., Chem. Commun.* **1995**, *135.*
16. McCullough, R. D.; Ewbank, P. C.; Loewe, R. S. *J. Am. Chem. Soc.* **1997**, *119,* 633.
17. Roncali, J. *Chem. Rev.* **1997**, *97*, 173.
18. Swager, T. M.; Marsella, M. J. *Mat. Res. Soc. Symp. Proc.* **1994**, *328*, 263.
19. Andersson, M.; Ekeblad, P. O.; Hjettberg, T.; Wennerström, O.; Inganäs, O. *Polym. Commun.* **1991**, *32*, 546.
20. Bäuerle, P.; Gaudl, K. U.; Wurthner, F.; Sariciftei, N. S.; Neugebauer, H.; Mehring, M.; Zhong, C.; Doblhofer, K. *Adv. Mater.* **1990**, *2*, 490.
21. (a) Patil, A. O.; Ikenoue, Y.; Basecu, N.; Colaneri, N.; Chen, J.; Wudl, F.; Heeger, A. J. *Synth. Met.* **1987**, *20*, 151. (b) Patil, A. O.; Ikenoue, Y.; Wudl, F.; Heeger, A. J. *J. Am. Chem. Soc.* **1987**, *109*, 1858. (c) Ikenoue, Y.; Saida, Y.; Kira, M.; Tomozawa, H.; Yashima, H.; Kobayashi, M. *J. Chem. Soc., Chem. Commun.* **1990**, 1694.
22. Aldissi, M. *Mol. Cryst. Liq. Cryst.* **1988**, *160*, 121.
23. Rasmussen, S. C.; Pickens, J. C.; Hutchison, J. E. *Macromolecules* **1998**, *31,* 933.
24. Casa, C. D.; Bertinelli, F.; Bizzarri, P. C.; Salatelli, E. *Adv. Mater.* **1995**, *7,* 1005.
25. Pohjakallio, M.; Sundholm, G. *Synth. Met.* **1993**, *55-57*, 1590.
26. Welzel, H.-P.; Kossmehl, G.; Schneider, J.; Plieth, W. *Macromolecules* **1995**, *28*, 5575.
27. Chen, F.; Mehta, P. G.; Takiff, L.; McCullough, R. D. *J. Mater. Chem.* **1996**, *6*, 1763.
28. (a) Andersson, M. R.; Pei, Q.; Hjertberg, T.; Inganäs, O.; Wennerström, O.; Österholm, J. E. *Synth. Met.* **1993**, *55-57*, 1227. (b) Pei, Q.; Inganäs, O.; Gustafsson, G.; Granstrom, M.; Andersson, M.; Hjertberg, T.; Wennerström, O.; Österholm, J. E.; Laakso, J.; Jarvinen, H. *Synth. Met.* **1993**, *55-57*, 1221.
29. Brédas, J. L.; Street, G. B.; Themans, B.; Andre, J. M. *J. Chem. Phys.* **1985**, *83*, 1323.
30. Souto Maior, R. M.; Hinkelmann, K.; Eckert, H.; Wudl, F. *Macromolecules* **1990**, *23*, 1268.
31. Zagórska, M.; Kulszewicz-Bajer, I.; Prón, A.; Firlej, L.; Brenier, P.; Galtier, M.; *Synth. Met.* **1991**, *45*, 385.

32. Lapkowski, M.; Zagórska, M.; Kulszewicz-Bajer, I.; Koziel, K.; Prón, A. *J. Electroanal. Chem.* **1991**, *310*, 57.
33. Zagórska, M.; Krische, B. *Polymer* **1990**, *31*, 1379.
34. Masuda, H.; Kaeriyama, K.; Suezawa, H.; Hirota, M. *J. Polym. Sci.: Part A: Polym. Chem.* **1992**, *30*, 945.
35. Inganäs, O. "Thermochromism and Solvatochromism in Substituted Polythiophenes" In *Handbook of Conductive Molecules and Polymers;* Nalwa, H. S., Ed.; John Wiley and Sons: New York, 1997; Vol. 3, Chapter 15.
36. Rughooputh, S. D. D. V.; Hotta, S.; Heeger, A. J.; Wudl, F. *J. Polym. Sci.: Part B: Polym. Phys.* **1987**, *25*, 1071.
37. Chen, S.-A.; Ni, J.-M. *Macromolecules* **1992**, *25*, 6081.
38. Inganäs, O.; Salaneck, W. R.; Österholm, J. E.; Laakso, J. *Synth. Met.* **1988**, *22*, 395.
39. Daoust, G.; Leclerc, M. *Macromolecules* **1991**, *24*, 455.
40. McCullough, R. D. *Adv. Mater.* **1998**, *10*, 93.
41. Brédas, J. L.; Elsenbaumer, R. L.; Chance, R. R.; Silbey, R. *J. Chem. Phys.* **1983**, *78*, 5656.
42. Henderson, P. T.; Collard, D. M. *Chem. Mater.* **1995**, *7*, 1979.
43. Bäuerle, P.; Segelbacher, U.; Maier, A.; Mehring, M. *J. Am. Chem. Soc.* **1993**, *115*, 10217.
44. Guay, J.; Kasai, P.; Diaz, A.; Wu, R.; Tour, J. M.; Dao, L. H. *Chem. Mater.* **1992**, *4*, 1097.
45. Bäuerle, P. *Adv. Mater.* **1992**, *4*, 102.
46. Havinga, E. E.; Rotte, I.; Meijer, E. W.; Ten Hoeve, W.; Wynberg, H. *Synth. Met.* **1991**, *41-43*, 473.
47. Bidan, G.; DeNicola, A.; Enée, V.; Guillerez, S. *Chem. Mater.* **1998**, *10*, 1052.
48. Xu, X.; Ishikawa, H.; Kobayashi, A.; Satoh, M.; Hasegawa, E. *Synth. Met.* **1993**, *55-57*, 4973-4978.
49. Pei, Q.; Inganäs, O. *Synth. Met.* **1992**, *46*, 353.
50. Sato, M.; Toshimi, S.; Yamauchi, A. *Synth. Met.* **1991**, *41-43*, 551.
51. Pei, Q.; Inganäs, O.; Österholm, J.-E.; Laakso, J. *Polymer* **1993**, *34*, 247.
52. Pei, Q.; Inganäs, O.; Gustafsson, G.; Granström, M.; Andersson, M.; Hjertberg, T.; Wennerström, O.; Österholm, J.-E.; Laakso, J.; Järvinen, H. *Synth. Met.* **1993**, *55-57*, 1221-1226.
53. Pomerantz, M.; Tseng, J. J.; Zhu, H.; Sproull, S. J., Reynolds, J. R.; Vitz, R.; Arnott, H. J. *Synth. Met.* **1991**, *41-43*, 825.
54. McCullough, R. D.; Jayaraman, M. *J. Chem. Soc., Chem. Commun.* **1995**, 135-136.
55. Chen, S.-A.; Ni, J.-M. *Macromolecules* **1993**, *26*, 3230-3231.
56. Wu, X.; Chen, T.-A.; Rieke, R. D. *Macromolecules* **1996**, *29*, 7671.
57. Roux, C.; Bergeron, J.-Y.; Leclerc, M. *Makromol. Chem.* **1993**, *194*, 869.
58. Zagórska, M.; Kulszewicz-Bajer, I.; Lapowski, M.; Laska, J.; Hasik, M.; Prón, A. *Synth. Met.* **1991**, *41-43*, 3009.
59. Leclerc, M.; Roux, C.; Bergeron, J.-Y. *Synth. Met.* **1993**, *55-57*, 297.
60. Adbou, M. S. A.; Holdcroft, S. *Macromolecules* **1993**, *26*, 2954.

61. Hotta, S.; Soga, M.; Sonoda, N. *Synth. Met.* **1988**, *26*, 267.
62. Mao, J.; Xu, B.; Holdcroft, S. *Macromolecules* **1993**, *26*, 1163.
63. Leclerc, M.; Daoust, G. *J. Chem. Soc., Chem. Commun.* **1990**, 273.
64. Daoust, G.; Leclerc, M. *Macromolecules* **1991**, *24*, 455.
65. Faid, K; Cloutier, R.; Leclerc, M. *Macromolecules* **1993**, *26*, 2501.
66. Chen, S.-A.; Tsai, C.-C. *Macromolecules* **1993**, *26*, 2234.
67. Rasmussen, S. C.; Pickens, J. C.; Hutchison, J. E. *Chem. Mater.* **1998**, In press.
68. Straw, B.; Rasmussen, S. C.; Hutchison, J. E. *Polym. Prepr., Am. Chem. Soc. Div. Polym. Chem.* **1998**, *39*, 87.
69. Narayan, S.; Desai, P.; Abhiraman, A. S.; Kowalik, J.; Tolbert, L. *Polym. Prepr., Am. Chem. Soc. Div. Polym. Chem.* **1998**, *39*, 183.
70. Gallazzi, M. C.; Castellani, L.; Zerbi, G. *Synth. Met.* **1991**, *41-43*, 495.
71. Gallazzi, M. C.; Castellani, L.; Marin, R. A.; Zerbi, G. *J. Polym. Sci.: Polym. Chem.* **1993**, *31*, 3339.
72. Ferraris, J. P.; Newton, M. D. *Polymer* **1992**, *33*, 391.
73. Wang, C.; Benz, M. E.; LeGoff, E.; Schindler, J. L.; Allbritton-Thomas, J.; Kannewurf, C. R.; Kanatzidis, M. G. *Chem. Mater* **1994**, *6*, 401.
74. Gill, R. E.; Malliaras, G. G.; Wildeman, J.; Hadziioannou, G. *Adv. Mater.* **1994**, *6*, 132.
75. van Hutten, P. F.; Gill, R. E.; Herrema, J. K.; Hadziioannou, G. *J. Phys. Chem.* **1995**, *99*, 3218.
76. Fell, H. J.; Mårdalen, J.; Samuelsen, E. J.; Hofsløkken, N. U.; Carlsen, P. H. J. *Synth. Met.* **1993**, *55-57*, 420.
77. Hotta, S. *Synth. Met.* **1987**, *22*, 103-113.

Chapter 22

Lowering the Band Gap of Ethylenedioxythiophene Polymers: Cyanovinylene-Linked Biheterocycles

Christopher A. Thomas and John R. Reynolds

Department of Chemistry, Center for Macromolecular Science and Engineering, University of Florida, Gainesville, FL 32611–7200

Abstract: Low band gap organic electrochromic polymers show promise for devices that can switch absorptive/transmissive states in both the infrared and visible spectral regions. Here we describe a rationale for lowering the band gaps of conducting polymers by incorporating donor-acceptor units along a polymer backbone. In this instance, we have employed 3,4-ethylenedioxythiophene (EDOT) and cyanovinylene linkages as electron rich and electron poor components respectively. The substitution pattern around the vinyl group has been varied resulting in polymers with bandgaps ranging from 1.1-1.6 eV when prepared via oxidative electropolymerization followed by subsequent charge neutralization. The polymers have been characterized by cyclic voltammetry and spectroelectrochemistry with coloration efficiencies and chronoabsorptometry results reported.

Varying the electron density along the main chain of conjugated polymers allows a high degree of control of numerous properties including redox switching potential, electrochromic transitions, luminescence energies and the ability to store charge in electrochemical capacitors and rechargeable batteries (*1*). Recently, EDOT based polymers have grown in importance based on their use as highly transmissive anti-static coatings with high conductivity and stability due to their low oxidation potentials (*2*). The high transmissivity of the conducting form is due to the electron rich nature of the polymer backbone yielding a relatively low electronic bandgap of 1.6 eV.

Our research group has been especially interested in the electrochromic properties of electropolymerized EDOT-based polymers (*3*). Due to its low bandgap, PEDOT is one of relatively few conjugated polymers that are cathodically coloring. As such, it is colored a deep-blue in the reduced state and switches to a highly transmissive neutral gray or sky blue in its conducting oxidized state. We find that varying the electronic character of the linking groups between thiophene rings in bis(EDOT)arylene monomers allows us to tune the band gap over a broad spectral range. For example poly(1,2-(2-(3,4-ethylenedioxythienyl))vinylene) (PBEDOT-V) exhibits a band gap of 1.4 eV (*3a*), while poly(3,6-(2-(3,4-ethylenedioxythienyl))carbazole) has a band gap of 2.5 eV (*3e*). By utilizing a combination of complementary anodically and cathodically coloring polymers, dual-polymer electrochromic devices were prepared that switch between a highly transmissive color-neutral state and a highly absorbing blue state (*4*). A benefit of using the low oxidation potential polymers in these devices is that they are highly stable to tens of thousands of deep double-potential switches.

Five factors have been identified which affect the band gap in conjugated polymers (*5*). These factors are interrelated and affect the overall band gap to varying degrees by summation of the individual contributions. Below, these factors are listed along with the specific requirements necessary in order to elicit band gap reduction.

1) *Bond length alternation*: The difference in length between single and double bonds in a conjugated polymer is related to the Peierls distortion that is responsible for the opening of a band gap at the Fermi level. Low bond length alternation leads to a smaller distortion which in turn lowers the band gap.
2) *Deviation from planarity*: Orbital overlap varies with the cosine of the twist angle and the band gap is directly related to the π overlap. Polythiophenes are known for rotational disorder, which must be minimized in order to ensure efficient orbital overlap.
3) *Resonance contributions*: In polymers derived from heterocyclic aromatic polymers there is a competition between electron density localized on the ring and delocalized over some larger region of the backbone.
4) *Donor-Acceptor substituent effects*: One approach to band gap reduction relies on alternating donor-acceptor units in a π conjugated polymer. There is an ideal strength for each of the donor and acceptor units (*6*).
5) *Interchain effects*: Organization of polymer chains into a solid and the transport differences that result play an important role in the properties of CPs (*7*).

In this work we explore the effect of varying the electron rich nature, and thus the donor strength, of a set of four monomer subunits. In this family we use EDOT and thiophene as the donor variants and cyanovinylene as the acceptor to prepare a family of polymers which have band gaps ranging from 1.1 to 1.6 eV.

Figure 1 schematically shows the expected relationship between the donor and acceptor frontier orbital energies of the monomer subunits and the energies of the polymer. We observe the HOMO energy to be related to the electron rich character of the heterocycle donor and the LUMO energy to be coupled to the acceptor unit. Consequently, the LUMO level is expected to remain constant over a series of polymers where the acceptor unit is identical.

Synthetic Approach

The cyanovinylene moiety is constructed by the Knoevenagel condensation of a suitable aldehyde with an aryl acetonitrile derivative using potassium *t*-butoxide. EDOT carboxaldehyde was prepared by the Vilsmeier formylation of EDOT with $POCl_3$ and DMF while EDOT acetonitrile was prepared by the $Ni(acac)_2$ catalyzed coupling of EDOT-ZnCl with bromoacetonitrile as shown below (*8*).

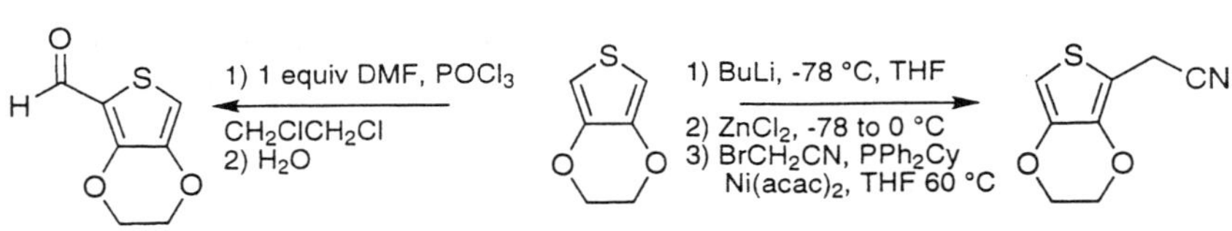

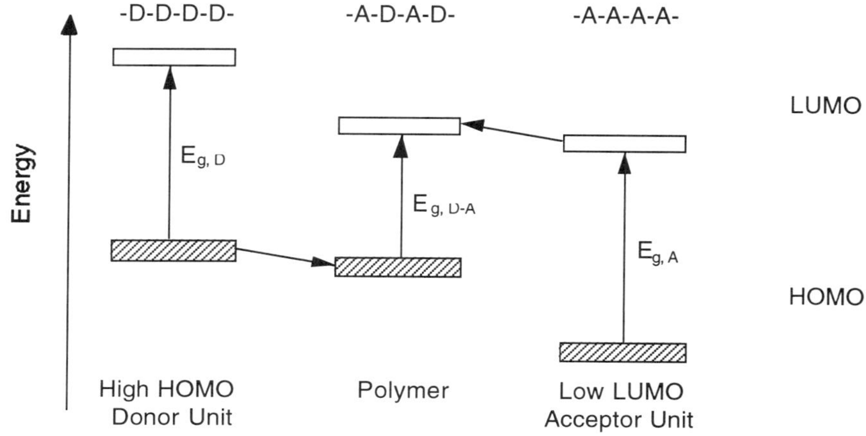

Figure 1: A donor is combined with an acceptor to yield a monomer with a hybrid electronic structure.

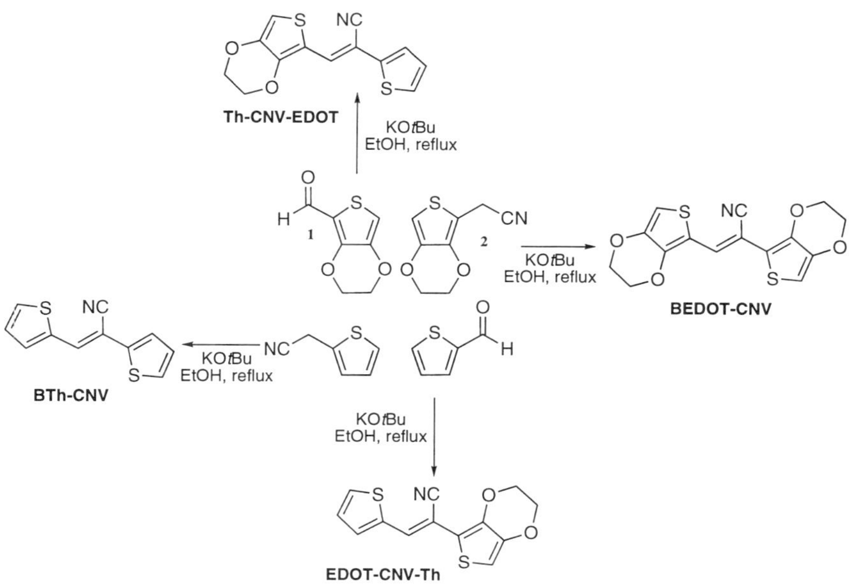

Scheme 1: Monomers were prepared by Knoevengel condensation of an aldehyde with an acetonitrile derivative.

2-Thiophenecarboxaldehyde and 2-thiopheneacetonitrile were obtained commercially from Aldrich. The Knoevenagel route is illustrated in Scheme 1. Yields range from 62-93% and all monomers gave satisfactory ^{1}H-NMR, ^{13}C-NMR and FAB-HRMS spectra. It should be noted that Th-CNV-EDOT was initially communicated by Roncali et al (*9*).

Electropolymerization. These monomers were anodically polymerized via repeated potential scanning on Pt button electrodes or deposited potentiostatically on ITO/glass at low potential to yield electrode confined films which are light blue. Table 1 lists some relevant properties of these materials. The onset is defined as the potential at which current starts to rise during the first scan of polymerization and low values are indicative of clean polymerization. As illustrated in Figure 2, this oxidation potential is ca. 250 mV

Table 1: Electrochemical properties of cyanovinylenes on a 0.73 cm^2 Pt button electrode vs Ag/Ag^+ reference electrode

Monomer	Monomer Properties		Polymer Properties		
Designation	E_{onset} /V	E_{peak} /V	$E_{1/2}$ anodic	$E_{1/2}$ cathodic	E_g /eV
BTh-CNV	0.18	1.10	(+0.5)	-1.6	1.6
BEDOT-CNV	0.58	0.70	(-0.1)	-1.67	1.1
Th-CNV-EDOT	0.78	0.90	(+0.15)	-1.52	1.3
EDOT-CNV-Th	0.75	>1.20	(+0.4)	-1.6	1.2

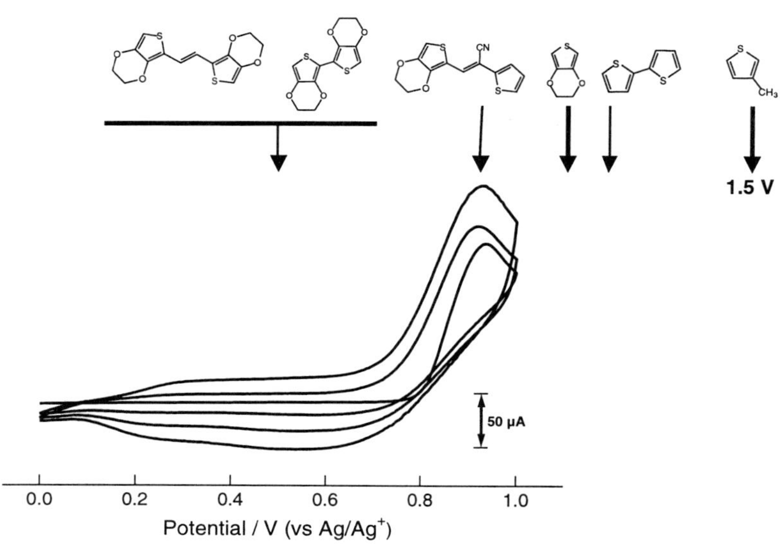

Figure 2: Repeated potential scanning electropolymerization of 10 mM Th-CNV-EDOT in 0.1 M $Bu_4N^+ClO_4^-$ (TBAP) at Pt. Scan rate was 100 mV/s.

lower than the parent EDOT and is in the range for very effective electropolymerization/deposition. The peak potential for Th-CNV-EDOT is noted along with the peak potentials for several other commonly used thiophene based monomers. Repeated scanning shows evolution of a polymer redox process at low potential coupled with an increased monomer peak current on sequential scans due to increasing surface area of the polymer modified electrode. A blue film appears on the electrode and the scan rate dependence of the polymer cyclic voltammetry (50-250 mV/s) indicates that the film is electrode confined, that is, the peak current for the oxidative doping process scales linearly with the scan rate.

Electronic and Electrochemical Properties

Polymer Electrochemistry. Cyclic voltammetry in monomer free tetrabutylammonium perchlorate (TBAP)/acetonitrile (ACN) indicates a broad oxidative p-doping process characteristic of this type of conducting polymer. Spectroelectrochemistry on ITO/glass in the UV-Vis-NIR region confirms the p-doping process. These spectra indicate a single peak for the neutral form of the polymer which gradually decreases in intensity with increased oxidative doping to a state which is highly absorbing at lower energies with a gradual tail into the visible region.

The onset for the oxidative doping process is taken to be the upper edge of the HOMO and increases systematically as the electron rich nature of the flanking heterocycle increases. The onset for the polymer reduction process is taken to be the LUMO and remains essentially the same as expected for a constant acceptor strength. As is characteristic of other reducible thiophene containing systems, this family of polymers displays a set of pre-peaks, small peaks at less extreme potentials than the main polymer redox processes. Disagreement persists over the origin of these peaks but in this system, the oxidative pre-peak grows in only after a polymer reduction and the reductive pre-peak grows in after polymer oxidation. Additionally, the pre-peaks are electrochemically coupled to one another, ruling out charge trapping as a possible explanation for them. Zotti has proposed that the pre-peaks arise from quinoid moieties formed from trace water at the 3 or 4 positions of a thiophene ring (*10*). In the case of BEDOT-CNV, a system with blocked 3 and 4 positions, the pre-peaks presumably arise from reaction at the vinyl group although we note pre-peaks in other compounds prepared in our laboratories with blocked 3 and 4 positions and no vinyl group.

Polymer reduction of Th-CNV-EDOT in ACN shows remarkably symmetrical peak currents when tetrabutylammonium is used as the electrolyte cation in an inert, water free atmosphere. The peak currents do not decay over *ca.* 50 scans and the shape of the redox process remains constant. In $LiClO_4$/ACN, the shape of the reduction process is not symmetrical and is highly broadened compared to the corresponding TBAP electrochemistry. Additionally, the peak currents decay rapidly over a period of only ten scans leaving little electroactivity in the resulting polymer. The difference in electrochemistry between the two supporting electrolytes is attributed to the small size of Li^+ pinning the negative charge on the backbone forming a tight ion pair with limited delocalization (*8*). In a general sense, the stability of reduced polymers has been examined and it is expected that with typical overpotentials, the $E_{1/2}$ for the reduction couple must be more positive than about *ca.* -500 mV for an air stable device to be formed (*9*).

Electrochromism. The E_g values for the polymers as determined by CV were confirmed using *in situ* spectroelectrochemical analysis. At an applied potential of –750 mV all of the polymers are in their neutral form allowing a band gap to be estimated as the onset of the π to π* transition (*e.g.* the smallest energy necessary to populate the π* orbital). This method agrees well (± 0.1 eV) with the bandgaps taken from CV. As the polymer film is oxidized to sequentially higher potentials, a lower energy valence to bipolaron transition is observed consistent with charge carrier formation. In the oxidized form of these polymers, there is an absorptive tail which extends into the visible region limiting the amount of optical contrast available between the neutral and oxidized forms at the absorbance maximum. This absorbance, which extends through the near IR and into the mid-IR, makes these materials useful as IR electrochromics.

IR and visible light electrochromic characterization of Th-CNV-EDOT was performed by employing an experiment where the potential was stepped rapidly between the completely oxidized or neutral (labeled "Red" in the figure) states. The percent transmittance at the λ_{max} (610 nm), 1064 or 1550 nm was measured simultaneously. The switching results for 610 and 1064 nm are shown in Figure 3. The relative position of the oxidized and neutral forms of the polymer are inverted in

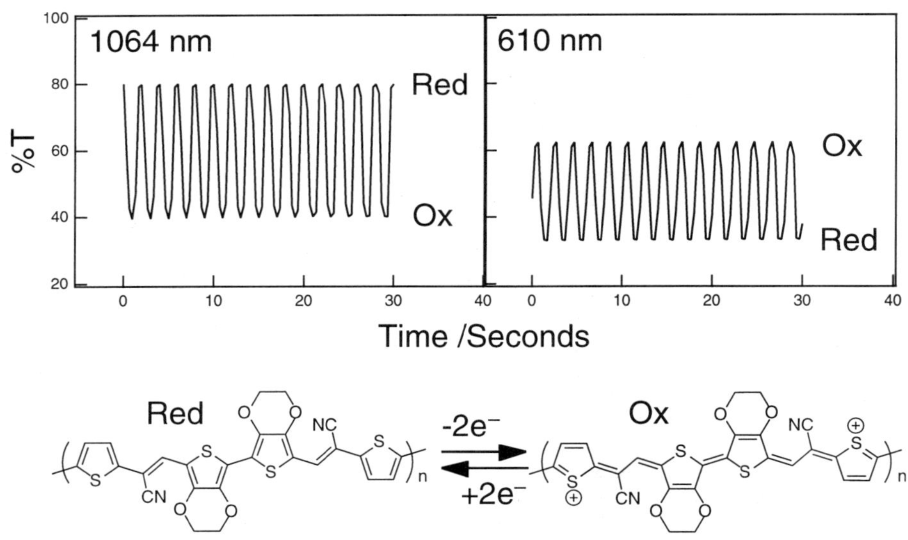

Figure 3. Chronoabsorptometry of Th-CNV-EDOT on ITO/glass in ACN/TBAP. Potentials were stepped between +1.0 V (Ox) and -750 mV (Red) at a frequency of 1 Hz.

the figure because the two wavelengths studied are on opposite sides of the isobestic point. Because of the absorptive tail of the oxidized form of the polymer at λ_{max}, the electrochromic contrast changed on the order $\Delta\%T_{ox}/\Delta\%T_{neutral}$: λ_{max} < 1064 nm < 1550 nm suggesting they will have further enhanced contrast at longer wavelengths. Coloration efficiency (*4*) is defined as the change in optical density per unit charge density and is a parameter used to compare electrochromic devices. For a 150 nm thick film of Th-CNV-EDOT on ITO/glass, the electrochromic contrast was 330 cm^2/C at 1064 nm and 400 cm^2/C at 1550 nm. For comparative purposes, electrochromic metal oxides (e.g. WO_3, IrO_2) have coloration efficiencies between 20-200 cm^2/C (*10*). As such, organic electrochromic materials based on EDOT have considerable promise in devices that operate over this regime.

Acknowledgements. We acknowledge early contributions to this work by Dr. Gregory A. Sotzing. Funding from the National Science Foundation (CHE 96-29854) and the AFOSR (F49620-96-1-0067) is greatly appreciated.

1 See for an extensive review the "Handbook of Conducting Polymers", Skotheim, T.A.; Elsenbaumer, R.L.; Reynolds, J.R., Marcel Dekker Inc., New York, NY, **1998**.

2 a) Jonas, F.; Heywang, G.; Schidtberg, W. Ger. Offen. DE 3,813,589, **1989**. b) Jonas, F.; Heywang, G.; Schidtberg, W.; Heinze, J.; Dietrich, M. Eur. Pat. App. EP 339,340, **1989**. c) Jonas, F.; Heywang, G.; Schidtberg, W.; Heinze, J.; Dietrich, M. U.S. Patent No. 5,035,926, **1991**. d) Heywang, G.; Jonas, F. *Adv. Mater.* **1992**, *4*, 116.

3 a) Sotzing, G. A.; Reynolds, J. R. *J. Chem. Soc., Chem. Commun.* **1995**, 703. b) Sotzing, G. A.; Reynolds, J. R.; Steel, P. J. *Chem. Mater.* **1996**, *8*, 882. c) Sankaran, B.; Reynolds, J. R. *Macromolecules* **1997,** *30*, 2582. d) Reddinger, J. L.; Sotzing, G. A.; Reynolds, J. R.; *J. Chem. Soc., Chem. Commun.* **1996**, 1777. e) Sotzing, G. A.; Reddinger, J. L.; Katritzky, A. R.; Soloducho, J.; Musgrave, R.; Steel, P. J.; Reynolds, J. R. *Chem. Mater.* **1997**, *9* , 1578. f) Sotzing, G. A.; Thomas, C. A.; Reynolds, J. R.; Steel, P. J.; Macromolecules, In Press.

4 Sapp, S. A.; Sotzing, G. A.; Reddinger, J. L.; Reynolds, J. R. *Adv. Mater.* **1996**, *8*, 808. Sapp, S. A.; Sotzing, G. A., Reynolds, J. R., *Chem. Mater.* **1998**, *10*, 2101.

5 Roncali, J. *Chem. Rev.* **1997**, *97*, 173-205.

6 Pagani, G.; Berlin, A.; Canavesi, A; Schiavon, G; Zecchin, S; Zotti, G. *Adv. Mater.* **1996**, 819.

7 Smith, R. C.; Fischer, W. M. ; Gin, D. L. *J. Am Chem. Soc.* **1997**, *119*, 4092-4093. PPV emission intensity is increased when PPV is formed as a single isolated chain in a polymerized LC matrix.

8 Torbjörn, F.; Klingstedt, T. *Synthesis* **1987**, *1*, 40.

9 Ho, H. A.; Brisset, H.; Frere, P.; Roncali, J. *J. Chem. Soc. Chem. Commun.* **1995**, 2309.

10 Zotti, G.; Schiavon, G.; Zecchin, S. *Synth. Met.*, **1995**, *72*, 275-281.

9 de Leeuw, D.M.; Simenon, M. M. J.; Brown, A. R.; Einerhand, R. E. F. *Synth. Met.*, **1997**, *87*, 53-59.

10 a) Faugnan, B. W.; Crandall, R. S.; Heyman, P. M. *RCA Rev.* **1975**, *36*, 177. b) Hitchman, M. J. *J. Electroanal. Chem.* **1977**, *85*, 135. c) Dautremont-Smith, W. C. *Displays I* **1982**, 3.

Chapter 23

Silylene-Tethered Divinylarene Copolymers: A New Class of Electroluminescent Polymer

Tien-Yau Luh [1], Ruey-Min Chen [1], Zhenbo Deng [2], and Shuit-Tong Lee [2]

[1] **Department of Chemistry, National Taiwan University, Taipei 106, Taiwan**
[2] **Department of Physics and Materials Science, City University of Hong Kong, Kowloon, Hong Kong**

A new class of electroactive copolymers of silylene-spaced conjugated segments is conveniently synthesized by hydrosilylation bis-vinylic silyl hydrides and bisalkynes. Flexible silylene-divinylbenzene copolymer exhibits strong intrachain aggregation at both ground and excited states leading to longer wavelength emission in the blue light region. More rigid polymers, on the other hand, shows compatible fluorescence spectra as those of the corresponding monomeric model compounds. Copolymers containing triphenylenevinylene-vinylene chromophore **8c** can serve as an emitting dopant for the fabrication of a blue-green organic light emitting diode (LED). The peak of the electroluminescence (EL) position of the LED device can be blue-shifted with increasing applied voltage. The present observation of the voltage-dependent EL emission suggests a new avenue for controlling the color of LEDs.

Organic electroluminescent devices have become a real possibility for applications as flat panel displays, since the first report of Tang and VanSlyke of an efficient double-layered device based on small organic molecules (*1*). A number of organic materials has now been found with improved emission spectral range, long-term stability and conversion efficiency (*2*). Conjugated polymers have been demonstrated to exhibit diverse electroactive properties. Poly(phenylene-vinylene) (PPV) was the first kind of such polymers to serve as an emission material in the light emitting diode (LED) (*3*) Since then, numerous conjugated polymers, copolymers, dye-doped polymers and metal complex polymers have been studied for LED applications (*2-7*).

Various model systems suggest that the photophysical properties of certain conjugated polymers can be represented by those of a short fragment of the corresponding chromophores (*8*). The wavelength and quantum efficiency of the emitted light of a conjugated moiety in an electroluminescence experiment may be determined by the conjugation length. Accordingly, introduction of spacers between well-defined chromophores in the polymeric chain can occasionally increase the processibility and, at the same time, the photophysical properties can be predicted

(*9,10*). In particular, such short conjugated segments could furnish desired HOMO-LUMO energy gap such that efficient multicolor display applications in the LED device could be achieved.

The use of aliphatic spacers has been well documented. Since the first discovery of the randomly segemented PPV-based conjugated polymer **1**, which exhibits the blue shifted light output relative to PPV (*11*), a more rational design by employing ether linkage to insulate the conjugated segments (e.g. **2** and **3**) has been reported. These latter copolymers exhibit blue light emission (*12*). Copolymers having aliphatic spacer **4** and **5** have also been prepared and their electroluminescent properties behave similarly (*13*). An alternative approach to tackle this problem is by attaching the chromophore(s) as pendant to a polymeric backbone (*14*).

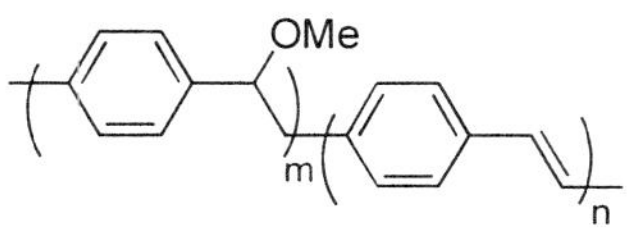

1

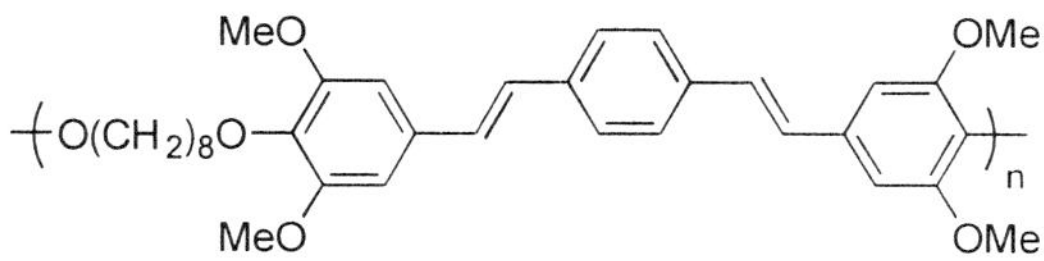

2

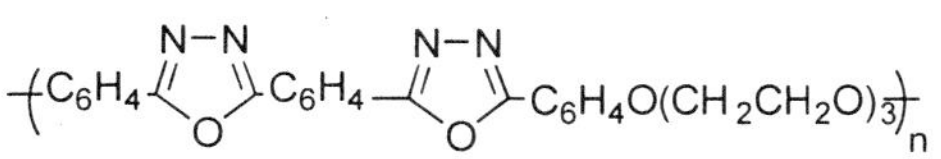

3

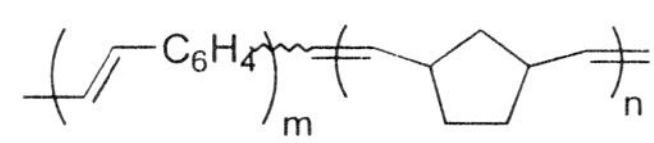

4

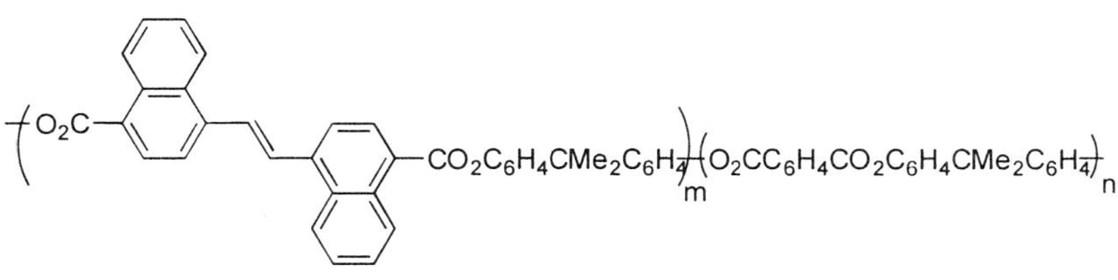

5

In this paper we summarize our recent investigations on the silylene-spaced conjugated copolymers.

Silylene Spacer

The presence of a silyl substituents on the conjugated system may change the band gap. Depending on the position of the silyl group, the band gap can be either increased (*15*) or reduced (*16*). There has been an increasing use of a tetrahedral silylene moiety as a bridge connecting chromophores in polymers (*16-20*). Wittig reaction of a dicarbonyl compound with preformed aryl-silane linkage provides a useful entry for the silylene-bridged copolymer **6** (*17,20*). Copolymers containing oligothiophenes and silanylene spacers **7** exhibit emission ranging from blue to red according to the number of thiophene and silane units (*18*).

More recently, we have employed the strategy using hydrosilylation of alkynes for the preparation of copolymers **8** for optoelectronic interests (*19*). The conjugated moiety in **9** are obtained conveniently by employing coupling reactions (eq 1) or Heck reaction (eq 2). The key idea of this hydrosilylation approach relies on our convenient procedure (*21*) for the synthesis of vinylic silyl hydrides **10** from the corresponding dithioacetals **9** (eq 3). Representative examples are summarized in Table 1. The corresponding monomers **11** are prepared in a similar manner for comparison.

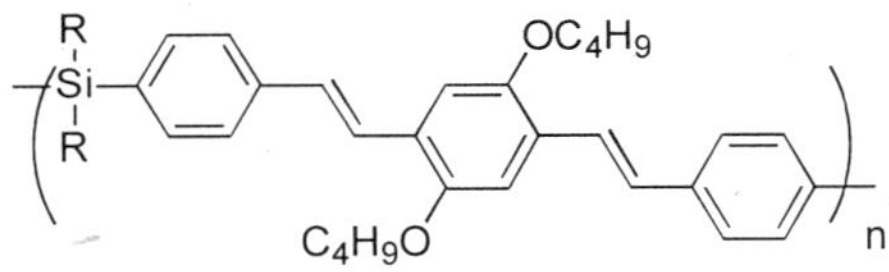

6

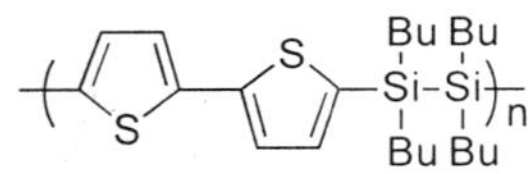

7

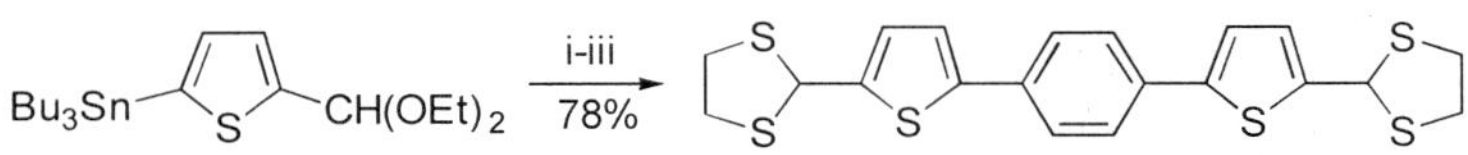

(1)

9b

(i) I-C_6H_4-I, cat.$PdCl_2(PPh_3)_2$, (ii) 10%HCl, (iii) 1,2-ethanedithiol, BF_3OEt_2

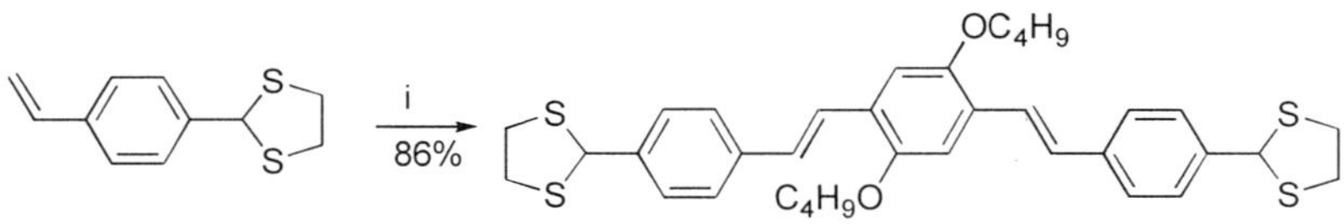

(2)

9c

(i) I-[2,5-$(BuO)_2C_6H_2$]-I, cat.$Pd(OAc)_2$, P(o-tolyl)$_3$, NEt_3, DMF

9 **10**

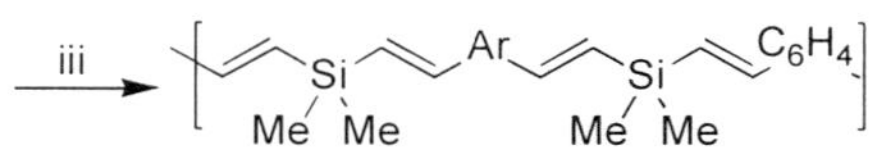

(3)

8

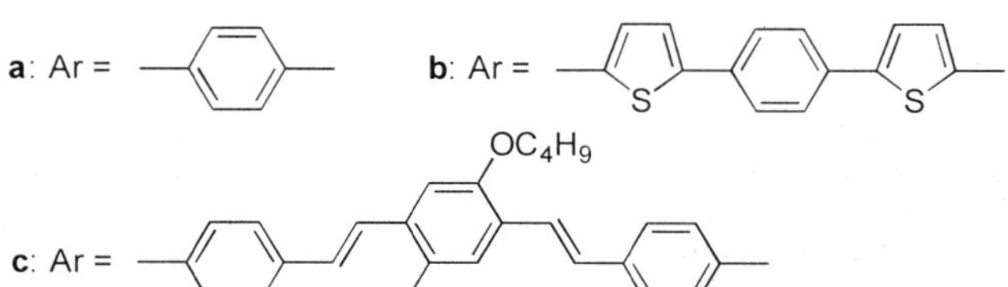

(i) (^{i}PrO)Me_2SiCH_2MgCl, $NiCl_2(PPh_3)_2$, (ii) $LiAlH_4$ (iii) $HC\equiv CC_6H_4C\equiv CH$, $RhCl(PPh_3)_3$

Table 1 Synthesis of Polymers **8**

9	%yield (**10**)	%yield (**8**)	Mn (PDI)
a	44	85	10667(3.5)
b	62	80	6167(2.0)
c	49	81	9759(1.6)

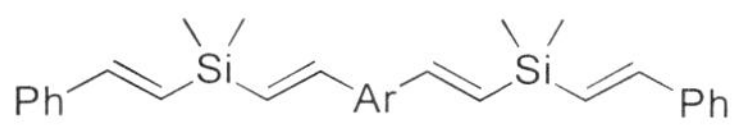

11

Photoluminescence

The fluorescence spectra for polymers **8** were examined and are compared with those of the corresponding monomers **11** (Figures 1-3). As shown in Figure 1, **8a** in solution exhibited dual fluorescence spectra. The profiles remains essentially unchanged with the solvent and with the concentration (10^{-5} to 10^{-7} M). The higher energy emission at ca 340 and 360 nm for **8a** is compatible with those for the monomer **11a**. The relative intensity of the emission in the blue light region (ca 420 nm) increases with the molecular weight of **8a**, such interaction becomes more important as the polymer becomes larger (*19*). Intramolecular exciplex formation between the chromophores in **8a** has been proposed. Hartree Fock (3-21G*) calculations on divinyl and distyrylsilanes suggested that the molecules are quite flexible (*19*). Accordingly, the opportunity for one chromophore unit in **8a** located proximal to the other in space would increase with the molecular weight. Related

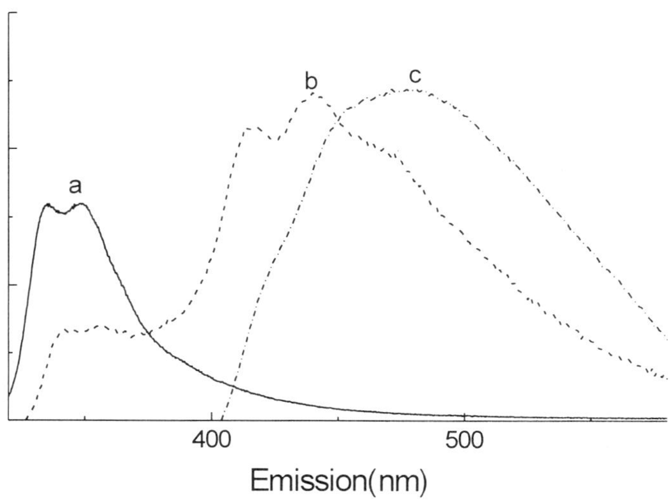

Figure 1 Emission spectra of a: **11a** (1×10^{-5} M in $CHCl_3$), b: **8a** (2×10^{-9} M in $CHCl_3$), and c: thin film of **8a**.

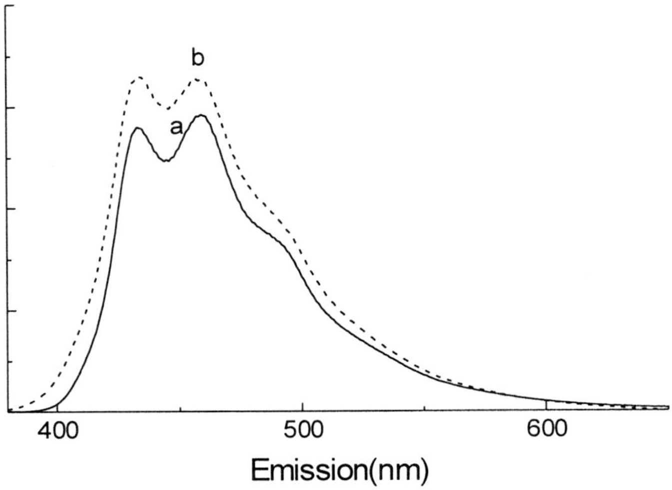

Figure 2 Emission spectra of a: **11b**(1 x 10^{-5} M in $CHCl_3$); b: **8b** (8 x 10^{-10} M in $CHCl_3$),.

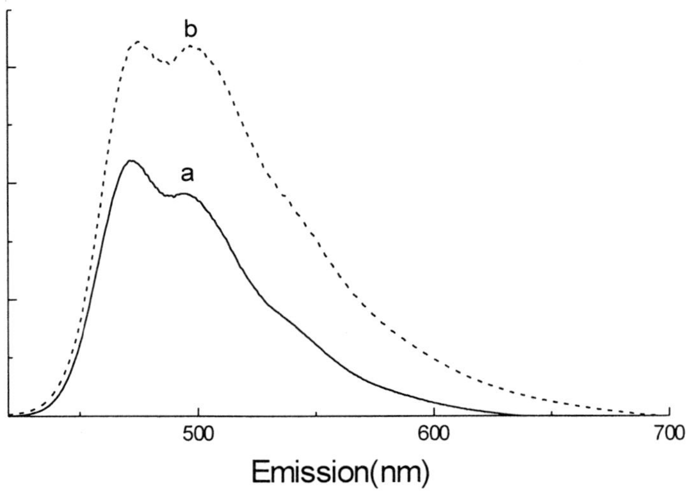

Figure 3 Emission spectra of **8c** (3 x 10^{-6} M in $CHCl_3$), **11c** (3 x 10^{-6} M in $CHCl_3$

photophysical behavior has also been observed in block copolymers obtained by ring opening metathesis polymerization of [2,2]paracyclophanene and norbornene (*13b*).

As shown in Figure 1, the vibronic fine structures ($\Delta\nu$ = 1507 cm^{-1}) for **8a** were also observed in this blue light region. This observation indicates that intrachain aggregation may also occur in the ground state. Intermolecular aggregation is known for polymers in solid film and therefore exhibits characteristic emission at longer wavelength (*22*). Indeed, the fluorescence of the thin film of **8a** appears at much longer wavelength than that of **8a** in solution (Figure 1). These results further suggest that intramolecular interaction between lumiphores in **8a** is responsible for the emission in solution at ca 420 nm. Because of such kind of interaction, divinylbenzene chromophore in these silylene-spaced copolymers will emit in the blue light region.

In contrast to the photophysical properties of **8a**, there is not much difference in the emission spectra between polymers **8b**, **c** and the corresponding monomers **11b**, **c** (Figures 2 and 3). This implies that the above-mentioned intrachain interaction may not occur in these polymers. Relatively speaking the conjugated moieties in **8b** and **c** are more rigid and therefore the steric requirement might prohibit the chromophores in these polymers in close proximity.

The fluorescence quantum yields of polymers **8** and the corresponding monomers in $CHCl_3$ solution are summarized in Table 2. Polymers **8b** and **c** were chosen for the electroluminescence (EL) investigation because of its high efficiency in photoluminescence (PL).

Table 2 Excitation wavelength (λex) and fluorescence quantum yield (Φ) of **8** and **11**

Substrate	λex (nm)	Φ
8a	300	0.026
8b	375	0.327
8c	407	0.562
11a	300	0.006
11b	375	0.217
11c	407	0.451

Electroluminescence

A blue-green emitting thin film electroluminescent device using **8c** was investigated. Universal device was employed for the LED studies (Figure 4). It is known that polymer blends of a hole-transporting polymer polyvinylcarbazole (PVK) with an emitting polymer such as polyphenylenephenylenevinylene (PPPV) will increase quantum efficiencies (*23*). In the present study, the hole-transport/ emitting layer containing a mixture of PVK doped with **8c** (1:0.2) was spin-coated onto the ITO glass. The thickness of the film measured with a profilometer (Alpha Step 500) was controlled by the speed of spin-coater and the concentration of the solution. The electron-transport layer, tris(quinoline-8-hydroxylate)-aluminum (Alq), and the cathode, Mg:Ag (10:1) were vacuum evaporated. Figure 5 shows the EL spectra of

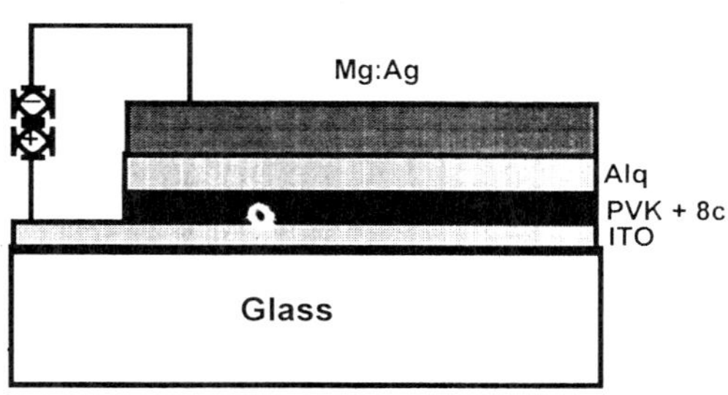

Figure 4 Structure of LED device

the device driven at 18 and 28 volt and the PL spectrum of the thin film of **8c**. The intensity of the EL peak varies with the applied voltage although the half-width remains at about 100 nm, and the position of the peak is at 516 nm (green) at the low voltage (18 V) but shifts to 496 nm (blue-green) at the high voltage (28 V).

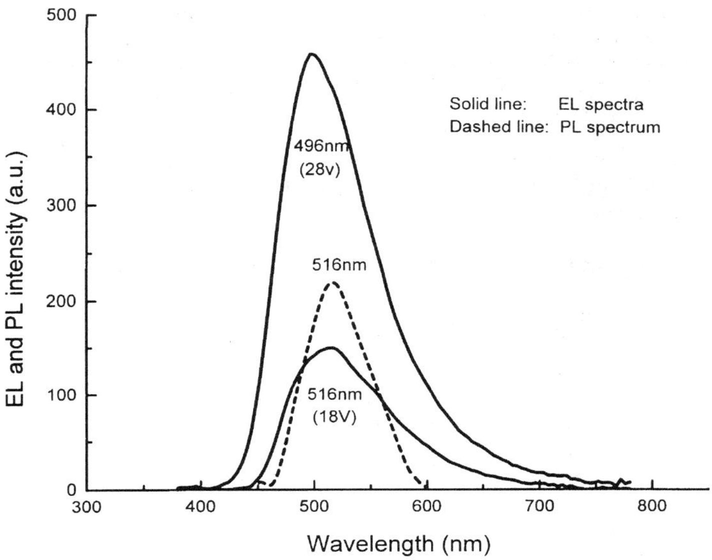

Figure 5 The PL spectrum of 8c and the EL spectra of the device

The brightness-current-voltage (B-I-V) curves of the device are shown in Figure 6. The threshold voltage was observed to be 12 V in the device. The EL brightness reached 800 cd/m^2 under 28 V and about 89 mA/cm^2. The EL peak emission of the device blue-shifted with increasing applied voltage. Figure 7 shows the peak position of EL versus the applied voltage. The wavelength of the peak position decreased by 24 nm (from 520 to 495 nm) as the applied voltage increased from 12 to 28 V.

As shown in Figure 5, the EL emission of the LED is similar to the PL spectrum of the thin film of **8c**, particularly when the device is driven at 18 V, indicating that the EL emission originates mainly from **8c**. This observation together with the absence of the blue emission from PVK suggested that there was an energy transfer from PVK to **8c** or the energy of the excitons was too low for PVK to emit. Under lower applied voltage, the EL is perhaps attributable to the emission from the Alq layer because the EL peak of the device is at 520 nm, which is close to the Alq emission at 524 nm. This suggests that at low electric field, the electrons are slowly moving so that electron-hole recombination occurs predominantly in the Alq layer. With increasing applied voltage, the electron mobility increases and the electron-hole recombination zone moves increasingly towards the PVK/**8c** layer. As a result, with increasing voltage the EL emission has an increasing contribution from **8c** and the EL peak shifts more to the blue. At still higher field, the recombination occurs near the ITO electrode from which reflection may become important and the effect of microcavity may set in, resulting in a further shortening of the wavelength to less than the PL peak of **8c**.

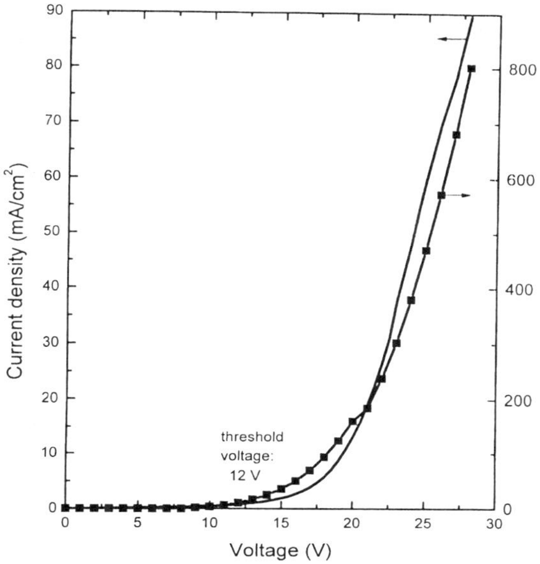

Figure 6 The brightness-current-voltage characteristics of the device

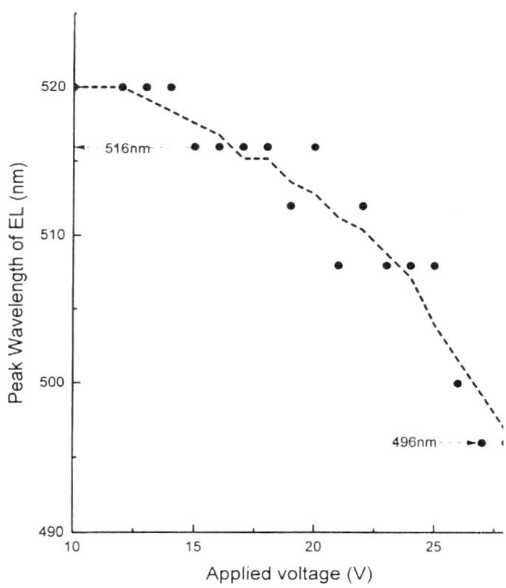

Figure 7 The change of EL peak position versus applied voltage (dot: experimental Value; dashed line: fitted curve of the data).

As shown in Figure 2, **8b** emits in the blue region. A similar device as shown in Figure 4 was made but without the alq electron transporting layer. This device gave blue light emission when the threshold voltage 15 V was applied.

Conclusion

In summary, we have demonstrated a new class of electroactive copolymers of silylene-spaced conjugated segments **8** which can be conveniently synthesized from the readily accessible starting materials. The photophysical studies indicate that **8a** exhibit strong intrachain aggregation at both ground and excited states leading to longer wavelength emission in the blue light region.

Polymer **8c** can serve as as an emitting dopant for the fabrication of a blue-green organic LED. The peak EL position of the LED device can be blue-shifted with increasing applied voltage. The present observation of the voltage-dependent EL emission suggests a new avenue for controlling the color of LEDs.

Acknowledgment. This work was supported by the National Science Council of the Republic of China (to TYL and RMC) and by the Strategic Research Grant (Project No. 7000771) of City University of Hong Kong (to ZBD and STL).

References

1. Tang, C. W.; Van Slyke, S. A. *Appl. Phys. Lett.* **1987**, *51*, 913.
2. For a recent review, see: Kraft, A.; Grimsdale, A. C.; Holmes, A. B. *Angew. Chem. Int. Ed. Engl.* **1998**, *37*, 402.
3. Burroughes, J. H.; Bradley, D. D. C.; Brown, A.R.; Marks, R.N.; Mackay, K.; Friend, R. H.; Burns, P. L.; Holmes, A. B. *Nature* **1990**, *347*, 539.
4. Gustafsson, G.; Cao, Y.; G.M. Treacy, G. M.; F.Klaveter, F.; N. Colaneri, N; Heeger, A. E. *Nature* **1992**, *357*, 477.
5. Hu, B.; Yang, Z.; Karasz, F. E. *J. Appl. Phys.* **1994**, *76*, 2419.
6. Kido, J.; K. Hongawa, K.; Okuyana, K.; Nagai, K. *Appl. Phys. Lett.* **1994**, *64*, 815.
7. Tao, X. T.; Suzuki, H.; Watanabe, T.; Lee, S. H.; Miyata, S.; Sasabe, H. *Appl. Phys. Lett.* **1997**, *70*, 1503.
8. Schumm, J. S.; Pearson, D. L.; Tour, J. M. *Angew. Chem. Int. Ed. Engl.* **1994**, *33*, 1360. Gebhardt, V.; Bacher, A.; Thelakkat, M.; Stalmach, U.; Meier, H.; Schmidt, H.-W.; Hannrer, D. *Synth. Met.* **1997**, *90*, 123.
9. (a) Ryu, M.-K; Lee, S.-M.; Zyung, T.; Kim, H. K. *Polym. Mater. Sci. Eng.* **1996**, *75*, 408. (b) Adachi, C.; Tautsui, T.; Saito, S. *Appl. Phys. Lett.* **1990**, *56*, 799. (c) Pope, M.; Swenberg, C. E. *Electronic Processes in Organic Crystals*; Oxford U: Oxford, 1982. (d) Kido, J.; Kohda, M.; Okuyama, K.; Nagai, K. *Appl. Phys. Lett.* **1992**, *61*, 761. (e) Yang, Z.; Sokolik, I.; Karasz, F. E. *Macromolecules* **1993**, *26*, 1188. (f) Kim, D. J.; Kim. S. H.; Lee, J. H.; Kang, S. J.; Kim, H. K.; Zyung, T.; Cho, I.; Choi, S. K. *Mol. Cryst. Liq. Cryst.* **1996**, *280*, 391. (g) Kim, H. K.; Ryu, M.-K.; Lee, S.-M. *Macromolecules* **1997**, *30*, 1236.
10. (a) Herrema, J. K.; Wildeman, J.; Wieringa, R. H.; Malliaras, G. G.; Lampoura, S. S.; Hadziioannou, G. *Polym. Prepr. (Am. Chem. Soc., Div. Polym. Chem.)* **1993**, *34*, 282. (**b**) Malliaras, G. G.; Herrema, J. K.; Wildeman, J.; Wieringa, R. H.; Gill, R. E.; Lampoura, S. S.; Hadziioannou, G. *Adv. Mater.* **1993**, *5*, 721. (c) Wildeman, J.; Herrema, J. K.; Hadziioannou, G.; Schomaker, E. *J. Inorg. Organomet. Polym.* **1991**, *1*, 567. (**e**) Herrema, J. K.; van Hutten, P. F.; Gill, R. E.; Wildeman, J.; Wierignga, R. H.; Hadziioannou, G. *Macromolecules* **1995**, *28*, 8102. (f) Brouwer, H. J.; Krasnikov, V. V.; Hilberer, A.; Hadziioannou, G. *Adv. Mater.* **1996**, *8*, 935.
12. Pei, Q.; Yang, Y. *Adv. Mater.* **1995**, *7*, 559. Yang, Z.; Sokolik, I.; Karasz, F. E. *Macromolecules* **1993**, *26*, 1188.

13. (a) Zhang, C.; von Seggern, H.; Kraabel, B.; Schmidt, H.-W.; Heeger, A. *J. Synth. Met.* **1995**, *72*, 185. (b) Bazan, G. C.; Miao, Y.-J.; Renak, M. L.; Sun, B. J. *J. Am. Chem. Soc.* **1996**, *118*, 2618.
14. Lee, J.-K.; Schrock, R. R.; Saigent, D. R.; Friend, R. H. *Macromolecules* **1995**, *28*, 1966. Hesemann, P.; Vestweber, H.; Pommerehne, J.; Mahrt, R. F.; Greiner, A. *Adv. Mater.* **1995**, *7*, 388. Li, X.-C.; Giles, C. M.; Gruner, J.; Friend, R. H.; Holmes, A. B.; Moratti, S. C.; Yong, T. M. *Adv. Mater.* **1995**, *7*, 898. Boyd, T. J.; Geerts, Y.; Lee, J.-K.; Fogg, D. E.; Lavoie, G. G.; Schrock, R. R.; Rubner, M. F. *Macromolecules* **1997**, *30*, 3553.
15. Höger, S.; McNamara, A.; Schricker, S.; Wudl, F. *Chem. Mater.* **1994**, *6*, 171.
16. Garten, F.; Hilberer, A.; Cacialli, F.; Esselink, E.; van Dam, Y.; Schlatmann, B.; Friend, R. H.; Klapwijk, T. M.; Hadziioannou, G. *Adv. Mater.* **1997**, *9*, 127.
17. Kim, H. K.; Ryu, M.-K.; Lee, S.-M. *Macromolecules* **1997**, *30*, 1236.
18. Malliaras, G. G.; Herrema, J. K.; Wildeman, J.; Wieringa, R. H.; Gill, R. E.; Lampoura, S. S.; Hadziioannou, G. *Adv. Mater.* **1993**, *5*, 721.
19. Chen, R.-M.; Chien, K.-M.; Wong, K.-T.; Jin, B.-Y.; Luh, T.-Y.; Hsu, J.-H.; Fann, W. *J. Am. Chem. Soc.* **1997**, *119*, 11321. Chen, R.-M.; Luh, T.-Y. *Tetrahedron* **1998**, *54*, 1197.
20. Miao, Y.-J.; Bazan, G. C. Macromolecules **1997**, *30*, 7414.
21. (a) Ni, Z.-J.; Yang, P.-F.; Ng, D. K. P.; Tzeng, Y.-L.; Luh, T.-Y. *J. Am. Chem. Soc.* **1990**, *112*, 9356. (b) For reviews, see: Luh, T.-Y. *Acc. Chem. Res.* **1991**, *24*, 257. *Pure Appl. Chem.* **1996**, *68*, 105.
22. So, Y.-H.; Zaleski, J. M.; Murlick, C.; Ellaboudy, A. *Macromolecules* **1996**, *29*, 2783 and references therein. Cornil, J.; dos Santos, D. A.; Crispin, X.; Silbey, R.; Brédas, J. L. *J. Am. Chem. Soc.* **1998**, *120*, 1289 and references therein.
23. Zhang, C.; von Seggern, H.; Pakbaz, K.; Draabel, B.; Schmidt, H.-W.; Heeger, A. J. Synth. etqls

Chapter 24

Electroactive Aniline Oligomers of Well-Defined Structures and Their Polymeric Derivatives

Yen Wei [1,6], Shuxi Li [1], Xinru Jia [1], Ming-Hsiung Cheng [1], Mat W. Mathai [1], Jui-Min Yeh [1], Wei Li [1], Susan A. Jansen [2], Zhi Yuan Wang [3], Chuncai Yang [1,3], Jian Ping Gao [3], Moshe Narkis [4], Arnon Siegmann [4], and Bing R. Hsieh [5]

[1] Department of Chemistry, Drexel University, Philadelphia, PA 19104
[2] Department of Chemistry, Temple University, Philadelphia, PA 19121
[3] Department of Chemistry, Carleton University, Ottawa K1S 5B6, Canada
[4] Department of Chemical Engineering, Technion-Israel Institute of Technology, Haifa 32000, Israel
[5] Xerox Corporation, The Wilson Center for Research and Technology, 800 Phillips Road, MS 114-39D, Webster, NY 14580

A general strategy for the synthesis of aniline oligomers with controlled molecular weight, narrow or unit molecular weight distribution and designable end-groups has been developed based on the theory of non-classical or reactivation chain polymerization. A series of oligomers have been prepared by oxidative polymerization of aniline in the presence of N-phenyl-1,4-phenylenediamine or 1,4-phenylenediamine as initiator. The molecular weight of the oligomers is controlled by varying the amount of initiator. Generally, lower oligomers can serve as the initiators to build higher oligomers. The oligomers with minimum 4 nitrogen atoms and 3 phenylene rings exhibit similar characteristic redox behavior and electroactivity as polyaniline. Electronic conductivity of the oligomers of 7 or 8 aniline units approaches that of polyaniline. Solubility of the oligomers is much improved over that of conventional polyaniline. Various functional groups can be introduced to the oligomers either by proper selection of starting materials or by post-synthesis modifications via common organic reactions. The functionalized oligomers undergo further polymerizations to afford a variety of new electroactive materials, including polyamides, polyimides, polyureas, polyurethanes, polyacrylamides and epoxy polymers. There are numerous potential applications for the oligomers and their polymeric derivatives.

Conductive polymers, such as polythiophene, polyaniline, polypyrrole, poly(p-phenylene vinylene), etc., have been studied extensively and explored for numerous potential applications because of their high conductivity and, probably more importantly, their electroactivity and other unique properties.[1-5] Recently, there have

[6] Corresponding author.

been many encouraging developments towards commercialization of the electroactive polymers for applications such as electroluminescent devices, corrosion resistant coatings and electrostatic dissipation coatings and fabrics.[5] However, one problem associated with large-scale commercial applications is the limited processibility of electroactive polymers. For instance, polyaniline solution in N-methyl-2-pyrrolidinone (NMP) often forms gels and unsubstituted polythiophene and polypyrrole are not soluble in common organic solvents. Furthermore, among the inherent drawbacks of the electroactive polymers, like most synthetic polymers, are the non-unity polydispersity in molecular weight and the existence of structure defects, which become obstacles for many electronic and optical applications that demand the materials of well-defined structures and high chemical purity.

In an effort to solve these problems, our research in the last several years has focused on the synthesis of electroactive oligomers of well-defined structures. The basic idea of our approach has been that the electroactive oligomers with the same repeating unit structures as their corresponding polymers can be prepared to have well-defined structures, controllable molecular weights, very narrow or unit polydispersity and designable functional end-groups. Because of the low molecular weights, these oligomers should have much enhanced solubility and, therefore, could be purified, thoroughly characterized and studied for their physical and chemical properties. On the other hand, since the end-groups can be readily varied to polymerizable functionalities, the oligomers function as monomers for further polymerization to yield a variety of new polymers, which possess both excellent mechanical properties of typical polymers and the electroactivity of the conducting polymers to a certain extent. It is particularly noteworthy that electroactive oligomers often exhibit similar or even improved physicochemical properties. As examples, the aniline oligomers appear to have good or even better anticorrosion properties.[6,7] The electronic properties of the aniline octamer were reported to be comparable to polyaniline.[8] Defect-free thiophene oligomers such as unsubstituted or α,ω-substituted sexithiophenes inherently possess the basic or much improved electronic and optical properties of polythiophene.[9] The p-phenylenevinylene oligomers have also been synthesized and investigated for electroluminescence applications.

In this article, we demonstrate our oligomer approach to the electroactive polymeric materials with oligoanilines as specific examples. The synthesis, properties and further polymerization of amino-terminated trianilines and other higher oligomers are reviewed and discussed. The synthesis of several new polymers including polyureas and polyurethanes derived from the α,ω-end-functionalized aniline oligomers is presented. Some of the potential applications of the oligomers and their polymeric derivatives are described.

A General Strategy for the Synthesis of Aniline Oligomers

There have been a number of methods reported for the preparations of aniline oligomers in the literature.[8,11-14] Most of them involved multi-step synthetic reactions or unstable reagents. Since later 1980s, Wei and coworkers have been interested in the development of fundamental understanding of the polymerization mechanism and, therefore, better methodologies for the synthesis of electroactive polymers including polyaniline[15-20], polythiophene[21,22] and polypyrrole[23]. A new general theory has been proposed as "non-classical chain polymerization" or "reactivated chain polymerization".[15-24] In this theory, the polymer chain propagation is accomplished by reactions of a reactive chain end (M_m*) with incoming monomer (M) leading to formation of a non-reactive or dormant product with higher molecular weight (M_{m+1}). This dormant chain is then reactivated chemically (e.g., oxidation or reduction, etc.), physically (e.g., heating or radiation, etc.) or biologically (e.g., enzymatic action) to become a reactive chain end (M_{m+1}*), which reacts with a monomer to complete another chain growth step yielding again a dormant chain (M_{m+2}):

$$M_{m+1} \xrightarrow[\text{Reactivation}]{} M_{m+1}^{*} \xrightarrow[\text{Chain growth}]{M} \longrightarrow M_{m+2}$$

This new class of polymerization process is neither a classical chain nor a classical step polymerizations. Examples of this polymerization could include many synthetical and natural polymerizations, such as the oxidative polymerization of anilines and other aromatic compounds[15-24], the recently emerged living radical polymerizations[25,26], in which dissociation of the capping groups can be considered as the reactivation step to allow the addition of one or more monomers before the re-capping, and biological polymerizations for the biosynthesis of nucleic acids, proteins and polysaccharides. Under proper conditions, the polymerization could be stopped at desired stages by not reactivating the dormant chains, implying that the non-classical chain polymerization could become a living polymerization. Application of this new theory has led to the development of a general, one-step synthesis of aniline oligomers of well-defined structures with controllable end-groups and molecular weights. In this method, various amounts of aromatic amine additives, such as 1,4-phenylenediamine (PDA) or N-phenyl-1,4-phenylenediamine (PPDA), were introduced into the aniline polymerization system in aqueous HCl solution with ammonium persulfate as oxidant at about -5 °C. The reaction conditions were very similar to conventional aniline polymerization[3], except for the presence of the additives. As illustrated in Scheme 1 with PPDA as example, the aniline dimer PPDA can be considered as a chain initiator or a "dormant" chain (M_m), which could be reactivated to reactive species such as nitrenium ions, iminium ions or others (M_m*) by oxidation, while aniline is considered as the monomer (M) because it has a higher oxidation potential than the M_m compound[16]. The highly reactive M_m* attacks the aniline monomer at the para position followed by deprotonation to afford a "dormant" aniline trimer (M_{m+1}), i.e., N-phenyl-4,4'-diaminodiphenylamine. Since the trimer has an even lower oxidation potential, it will be oxidized to form the reactive M_{m+1}*, which again reacts with another aniline monomer resulting in M_{m+2}. The process repeats to yield higher oligomers and polymers[16].

The aniline dimer PPDA plays a similar role as the chain initiator in a classical chain polymerization. Therefore, the amount of PPDA added to the aniline polymerization should significantly affect the molecular weight of polymers. Indeed, we have found that the molecular weight decreases as the amount of PPDA is increased (Fig. 1). At 20 mol-% PPDA, the number-average molecular weight (M_n)

Scheme 1

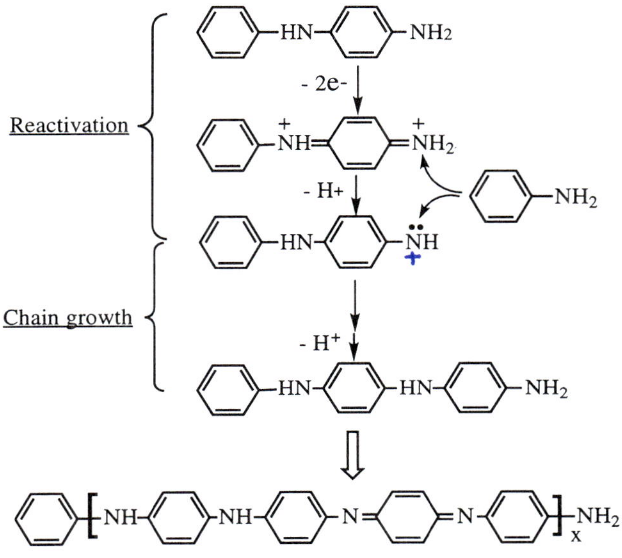

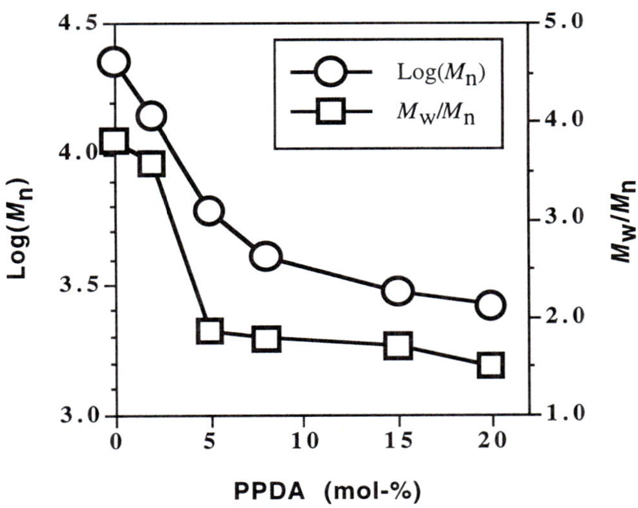

Fig. 1. Effect of the amount of N-phenyl-1,4-phenylenediamine (PPDA) on the molecular weight (M_n) and molecular weight distribution (M_w/M_n) of aniline polymer or oligomers.

was only about 2600 as determined by gel-permeation chromatography (GPC) with polystyrene calibration[27]. The product can be best described as an oligomer. Furthermore, there is a clear trend, as shown in Fig. 1, that the molecular weight distribution (i.e. M_w/M_n) becomes narrower as the amount of PPDA is increased. Similarly, in the system with PDA as the initiator, both the molecular weight and molecular weight distribution also decrease with the increase in the PDA concentration. The oligomers were found to have two amino end-groups. It should be noted that with either PDA or PPDA as the initiator at 2-5 mol% or higher, the GPC curve of the product changes from a bimodal pattern to a single narrow peak.

Scheme 2

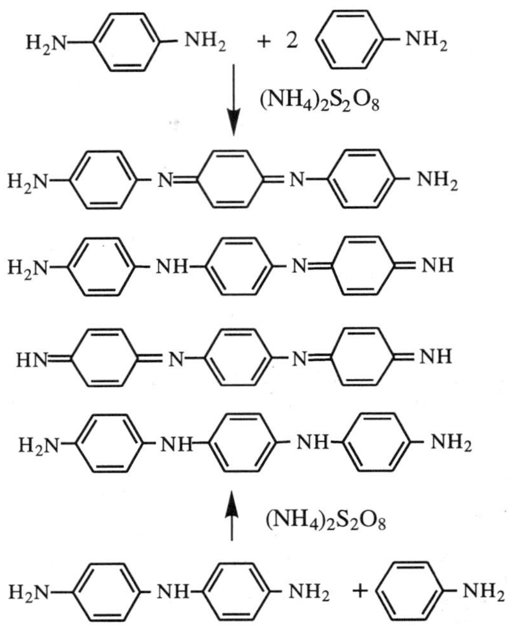

1

2

3

4

When a very high concentration of PDA (i.e. at the molar ratio of PDA to aniline of 1:2), an amino-capped aniline trimer can be synthesized in one step as shown in Scheme 2.[28] There are four possible structures (**1-4**) for the trimeric product depending on the oxidation state and the position of the quinodal diimine unit. Under the present experimental conditions, the product typically contains the isomer **1**, i.e., N,N'-bis(4'-aminophenyl)-1,4-quinonenediimine, as the major component and relatively small amount of isomer **2**. Upon chemical or electrochemical reduction, a single compound (**4**), i.e., N,N'-bis(4'-aminophenyl)-1,4-phenylenediamine, was obtained, which could be recrystallized to give needle-shaped crystals. Various C- or N-substituted anilines could also be employed to yield a series of the trimeric derivatives[28]. The trimeric products were also prepared by oxidative reactions of an amino-terminated aniline dimer, i.e. 4,4'-diaminodiphenylamine, with aniline monomer at 1:1 molar ratio in the presence of an oxidizing agent such as ammonium persulfate (Scheme 2) [29,30]. In general, the oligomers with relatively lower oxidation potentials than the monomers can be considered as the initiators to build higher oligomers. Using this general strategy with the trimer as the starting material to react with monomeric or dimeric anilines, a series of higher oligomers (e.g., tetramers to heptamers, Scheme 3) with one or two amino end-groups have been successfully prepared and isolated at a number of oxidation states[31]. The reduced form of the trimer (**4**) as the starting material appears to give better yields when the persulfate was used as the oxidant. By choosing appropriate starting materials and reactant stoichiometry, the oligomers containing both odd and even numbers of nitrogen atoms could be synthesized[28,29,31].

Scheme 3

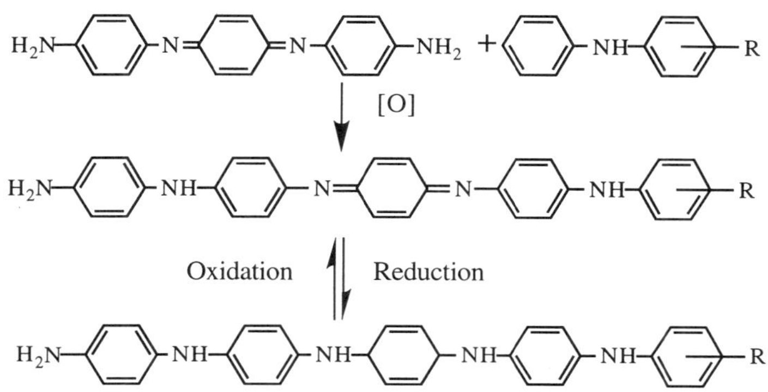

R = -H, *m* -Me, *m* -OMe, *p* -NH_2, -NH-Ph, or *p*-NH-Ph-NH-Ph

Redox Reactions and Electronic Properties of the Oligomers

All the oligomers undergo similar redox reactions as polyaniline (Scheme 4). Their oxidation states can be changed chemically or electrochemically[28-31]. Chemical reduction can be achieved by treatment with reducing agents such as phenyl hydrazine, hydrazine, boron hydrides, etc., or by hydrogenation in the presence of catalysts such as platinum oxide, while oxidation takes place upon exposure to air or treatment with oxidizing agents such as hydrogen peroxide, ammonium persulfate, tetraacyl lead, etc. The redox reaction can also be carried out electrochemically. Cyclic voltammograms (CV) of the oligomers with at least four nitrogen atoms and three phenylene rings (e.g., the trimer **4** or higher oligomers) have a pattern of at least two redox pairs, which is similar to that of polyaniline although the specific values of redox potentials are different. It is noted that the electrochemical behavior (i.e., CV patterns) of the lower oligomers such as 1,4-phenylenediamine and the amino-terminated dimer **5** differs significantly from that of polyaniline. Both the theoretical calculations and experimental results suggest that the minimum length of the oligomer should be four nitrogen atoms and three phenylene rings in order to have maintain the essential electrochemical characteristics of polyaniline[32-34]. Such as minimum structure would allow all the typical variations of oxidation state, e.g., leucoemeraldine (**4**), emeraldine (**1** and **2**) and pernigraniline (**3**), as in polyaniline (Scheme 4). Electronic conductivity of the oligomers upon doping with 1.0 M HCl was measured by the standard 4-probe technique. As shown in Fig. 2, the logarithm of conductivity seems to increase linearly with the length (i.e., the number of aniline units) of the oligomer. The heptamer exhibited a conductivity close to 0.1 S/cm, which is approaching the value (1-10 S/cm) for polyaniline under identical doping and measurement conditions. The electronic absorption λ_{max} for the benzenoid to quinoid excitonic transition in the emeraldine base (EB) form of the heptamer was

Scheme 4

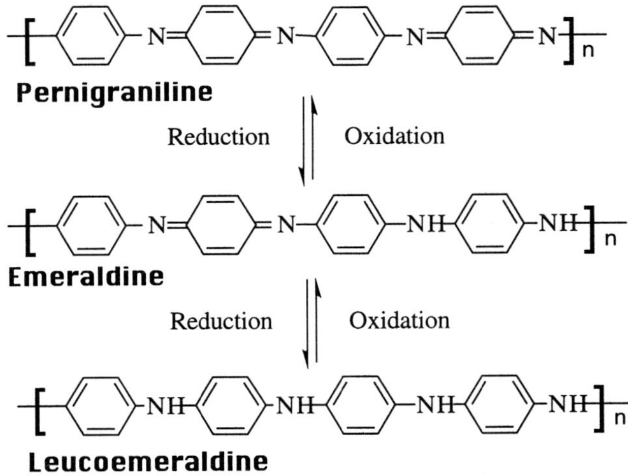

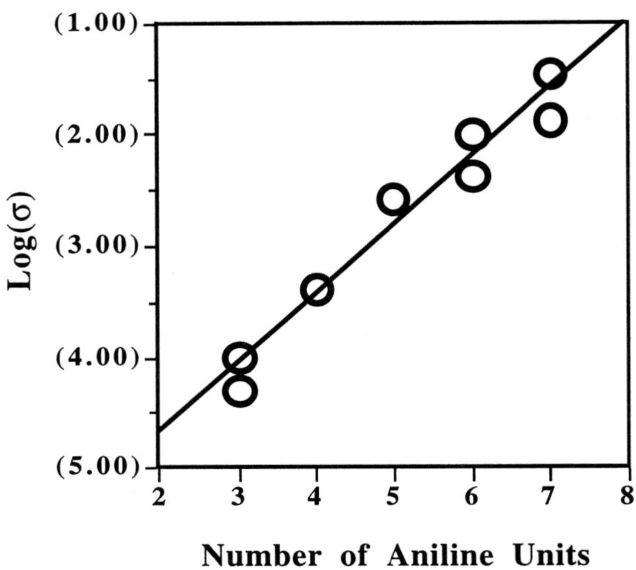

Fig. 2. Plot of Log(conductivity) vs. number of anilino unit in the aniline oligomers, doped with 1 M HCl. Unit of conductivity (σ): S/cm.

about 620 nm, again approaching that (~636 nm) for the EB of polyaniline[35]. These observations are in line with those reported by Wudl et al. for the aniline octamer[8]. It is hence reasonable to conclude that the oligomers with 7-8 aniline units should be sufficient to reach the conductivity level of HCl-doped polyaniline in a largely amorphous state and to exhibit UV-vis characteristics comparable to polyaniline. Recently, Buchwald et al. reported similar spectral results on the oligomers prepared via a clever synthesis in which the secondary amines (-NH-) on the main chain are protected as a *t*-butyl carbamate (BOC) during the palladium-catalyzed coupling reactions[36].

Functionalization of the Aniline Oligomers

In addition to the redox chemistry, the aniline oligomers may undergo a variety of reactions on the amino, phenylene and/or quinonediimine groups of the main chain and, more importantly, on the end functional groups. As mentioned in the previous sections, one or two primary or N-substituted amino end-groups were obtained directly by using appropriate reactants in the oxidative reactions. The amino end-groups are very versatile and could be readily transformed to many other functional groups through conventional organic reactions of aromatic amines, such as azide and diazonium chemistry, Schiff base formation, reductive amination, N-alkylation or acylation, etc. [30,37] Scheme 5 shows some examples of compounds (**5-8**) that could be derivatized from the reactions of trimer **1** or **4**. Since numerous possible functional groups could be introduced to the aniline oligomers, we should be able to build a library of new electroactive compounds with designable end-functionality for specific applications, including electrochemically switchable surfactants, electroactive monolayer or multilayer molecular and supramolecular assemblies, and further polymerizations to afford new organic or organic-inorganic hybrid polymers with electroactive oligomer building blocks. Furthermore, preliminary results of spectroscopic and electrochemical studies show that the aniline oligomers have interesting interactions with metals, metal ions and metal oxide clusters as well as with organic compounds such as tetracyanoquinodimethan (TCNQ)[38]. There have also been reports that the aniline trimers could be employed to build electroactive dendrimers[39].

Oligomers As Building Blocks for Further Polymerizations

In comparison with polyaniline, unsubstituted oligomers generally exhibit much improved solubility in organic solvents including NMP and the solutions are stable for a longer period of time (≥7 days) without any visible sign of gelation. The oligomers upto tetramers and some substituted pentamers show good solubility in common organic solvents such as tetrahydrofuran (THF), N,N-dimethylformamide (DMF), dimethyl sulfoxide (DMSO), etc. With appropriate C- or N-substitutions (e.g. -RCOOH and -RSO_3H) the oligomers could be made water-soluble. Hence, numerous new electroactive compounds and derivatives could be synthesized by conventional solution reactions. As examples, acrylamides with the aniline oligomer side groups were prepared by reacting the reduced form of the aniline tetramer, $H[C_6H_4NH]_3$-$C_6H_4NH_2$, with acryloyl chloride, which could then be polymerized via radical polymerization to afford polyacrylamides with electroactive side chains[29]. The amino-capped aniline oligomers could be employed as difunctional monomers in conventional step polymerizations. In this capacity, the electroactive amino-terminated trimer (e.g., **1**) is of particular importance because it represents the minimum structure unit of polyaniline and it can be readily synthesized in large quantity from a one-step reaction of inexpensive materials[28-31]. Availability of the trimer allowed us to exploit a variety of potentially electroactive polymers. Thus, polyamides (**9**)[29], polyimides (**10, 11**)[29,30,40], polyureas (**12**)[40], polyazomethines (**13**)[37], polyurethanes (**14**)[40] and epoxy polymers (**15**)[6,27] were obtained from step polymerization of the trimer (**1** or **4**) with dicarboxylic acid chlorides, dianhydrides, diisocyanates, dialdehydes, dioxyacid chlorides, and epoxy resins, respectively, as depicted in Scheme 6. Except for the crosslinked products such as epoxy polymers, the use of the reduced trimer (**4**) as the monomer appeared to result in higher molecular weight linear polymers than that of the oxidized trimer **1**. This can be attributed probably to (a) a decrease in the reactivity of the amino end-groups because of the electron-withdrawing effect of the quinonenediimine group and (b) the tautomerization of **1** to **2**, which has a less reactive imino (>C=NH) end-group[30].

Scheme 5

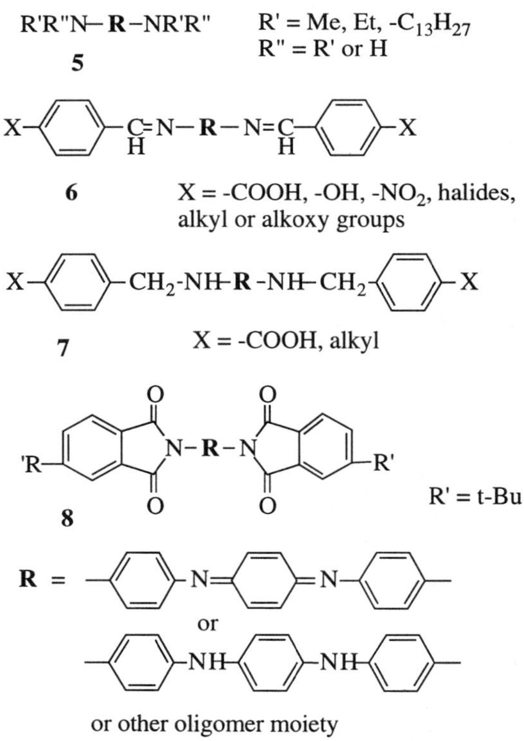

Scheme 6

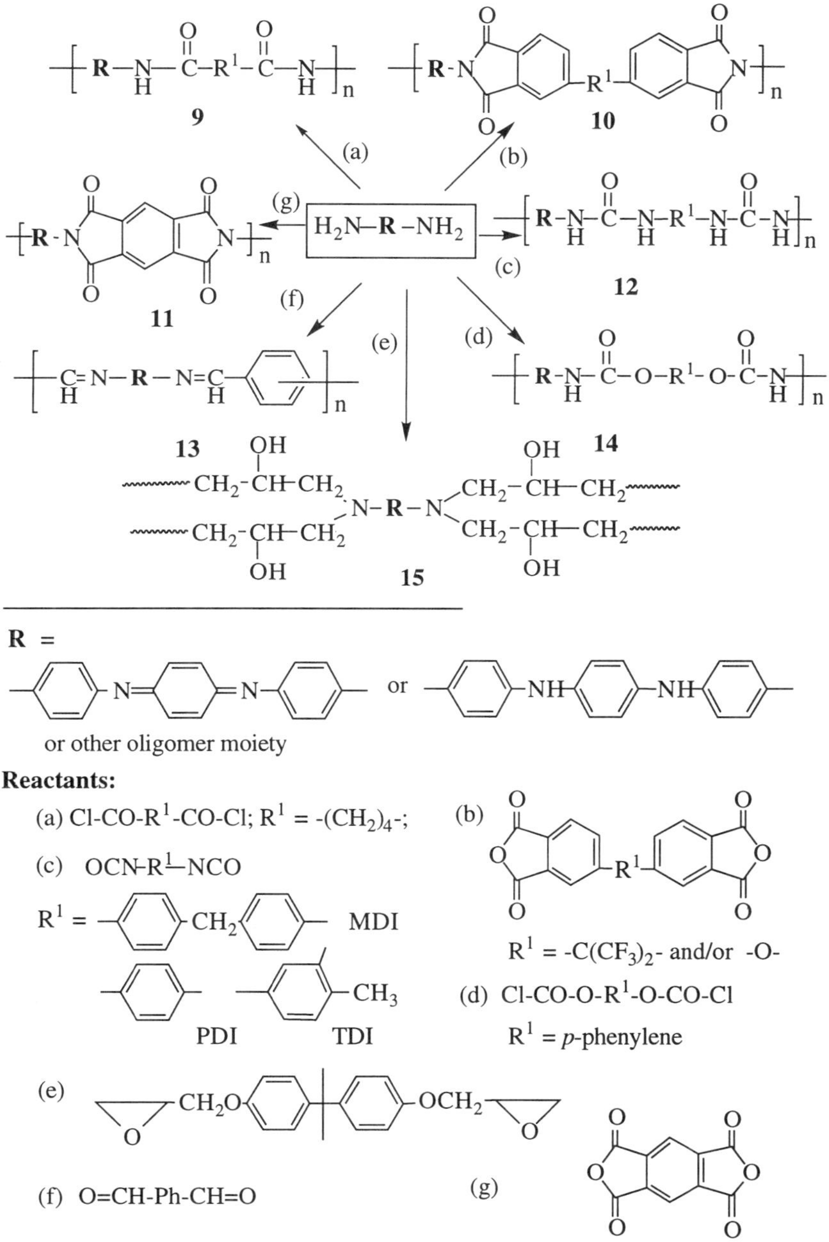

As a specific example of the synthesis of polymers with the aniline oligomer building blocks, various polyureas (**12**) were prepared from the polymerization of the amino-capped aniline trimer (**1**) with 4,4'-methylenebis(phenyl isocyanate) (MDI), toluene-2,4-diisocyanate (TDI) and 1,4-phenylene diisocyanate (PDI) in good yields (90-96%). The polymerizations were carried out at 100 °C for 3 to 4 hours in NMP solution in the presence of 1,4-diazabicyclo[2,2,2]-octane (Dabac) catalyst. The structures of the resultant polyureas were supported by elemental analysis, ^{1}H-NMR, FTIR and UV-visible spectroscopy. All the polyureas were of high molecular mass as evidenced by their large intrinsic viscosity values (≥1.0 dL/g), which were determined in dilute NMP solutions containing 5 wt% of LiCl at 25.0 °C. The readily attainable high molecular weights in this system could be attributed to the high reactivity of the isocyanate groups. The electronic properties and characteristics of both chemical and electrochemical redox reactions of the polyureas were found to be quite similar to those of the aniline trimers.

Engineering and Potential Technological Applications of the Oligomers

There are many possible technological applications for the aniline oligomers and both of their small molecular and polymeric derivatives. Like polyaniline, the oligomers and their derivatives could be used, for instance, as electroactive materials in fabricating electrochromic, electroluminescent, biosensor, and electroactuator devices, chemically or electrochemically tunable gas-separation membranes, anticorrosion and antistatic coatings, rechargeable batteries, etc.[1-5] Taking advantage of their well-defined structures and designable end-groups, properly functionalized oligomers could form a variety of molecular or supramolecular assemblies for potential electronic and optical applications. By introducing silane or alkoxysilyl groups in the oligomers, one should be able to prepare electroactive organic-inorganic hybrid or nanocomposite materials via the sol-gel reactions[41]. The oligomers or their metal complexes could find potential catalysis applications. Like many aromatic amines, the aniline oligomers in their reduced forms may serve as antioxidants and free-radical absorbents. The oligomers may also serve as a component in conventional or conducting polymer blends to renter the blends electroactive or more processible. It has been demonstrated by electrochemical[7,27] and X-ray photoelectron spectroscopic studies[34] that the anticorrosion performance of the aniline oligomers is as good as, and sometimes better than, that of polyaniline[42]. Besides the electroactivity, the polymers containing the oligomer building blocks should have many other desired chemical and physical properties, such as good mechanical strength and thermal stability, for applications as structural materials. Many of the above-mentioned applications have been explored or are currently under investigation in our laboratories.

Summary and Remarks

We have described a general methodology, which has been developed based on the theory of non-classical or reactivation chain polymerization, for the synthesis of aniline oligomers with controlled molecular weight, narrow or unit molecular weight distribution and designable end-groups. A series of oligomers were prepared by oxidative polymerization of aniline in the presence of N-phenyl-1,4-phenylenediamine or 1,4-phenylenediamine as initiator. The molecular weight of the oligomers is controlled by varying the amount of the initiator. As the initiator concentration is increased, the molecular weight decreases and the molecular weight distribution becomes narrower and approaches unity. Generally, lower oligomers can serve as the initiators to build higher oligomers. The oligomers with minimum 4 nitrogen atoms and 3 phenylene rings exhibit similar characteristic redox behavior and electroactivity as polyaniline. Electronic conductivity increases with the number of aniline units in the oligomers and approaches the value for polyaniline as the number is about 7 or 8. Solubility of the oligomers is much improved over that of conventional polyaniline. Various functional groups can be introduced to the oligomers either by proper selection of starting materials in the oxidative reactions or by post-synthesis modifications via conventional organic transformations. The functionalized oligomers can also undergo further polymerization to afford a variety of new electroactive materials, including polyamides, polyimides, polyureas, polyurethanes, polyacrylamides, and epoxy polymers. The electroactive conjugated oligomers of well-defined structures have many significant advantages over their corresponding polymers. There are also numerous potential applications for the oligomers and their polymeric derivatives.

Experimental Section

The synthesis of various substituted or unsubstituted aniline oligomers was reported in previous publications[27-31]. Preparation of new polyacrylamides[29], polyamides[29], polyimides[29,30,40], polyureas[40], polyazomethines[37], polyurethanes[40] and epoxy polymers[6,27] with electroactive oligomer building blocks was also reported. The synthesis of electroactive polyureas containing the aniline trimer (**1**) was achieved by following a modified procedure for the preparation of high molecular weight polyureas[43]. As a typical procedure, 0.581 g (2.0 mmol) of the aniline trimer (**1**)[28] and 2 mg of the catalyst 1,4-diazabicyclo[2,2,2]-octane (Dabac, 98%, Aldrich) were dissolved in 10 mL anhydrous N-methyl-2-pyrrolidinone (NMP). The blue solution was heated to 100 °C under stirring and nitrogen, to which a solution of 0.553g (2.2 mmol) of freshly distilled 4,4'-methylenebis(phenyl isocyanate) (MDI, Aldrich) in 11 mL NMP was added dropwise in a period of 30 min. The reaction was allowed to proceed at 100 °C for additional 3 hours. Occasionally, small amounts of NMP solvent were added to the system if the viscosity became too high. After the reaction was completed, the solution was poured into distilled water under stirring. A dark-violet precipitate was collected by filtration. After washing with water and ethanol several times, the filtration cake was cooled in liquid nitrogen and was crushed and ground into fine powders, which were then extracted with acetone in a Soxhlet apparatus for 48 hours. Upon drying overnight at 80 °C under vacuum, the polyurea (**12**, R' = -Ph-CH_2-Ph-) was obtained in 94 % yield. Polymerizations of the aniline trimer with toluene-2,4-diisocyanate (TDI, Aldrich) and 1,4-phenylene diisocyanate (PDI, Aldrich) were carried out following the similar procedures to afford corresponding polyureas in 90% and 96% yields, respectively.

Acknowledgments. We wish to dedicate this article to Prof. Alan G. MacDiarmid for his pioneering contributions to the field of conducting polymers. This work was supported in part by Akzo-Nobel Corporation (YW), by Harry Stern Foundation via a grant to the Drexel-Technion collaborative research projects (YW, MN and AS), and by the Natural Sciences and Engineering Research Council of Canada (ZYW).

References

1. (a) Skotheim, T.A.; Elsenbaumer, R.L.; Reynolds, J.R., Eds. *Handbook of Conducting Polymers*, 2d ed., Marcel Dekker: New York, 1997. (b) Skotheim, T. A., Ed. *Handbook of Conductive Polymers*, Vols. I and II, Marcel Dekker: New York, 1986.
2. (a) MacDiarmid, A.G.; Epstein, A.J. *Faraday Discussion, Chem. Soc.* **1989**, *88*, 317. (b) Genies, E.M.; Boyle, A.; Lapkowski, M.; Tsintavis, C. *Synth. Met.* **1990**, *36*, 139. (c) Roncali, J. *Chem. Rev*. **1992**, *92*, 711.
3. MacDiarmid, A. G.; Chang, J. C.; Richter, A. F.; Somasiri, N. L. D.; Epstein, A. J. In *Conducting Polymers*; Alcacer, L., Ed., Reidel Publishing Co.: Holland, 1987.
4. Sheikh-Ali, B.M.; Wnek, G.E. In *Chemistry of Advanced Materials*, Interrante, L.V.; Hampden-Smith, M.J., Eds., Wiley-VCH: New York, 1998, pp. 73-98.
5. Roberts, S.; Ondrey, G. *Chem. Eng.* July 1996, p. 44.
6. Wei, Y.; Yang, Y.; Ding, T.; Yeh, J.-M.; Wei, G. *Polym. Mater. Sci. Eng.* **1996**, *74*, 209.
7. Wei, Y.; Yang, C.; Yeh, J.-M.; Ding, T.; Wei, G.; Jin, D.; Wang, J.; Jia, X.; Jansen, S.A. In *Corrosion Control by Organic Coatings*, (ACS Symp. Ser. 689), Bierwagen, G., Ed., Am. Chem. Soc.: Washington, DC, Ch. 31, pp. 382-395, 1998.
8. Lu, F.-L.; Wudl, F.; Nowak, M.; Heeger, A.J. *J. Am. Chem. Soc*. **1986**, *108* , 8311.
9. (a) Garnier, F.; Horowitz, G.; Peng, X.Z.; Fichou, D. *Adv. Mater*. **1990**, *2*, 592. (b) Garnier, F.; Hajlaoui, R.; Yassar, A.; Srivasyava, P. *Science* **1994**, *265*, 1684.
10. Wei, Y.; Yang, Y.; Yeh, J.-M. *Chem. Mater.* **1996**, *8*, 2659.
11. (a) Honzl, J.; Ulbert, K.; Hadek, V.; Tlustakova, M. *Chem. Commun*. **1965**, 440. (b) Honzl, J.; Tlustakova, M. *J. Polym. Sci.: Part C*, **1968**, *22*, 451.
12. (a) Shacklette, L. W.; Wolf, J. F.; Gould, S.; Baughman, R. H. *J. Chem. Phys.* **1988**, *88*, 3955. (b) Cao, Y.; Li, S.; Xue, Z.; Guo, D. *Synth. Met.* **1986**, *16*, 305.
13. (a) Feng, J.; Zhang, W. J.; MacDiarmid, A. G.; Epstein, A. J. *Proc. SPE/ANTEC'97*, **1997**, *55*(2), 1373. (b) Wienk, M. M.; Janssen, R. A. *J. Am. Chem. Soc.* **1996**, *118*, 10626.
14. Bebert, P. H.; Batich, C. D.; Tanner, D. B.; Herr, S. L. *Synth. Met.* **1989**, *34*, 3139.

15. (a) Wei, Y.; Tang, X.; Sun, Y.; Focke, W.W. *J. Polym. Sci., Part A, Polym. Chem.* **1989**, *27*, 2385.
16. Wei, Y.; Jang, G.-W.; Hsueh, K.F.; Chan, C.-C.; Hariharan, R.; Patel, S.A.; Whitecar, C.K. *J. Phys. Chem.* **1990**, *94*, 7716.
17. Wei, Y.; Hsueh, K.F.; Jang, G.-W. *Polymer* **1994**, *35*, 3572.
18. Wei, Y.; Sun, Y.; Tang, X. *J. Phys. Chem.* **1989**, *93*, 4878.
19. Wei, Y.; Ramakrishnan, H.; Patel, S.A. *Macromolecules* **1990**, *23*, 758.
20. Wei, Y.; Sun, Y.; Jang, G.-W.; Tang, X. *J. Polym. Sci., Part C* **1990**, *28*, 81.
21. Wei, Y.; Chan, C.-C.; Jang, G.-W.; Tian, J. *Chem. Mater.* **1991**, *3*, 888.
22. (a) Wei, Y.; Jang, G.-W.; Chan, C.-C. *J. Polym. Sci., Part-C* **1990**, *28*, 219. (b) Wei, Y.; Tian, J. *Macromolecules* **1993**, *26*, 457. (c) Wei, Y.; Tian, J.; Glahn, D.; Wang, B.; Chu, D. *J. Phys. Chem.* **1993**, *97*, 12842.
23. Wei, Y.; Tian, J.; Yang, D. *Makromol. Chem., Rapid Commun.* **1991**, *12*, 617.
24. (a) Wei, Y. "Lecture Notes on Polymer Chemistry", Drexel University, 1991. (b) Wei, Y. *ACS Symp. on Electrochemistry of Conductive Polymers* (Div. Colloid & Surface Chem.) Washington, D.C., Abstr. No. 117, 1992. (c) Wei, Y. In *Proc. Symp. Frontiers of Chemistry* (CWCYC-2); Wu, Y.-D.; Yan, Y.-J., Eds., Hong Kong U. Sci. Tech. Press: Hong Kong, pp. 221-222, 1997. (d) Wei, Y.; Jia, X.; Jin, D.; Yeh, J.-M.; Narkis, M.; Siegmann, A. *Polym. Prepr. (Am. Chem. Soc., Div. Polym. Chem.)* **1998**, *39(1)*, 115-116.
25. (a) Webster, O.W. *Science* **1991**, *251*, 887. (b) Patten, T.E.; Xia, J.; Abernathy, T.; Matyjaszewski, K. *Science* **1996**, *272*, 866. (c) Georges, M.K.; Veregin, R.P.N.; Kazmaier, P.M.; Hamer, G.K. *Trends Polym. Sci.* **1994**, *2*, 66.
26. (a) Wei, Y.; Connors, E.J.; Jia, X.; Wang, B. *Chem. Mater.* **1996**, *8*, 604. (b) Wei, Y.; Connors, E.J.; Jia, X.; Wang, C. *J. Polym. Sci., A: Polym. Chem.* **1998**, *36*, 761.
27. (a) Wei, Y.; Yeh, J.-M.; Wang, J.; Jia, X.; Yang, C.; Jin, D. *Polym. Mater. Sci. Eng.* **1996**, *74*, 202. (b) Yeh, J.-M. Ph.D. dissertation, Drexel University, 1996.
28. Wei, Y.; Yang, C.; Ding, T. *Tetrahedron Lett.* **1996**, *37*, 731.
29. Wei, Y.; Yang, C.; Chen, M.-H.; Li, W. *Proc. SPE/ANTEC* **1997**, *55(2)*, 1369.
30. Wang, Z.Y.; Yang, C.; Gao, J.P.; Lin, J.; Wei, Y.; Li, S. *Macromolecules* **1998**, *31*, 2702.
31. Wei, Y.; Yang, C.; Wei, G.; Feng, G. *Synth. Met.* **1997**, *84*, 289.
32. Sein, Jr., L.T.; Kolla, S.; Pasupuleti, P.; Patel, K.; Jansen, S.A.; Wei, Y. *Polym. Prepr. (Am. Chem. Soc., Div. Polym. Chem.)* **1998**, *39(1)*, 117.
33. (a) Vallerio, R.; Keyer, R.; Grabania, S.; Landis, M.; Jansen, S.A.; Wei, Y. *Mater. Res. Soc. Symp. Proc.* **1996**, *413*, 523. (b) Sein, Jr. L.T.; Levarity, L.; Keyer, R.; Jansen, S.A.; Wei, Y. In *Electrical and Optical Polymer Systems: Fundamentals, Methods and Applications*; Wise, D.L.; Wnek, G.E.; Trantolo, D.J.; Cooper, T.M.; Gresser, J.D., Eds., Marcel Dekker:New York, Ch. 1, 1998.
34. Fahlman, M.; Guan, H.; Li, S.; Wei, Y.; Epstein, A.J. *Chem. Mater.* **1998**, submitted.
35. Wei, Y.; Hsueh, K.F.; Jang, G.-W. *Macromolecules* **1994**, *27*, 518-525.
36. Singer, R.A.; Sadighi, T.P.; Buchwald, S.L. *J. Am. Chem. Soc.* **1998**, *120*, 213.

37. (a) Wei, Y.; Li, S.; Shan, C.; Feng, Q.; Lin, M.; Guterman, E. Results to be published. (b) Mathai, M.W. MS Dissertation, Drexel University, 1998.
38. Jansen, S.A.; Pasupuleti, P.; Patel, K.; Li, W.; Li, S.; Wei, Y. Results to be published.
39. Viswanathan, T.; Feng, Q.; Toland, A. *Polym. Prepr. (Am. Chem. Soc., Div. Polym. Chem.)* **1997**, *38(1)*, 141.
40. (a) Cheng, M.-H. MS Dissertation, Drexel University, 1997. (b) Wei, Y.; Li, S.; Cheng, M.-H.; Li, W. Results to be published.
41. (a) Wei, Y.; Wang, W.; Yeh, J.-M.; Wang, B.; Yang, D.; Murray, Jr., J.K.; Jin, D.; Wei, G. In *Hybrid Organic-Inorganic Composites*, ACS Symp. Ser. No. 585; Mark, J.E.; Lee, C.Y.-C.; Bianconi, P.A., Eds., Am. Chem. Soc.: Wanshington DC, Ch. 11, pp. 125-141, 1995. (b) Jang, G.-W.; Chen, C.; Gumbs, R.W.; Wei, Y.; Yeh, J.-M. *J. Electrochem. Soc.* **1996**, *143*, 2591.
42. Wei, Y.; Wang, J.; Jia, X.; Yeh, J.-M.; Spellane, P. *Polymer* **1995**, *36*, 4535.
43. Li, S. Ph.D. Dissertation, Department of Chemistry, Stuttgart University, Germany, 1988.

Chapter 25

Vapor Phase Molecular Epitaxy via Self-Assembly Reactions

Vladimir Burtman [1], Alexander Zelichenok [1], Aharon Yakimov [2], and Shlomo Yitzchaik [1,3]

[1] Inorganic and Analytical Chemistry Department, and [2] The Racah Institute of Physics, The Hebrew University of Jerusalem, Jerusalem 91904, Israel

In this paper we introduce the Molecular Layer Epitaxy (MLE) method for epitaxial growth of covalently-linked organic multilayered structures. This method combines vapor phase and solution based multilayers assembly techniques in a unified concept. The MLE approach was proved fruitful by applying Chemical Vapor Deposition (CVD) techniques. The resulting MLE-derived low-dimensional multilayered structures exhibited high structural regularity and thermal stability. The above vapor phase assembly technique led to the formation of organic multiple quantum wells (OMQW) structures, where the solid-state electronic properties are governed by finite size effects. The various emitting species in the solid state were studied by modeling intermolecular interactions in solution. The strong tendency for π-aggregation in model compounds is evident in their crystal structure as well. This driving force for in-plane stacking also enhances the mobilities of electrons with-in this layer leading to unique electroluminescent properties. The suggested MLE approach for multilayered thin film deposition should enable the future advance in areas of material science connected with nano-technologies, and better understanding of fundamental issues in quantum mechanic and solid state physics.

The study of electronic processes in organic semiconducting heterostructures has been a subject of recent extensive research efforts. Of particular importance are the efforts to fabricate and study organic multiple quantum wells (OMQW's) structures, in which the charge carriers are confined in one direction to a characteristic length which is smaller than their De-Broglie wavelength. These efforts are motivated by both the

[3] Corresponding author.

naphthalene-based compounds shown that an electron mobility in range of 10^{-6} - 10^{-3} cm^2 V^{-1} s^{-1} are achievable in molecular electronic devices (4-7). In addition, completely new and applicable photophysical phenomena of organic superlattices have been predicted, including enhancement of optical nonlinearities (8-11) and photoelectric transformations (12,13). Solution and vapor phase deposition techniques, however have already proved the capability of growing multilayered heterostructures. (see Figure 1).

State-of- the-Art in Organic Multilayer Thin Film Deposition Techniques.

Solution-derived techniques for the growth of organic multilayers include the Langmuir-Blodgett (LB) technique and Self-Assembled Monolayers (SAM) technique. The LB technique, a physisorption process, yields films exhibiting anisotropic electron transport (14) and tunneling, (15-19) again suggesting OMQW behavior. While the LB method is useful in achieving 2D multilayered physisorbed structures, LB films suffer from low chemical and thermal stability and cannot incorporate large chromophores without phase-segregation and micro-crystal formation with time.

The solution-derived SAM methods were found to be useful in the construction of covalently bonded organic self-assembled multilayers and superlattices (20-32). Chromophoric self-assembled superlattices provide the advantages of strong chemisorption through Si-O bonds, chemical and thermal stability, and the ability to form noncentrosymmetric structures (23,33). However structural regularity achieved by SAM is not process-controlled and include the application of a capping layer, a synthetic step that forms siloxane-based cross-linked network. This capping reaction was found to be essential for constructions of superlattices since it gaps through pinholes in the underlayer and thus, provide a route to defect annealing mechanism. The resulting superlattices are characterized with very smooth interfaces in between the various layers and the periodic structure was manifested by the appearance of a Brag-diffraction peak in x-ray reflectivity studies (34).

Electrostatic SAM (EL-SAM) method make use of polyelectrolytes of opposite charge to produce layered structures in which various charged organic material may be introduced in each layer by simple neutralization process, thus enabling a quick way to the formation of organic multilayered structures (35). A number of research groups have been exploring the coulombic interactions that produce polyelectrolyte-surfactant complexes to build ordered thin film structures using layer-by-layer assembly. Conducting films (36-43) and electroluminescent devices (44) have been produced in such a fashion. It was also shown the pertinence of EL-SAM methods to incorporate nanocrystals in multilayered structures (45). However the electrostatic nature of EL-SAM method produces rough interfaces in the resulting multilayered structures, inducing defects accumulation mechanism with the increase of the layers number. Thus the structural regularity and stability of EL-SAM multilayers is still inferior with the same structures achieved by classical SAM routes.

The main disadvantages of LB and SAM solution derived methods is the fact that these methods are practiced with large volumes of high purity solvents for each deposition cycle. Thus, LB and "classical" SAM are too time-consuming methods, which are non-

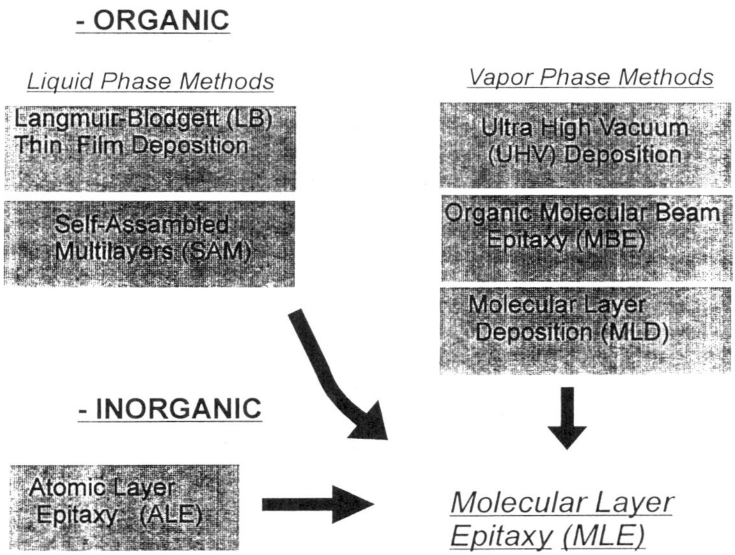

Figure 1. Assembly methods for organic/inorganic multilayered structures.

compatible with current semiconductors production lines and are generally involved with equipment such as Schlenk-line, glove-box and vibration isolated setups. In addition, special effort have to be applied (as synthesis of soluble precursor materials or complicated reaction conditions) in order to use a wide-range of organic materials, exhibiting attractive physical properties, as for example naphthalene and perylene derivatives. EL-SAM is more rapid and simple technology, however this method suffers from very rough interfaces that prevents the assembly of the ideal quantum confined monolayer in multilayered structure. This method is also limited to only charged water soluble polymeric precursors. Beside that all solution-derived methods are susceptible to surface contamination by the glassware and solvent manipulation techniques.

The disadvantages of wet methods are eliminated when ultra high vacuum (UHV) vapor phase growth methods are used for assembly of organic multilayered thin-films such as: organic UHV deposition (47), organic molecular beam epitaxy (MBE) (46-48) and molecular layer deposition (MLD) (49, 50). These vapor phase methods use standard semiconducting apparatus setup, believed to provide both layer thickness control and automatic thermally cleaned substrate and environment. A number of interesting optical and photophysical phenomena have already been found in OMBE derived OMQW, including the observation of exciton confinement in photoluminescence (PL), (51, 52) and electroluminescence (EL), (53, 54) and electric field modulation of PL (53, 55). Preparation of crystalline thin organic films by the UHV methods relies on the bonding of molecular layers *via* weak van der Waals (vdW) forces to achieve and preserve quasi-epitaxial structures (56). Thus, perfect monolayers without step edges are difficult to achieve and the lower limit is an average of three 'monomolecular' layers with exception of MLD that relies on hydrogen-bonding. The vdW stacking growth mechanism practiced in all UHV methods limit the full potential of assembling heterostructures since vdW stacking relies on lattice matching of the various precursors. The main conditions, which thin films have to satisfied in order to overcome this obstacle is the combination of epitaxial growth with covalent bonding. Some organic UHV and MBE methods succeed in epitaxial growth (46, 57) and solution-derived SAM (58) enable to form interlayer covalent bonds. However, reviewing the situation for organic multi-layered thin film deposition techniques it is possible to state that, there is no heterostructures fabrication method enabling the simultaneous unification of "epitaxy" and "covalent bonding" concepts. The inorganic atomic layer epitaxy (ALE) (59) is the only vapor phase derived method that can build chemically bonded epitaxial multilayers in a layer-by-layer fashion.

ALE approach utilizes the difference between chemical and physical adsorption. When the first layer of atoms of reactive molecular species reaches a solid surface where a chemical reaction (generally condensation) takes place, while subsequent physisorbed layer tend to interact much less strongly. If the substrate surface is heated sufficiently one can achieved a condition such that only the chemisorbed layer remains attached. Repeating this reaction cycle with different compounds leads to a controlled layer-by-layer growth (60). ALE method was successfully applied to epitaxial growth of "conventional" semiconducting inorganic materials such as II-VI, III-V compounds and silicon and silicon related compounds (61).

In this contribution we introduce a new method for organic multilayered hetero-structures deposition, Molecular Layer Epitaxy (MLE), that in a sense is widering the scope of ALE concept for new class of organic and organic/inorganic multilayers containing materials. Methodologically MLE encompass best achievements of organic and inorganic multilayered deposition methods in an hybrid straightforward concept. We show in this work the first application of MLE method for organic superlattices growth. MLE method may also include elements of ALE technique as an inorganic monolayers are encapsulated in organic multilayered matrix or *vise versa* - organic monolayers integration in inorganic superlattices. Thus we expect the MLE approach to enable the future advances in nano-chemistry and -technology, low-dimensional physics and applied sciences such as molecular electronic and photonics.

Methodology of Molecular Layer Epitaxy.

In the new method for organic multilayers deposition, molecular layer epitaxy (MLE), the epitaxial growth is achieved by self-limited vapor-phase reactions on a templated surface. As in ALE method, MLE is governed by covalent bonding at the intermolecular level that leads to ideal monomolecular growth. The chemical principles of the MLE method can be divided into four levels: *(i)* a template layer, *(ii)* self-restricted vapor phase reactions, *(iii)* covalent ("c-axis") interlayer bonding and *(iv)* π-stacking in x-y plane.

The first concept, the formation of a *template layer,* is suggested to enhance adsorption of the first monomer onto the surface as a single monolayer. A template layer is a self-assembled monolayer deposited from solution, which exposes a reactive functionality toward the first monomer, prior to introduction of the substrates to vapor phase assembly. This template layer dictate epitaxial growth and ensure the second concept of *self-limiting growth* of only one monolayer. Therefore, this layer bridges solution and vapor phase growth m*echanisms in a unified concept.*

The *covalent ("c-axis") interlayer bonding* and *π-stacking in x-y plane*, which are the third and forth level of the MLE concept, are achieved by the use of bi-functional rigid and flat molecular precursors material and the appropriate thermodynamic and kinetic deposition conditions. *In situ* formation of covalent c-axis interlayer bonding was achieved by chemical reactions of reactive precursor's groups, that do not allow self-condensation reactions in the vapor phase, with activated surface radicals at appropriate thermodynamic condition of the vapor phase reaction. The driving force for *face-to-face in x-y plane packing* is π-stacking (P_z orbitals overlap) of disc shaped π-conjugated units and hydrophobic interactions. These conditions are achieved by sufficient concentration of reactive precursors, which have a strong intramolecular π-conjugation in the plane of the rigid heterocyclic rings, the needed alignment for the surface reaction. The positioning of the two reactive units of the precursor on the two poles of the molecule ensure anchoring to the surface by only one reactive group forcing the molecule to align perpendicular to the surface exposing the second functional group to the new emerging interface and thus align the P_z orbital in an optimal position for in-plane stacking.

The general strategy of MLE towards formation of OMQW's is illustrated in Figure 2, showing the continuous pulse-mode multilayered growth process. The template layer (T) is formed by usage of coupling agent in solution. A, B, C which are active

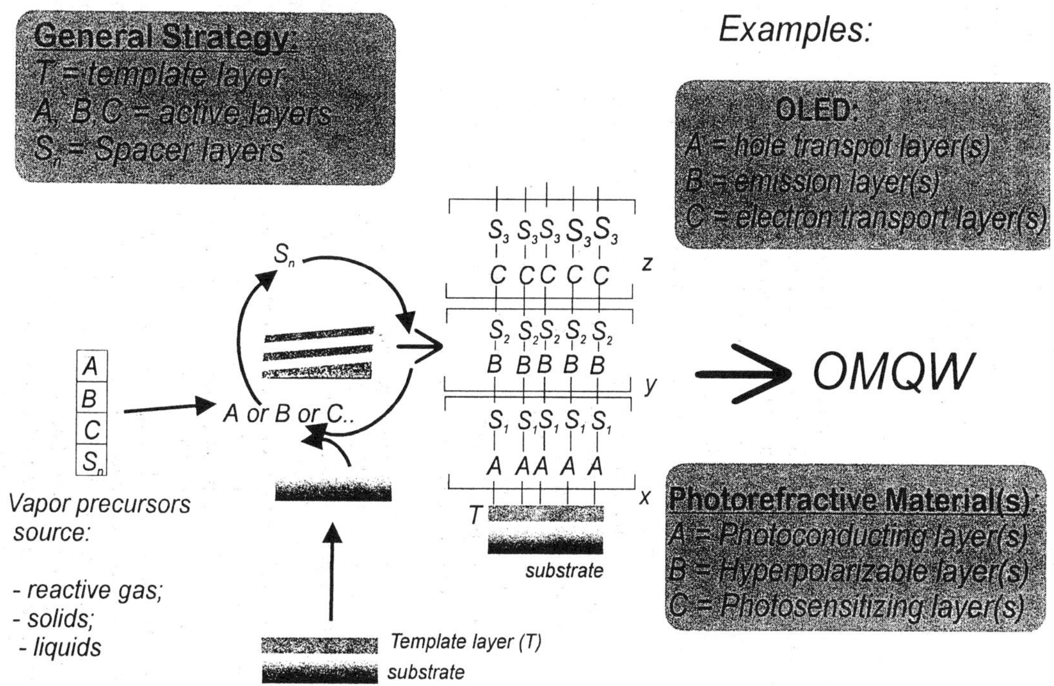

Figure 2. Molecular layer epitaxy (MLE) conceptual approach for the formation of organic multiple quantum well (OMQW) structures.

layers and S_n which are different spacer layers (n = aromatic or aliphatic or metalloorganic spacers) can be deposited repeatedly in a layer-by-layer fashion from the vapor phase, using selected functionalities of varies precursors to form the OMQW's structures. It is shown also a principle construction of MLE-derived useful molecular opto-electronic devices such as OLED's and photorefractive materials (PR).

Molecular Layer Epitaxy *via* Chemical Vapor Deposition routes.

MLE approach is applicable to all vapor-phase deposition techniques. We show the first principle application of MLE concept by chemical vapor deposition (CVD) routes. CVD deposed coatings and films are used extensively in the inorganic semiconductor and related industries for making single devices, integrated circuits, microwave hybrid integrated circuits, compact discs, solar reflective glazing, fiber optics, photovoltaic cells, sensors, displays, and many other electro-optic products (60). CVD is believed to offer potential advantages over physical deposition techniques such as ultrahigh vacuum (UHV) deposition, molecular beam epitaxy (MBE), sputtering etc. i.e. the fast growth speed, the ability of coating complex shapes and of large area deposition and the pertinence for large-scale mass production. Thus, we expect that our CVD method for organic multilayered structures deposition can be very easily integrated to existing production lines. We designed a low pressure CVD laminar flow ($Re < 250$) reactor with base pressure of 10^{-5} Torr and molecular gas pressure of 10^{-3} Torr. A multiwavelength ellipsometer (M-44, Woollam Co.) was used for *in-situ* monitoring of the molecular layer assembly. The scheme of the CVD apparatus for MLE is shown on Figure 3. The carrier-gas (Ar) assisted CVD setup enable the use of solid, liquid and gas precursors in sublimers, bubbles and gas lines respectively in a pulsed mode. The total pressure was controlled by baratron connected with mass-flow controllers. The temperature controlling setup was divided for the various precursor reservoirs, feadthrough lines into reactor, three major zones in the laminar flow reactor, the susceptor zone and the lines getting out of the reactor. Pulses of N_2 gas were used for the cleaning steps in between pulses of reactive precursors. The completion of each synthetic step was verified by in-situ ellipsometry.

The surface chemistry principles for assembling of covalently linked interlayers organic superlattices via the MLE method are illustrated in Figure 4. Float glass (Kaufman Glass, Wilmington), silicon wafers and ITO coated slides (Delta technologies, Stillwater) were cleaned in aqueous detergent, rinsed copiously with deionized (DI) water, cleaned for 1 h in an ultrasonic bath in hot (90 °C) "piranha" solution (H_2SO_4:H_2O_2 70:30 v/v), and then allowed to cool to room temperature over the course of 1 h while sonicating. The cleaned substrates were rinsed then with DI water. Further cleaning was carried out using an $H_2O/H_2O_2/NH_3$ (5:1:1) cleaning procedure. After subsequent washing with DI water, the substrates were immersed for 5 min in pure methanol, then in methanol/toluene (1:1), and finally in dry toluene.

From the last solvent the substrates were directly transferred under inert atmosphere to a solution of the silanizing reagent (3-aminopropyltrimethoxysilane) in dry toluene (3mM, toluene, 90° C for 16 hours under N_2 atmosphere), (Figure 4, step *i*). At the

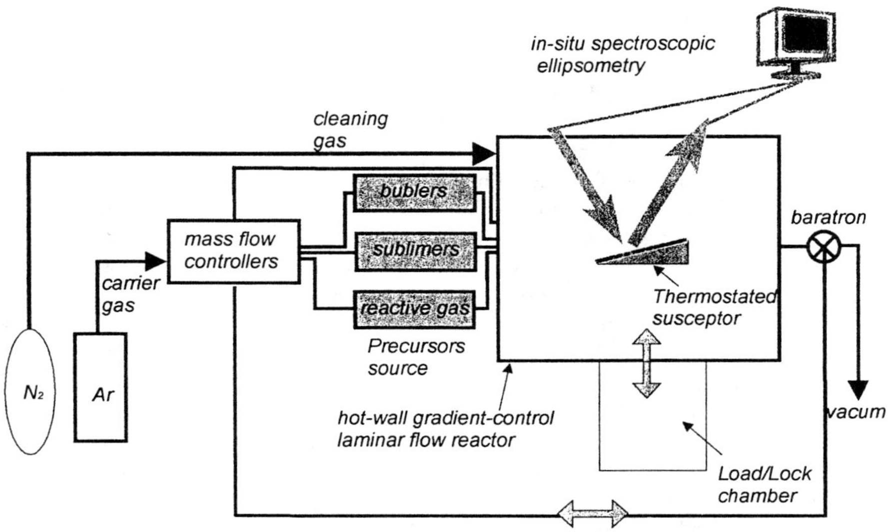

Figure 3. MLE apparatus: carrier gas assisted pulsed Chemical Vapor Deposition (CVD) reactor scheme.

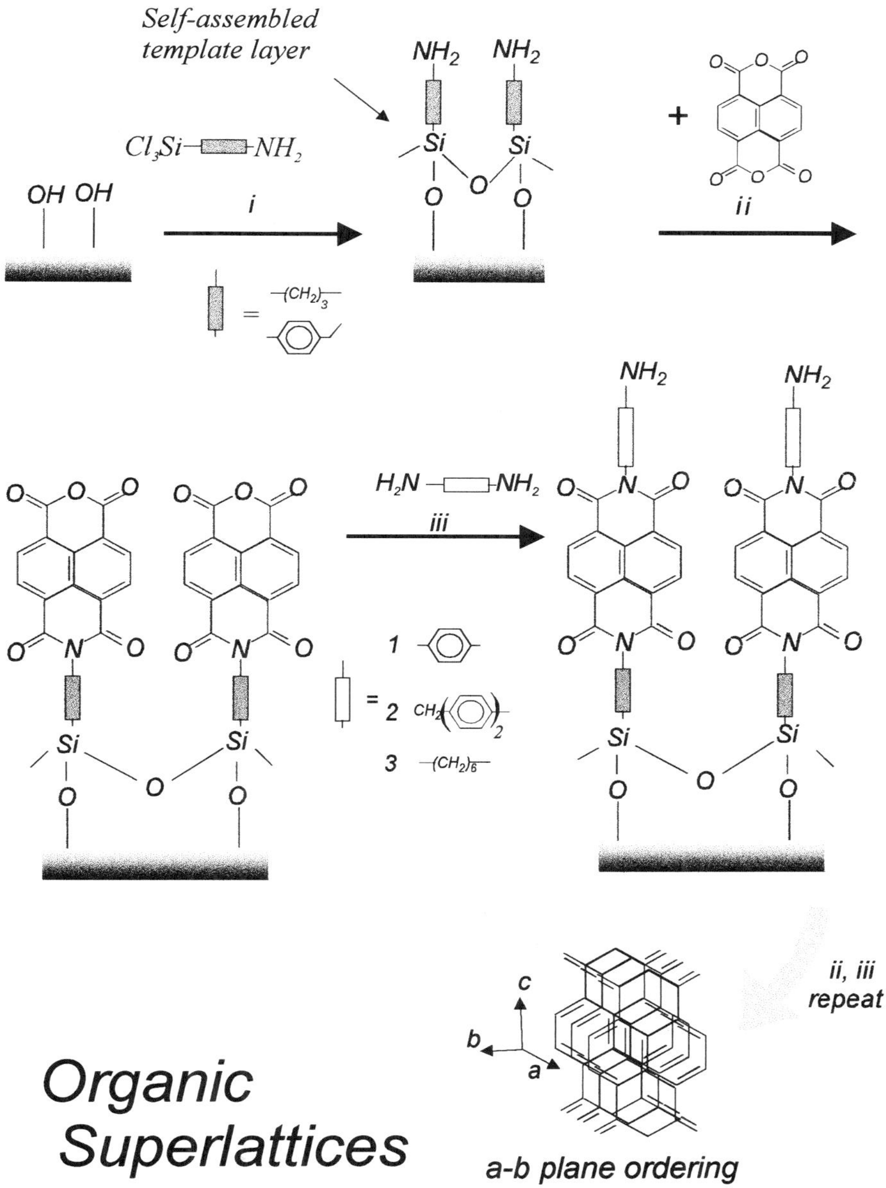

Figure 4. Synthetic route to organic superlattices via MLE.

completion of the silanization, the substrates were removed from reaction solution and rinsed twice for 10 min each with toluene, toluene/ acetone, and acetone. The substrates were cured in a convection oven at 115 °C for 15 min to complete the siloxane network formation. After the *template layer* is deposited all of the following film growth steps were conducted in the CVD reactor. In step *ii*, 3,4;7,8-naphthalenetetracarboxylic-dianhydride, NTCDA, is vaporized to react with the surface bound amine to form imide linkages. Reaction with vaporized aliphatic- or aromatic-diamine (Figure 4, step *iii*) then regenerates an aminated surface ready for the next dianhydride insertion.
The following conditions for CVD deposition were used: NTCDA precursor was vaporized at T = 200-250 °C, 1,6-diaminohexane (DAH) and 4-4'-dianilinomethane (DAM) and 4, 4'-Methylenedianeline (MDA) precursors were vaporized at T = 35-40 °C, 55-75 °C, and 90-110 °C correspondingly to react with the surface bound amine or anhydride to form imide linkages on growing interface at substrate temperature of 200-290 °C for 15 -25 min for each deposition step at total pressure of 0.05 Torr. After formation of each consisted layer, the reaction zone was cleaned by vacuo evacuation of unreacted precursors under N_2 flow for 5 min. The walls of the reactor were independently heated for temperatures 280-380 °C. Repeating steps *ii-iii* several times leads to the formation of naphthalenetetracarboxylic-diimide (NTCDI) based superlattices.
The subsequent step in fitting MLE concept to CVD condition is the study of particular conditions enabling surface chemistry on each step. The study by differential scanning calorimetry (DSC) of reactive precursors was used to obtain the temperature dependent imidization reaction for the MLE process. The DSC thermograms for modeling the elemental building step is shown in Figure 5. In both cases DSC thermograms indicate a two-step process of chemical bonds formation. The first step is the formation of naphthalene amidodicarboxylic acid. The second step is the cyclization reaction of the acid-amide to yield the naphthalene-tetracarboxylicdiimid. The cyclization reaction takes place at 350 °C for the hexyl spacer and at 500 °C for the phenyl insert under ambient conditions. Vapor phase methods generally enable to lower the deposition temperatures on a factor by 1/3 - 1/4 (62). We find that the same reaction under CVD reactor condition take place at 220 °C and 280 °C for hexyl and phenyl spacers respectively. Such lowering of the reaction temperatures is attributed to vapor phase reaction, where the reactive groups are aligned in an optimal way for condensation reaction and simultaneous removal of water vapor from the reaction zone, which shift the reaction equilibrium towards the imide-product side. It can also be observed that just a simple change in spacer from aliphatic to aromatic in NTCDI superlattice induce substantial changes in synthetic conditions for imide bridge formation. It can be also guessed that the physical and structural properties of resulting organic superlattices with a different spacers should be deviant.

MLE-derived Organic Superlattices.

The structural regularity of the resulting NTCDI-based ultra-thin films was tested by variable angle spectroscopic ellipsometry (VASE, Woollam Co.) at 75°. Figure 6

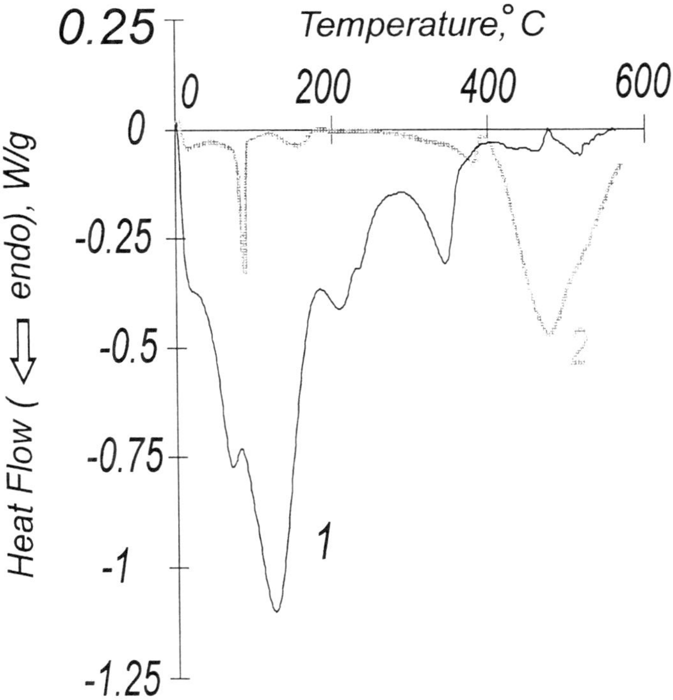

Figure 5. DSC thermograms of the reactive MLE precursors: (1) NTCDA+DAH and (2) NTCDA+DAM.

shows measured and fitted ellipsometric data for a superlattice with 4 bi-layers, NTCDI-hexamethylene (NTCDI-HM). Each bi-layer contains one semiconductive monolayer of NTCDI and a dielectric one, hexamethylene, oriented along the c-axis. Molecular c-axis interplanar spacing are: 8.19 Å for NTCDI and 6.22 Å for the hexamethylene spacer. In order to verify dimension parameters of resulting OMQW structures and spectroscopic data we synthesized low-molecular weight model-compound N, N'-dihexyl-NTCDI. The length of heterocyclic molecular part and aliphatic chain of this model compound reproducing the same NTCDI-HM superlattice interlayer spacing on the molecular level are 8.5 Å for NTCDI and 6.66 Å for the hexamethylene spacer (63). There is a reasonable fitting between ellipsometry and crystallographic data on model compound.

Contact angle was measured after every deposition step. Contact angles were: 17.5°, 45°, 92° and 60° for SiO2/Si and after steps: *i -iii* respectively. Values of contact angle for $(NTCDA\text{-}HM)_n$ thin films for steps ii-iii were repeated with diversity of ± 10 % up to n = 9. The growth of the contact angle value corresponds to hydrophobic changes on surface of growing interface, where the exposed anhydride is more hydrophobic than the exposed amines.

The absorption spectra of multilayered structure with the aromatic spacer, diphenylmethane $(NTCDI/DPM)_n$, n = 4 is shown in the Figure 7. The baseline was corrected before each and every measurement. As in the case of $(NTCDI/HM)_n$ multilayers (62), the optical density in the UV region peaks at about 360 nm and 390 nm and rise monotonicaly with the number of deposited bi-layers. However, in differ to multilayers with hexyl spacer (64), in case of $(NTCDI/DPM)_n$, there is a new very broad absorption band in the near infra-red (NIR) spectral range, starting with 520 nm and extended over 1750 nm. Moreover, on top of this band a narrower new band starting from about 550 nm until 1350 nm appear in the spectra beginning from the 4th bi-layer. This kind of absorption spectrum points out the existence of thermally excited electrons, which have an opportunity to "travel" within the potential well's structure (65). The large number of charge carriers located on the π-conjugated molecular plane in multilayered structure, seems to be responsible to such low energy intermolecular charge-transfer transition. The UV-Vis $\lambda_{max}(\varepsilon)$ and FTIR $\nu_{C=O}$ spectral bands of NTCDA precursor are: 369 nm ($\varepsilon = 2.1 \times 10^4$ L/mol×cm) and 1779 cm^{-1}. Upon imide formation in model compounds these bands were shifted to 381 ($\varepsilon = 2.6 \times 10^4$ L/mol×cm) and 1655.9 cm^{-1} for dihexyl-NTCDI and to 380 ($\varepsilon = 2.3 \times 10^4$ L/mol×cm) and 1655.5 cm^{-1} in case of diphenyl-NTCDI. The same shifts are observed also in the MLE-derived films where additional thickness-dependent factor influence the spectra, as is demonstrated next.

It was suggested that excitons are responsible for finite size effects in the absorption spectra of organic multilayered structures (66). The size-dependent characteristic in NTCDI-based nano-structures were investigated by optical UV-Vis-NIR absorption signature. The blueshift in absorption peak, increase in transition energy ΔE, predicted as a function of multilayers thickness decrease observed for organic semiconducting PTCDA/NTCDA organic MBE-derived structure (64).

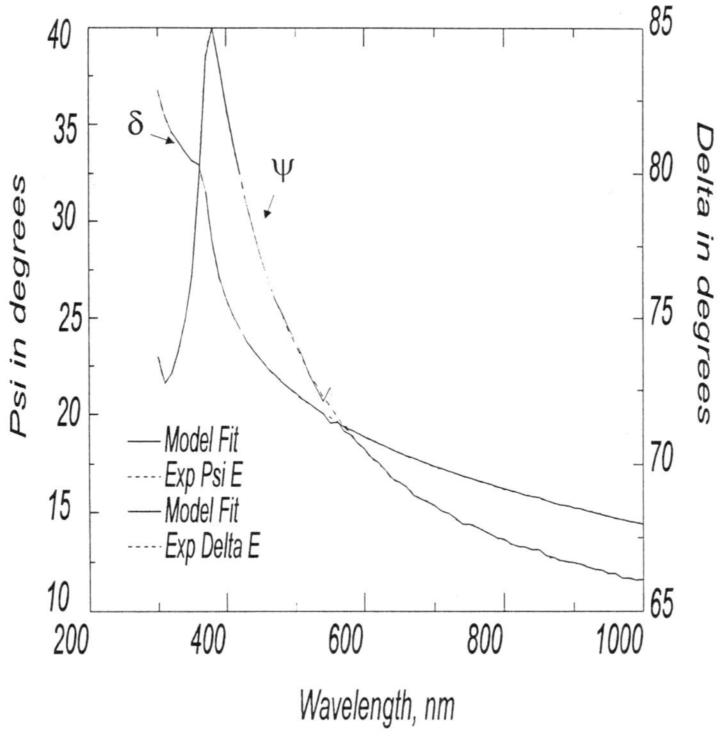

Figure 6. Spectroscopic ellipsometry experimental and theory for 4 bi-layers structure of (NTCDI-HM) superlattice, assembled by the MLE method.

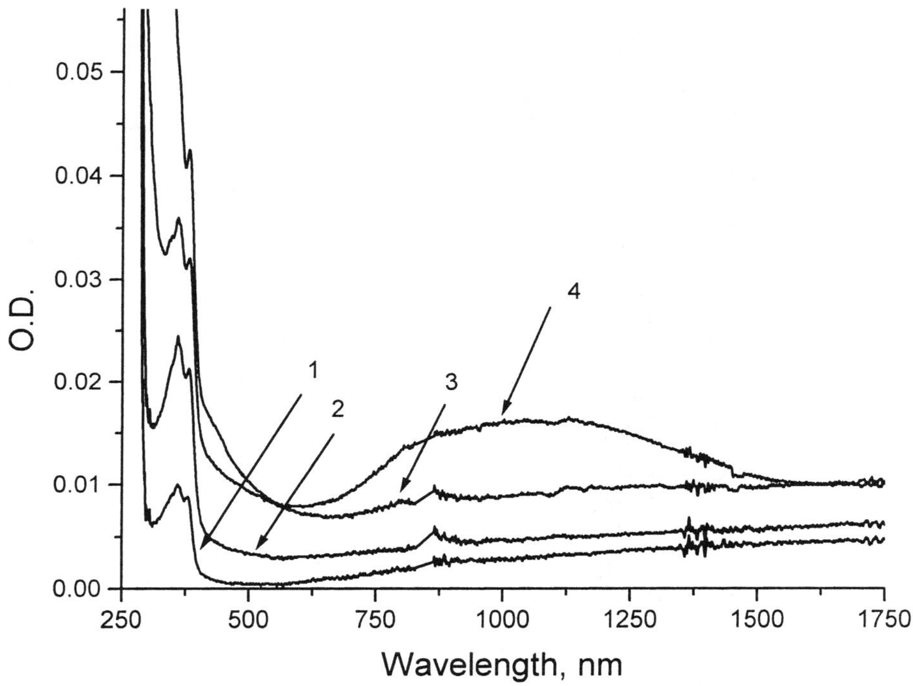

Figure 7. Absorption spectra of (NTCDI-DPM)n (n is number of deposited bi-layers) ultrathin film, where n = 1 - 4.

The same trend in shift of the lowest-energy absorption peak of the NTCDI in the naphthalene-imide based superlattices is observed. Energy blueshift of about 12 meV (see Figure 8) is observed for the fewer layers containing NTCDI multilayered structure with aliphatic hexamethylene (HM) and two aromatic spacer: 4, 4'-diphenylmethane (DPM) and p-phenylene (pP). It may be observed that the biggest shift of low energy peak is observed in OMQW's structure with DPM spacer (ca. 25 meV). The detailed Hamiltonian fitting for blue-shifted absorption due to exciton confinement in the OMQW structure will be discussed elsewhere (63).

OMQW's-based LED Devices.

We used MLE-derived multilayered structures to form organic light emitting diodes (OLED). The MLE-derived OLED exhibit EL lifetime more than 100 hours (RT, ambient environment, U = 4 V). The electroluminescence (EL) spectra of OLED formed by introduction 2 bi-layered NTCDI structure with DPM spacer in SiO_2/ITO/OMQW/Al configuration at 8 Volts is shown in Figure 9. The voltage depend EL of 4 bi-layered NTCDI structure with HM spacer at 3 and 6 Volt is shown in the insert of Figure 9. It is clearly observed that two types of emission bands appears. The first one (marked as 1) is attribute to EL of NTCDI isolated chromophore and the second (marked as 2) to EL NTCDI of π-aggregates. The same trend in appearing of new EL peak due to epitaxial ordering was observed in perylene-based OLED (54). In order to verify such assignment we study the spectroscopy of models compound. The PL spectra of dihexyl NTCDI model compounds exhibited similar behavior. The emission spectra of dihexyl NTCDI in toluene (spectra I) and in mixture of toluene with methanol (1:19) (spectra II) is shown on Figure 10. In the case of polar solvent, which provided best conditions for aggregation, two different peaks of PL were clearly observed: one in region 400-430 nm and second broad peak at about 570 nm, those were attributed to monomer and aggregate PL respectively in differ to PL spectra in toluene, where only monomers emission was observed. The induced aggregation in solution and the appearance of red shifted absorption and emission bands clearly modeled the red shifted absorption and EL bands in the solid-state.

Conclusions.

In this contribution we introduced the MLE method - a new synthetic and instrumental approach for deposition of organic multilayered structures. Vapor phase MLE approach enable (a) simple multilayers deposition method, usage of wide-range class of organic materials that are difficult to assemble by solution-derived methods, (b) chemical process-control of monolayer growth in differ from UHV, MBE and MLD deposition techniques and (c) chemical bonds on interlayer. We demonstrated the first realization of MLE approach on gas-assisted CVD setup. We exemplified the formation of a new type of organic superlattices, unachievable by other deposition techniques. The strong

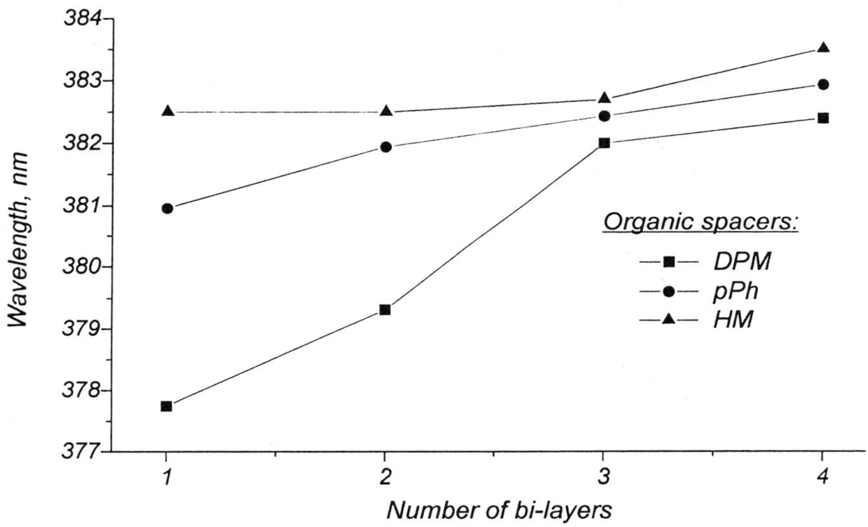

Figure 8. UV absorption maximum in NTCDI-based multilayered structures as a function of the bi-layers number with various organic spacers: HM, hexamethylene ; pPh, para-phenylene; DPM, diphenylmethane.

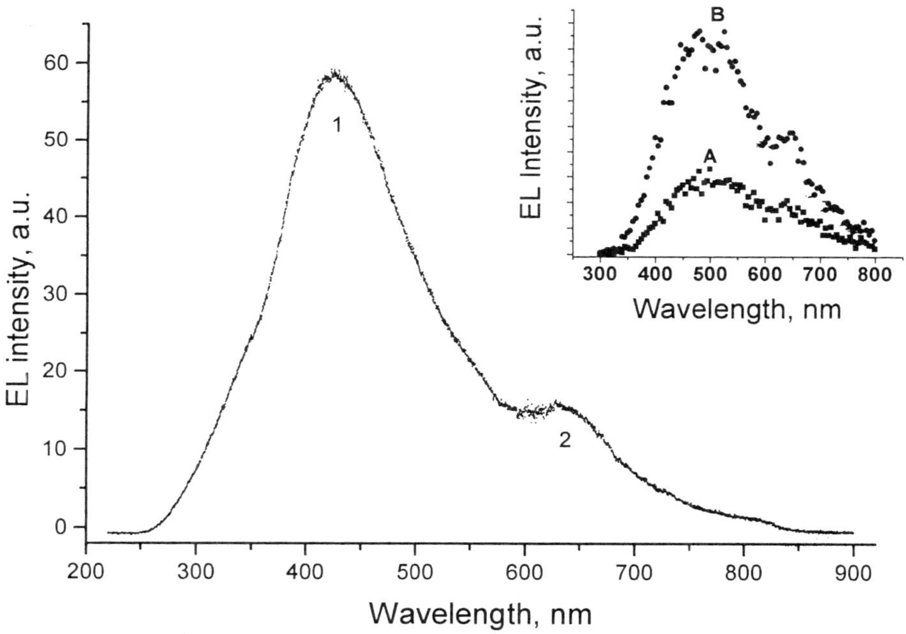

Figure 9. The electroluminescence (EL) spectra of OLED in SiO_2/ITO/(NTCDI-DPM)$_2$/Al configuration at 8 Volts. Peak 1 and 2 correspond to EL of monomers and aggregates respectively. Inset: voltage dependence EL at (A) 3 V and (B) 6 V of SiO_2/ITO/(NTCDI-HM)$_4$/Al OLED-configuration.

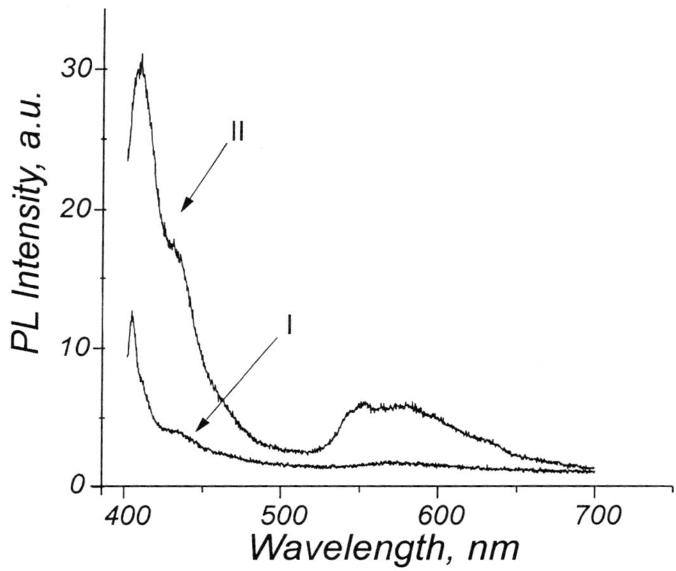

Figure 10. The emission spectra of dihexyl- NTCDI model compound in (I) toluene and in (II) mixture of MeOH/toluene (19:1). The emission spectra is recorded with excitation at 361 nm.

tendency of NTCDI molecular building block to assemble face-to-face structures was verified for the model-compounds in solution and dominated the solid-state's optical and electrical properties of OMQW structures. This assembling manner can also explain the EL bands observed in NTCDI-based structures, where two types of emission are clearly observed. We expect this MLE method to be useful in submicron lithography and the production of novel organic heterostructures for molecular electronic and photonic applications.

Acknowledgments

S.Y. Acknowledges support for this research by the Israel Science Foundation (ISF) under grand # 651/95-2, US-Israel binational science foundation (BSF) under grand # 95-0085 and the Israel Academy of Science and Humanitarians. V.B. thanks the Berg Foundation for Post Doctoral Fellowship.

References

1. Chen, Z.; Burrows, P. E.; Bulovic, V.; Forrest, S. R, Tompson, M. E. *Science*, **1997**, *276*, 2009.
2. Sheats, J. R. *Science* **1997**, *277*, 191.
3. Vaeth, K. M.; Jensen, K. F.; *Adv. Mater.* **1997**, *9*, 490.
4. Laquindanum, J. G.; Katz, H. E.; Dodabalapur, A.;. Lovinger, A. J. *J. Am. Chem. Soc.* **1996**, *118*, 11331.
5. Ostrick, J. R.; Dodabalapur, A.; Torsli, L.; Lovinger, A. J.; Kwock E. W., Miller, T. M., Galvin M., Berggren, M., and Kaz, H. E. *J. Appl. Phys.* **1997**, *81*, 6804.
6. Horovitz, G., Kouki, F.; Spearman, P.; Fishou, D.; Nogues, C.; Pan, X.; Garnier, F. *Adv. Mater.* **1996**, *8*, 242.
7. Bao, Z.; Lovinger, A. J.; Dodabalapur, A. *Adv. Mater.* **1997**, *9*, 42.
8. Lam, J.F.; Forrest, S. R.; Tangonan, G. L. *Phys. Rev. Lett.* **1991**, *60*, 1614.
9. Agranovich, V. M. and Dubovsky, O. A. *Chem. Phys. Lett.* **1993**, *210*, 458.
10. Tran-Thi, T-H.; Fournier, T.; Sharonov, A. Yu.; Tkachenko, N.; Lemmetyinen, H.; Grenier., P.; Truong, K-D.; Houde, D. *Thin Solid Films* **1996**, *273*, 8.
11. Ashwell, G. J.; Walker, T. W.; Leeson, P.; Grummt, U.-W.; Lehmann, F. *Langmuir* **1998**, *14*, 1525.
12. Zakhidov, A. A. and Yoshino, K. *Synthetic Metals* **1994**, *64*, 155.
13. Sakomura, M.; Fujihira, M. *Thin Solid Films* **1996**, *273*, 181.
14. Donovan, K. J.; Scott, K.; Sudiwala, R. V.; Wilson, E. G.; Bonnett, R.; Wilkins, R. F.; Paradiso, R.; Clark, T. R.; Batzel, D. A.; Kenney, M. E. *Thin Solid Films* **1993**, *232*, 110.
15. Donovan, K. J.; Scott, K.; Sudiwala, R. V.; Wilson, E. G.; Bonnett, R.; Wilkins, R. F.; Paradiso, R.; Clark, T. R.; Batzel, D. A.; Kenney, M. E. *Thin Solid Films* **1994**, *244*, 923.

16. Donovan, K. J.; Elliott, J. E.; Scott, K.; Wilson, E. G; Batzel, D. A.; Clark, T. R.; Kenney, M. E. *Thin Solid Films* **1996**, *273*, 229.
17. Fisher, C. M.; Burghard, M.; Roth, S.; von Klitzing, K. *Surface Science*, **1996**, *361/362*, 905.
18. Burghard, M.; Fisher, C. M.; Roth, S.; Schlick, U.; Hanack, M. *Synthetic Metals*, **1996**, *76*, 241.
19. Wilson, E. G. *Thin Solid Films* **1996**, *273*, 1.
20. Maoz, R.; Maltis, S.; DiMasi, E.; Ocko, B. M.; Sagiv, J. *Nature*, **1996**, *384*, 150.
21. Tillman, N., Ulman, A.; Schildkraut, J. S.; Penner, T. L. *J. Am. Chem. Soc.*, **1988**, *110*, 6136.
22. Yitzchaik, S.; Lin, W.; Marks, T.J.; Lin, W.; Wong, G.K. *Poly. Mat. Sci. Eng.* **1995**, *71*, 271
23. Yitzchaik, S.; Marks, T. J. *Acc. Chem. Res.* **1996**, *29*, 197.
24. Ulman, A. *Chem. Rev.* **1996**, *96*, 1533.
25. Ulman, A., *An Introduction to Ultrathin Organic Films*, Academic Press: New York, 1991, Part 3.
26. Ulman, A., *Organic Thin Films and Surfaces:* Directions for the Nineties, Academic Press: New York, 1995.
27. Keller, S. W.; Kim, H.-N.; Mallouk, T. E., *J. Am. Chem. Soc.*, **1994**, *116*, 8817.
28. Lee, H.; Kepley, L. J.; Hong, H.-G.; Akhter, S.; Mallouk, T. E. *Phys. Chem.*, **1988**, *92*, 2597.
29. Katz, H. E., *Chem. Mater.* **1994**, *6*, 2227.
30. Katz, H. E.; Wilson, W. L.; Scheller, G. *J. Am. Chem. Soc.* **1994**, *116*, 6636.
31. Katz, H. E.; Scheller, G.; Putvinski, T. J.; Schilling, M. L.; Wilson, W. L.; Chidsey, C. E. D., *Science*, **1991**, *254*, 1485.
32. Whitesides, G. M. *Proceeding of the SPIE-The Int. Soc. for Optical Engineering* **1996**, *2716*, 307.
33. Li, D.; Ratner, M. A.; Marks, T. J.; Zhang, C., Yang, J., Wong, G. K. *J. Am. Chem. Soc.* **1990**, *112*, 7389.
34. Lin, W.; Yitzchaik, S.; Lin, W.; Malik, A.; Durbin M. K.; Richer, A. G.; Wong, G. K. Dutta, P.; Marks, T.J.; *Angew. Chem. Int. Eng. Ed.* **1995**, *34*, 1497.
35. Ober, C. K.; Wegner, G. *Adv. Mater.* **1997**, *9*, 17.
36. Decher, G.; Hong, J. -D. *Makromol. Chem., Macromol. Symp.* **1991**, *46*, 321.
37. Lvov, Y. M.; Decher, G. *Crystallography Reports* **1994**, *39*, 628.
38. Lvov, Yu.; Decher, G.; Haas, H.; Möhwald, H.; Kalachev, A. *Physica B* **1994**, *198*, 89.
39. Decher, G.; Lvov, Y.; Schmitt, J. *Thin Solid Films* **1994**, *244*, 772.
40. Lvov, Y.; Haas, H.; Decher, G.; Möhwald, H.; Kalachev, A. *J. Phys. Chem.* **1993**, *97*,12835.
41. Cheung, J. H.; Fou, A. F.; Rubner, M. F. *Thin Solid Films* **1994**, *244*, 985.
42. Ferreira, M.; Cheung, J. H.; Rubner, M. F. *Thin Solid Films* **1994**, *244*, 806.
43. Surhorukov, G. B.; Möhwald, H.; Decher, G.; Lvov, Y. M. *Thin Solid Films* **1996**, *284-285*, 220.

44. Tarabia, H.; Hong, H.; Davidov, D.; Kirstein, S.; Steitz, R.; Neumann, R.; Avny, Y. *J. Appl. Phys.* **1998,** *83*, 725.
45. Schmitt, J.; Decher, G.; Dressick, W. J.; Brandow, S. L.; Geer, R. E.; Shashidhar, R., Calvert, J. M. *Adv. Mater.* **1997**, *9*, 61.
46. Forrest, S. R.; Burrows, P. E.; Haskal, E. I.; So, F. F. *Phys. Rev. B* **1994,** *49*, 11309.
47. Imanachi, Y.; Hattori, S.; Kakuta, A.; Numata, S.; *Phys. Rev. Lett.*, **1993,** *71*, 2098.
48. Imanishi, Y.; Hattori, S.; Hamada, T.; Ishihara, S.; Nalwa, H.S.; Kakuta, A., *Nonlinear Optics, Principles, Materials, Phenomena, and Devices* **1995,** *13*, 63.
49. Yoshimura, T.; Tastuura, S.; Sotoyama, W. *Appl. Phys. Lett.* **1991**, *59*, 482.
50. Yoshimura, T.; Tastuura, S.; Sotoyama, W.; Matsuura, A.; Hayano, T. *Appl. Phys. Lett.* **1992**, *60*, 268.
51. So, F.F.; Forrest, S. R . *Phys. Rev. Lett.* **1991**, *60*, 2649.
52. Haskal, E. I.; Zang, Y.; Burrows, P. E.; Forrest, S. R. *Chem. Phys. Lett.* **1994**, *219*, 325.
53. Ohmori, Y.; Fujii, A.; Uchida, M.; Morishima, C.; Yoshino, K. *Appl. Phys. Lett.* **1993**, *63*, 1871.
54 Toda, Y.; Yanagi, H. *Appl. Phys. Lett.* **1996**, *69*, 2315.
55. Tsuzuki, T.; Hirota, N., Noma, N.; Shirota, Y. *Thin Solid Film*, **1996**, *273*, 177.
56. Forrest, S. R.; Burrows, P. E.; Haskal, E. I.; So, F. F. *Phys. Rev. B* **1994**, *49*, 11309.
57. Fenter, P.; Burrows, P. E.; Eisenberger, P.; Forrest, S. R. *J. Crystal Growth*, **1995,** *152*, 65.
58. Yitzchaik, S.; Roscoe, S. B.; Kakkar, A. K.; Allan, D. S.; Marks, T. J.; Xu, Z.; Zhang, T.; Lin, W.; Wong, G. K. *J. Phys. Chem.* **1993**, *97*, 6958.
59. M. Leskelä and L. Niinistö *Atomic Layer Epitaxy*, ed. T. Suntola and M. Simpson, Blackie and Son Ltd, **1990.**
60. Goodman, C. H. L.; Pessa, M. V. *J. Appl. Phys.* **1986**, *60*, R65.
61. Itoh, A.; Matsunami, H. *Critical reviews in Solid State and Material Science*, **1997,** *22*, 111.
62. *The Chemistry of the Semiconductor Industry* **1987**, Ed. Moss., and Ledwith.
63. Burtman, V.; Zelichenok, A.; Yitzchaik, S. submitted.
64. Burtman V.; Zelichenok A.; Yitzchaik, S. *Polymer Preprints*, **1998**, *38,* 167.
65. *Physics of Nanostructures* **1992**, Ed. Davies, J. H. and Long, A. R.
66. Shen, Z.; Forrest, S. R. *Phys. Rev.* **1997,** *55*, 10578.

Chapter 26

Multilayer Growth, Optical Properties, and Film Uniformity of Zinc–Bisquinoline Assemblies

D. L. Thomsen, III, T. Phely-Bobin, and F. Papadimitrakopoulos [1]

Department of Chemistry, Polymer Science Program, Nanomaterials Optoelectronics Laboratory, Institute of Materials Science, University of Connecticut, Storrs, CT 06269–3136

Polymeric zinc-bisquinoline grown by the reactive self-assembly of 8,8'-dihydroxy-5,5'-biquinoline (bisquinoline) and diethyl zinc has been reported to yield light emitting diodes (LEDs). The potential of this method to produce insoluble and intractable structures of controllable supramolecular architecture suitable for semiconducting applications has stimulated an in-depth investigation of the growth mechanism of these polymeric chelates. Ellipsometric and quartz-crystal microbalance measurements indicate a slow initial growth, reaching steady state after the sixth successive dipping cycle. A refractive index of 1.69 ± 0.01 at 633 nm for these films, as determined by variable angle spectroscopic ellipsometry, was found to be among the highest reported values for a self-assembled metallorganic multilayer.

Metal chelates of 8-hydroxyquinoline (8-Hq), and in particular aluminum tris-(8-Hq) (Alq_3), have received considerable attention toward the development of organic light emitting diodes (OLEDs).(*1,2*) In contrast to traditional inorganic light-emitting devices, OLEDs based on sublimed organic glasses or semicrystalline films are naturally limited by morphological changes at elevated temperatures and high current densities.(*3-5*) In order to suppress crystallization and densification in OLEDs, materials have been engineered with increased molecular asymmetry, hyperbranching, ring-puckering and entropic disorder via molecular doping.(*6,7*) While there is considerable difficulty in achieving stable disordered systems with these molecular compounds, polymeric materials offer a more traditional route to inhibit large morphological reorganizations.

[1] Corresponding author.

The incorporation of polymers in light emitting devices (LEDs) is currently challenged by purity and processability issues. The ability to purify macromolecules devoid of defects requires tight control of side reactions.(*8*) Furthermore, when traditional thin-film spin or dip casting techniques are utilized to fabricate multilayer LEDs, interlayer solubility, thickness variations, and pinhole defects may commonly occur. Self-assembly of thin films has introduced ways to improve film uniformity and layer architecture at the molecular level.(*9*) Rubner *et al.*(*10*) have constructed pinhole-free poly(*p*-phenylenevinylene)-based LEDs as thin as 300 Å. Although the versatility of a polycation/polyanion layering approach shows increasing promise as an alternative fabrication method, it is inapplicable for molecular-based devices which lack ionic functionalities, such as Alq_3.

Self-assemblies of small molecules on glass and metal surfaces have been demonstrated in numerous motifs with alkanol and alkanoic acids,(*11*) alkylsilanes,(*12*) disulfides and thiols,(*13,14*) and isonitriles.(*15*) Zirconium phosphonates represent an interesting class of self assemblies. The aqueous or highly polar (e.g., dimethyl sulfoxide (DMSO)) solvents necessary for solubilizing the $ZrOCl_2$ and the bifunctional phosphonic acid terminated organics result in heavily hydrated assemblies which develop pores upon drying.(*16,17*) Transition metal coordination with bifunctional isonitriles in polar solvents has also been used to fabricate multilayer self-assemblies.(*15*) Currently, only a limited number of coordinating transition metals have been suitable for multilayer fabrication, and the substantial film shrinkage upon drying (ca. 35%) are limiting factors in device fabrication.

The fabrication of polymer analogs of 8-hydroxyquinoline-based metal chelates (such as Alq_3, etc.) for electroluminescence applications has been a challenging task. These metal chelate polymers are nontraditional polymers and are usually associated with considerable handling difficulties. Their major intricacy arises from complexation-decomplexation dynamics, which are very sensitive to the pH and ionic strength of the solute.(*18*) For linear metal chelate polymers, solubilization typically occurs only in polar aprotic solvents,(*19*) which are difficult to remove from spun films. The insoluble and intractable nature of these polymers makes them amenable to the self-assembly growth which is the subject of this paper.

Experimental

Self Assembly Growth of Zinc-bisquinoline. Monomer Synthesis and purification was reported elsewhere.(*20,21*) Single polished silicon wafers with native oxide were cleaned in 1/1 H_2SO_4/30%H_2O_2 for 20 minutes followed by washing with distilled H_2O, and rinsing with MeOH. Silicon and silver coated quartz microcrystals were cleaned by subsequent rinsing in dilute HCl, distilled water, ultrasonically cleaned in ethanol for five minutes and air dried. The concentrations of diethyl zinc in THF was 1.5 x 10^{-5} M and the bisquinoline in THF was 3 x 10^{-4} M from THF. Hydroxy-functionalized surfaces were first dipped in the THF solution of $ZnEt_2$ for two minutes, followed by a rinse in the THF bath for an equal amount of time. Subsequently, the film is dipped into the THF solution of bisquinoline for one to two minutes followed by another THF rinse to form a polymer repeat unit of zinc-bisquinoline. In this fashion, the layer thickness increases by cycling through the dipping sequences. We define a dip cycle (d.c.) as a successive $ZnEt_2$ / THF wash / bisquinoline / THF wash sequence of 8 minutes duration. Films were prepared in thicknesses from 0 Å to 1275 Å.

Ellipsometry. Measurements were performed on a Variable Angle Spectroscopic Ellipsometer (VASE) manufactured by J. A. Woollman Co. equiped with PTI monochromator having a 4 nm bandwidth and scanned between 600 - 1000 nm at incident angles of 70, 72, 74, 76 and 78° for films between 0 Å and 1275 Å. Data were collected every 10 nm at 30 revs./meas. for films under 30 nm and at 10 revs./meas. for films over 30nm. Extrapolation for Psi and Del values at 633 nm at 70 degrees incidence had a calculated error due to finite monochromator slit width. For films under 2, 5, 20 and 30 as well as for over 30 nm Psi errors were ± 0.04, ± 0.04, ± 0.04, ± 0.04 and ± 0.22 while Del errors were ± 1.14, ± 0.5, ± 0.3, ± 0.15 and ± 0.50 respectively. The isotropic index of refraction was calculated from the ellipsometric Psi and Del values at multiple angles of incidence of seven films of zinc-bisquinoline, between 1080 Å to 1275 Å respectively, where Psi is near its maximum. With (VASE) measurements at 60°, 64°, 68°, 72°, and 76° incident angles scanned between 600 nm and 1000 nm, the isotropic index of refraction *(n)* was calculated from the wavelength *(λ)* fit of the Cauchy dispersion model, $n(\lambda) = A + B/\lambda^2 + C/\lambda^4$, using optical constants: A=1.5474 ± .00864, B= 0.076763 ± .00862, C = -0.0071004 ± .00208, Mean Square Error (MSE) = 33.48. The resulting index of refraction ranges from 1.71 to 1.62 between 600 and 1000 nm and is $n = 1.69 \pm 0.01$ extrapolated at 633 nm (He-Ne laser line).

Quartz Crystal Microbalance (QCM). Experiments were performed on 9 MHz silver coated quartz resonators from USI-Systems.

Results and Discussion

Scheme I illustrates the self-assembly growth of zinc-bisquinoline from 8,8'-dihydroxy-5,5'-biquinoline (bisquinoline). The reaction of bisquinoline with $Zn(Et)_2$ is almost instantaneous. Figure 1 illustrates the ellipsometrically determined film thicknesses as a function of dip cycle. The two markedly different regimes (Stage I and Stage II) with growth rates of 8.3 and 26.5 Å/dip cycle (d.c.) respectively, indicate a complex film growth. Films thicker than 250 Å maintain the 26.5 Å/d.c. growth rate as far as we have measured (c.a. 1,275 Å).

Figure 2 displays the quartz crystal microbalance (QCM) frequency shifts as a function of dip cycle. A similar trend is observed in relation the ellipsometrically determined film thicknesses of Figure 1. The QCM frequency shift ΔF (Hz) is related to the mass increase, M (g), and surface area of the resonator A, through the following equation (1):*(22)*

$$\Delta F = -1.832 \times 10^8 \, M/A \quad (1)$$

Maintaining constant the area and the surface roughness of the resonator, the QCM frequency shift is proportional to the increased mass. By the same formalism and assuming that the growth on silver substrates (on the thin oxide that covers the silver layer) is similar to that of silicon substrates, the ratio of QCM frequency shifts over the ellipsometrically determined film thicknesses is proportional to the film density (see Figure 2). The rapid increase of film density in Stage I followed by the apparent leveling in Stage II correlates this transition with a densification process.

A dense packing arrangement of zinc-bisquinoline as determined from molecular mechanics minimization in periodic boundary conditions suggests a molecular repeat of 12.3 Å. When this length, which represents the expected film growth for each dip cycle, is compared with the experimentally derived thickness growth at both stages, a number of observations can be inferred:

(i). In Stage I, the first layer growth appears to be exceedingly high (21.3 Å versus the 8.3 Å of subsequent layers). Taking into account the ellipsometric error at diminished thicknesses, and the apparent small frequency shift of the QCM resonator, we could assign such behavior on either high surface roughness or highly porous oxide structures formed by the adsorbed moisture when it comes in contact with the $ZnEt_2$ solution.

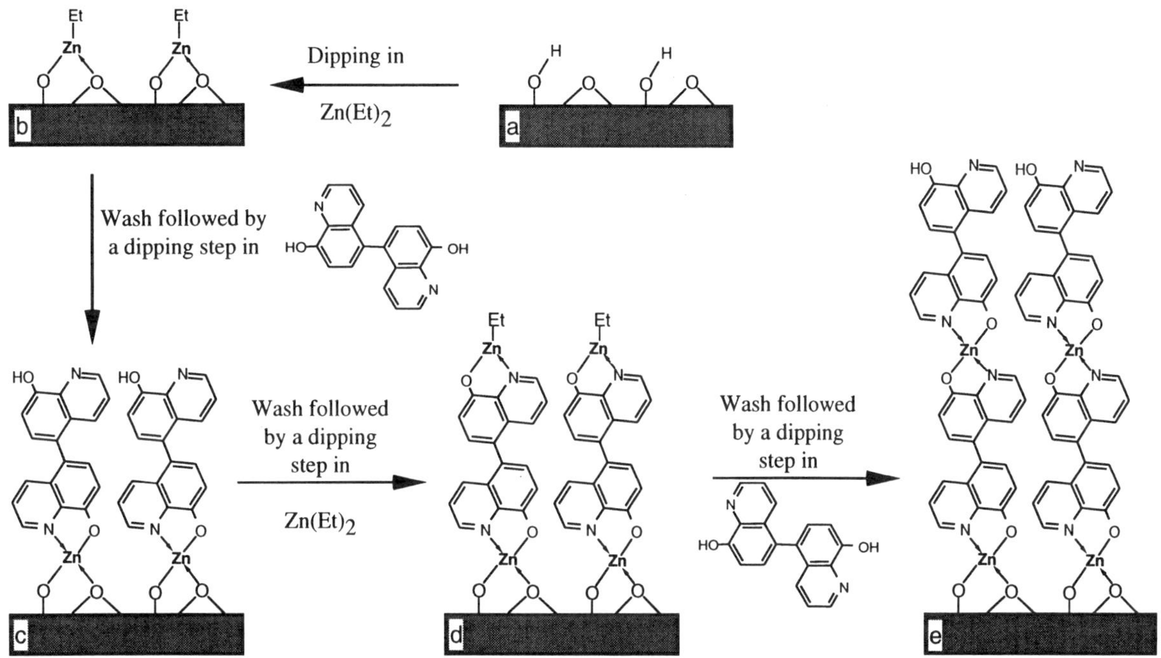

Scheme 1. Schematic representation of the self-assembly growth of poly(zinc-bisquinoline).

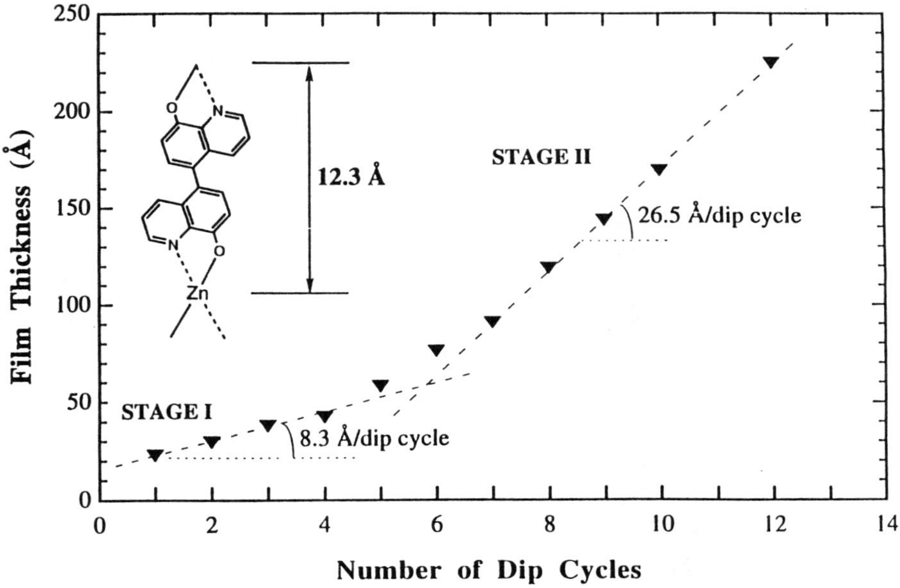

Figure 1. Ellipsometrically determined thickness versus successive dip cycles for the poly(zinc-bisquinoline) self-assembly process.

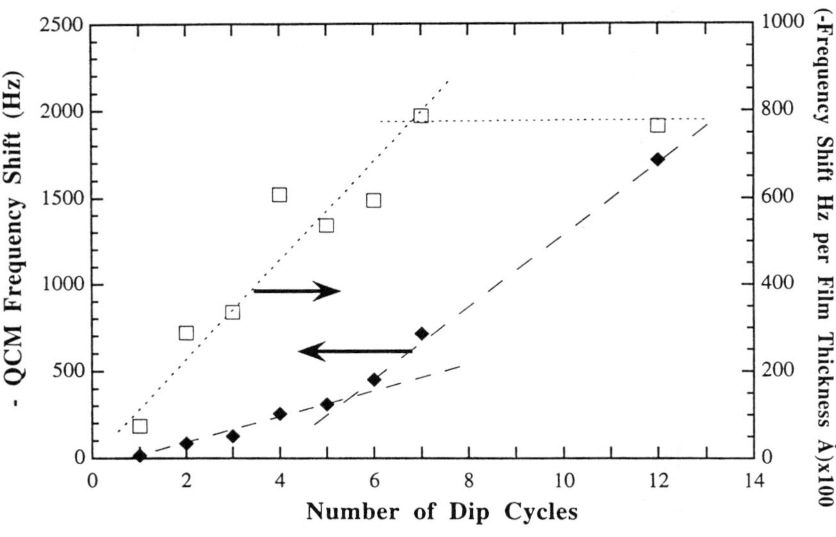

Figure 2. Quartz crystal microbalance frequency shift and its ratio versus the ellipsometrically determined film thickeness growth of as a function of successive dip cycles.

(ii). The smaller growth rate (8.3 Å/d.c.) versus the expected one (12.3 Å/d.c.) along with the densification behavior observed in Stage I (from Figure 1 and 2 respectively) could be explained by either surface-roughness healing, or(and) by improvement in chain orientation. Also, an initial surface coverage effect has been seen with a number of polyanion/polycation self-assemblies, where at early stages the growth on unfunctionalized surfaces was reduced.(*22,23*)However, considerable work is still needed to understand this behavior.

(iii). The nearly doubling of growth rate (26.5 Å/d.c.) in Stage II along with the leveling of the apparent density could arise from a high degree of chain orientation that cooperatively stabilize further growth through packing and the asscociative behavior of quinolinols with diethly zinc. Alcohols and phenols have been shown to form dimeric and trimeric species in the presence of diethly zinc in anhydrous conditions.(*24*) The excess reactant (e.g. $ZnEt_2$) could be stored within the aligned chains (acting in a sense as a plasticizer, resisting removal by the wash step) only to be extracted out by its diminishing concentration gradient at the surface where bisquinoline is coordinated in an aligned fashion.

The film uniformity of zinc bisquinoline films was shown by and spectroscopic ellipsometry. The films over a range of thicknesses, 0 Å to 1275 Å exhibit similiar refractive indices between n = 1.65 and n = 1.70. The optical uniformity of zinc bisquinoline was demonstrated by plotting the ellipsometric trajectory of films with thicknesses from 0 Å to 1275 Å, grown on silicon wafers. These data correspond to the extrapolated Psi and Del values at 633 nm and 70 degrees incidence, and comparing to model trajectories of 1.65, 1.70, and 1.75 refractive indices (see Figure 3). The colors from interference effects of these films on silicon ranged from light pink, brown, gold, purple, navy blue, royal blue, and blue-grey as the films increased from 200 Å - 1275 Å. Figure 4 illustrates the isotropic index of refraction for zinc bisquinoline in its transparent region between 600 nm and 1000 nm determined by spectroscopic ellipsometry using the Cauchy dispersion model. (EQN 1) The isotropic index of refraction was calculated from the ellipsometric Psi and Del values (see Figure 2) at multiple angles of incidence of seven films of zinc-bisquinoline, between 1080 Å and 1275 Å respectively, where Psi is near its maximum,(*25-27*) based on the isotropic Cauchy fit model and its constants given in equation 2:

$$n(\lambda) = A + B/\lambda^2 + C/\lambda^4 \qquad (2)$$

with: A=1.5474 ± .00864, B= 0.076763 ± .00862, C = -0.0071004 ± .00208,
Mean Square Error (MSE) = 33.48

When this is extrapolated at 633 nm a refractive index of 1.69 ± 0.01 is obtained which has been found to be among the highest reported values for a self-assembled metallorganic multilayer. To the best of our knowledge, the current index of refraction (n = 1.69 ± 0.01) is the highest of the reported values of multilayer assemblies of conjugated chromophores.

The crystalline packing arrangement of poly(zinc-bisquinoline) has been a topic of considerable interest. Up to the present, evidence of crystalline order has been witnessed only from the solution-synthesized oligomers of poly(zinc-bisquinoline)(*28*). This is in good agreement with the behavior of its low molecular analog (zinc-bis-(8-hydroxyquinoline)) where its dihydrate forms highly ordered single crystals. The single crystal of zinc quinoline dihydrate exhibits a density of 1.68 g/cm3 and refractive indices, as determined by immersion methods to be, n_α = 1.65, n_β = 1.78, and n_γ > 1.82.(*29*) In light of the fact that wide-angle X-ray diffraction of zinc bisquinoline assemblies have not shown clear signs of crystallinity, refractive index correlation can provide sufficient insight of the molecular packing in such films.(*27*) Thin films, thick enough for accurate optical constant determination,

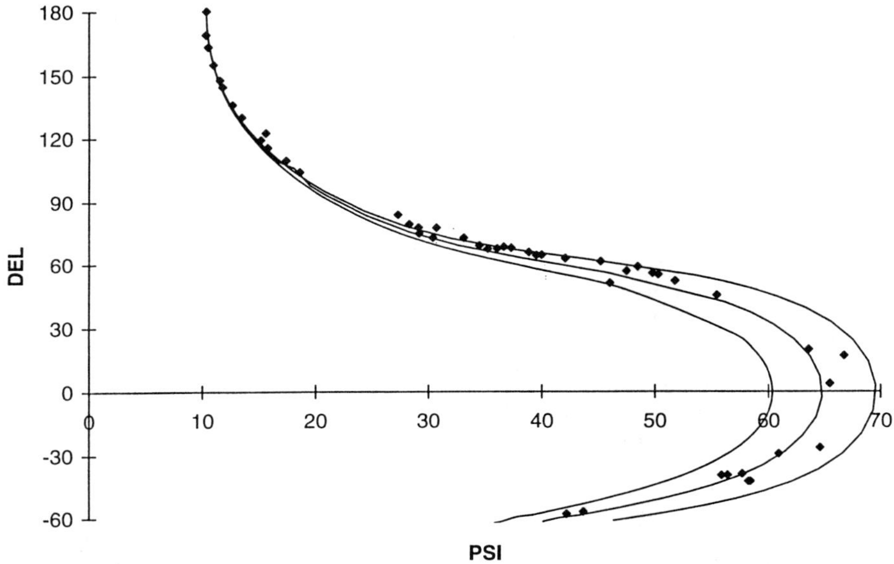

Figure 3. PSI versus DEL trajectory plot of poly(zinc-bisquinoline) films ranging in thicknesses from 0 Å to 1275 Å, at 633nm and 70° incidence. From left to right the solid lines represent the model trajectories of $n = 1.75$, 1.70, and 1.65 respectively.

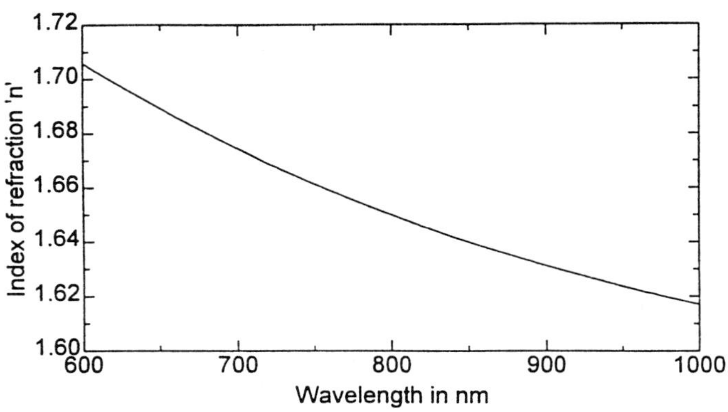

Figure 4. Index of refraction of poly(zinc-bisquinoline) from the Cauchy fit model.

can be directly correlated with bulk refractive indices.(*30-32*) The measured bulk values of zirconium phosphonates (Zn is much lighter than Zr) appear to range between 1.52 - 1.54(*16,33-35*) where polyelectrolyte self-assemblies have shown refractive indices ranging from 1.47 - 1.60.(*36*) Considerable amount of effort has been exerted to characterize the order in self-assembled monolayers and multilayers.(*32*) Self-assembled monolayers and multilayers of alkyl silanes,(*37*) alkanoic acids(*38*) and thiols(*14*) have demonstrated order as a result of chain packing. The high refractive index of the zinc bisquinoline self-assembled films has been attributed to the packing.

Conclusion

The insoluble and intractable nature of the zinc bisquinoline chelates, along with their improved processability, film forming quality and increased purity enables their usage in a variety of semiconducting applications.(*21*) Zinc bisquinoline films were determined to have a refractive index of 1.69 ± 0.01 at 633 nm. The uniformity of films was demonstrated by the ellipsometric trajectory over a wide range of thicknesses. These films possess a high refractive index, indicative of dense films, and film uniformity. A better understanding of the underlying structure and orientation of zinc bisquinoline self-assemblies will provide the necessary insight for improving the photoluminescence and electroluminescence behavior of these films.

Acknowledgments

The authors would like to thank Dr. Y. Lvov for helping with the QCM measurements. Financial support from NSF CAREER Grant DMR-9702220 and the Critical Technologies Program through the Institute of Materials Science, University of Connecticut are greatly appreciated.

Literature Cited

(1) Tang, C. W.; VanSlyke, S. A.; Chen, C. H. *J. Appl. Phys.* **1989**, *65*, 3610 - 3615.
(2) Sheats, J. R.; Antoniadis, H.; Hueschen, M.; leonard, W.; Miller, J.; Moon, R.; Roitman, D.; Stocking, A. *Science* **1996**, *273*, 884 - 888.
(3) Do, L.-M.; Han, E. M.; Niidome, Y.; Fujihira, M.; Kanno, T.; Yoshida, S.; Maeda, A.; Ikushima, A. J. *J. Appl. Phys.* **1994**, *76*, 5118 - 5121.
(4) Han, E.-M.; Do, L.-M.; Yamamoto, N.; Fujihira, M. *Thin Solid Films* **1996**, *273*, 202 - 208.
(5) Higginson, K. A.; Zhang, X.-M.; Papadimitrakopoulos, F. *Chem. Mater.* **1997**, Accepted for Publication.
(6) Shirota, Y.; Kuwabara, Y.; Inada, H.; Wakimoto, T.; Nakada, H.; Yonemoto, Y.; Kawami, S.; Imai, K. *Appl. Phys. Lett.* **1994**, *65*, 807 - 809.
(7) Chen, C. H.; Shi, J.; Tang, C. W. *A.C.S. Polymer Preprints* **1997**, *38*, 317.
(8) Papadimitrakopoulos, F.; Yang, M.; Rothberg, L. J.; Katz, H. E.; Chandross, E. A.; Galvin, M. E. *Mol. Cryst. Liq. Cryst.* **1994**, *256*, 663 - 661.
(9) Decher, G.; Hong, J. D.; Schmitt, J. *Thin Solid Films* **1992**, *210/211*, 831-835.
(10) Fou, A. C.; Onitsuka, O.; Ferreira, M.; Rubner, M. F.; Hsiesh, B. R. *J. Appl. Phys.* **1996**, *79*, 7501.
(11) Allara, D. L.; Nuzzo, R. G. *Langmuir* **1985**, *1*, 52-66.
(12) Sagiv, J. *J. Am. Chem. Soc.* **1980**, *102*, 92-98.
(13) Nuzzo, R. G.; Allara, D. L. *J. Am. Chem. Soc.* **1983**, *105*, 4481-4483.
(14) Porter, M. D.; Bright, T. B.; Allara, D. L.; Chidsey, C. E. D. *J. Am. Chem. Soc.* **1987**, *109*, 3559-3568.

(15) Ansell, M. A.; Zeppenfeld, A. C.; Yoshimoto, K.; Cogan, E. B.; Page, C. J. *Chem. Mater.* **1996**, *8*, 591.
(16) Katz, H. E. *Chem. Mater.* **1994**, *6*, 2227.
(17) Feng, S.; Bein, T. *Nature* **1994**, *368*, 834-836.
(18) Yeh, Y. S. *J. Polym. Sci.* **1990**, *28*, 545.
(19) Thomsen, D. L.; Phely-Bobin, T.; Papadimitrakopoulos, F. *JACS* **1998**, Submitted.
(20) Thomsen, D. L.; Papadimitrakopoulos, F. *Macromol. Symp.* **1997**, *125*, 143.
(21) Thomsen, D. L.; Higginson, K. A.; Papadimitrakopoulos, F. *Proc. Inter. Soc. Opt. Eng.* **1997**, *3148*, In Print.
(22) Lvov, Y.; Ariga, K.; Ichinose, I.; Kunitake, T. *J. Am. Chem. Soc.* **1995**, *117*, 6117-6123.
(23) Lvov, Y.; Decher, G. *Crystallography Reports* **1994**, *39*, 628.
(24) Boersma, J.; Noltes, J. G. *J. Organometallic Chem.* **1968**, *13*, 291 - 299.
(25) Tompkins, H. G. *A User's Guide to Ellipsometry*; Academic Press: San Diego, CA, 1993.
(26) Arwin, H.; Aspnes, D. E. *Thin Solid films* **1984**, *113*, 101 - 113.
(27) Aspnes, D. E. *The Accurate Determination of Optical Properties by Ellipsometry*; Academic Press: New York, 1985.
(28) Berg, E. W.; Alam, A. *Anal. Chim. Acta* **1962**, *27*, 454.
(29) Merritt, L. L.; Cady, R. T.; Mundy, B. W. *Acta. Cryst.* **1954**, *7*, 473.
(30) Tillman, N.; Ulman, A.; Schildkraut, J. S.; Penner, T. L. *J. Am. Chem. Soc.* **1988**, *110*, 6136-6144.
(31) Tillman, N.; Ulman, A.; Penner, T. L. *Langmuir* **1989**, *5*, 101-111.
(32) Ulman, A. *An Introduction to Ultrathin Organic Films from Langmuir-Blodgett to Self-Assembly*; Academic Press: Boston, MA, 1991.
(33) Katz, H. E.; Scheller, G.; Putvinski, T. M.; Schilling, M. L.; Wilson, W. L.; Chidsey, C. E. D. *Science* **1991**, *254*, 1485-1487.
(34) Katz, H. E.; Schilling, M. L.; Chidsey, C. E. D.; Putvinski, T. M.; Hutton., R. S. *Chem. Mater.* **1991**, *3*, 699-703.
(35) Lee, H.; Kepley, L. J.; Hong, H.-G.; Mallouk, T. E. *J. Am. Chem. Soc.* **1988**, *110*, 617.
(36) Ramsden, J. J.; Yu. M. Lvov; Decher, G. *Thin Solid Films* **1995**, *254*, 246-251.
(37) Gun, J.; Iscovici, R.; Sagiv, J. *Journal of Colloid and Interface Science* **1984**, *101*, 201-213.
(38) Allara, D. L.; Nuzzo, R. G. *Langmuir* **1985**, *1*, 45-52.

Chapter 27

The Construction of Circuitry Using Spatially Coupled Bipolar Electrochemistry

Jean-Claude Bradley, Zhongming Ma, Shanthi Christaffer, Jeffrey Crawford, Karima Ernazarova, and Samuel G. Stephens

Drexel University, Department of Chemistry, 32nd and Chestnut Streets, Philadelphia, PA 19104

The formation of electrical contacts between metal structures and devices is an integral aspect of circuit construction at all scales of commercial importance. Currently photolithographic, screen printing and microsoldering techniques are the methods of choice to establish connections. However, these approaches require masks, templates or intimate physical contact with the components. Spatially Coupled Bipolar Electrochemistry (SCBE) is a novel technique which makes use of electric fields to create electrical connections between components, which not only avoids physical contact but is also applicable in principle to the formation of three dimensional circuitry. The SCBE technique has been developed to a point where the construction of functional robust circuits has been achieved. Preliminary data demonstrating the application of this approach to the formation of polypyrrole bridges between isolated gold structures is also presented.

There is a recent trend in materials science attempting to establish electrical contacts between conductive components without resorting to photolithographic techniques. One such approach involves electro-crystallization[1] or electropolymerization[2-6] from adjacent electrodes until an electrical connection is achieved by random physical contact of the growing conductive polymer or salt. This has opened up a new avenue where it is possible to create conductive polymer based diodes, transistors and signal amplifiers[2-3]. Other researchers have employed template [7-12] and localized thermal plating[13] strategies to construct conductive paths. The advantage of these approaches over photolithographic techniques is the possibility of construction of circuits without being confined to 2-D architectures.

Theory of Bipolar Electrochemistry

When a conductive particle is exposed to an electric field, it causes the particle to polarize. As a consequence an overpotential η varying according to a cosine law is induced at the surface of the particle (Eq. 1, Fig. 1)[14,15]. Thus, a maximum potential difference will occur at opposite poles of the particle. In order to carry out electrochemistry at the surface of the particle a critical voltage difference V_c corresponding to the sum of two half-cell reactions must be reached. Thus, for a given particle of radius r and an applied electric field E there will exist two polar regions defined by a critical angle θ_c within which electrochemistry will occur. (Eq. 2, Fig. 2). This forms the theoretical basis of toposelective electrodeposition.

$$\eta_x = Er\cos(\theta) \qquad (1)$$

$$\theta_c = \cos^{-1}(V_c/2Er) \qquad (2)$$

Bipolar electrochemistry[14-26] is the term used to describe this process since the particle serves as both anode and cathode. This phenomenon has been investigated using fluidized or packed bed electrodes for applications in metal recovery[16-19], electrosynthesis[20-23] and ultramicroelectrode[24-26] studies.

Spatially Coupled Bipolar Chemistry (SCBE)

We have recently developed a novel technology to complement existing methods of forming interconnects: Spatially Coupled Bipolar Chemistry (SCBE)[27-30] (Fig.3). In SCBE, toposelective electrodissolution and electrodeposition are spatially coupled to generate copper wires at predictable and highly selective locations. Electrodeposition in the absence of supporting electrolyte leads to the formation of highly ramified metallic structures which serve as conductive wires in our applications. The ability to connect structures not in physical contact with an external voltage source contributes a unique flexibility not only to create interconnects but also to the formation of freestanding metallic structures in general.

This concept has been extended from the formation of hanging wires between copper spheres[27] to the formation of robust and adherent electrical connections between pairs of copper rings on sections of commercial circuit boards.[28] In this case, an initial ramified copper deposit is grown using the SCBE approach followed by an electroless copper plating step to create much more robust connections. The use of

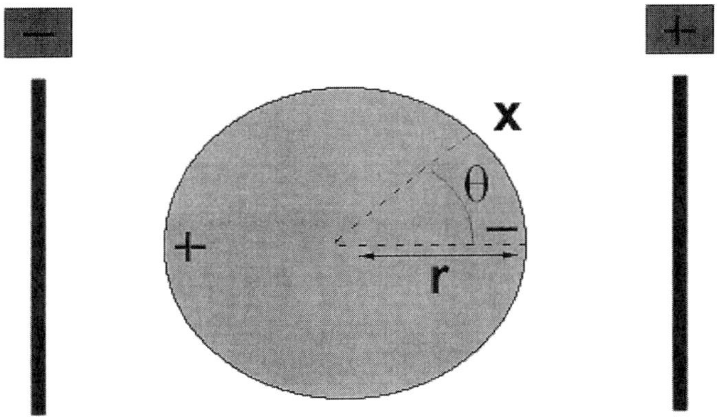

Figure 1. The polarization of a spherical metal particle in an electric field. The overpotential η at position x can be calculated according to Equation 1 (see text).

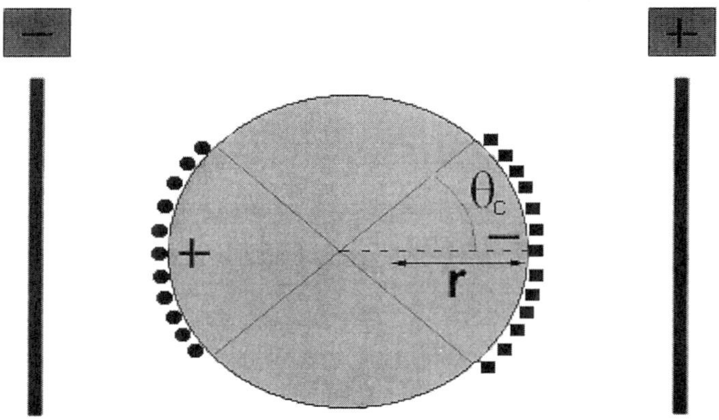

Figure 2. The theoretical areas within which electrochemistry can occur on the surface of a polarized metal sphere, calculated according to Equation 2 (see text).

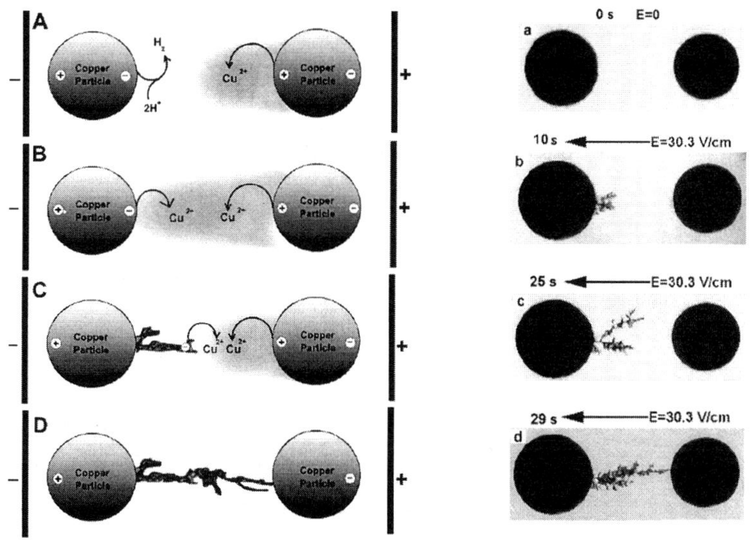

Figure 3. Schematic (left) and experimental observation (right) of wire formation between two particles under bipolar conditions.

A. Two copper particles are placed in an aqueous environment and an electric field is applied. The polarization of each particle is shown. Initially the particle on the right liberates copper ions while the particle on the left reduces water. The shaded area represents a hypothetical distribution of the ionic cloud. (For clarity only the phenomena in the interparticle region are shown).

B. When the copper ion concentration near the particle on the left is high enough electrodeposition occurs and the wire begins to grow facing the other particle.

C. Electrodeposition occurs preferentially at the wire tip where cathodic polarization is expected to be highest.

D. When the wire reaches the particle on the right, electrical contact is made. At this point there is no potential difference between the particles and elctrochemical processes in the interparticle region cease. Experimental conditions are described in detail elsewhere.[27] (Reproduced with permission from reference 27. Copyright 1997 Macmillan Magazines, Ltd.)

organic solvents such as toluene/acetonitrile mixtures instead of aqueous media permits the growth of initially much thicker deposits thus substantially lowering the electroless plating time necessary to achieve a robust electrical contact. In addition, non-aqueous media are conducive to the application of fields on the order of several kV/cm which has allowed growth of wires as small 0.3 micrometers on copper structures on the order of a few micrometers.[29] As shown in Figure 4, the application of electric fields within selected areas of a circuit board can be used to create selective circuitry. These results will be detailed in a subsequent publication.[30]

The Ramified Growth of a Conductive Polymer on Isolated Metal Substrates

Bipolar electrochemistry requires a medium of sufficiently low conductivity to avoid excessive parasitic current leakage at the feeder electrodes. In this regard the bipolar electropolymerization of conductive polymers appears to be an attractive additional method to create interconnects between isolated metal components because the neutral monomer does not significantly contribute to conductivity and can thus be introduced in relatively high concentrations. Furthermore, as opposed to the examples of SCBE described above using copper structures, it is not necessary for the metal component to electrodissolve to create an interconnect.

Figure 5 provides an example of bipolar electropolymerization of polypyrrole onto an isolated gold particle. A mixture of toluene and acetonitrile is used to provide relatively high resistivity. A small amount of sodium p-toluenesulfonate is necessary to ensure the ramified growth of the conductive polymer. If the additive is omitted only a film forms on the gold particle. When two gold particles are aligned with the direction of the applied electric field, an apparent connection can be seen to form within minutes. If a single field direction is used a large variation in the thickness of bridging polymer is observed (see Figure 6). A more symmetrical linkage can be obtained by alternating the direction of the applied field every few minutes (see Figure 7). Whether an actual electrical connection has formed is yet to be determined. If copper is used instead of gold, a metallic wire will form much more quickly than the polypyrrole growth. This can be ascertained readily from the direction of growth. Copper deposits grow from the cathodically polarized region while polypyrrole grows from the anodically polarized side of the particles.

Conclusion

We have demonstrated a novel method to create either freestanding or surface-bound wires capable of establishing electrical contacts between isolated metal structures. The technique relies on spatially coupled electrodissolution and electrodeposition mediated by an electric field. Such bipolar conditions avoid the need to physically contact the metallic components in the system, which should prove to be a considerable advantage when dealing with far smaller components. In the case of wire formation on circuit boards, the metallic paths laid down serve as templates for an electroless deposition step, which yields robust and adherent electrical connections. In principle, the method is not limited to planar geometries. The application of electric field vectors in selected directions through an array of conductive structures in a porous medium may represent a powerful approach to the formation of three-dimensional circuitry. Furthermore the use of a conductive polymer allows the use of metal structures which do not electrodissolve in the medium.

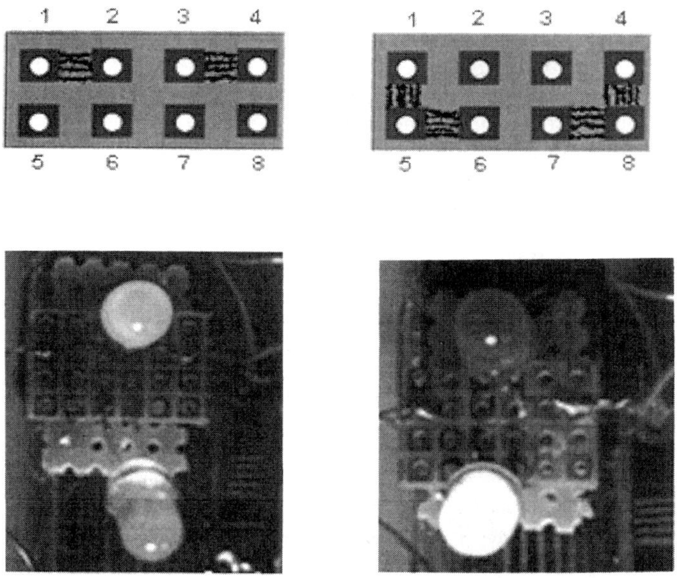

Figure 4. Demonstration of the programmed growth of circuitry by applying electric fields to form pattern 1 (A) and pattern 2 (B). The resulting circuits are shown in micrographs C and D formed after the application of patterns 1 and 2, respectively. Prior to the application of electric fields, leads were soldered at positions 1 and 4 and two diodes across positions 2-3 and 6-7, respectively. In C the growth of pattern 1 leads to the lighting of the top diode whereas in D the growth of pattern 2 leads to the lighting of the lower diode. (Reproduced with permission from reference 30. Copyright 1999 Electrochemical Society.)

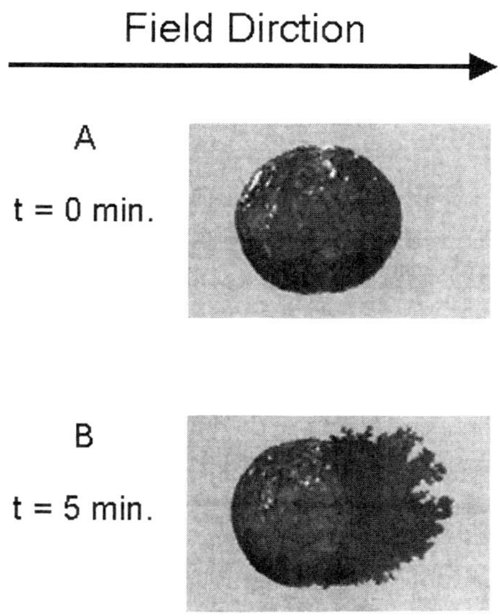

Figure 5. Bipolar electropolymerization of polypyrrole onto a single gold particle.

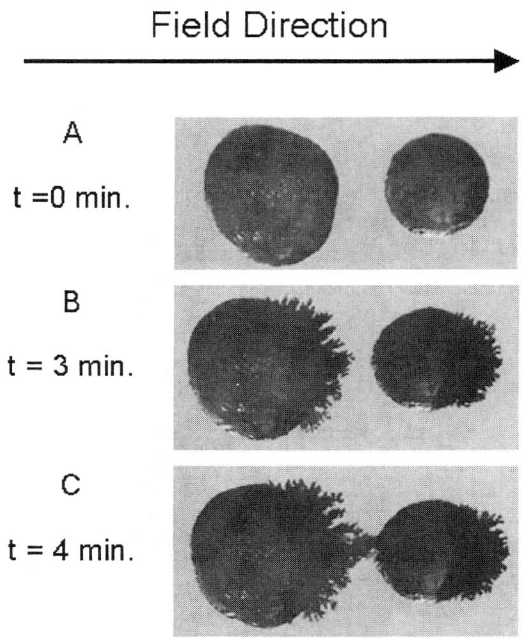

Figure 6. Formation of an apparent interconnection between two isolated gold particles by bipolar electropolymerization on polypyrrole using a single field direction.

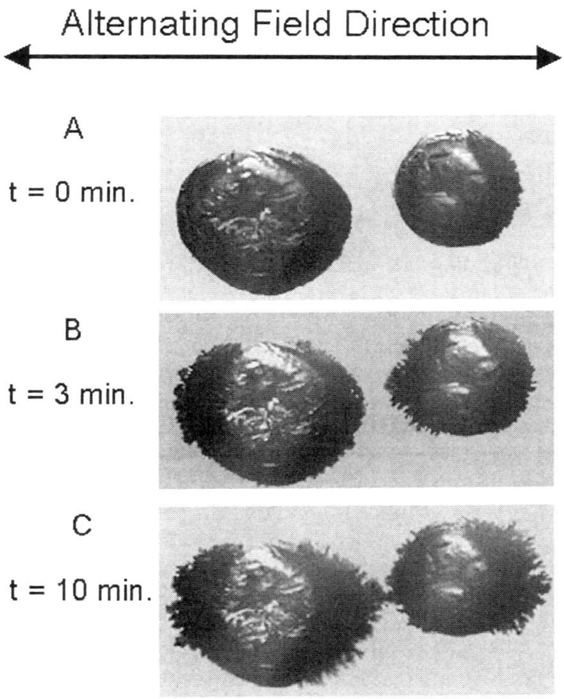

Figure 7. Formation of an apparent interconnection between two isolated gold particles by bipolar electropolymerization on polypyrrole using an alternating field direction.

Experimental:

Circuit Board Growth

Circuit growth was carried out on printed circuit boards (Radioshack, part number 276-158B) consisting of 2mm x 2 mm copper squares 0.5mm apart with 1mm diameter unplated through holes. The boards were cut down to sections 4 squares wide by 6 squares long and placed on an array of platinum pins (2x4), the details of which are reported elsewhere.[30] The electric field application was carried out in a solution of 60/40 toluene/acetonitrile. A power supply (Bertan Associates Inc. model PMT-10A/option 3, rated for 0-1000v @ 4 mA DC) was used to deliver the field across the desired pins. After growth of the wire, the circuit boards were subjected to an electroless copper plating solution (Technic Inc. EC-70) for 3 hours.

Conductive Polymer Growth

Gold particles (2.7 and/or 3.8 mm) were used. 1M pyrrole in 1:1 toluene/acetonitrile containing 1mM sodium p-toluenesulfonate was used throughout. The electric field was applied through two platinum wires (1mm in diameter) positioned vertically in the solution in such a manner that the particle(s) was (were) within the general area where the electric field was expected to be maximum. When two particles were used, they were aligned in the direction of the electric field. The electrodes were positioned 1.5 cm apart, with voltages applied in the range of 250-300 V resulting in fields of 167-200 V/cm.

Case #1: Electropolymerization onto a single gold particle:

With the conditions described above polypyrrole will form from the positively polarized region of the gold particle. See Fig. 5. The voltage was 251 V.

Case #2: Electropolymerization between two gold particles using a single field direction:

An apparent contact is shown in Fig. 6C (time=4 min.) after exposing the particles in Fig. 6C (time= 0 min.) to an electric field in the direction shown. The voltage was 300 V and the particle separation was 0.68 mm.

Case #3: Electropolymerization between two gold particles using an alternating field direction:

An apparent contact is shown in Fig. 7C (time= 10 min.) after exposing the particles in Fig. 7A (time= 0 min.) to an alternating electric field in the direction shown. The direction was alternated with sequential exposure times of approximately 2 minutes. The voltage was 250 V and the separation was 1.0 mm.

References

1. Gutner, C. and Sailor, M. J. *Adv. Mater.* **1996**, *8*, 897.
2. McCoy, C.H. and Wrighton, M.S. *Chem. Mater.* **1993**, *5*, 914.
3. Jones, E.T.T.; Chyan, O.M. and Wrighton, M.S. *J. Am. Chem. Soc.* **1987**, *109*, 5526.
4. Sailor, M.J. and Curtis, C.L. *Adv. Mater.* **1994**, *6*, 688.
5. Fujii, M.; Arii, K. and Yoshino, K. *Synthetic Metals* **1995**, *71*, 2223.
6. Curtis, C.L.; Ritchie, J.E. and Sailor, M.J. *Science* **1993**, *262*, 2014.
7. Hidber, P.C.; Nealey, P.F; Helbig, W. and Whitesides, G.M. Langmuir **1996**, *12*, 5209.

8. Chou, S.Y.; Krauss, P.R. and Renstrom, P.J. *Science* **1996**, *272*, 85.
9. Wang, L.; Yu-Zhang, K; Metrot, A; Bonhomme, P and Troyon, M. *Thin Solid Films* **1996**, *288*, 86.
10. Masuda, H. and Fakuda, K. *Science* **1995**, *268*, 1466.
11. Martin, C.R. *Science* **1994**, *266*, 1961.
12. Huber, C.A., et al. *Science* **1994**, *263*, 800.
13. Von Gutfeld, R.J. and Vigliotti, D.R. *Appl. Phys. Lett.* **1990**, *56*, 2584.
14. Eardley, D.C.; Handley, D.; Andrew, S.P.S. *Electrochim. Acta* **1973**, *18*, 839.
15. Fleischmann, M.; Ghoroghchian, J.; Pons, S. *J. Phys. Chem.* **1985**, *89*, 5530.
16. Lee, J.K.; Shemilt, L.W. and Chun, H.S. *J. Appl. Electrochem.* **1989**, *19*, 877.
17. Kusakabe, K; Kimura, T; Morroka, S and Kato, Y. *J. Appl. Electrochem.* **1987**, *17*, 724
18. Plimley, R.E. and Wright, A.R. *Chem. Eng. Sci.* **1984**, *39*, 395
19. Handley, D and Eardley, D.C. *Chemistry and Industry* **1975**, 330
20. Sudoh, M.; Kodera, T and Ichino, T. *J. Chem. Eng. Jpn.* **1991**, *24*, 165
21. Manji, A and Oloman, C.W. *J. Appl. Electrochem.* **1987**, *17*, 532
22. Sudoh, H; Kodera, T.; Hino, H and Shimamura, H. *J. Chem. Eng. Jpn.* **1988**, *21*, 198
23. Kim, H.J.; Kusadabe, K.; Hokazono, S.; Morroka, S. and Kato, Y. *J. Appl. Electrochem.* **1987**, *17*, 1213
24. Ghoroghchian, J.; Pons, S and Fleischmann, M. *J. Electroanal. Chem.* **1991**, *317*, 101
25. Rolison, D.R. *Chem. Rev.* **1990**, *90*, 867
26. Fleischmann, M.; Ghoroghchian, J.; Rolison, D. and Pons, S. *J. Phys. Chem.* **1986**, *90*, 6392
27. Bradley, J.-C.; Chen, H.-M.; Crawford, J.; Eckert, J; Ernazarova, K; Kurzeja, T; Lin, T; McGee, M; Nadler, W. and Stephens, S.G. *Nature* **1997**, *389*, 268.
28. Bradley, J.-C.; Crawford, J.; Ernazarova, K.; McGee, M. and Stephens, S.G. *Adv. Mater.* **1997**, *9*, 1168
29. Bradley, J.-C.; Crawford, J.; McGee, M.; Stephens, S.G. *J. Electrochem. Soc.* **1998**, *145*, L45.
30. Bradley, J.-C.; Ma, Z.; Clark, E.; Crawford, J., Stephens, S.G *J. Electrochem. Soc.* in press.

Author Index

Author Index

Subject Index

Subject Index

C

D

E

F

G

H

I

K

L

M

N

O

P

U

V

W

Z